I0483484

GPT Meets Game Theory

Game theory systems can be seen as players working together or competing to achieve goals. *GPT Meets Game Theory* explores a new way to understand and employ neural networks through the lens of game theory. Focusing on transformers, the engines behind today's most advanced AI, it explains key mathematical concepts and strategies in a clear, accessible way.

As AI models are growing larger and taking on more data, *GPT Meets Game Theory* draws from biology, physics, as well as game theory, to help readers understand how we can interpret and guide the models' behavior. It also looks at how these ideas apply to "mean-field" models and how they can be used in situations like federated learning, where many devices work together to train an AI system. The book shows how choosing the right AI design and training method is like making strategic moves in a game - especially when multiple AI agents are involved.

GPT Meets Game Theory offers an illuminating read for computer science, engineering, and mathematics researchers interested in the mathematical underpinnings of deep learning models, particularly transformers, and also for those who are curious about how game theory can apply to the training and optimisation of these models.

GPT Meets Game Theory

Training and Optimizing Generative AI Models

Hamidou Tembine

CRC Press
Taylor & Francis Group
Boca Raton London New York

CRC Press is an imprint of the
Taylor & Francis Group, an **informa** business
A CHAPMAN & HALL BOOK

Front cover image: afe207/Shutterstock

First edition published 2026
by CRC Press
2385 NW Executive Center Drive, Suite 320, Boca Raton FL 33431

and by CRC Press
4 Park Square, Milton Park, Abingdon, Oxon, OX14 4RN

ISBN: 978-1-041-12409-2 (hbk)
ISBN: 978-1-041-12407-8 (pbk)
ISBN: 978-1-003-66458-1 (ebk)

DOI: 10.1201/9781003664581

Typeset in CMR10 font
by KnowledgeWorks Global Ltd.

Publisher's note This book has been prepared from camera-ready copy provided by the author(s)/editor(s).

To Yandai, Pama, Marie-Claire,
Jean-Pierre and Florence
for their unconditional support

Contents

Preface

Machine intelligence is rapidly evolving from an academic curiosity to a cornerstone of modern technologies of high societal impact. At the heart of this evolution is co-learning, specifically, machine learning including deep learning, reinforcement learning and co-opetitive learning. In parallel, game theory, a framework for understanding strategic interactions, has long been instrumental in fields as varied as engineering, computer science, economics, biology, finance, language, ecology, and psychology. But what happens when these two powerful disciplines meet?

This book on Generative Transformer Meets Game Theory tackles this question. It offers readers an exploration of how deep learning, particularly transformer models, can be enriched through the strategic lens of game theory. This book opens up a conversation: what if we could frame the learning process of neural networks as games? What if we could treat the interactions within these models, between layers, nodes, and agents , as a form of strategic play?

This is not just a theoretical exercise. By viewing machine intelligence co-development through the framework of strategic decision-making problems, we invite you to rethink the way we approach consistency check, optimization, training, and scaling in machine intelligence systems. For example, the interplay of different components in a deep neural network can be seen as decision-makers in a game, each with its own incentives, constraints, and strategies. Such a perspective has the potential to enable new insights, driving the next generation of machine intelligence advancements.

As the challenges of building increasingly large machine intelligence models grow, this book offers a timely and much-needed approach to tackling issues such as consistency, robustness, efficiency, multi-criteria performance, and fairness. The application of game theory can help address these challenges by offering new solutions to variational inequality problems, training stability, and coordination among various model components. Whether we are dealing with transformers, federated training systems, or multi-agent environments, the strategic insights provided by game theory could hold the key to solving some of machine intelligence's most pressing problems. But the importance of this book goes beyond its immediate technical contributions. It represents a broader trend in scientific inquiry: the blending of disciplines to solve real-world problems. As fields like machine intelligence, engineering, mathematics, and economics converge, we are seeing new frameworks and methodologies emerge, transforming not only how we build intelligent systems but also how we think about intelligence itself. This book is a bold step into a future where machine co-intelligence is not just about algorithms, data, and models, but also about strategy, interaction, incentives, coopetition and collaboration. We have laid out a path that is both mathematically rigorous and conceptually innovative, inviting you to think critically about how intelligent systems co-learn, co-adapt, and interact. This work challenges us to consider machine intelligence not just as a set of tools, but as a dynamic system of strategic actors, each playing a part in the game of collective intelligence.

The fusion of machine intelligence and game theory has the potential to transform how we think about problem-solving in our increasingly knowledge-and-data-driven world. We present a new frontier where two powerful disciplines, deep learning exemplified by transformer architectures and game theory, come together to develop innovative methods for training, optimizing, and analyzing machine intelligence models. The rise of transformers, particularly in generative models, has been an interesting development in machine learning. In the last months we have had several limitations of the initial architectures of transformers. These models, with their ability to handle vast amounts of data and relationships, are influencing content generation. However, as these models grow larger and with more variables, the challenges of consistency check, training, tuning, and ensuring their efficiency become more pronounced. Game theory offers a framework to tackle these challenges. In a world where decisions are not made in isolation, game theory teaches us how to think about interactions, competition, coopetition and cooperation, ideas that are surprisingly relevant to the consistency check, coherence and training of neural networks. Viewing the consistency check of transformers as a game opens up new possibilities for improving their performance and addressing their limitations. This book is intended for readers from diverse backgrounds who share an interest in both the practical applications of machine intelligence and the mathematical theories that support it. Whether you are a machine intelligence researcher, data scientist, mathematician, or someone with a keen interest in the future of intelligent systems, this text will offer you a novel perspective. While the material is rigorous and technical in places, our goal is to provide a bridge between theory and real implemented architectures, showing how the seemingly abstract concepts of game theory can be applied to real-world machine intelligence models. In writing this book, we aim to spark a new conversation in the field of machine intelligence. As technology continues to evolve, it is important to look for synergies between different fields of study. We hope this exploration into the intersection of deep learning and game theory will inspire further research and new innovations in the machine intelligence community. We also acknowledge that the ideas presented here are part of an ongoing journey. As deep learning architectures grow ever more sophisticated and the data becomes even more interdependent, the strategic dimension of machine intelligence, how models make decisions, learn from their environment, and interact with one another, will only become more critical. It is our hope that this book provides readers with both a deeper understanding of the inner workings of machine intelligence and new tools for pushing the boundaries of what machine intelligence can achieve.

About the Author

Hamidou Tembine is a professor of machine intelligence at the University of Quebec in Trois-Rivieres, Canada, and the co-founder of Timadie, which is a platform of platforms that brings together companies, laboratories, and professional associations. It includes various platforms such as Guinaga for blockchain in agriculture, Grabal for livestock and poultry, SK1 Sogoloton for Distributed Information, WETE for women's empowerment through entrepreneurship, the knitting and crocheting club, women drone pilots, CI4SI for Collective Intelligence for Societal Impact, AI Mali, LABINCO, and more. Since 2010, Tembine has been the founding director of the Learning & Game Theory Laboratory and, since 2020, one of the principal investigators of the Stability Research Center. He has published over 300 scientific publications and has received over 7000 citations and an h-index of 42 to date. Tembine has received more than 10 awards for best scientific articles. He is also the recipient of the Simons Prize in 2020, the Next Einstein Forum Fellow in 2018, and the Best Young Researcher Award in 2014 from the IEEE Communication Society. He has authored five books on game theory and learning, including his publication on *Distributed Strategic Learning* in 2012 by CRC Press, Taylor & Francis. He is a co-author of the book on *Machine Intelligence in Africa in 20 questions*, published in June 2023. He graduated in Applied Mathematics from Ecole Polytechnique (Palaiseau, France) and earned a Ph.D. degree in Computer Science from INRIA and University of Avignon, France. He further received a master's degree in game theory and economics. His main research interests are learning, evolution, and games. In 2019, he was certified by the MIT Sloan School of Management on Business Blockchain Technologies. Tembine has been co-organizer of several scientific meetings on game theory in agriculture, water, food, environment, networking, wireless communications, transportation systems, and smart energy systems. He has been a visiting researcher at the University of California at Berkeley (US, 2011-2012), University of McGill (Montreal, Quebec, Canada, 2010), University of Illinois at Urbana-Champaign (UIUC, US, 2007-2010), Ecole Polytechnique Federale de Lausanne (EPFL, Switzerland, 2008-2009) and, University of Wisconsin (Madison, US, 2009).

Symbols

Symbol Description

$\mathbb{NN}$	Neural Network	$\|.\|$	Norm operator on the Hilbert space
L	Number of layers	λ_l	Learning rate
H	Number of Heads	$\mathcal{R}_\alpha$	Risk-measure
Q	Query	R_l	Activation function
K	Key	r_l	Activation function in one dimension or entrywise
V	Value		
W	Attention Weight	$\mathbb{P}$	Mean-field term $\mathbb{P}(dx_0 dy_L)$
W_1	Weight	$\phi\#\nu$	Pushforward measure of μ under the map ϕ
b_1	Bias		
W_2	Weight	γ	Operator averagedness level
b_2	Bias	$\partial\phi$	Subgradient
x_0	Initial signal	$ReLU$	$[.]_+ = \max(0, .)$ maximum between 0 and the argument.
μ_0	initial data distribution		
ν^l	Distribution at layer l		
y^L	Output signal	σ	Sigmoid
$\mathcal{H}_l$	Non-zero Hilbert space	S	Boltzmann-Gibbs distribution: softargmax
$\mathcal{O}_{l,nn}$	Normalization Operator of layer l		
$\mathcal{O}_{l,ff}$	2-Layered Feedforward Operator of layer l	S	Boltzmann-Giibbs distribution: softargmin
$\mathcal{O}_{l,att}$	Attention Operator of layer l	$\mathcal{L}$	$\mathcal{L}(\mathcal{H}^d, \mathcal{H}^k)$ Linear operators from $\mathcal{H}^d$ to $\mathcal{H}^k$
Id_l	Identity Operator		
$\mathcal{O}_l$	Operator of layer l	D	sequence/memory/data size

Example of transformer used in the book:

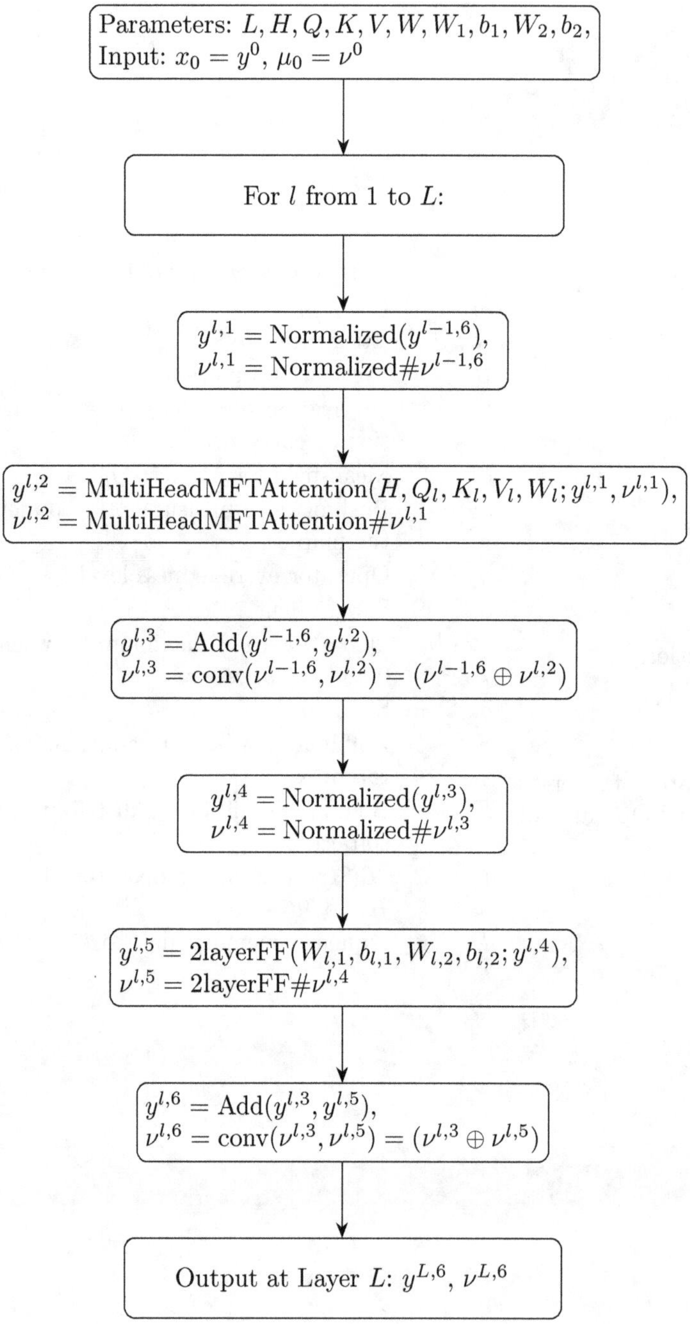
Parameters: $L, H, Q, K, V, W, W_1, b_1, W_2, b_2$,
Input: $x_0 = y^0$, $\mu_0 = \nu^0$
For l from 1 to L:
$y^{l,1} = \text{Normalized}(y^{l-1,6})$,
$\nu^{l,1} = \text{Normalized}\#\nu^{l-1,6}$
$y^{l,2} = \text{MultiHeadMFTAttention}(H, Q_l, K_l, V_l, W_l; y^{l,1}, \nu^{l,1})$,
$\nu^{l,2} = \text{MultiHeadMFTAttention}\#\nu^{l,1}$
$y^{l,3} = \text{Add}(y^{l-1,6}, y^{l,2})$,
$\nu^{l,3} = \text{conv}(\nu^{l-1,6}, \nu^{l,2}) = (\nu^{l-1,6} \oplus \nu^{l,2})$
$y^{l,4} = \text{Normalized}(y^{l,3})$,
$\nu^{l,4} = \text{Normalized}\#\nu^{l,3}$
$y^{l,5} = \text{2layerFF}(W_{l,1}, b_{l,1}, W_{l,2}, b_{l,2}; y^{l,4})$,
$\nu^{l,5} = \text{2layerFF}\#\nu^{l,4}$
$y^{l,6} = \text{Add}(y^{l,3}, y^{l,5})$,
$\nu^{l,6} = \text{conv}(\nu^{l,3}, \nu^{l,5}) = (\nu^{l,3} \oplus \nu^{l,5})$
Output at Layer L: $y^{L,6}$, $\nu^{L,6}$

Introduction

Machine intelligence (MI, [72, 71]) focuses on the creation of models, evolutionary dynamics, and algorithms that enable machines or software to co-learn from data and improve their performance over time. Machine intelligence is therefore an advanced computer science that allows a machine, device, software, program, code, or algorithm to interact intelligently with its environment, which means it can collect data, make decisions, perform actions, and develop strategies to maximize its chances of successfully achieving its preferences and objectives [42]. A neural network is a computational model inspired by the structure and function of the neural units, comprised of interconnected nodes or machine/artificial neural units that process and transform information to solve various machine learning and pattern recognition tasks. A large language model is a type of machine intelligence model that uses a vast amount of data to understand and generate human-like text or language responses. The methodology has been extended to large learning models that go beyond texts. A large learning model typically refers to a learning model, often a neural network, that is trained on a substantial amount of data to perform various tasks or make risk-aware predictions with high accuracy. In recent years, both deep learning and game theory have emerged as powerful tools in various fields, from machine intelligence to engineering to economics. We provide the intersection of these two domains, offering insights into how they complement and enhance one another in solving societal problems:

- Machine learning is part of machine intelligence.
- Deep learning is part of machine learning.
- Generative pretrained transformer is part of deep learning.
- Generative pretrained transformer is a specific large learning model.
- Deep learning is a game.
- Generative pretrained transformer outcomes are Nash equilibria of non-zero sum hierarchical game.
- Pre-training generative transformer is a game (coalitional, cooperative, non-cooperative depending on the implementation).
- Generative models focus on the joint distribution of input-output. There are many other generative machine intelligence tools that differ from pre-trained generative transformers. It includes Long Short-Term Memory, Gated Recurrent Units, Convolutional Neural Networks, Variational Encoders, Generative Adversarial Networks, Neural ODEs, Diffusion Models, Sub/Super Diffusion Models, Sparse Mixture-of Experts, Gaussian Mixture Models, Hidden Markov Models, Distributional Reinforcement Learning, etc.
- There are also other machine intelligence methods used in discriminative tasks such as classification, identification, etc.

- Classifying methods strictly as generative or discriminative can be technically misleading because many methods have dual capabilities. A movie model trained on predicting the next image/scene (discriminative) can generate videos by iteratively applying this discriminative task. Some tasks require both generation and discrimination. Q-reinforcement learning uses discriminative aspects from estimated rewards and generation from the transition to the next state. Generative adversarial networks use co-training from a generator and discriminator. Joint Embedding Predictive Architecture uses self-supervised learning, and combine both discriminative and generative tasks.
- Coalitional training of transformers is a game.
- Federated training is a game.
- Diffusion-Transformer is a Mean-Field-Type Game.
- Self-Attention Mechanism is a Mean-field-Type Game (see Figure 0.1).

This book is designed for readers who are interested in the mathematical underpinnings of deep learning models, particularly transformers, and those who are curious about how game theory can be applied to training and optimizing these models. The opening chapter, Deep Learning Meets Game Theory [191, 122, 84], introduces the foundational architectures of deep learning, focusing on the significance of activation functions and their properties. It presents how learning processes in neural networks can be framed as games, exploring various game structures such as single-leader, single-follower setups, and hierarchical games. This perspective brings a fresh angle to understanding deep learning, viewing it as a competitive or cooperative dynamic between different components of a model.

As the book progresses into Chapter 2, Mathematics of Transformers [64], it zooms in on the transformer architecture, which is applied to natural language processing and generative machine intelligence. The chapter explores key mathematical concepts that define transformers, including attention mechanisms, normalization challenges, and tensor-graph neural networks. The idea that transformers can be understood through the lens of game theory provides a novel framework for tackling optimization and performance issues.

Moving on to Extremely Large Transformers in Chapter 3, the book introduces the concept of mean-field limits, a crucial consideration when dealing with vast datasets and large-scale models. It addresses the asymptotics and convergence behaviors of transformers, providing insights into their scalability and performance in real-world applications.

Chapter 4, Mean-Field-Type Transformers [200], extends the discussion of game-theoretic models to a more specialized class of transformers, focusing on mean-field-type self-attention mechanisms and their applications in federated learning. The connection to game theory is evident in how these models manage distributed training across multiple agents, reflecting strategic interactions in multi-agent systems.

Chapter 5, Mean-Field-Type Learning, focuses on distribution-dependent adjustments towards the target.

The final chapter, Strategic Deep Learning [189, 184, 178, 121], brings everything together by discussing the strategic choices involved in mean-field-type deep learning, especially when multiple intelligent agents are involved. It introduces key concepts related to architecture selection, decision-making, and mean-field-type learning strategies in environments where machine learning agents interact with one another.

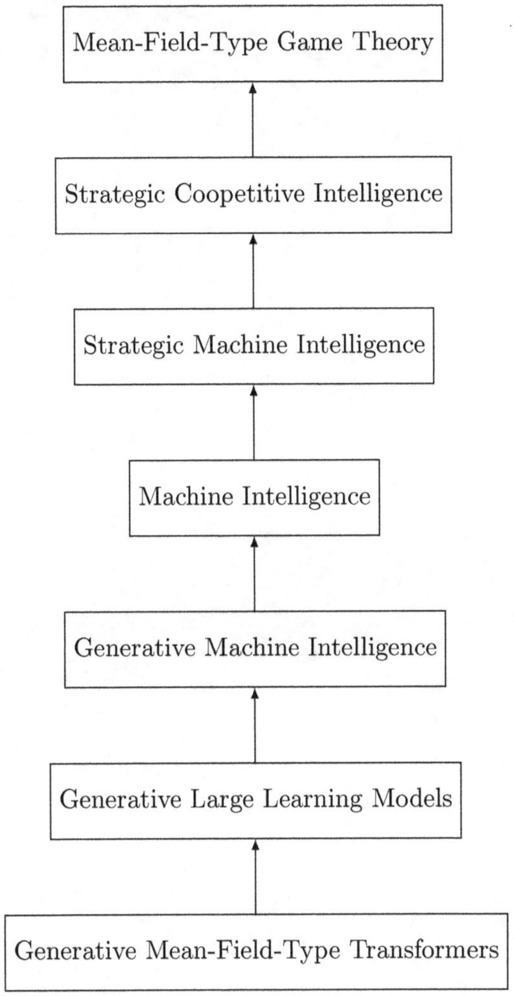

FIGURE 0.1
Connection Between Mean-Field-Type Game Theory and Transformers

The content of this book is an advanced text that challenges the reader to think of deep learning not only as a computational problem but as a strategic one. It is ideal for those who seek to deepen their understanding of the mathematics behind neural networks and the strategic interactions that govern their training and optimization. Whether you are a researcher, practitioner, or student, this book will equip you with the tools to explore the future of machine intelligence from a game-theoretic perspective.

1

Deep Learning Meets Game Theory

In recent years, two fields that were once largely independent: strategic deep learning and game theory, have begun to converge in profound ways. This merger has led to the rise of novel frameworks that blend strategic decision-making with the complexity of modern strategic-learning-enabled machine learning architectures. At the forefront of this convergence is the generative transformer model, which has sparked significant advancements in the applications of machine intelligence. In this chapter, we explore how these powerful deep learning models intersect with game theory. The surge of machine intelligence is reshaping industries, from finance and healthcare to logistics and entertainment. At the heart of this revolution is deep learning, a subset of machine learning and subset of machine intelligence that powers systems like chatbots, recommendation engines, and predictive analytics. But as these neural networks become more complex, optimizing and training them has become a strategic challenge, one that increasingly resembles the high-stakes dynamics of competitive markets or diplomatic negotiations. That is where game theory comes in. Typically used to analyze interactive scenarios in engineering, computer science, biology, economics and business, game theory offers a powerful framework to understand interactions between different agents or systems, each trying to achieve their own goals. We look at how these seemingly distinct worlds, deep learning and game theory, can converge to offer fresh perspectives on machine intelligence development.

The chapter opens by breaking down the architecture of deep learning models, explaining the role of bounded linear operators and activation functions, those mathematical operations that guide how a neural network processes information. But what is particularly intriguing is the proposal to view deep learning outcomes as games, where the components of a neural network, its layers, nodes and neurons, interact much like decision-makers in a strategic competition. The choices made by one part of the network influence the others, much like how moves in a game affect other participants. From single-leader, single-follower structures to multi-layer hierarchical games, we map out how every neural network can be interpreted as a game. This game-theoretic perspective is not just an academic exercise. It opens new doors for improving how machine intelligence systems are trained and optimized. By considering the interactions within a network as strategic decisions, developers can potentially enhance both the performance and stability of their models. It introduces the foundational elements of deep learning and game theory. It explores how activation functions in neural networks, such as HoloNorm, ReLU, SiLU, GELU, Swish or Sigmoid, can be reinterpreted as strategies in a game. Each layer of the network can be viewed as an agent in a hierarchical game, where decisions made by one layer influence the others. Concepts like averaged non-expansive activation functions and maximally cyclically monotone operators are introduced as ways to mathematically describe these interactions. The chapter also addresses critical challenges facing machine intelligence training today, from gradient descent inefficiencies to the structures of activation function design. We show that a deeper understanding of game dynamics can lead to breakthroughs in training algorithms, offering new strategies for refining deep learning models in ways that conventional methods have struggled to achieve. As the chapter unfolds, readers are guided through real-world examples, including the practical applications of residual networks and the cutting-edge transformer architectures that have transformed text and image processing. The message is clear: *machine intelligence is no longer just about raw computation and massive datasets. It is also about strategy, interaction, risk-awareness and optimization, elements that game theory can help unlock.* As machine intelligence continues to reshape global markets and industries, the insights offered in this chapter suggest that the next leap forward may come not just from better data or bigger models, but from smarter, more strategic training and learning approaches.

1.1 Deep Learning Architectures

This section formulates deep neural network architectures layer by layer and uses operator theory to characterize the outcomes of these networks.

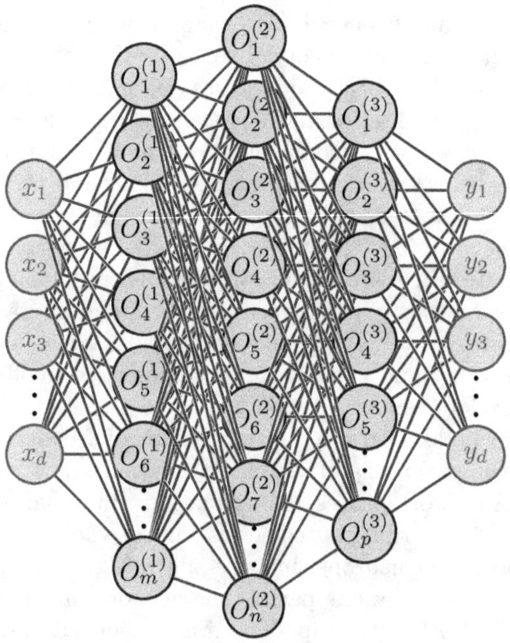

FIGURE 1.1
A schematic representation of the deep learning architecture considered in this chapter. $O_k^{(l)} := O_{l,t}(k)$. The integers m, n, p, d may be different.

We consider a neural network (see Figures 1.1, 1.2, 1.3) with $L \geq 1$ layers. The layer operations run over $\{\mathcal{H}_l\}_{0 \leq l \leq L}$ non-zero Hilbert spaces. For every layer $l \in \{1, \ldots, L\}$, let $W_l : \mathcal{H}_{l-1} \to \mathcal{H}_l$ be a bounded linear operator, and consider the family of activation operators $R_l : \mathcal{H}_l \to \mathcal{H}_l$. The neural network is defined by the composition of operators

$$R_L \circ (W_L \cdot + b_L) \circ R_{L-1} \circ (W_{L-1} \cdot + b_{L-1}) \circ \ldots \circ R_1(W_1 \cdot + b_1), \tag{1.1}$$

where W_l is the weight operator and b_l captures the bias parameter. The number L is referred to as the depth of the neural network.

- When L is large enough, it is called a deep neural network, and the associated learning algorithm is called deep learning.
- When L is relatively small it is a shallow learning.

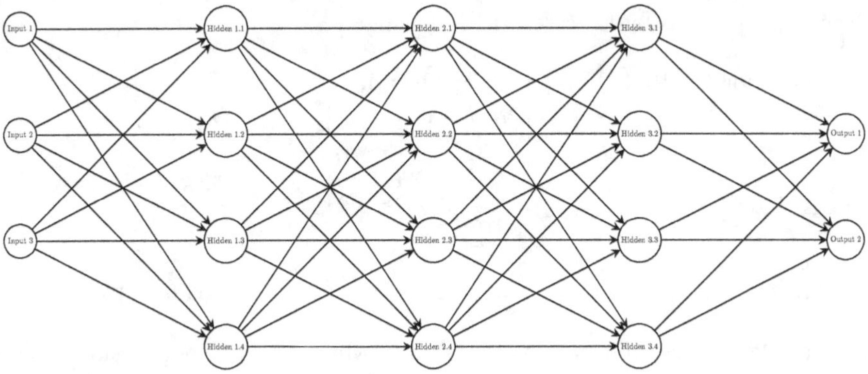

FIGURE 1.2
An example of neural network: 3 input neurons, 2 output neurons, three hidden layers, each with four neurons, between the input and output layers.

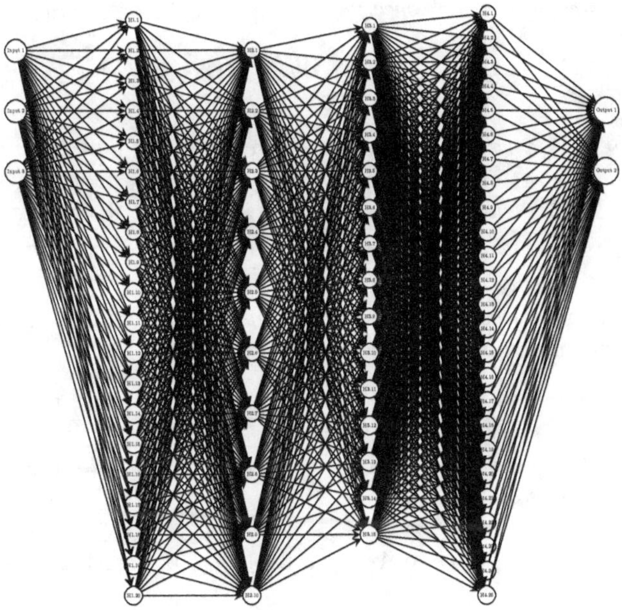

FIGURE 1.3
An example of a dense neural network: 3 input neurons, 20 neurons in the first hidden layer, 10 neurons in the second hidden layer, 15 neurons in the third hidden layer, 25 neurons in the fourth hidden layer, 2 output neurons.

Note, however, that the threshold and border between deep learning and shallow learning are unclear because it has been moving in the last four decades.

We look at a consistency check of a neural network by iterating it. Let $x_0 \in \mathcal{H}_0$ be an initial signal. This signal goes to the input layer. Let $\{\lambda_t\}_{t\geq 0}$ be a non-negative sequence, serving as the learning rate of the algorithm. At layer $l \geq 1$, consider $\hat{O}_l$ the map defined by

$$\hat{O}_l : \left| \begin{array}{l} \mathcal{H}_{l-1} \to \mathcal{H}_l \\ x \mapsto R_l(W_l x + b_l). \end{array} \right.$$

Iterate layer by layer and by timestep the following:

$$\text{for } t \in \{0,1,2,\ldots\} \left| \begin{array}{l} y_{1,t} = \hat{O}_1(x_t), \\ \text{for } l \in \{2,\ldots,L\},\ y_{l,t} = \hat{O}_l(y_{l-1,t}), \\ x_{t+1} = x_t + \lambda_t(y_{L,t} - x_t). \end{array} \right. \tag{1.2}$$

which means that

$$\begin{array}{l} x_0 \in \mathcal{H}_0, \\ \text{For } t \in \{0,1,2,\ldots\}:\ \ x_{t+1} = x_t + \lambda_t(\hat{O}_L \circ \hat{O}_{L-1} \circ \ldots \circ \hat{O}_1(x_t) - x_t). \end{array} \tag{1.3}$$

Equation (1.2) defines an iterative algorithm in which an averaged version of the output from the last layer is fed back as an input signal, and the process is repeated. For consistency of the operation, the output of $\hat{O}_L$ should be in the same space as the input, i.e., $\mathcal{H}_L \subseteq \mathcal{H}_0$. As we want to examine situations in which time goes to infinity, the cumulative step-size $\sum_{t=1}^{T} \lambda_t \to +\infty$ as T goes to infinity.

Definition 1. *The neural network specification is given by the family*

$$\mathbb{NN} := (x_0, L, \{R_l, W_l, b_l, \mathcal{H}_l\}_{1\leq l\leq L}, \lambda).$$

Problem 1. *What does a well-trained Deep Neural Network do? To answer this question, we aim to find and characterize the behavior of the neural network for a large class of architectures* $\mathbb{NN}$*. To this end, we consider algorithm in (1.2) and examine the possible limit (if any) of* $(x_t, y_{1,t}, \ldots, y_{L,t})$ *as t goes to infinity.*

Problem 2 (Training & Design)**.** *Given an input-output data set* $\{(x_i, d_i = y_{L,i}),\ i \in \{1,\ldots,D\}\}$*, find* $\mathbb{NN}$ *such that*

$$\hat{O}_L \circ \hat{O}_{L-1} \circ \ldots \circ \hat{O}_1(x_i) = d_i,\ \forall i \in \{1,\ldots,D\},$$

which is reformulated in a weak sense as evolving equalities: Find a vector $\theta^* := (W^*, b^*) = (W_l^*, b_l^*)_{l\in\{1,2,\ldots,L\}}$ *such that*

$$r_L \circ \tilde{A}_{L,\theta_L^*} \circ \ldots \circ (r_1 \circ \tilde{A}_{1,\theta_1^*})(x_i) - y_{L,i} = 0,\ \ i \in \{1,\ldots,D\}, \tag{1.4}$$

where $\tilde{A}_l = \tilde{A}_{l,\theta_l} :\ x_{l,i} \mapsto W_l x_{l,i} + b_l$ *and* $A_{l,i} = A_{l,x_{l,i}} :\ \theta_l \mapsto W_l x_{l,i} + b_l$*, which is linear and non-zero, and* $A_{l,i}^{\dagger}$ *is the adjoint operator of* $A_{l,i}$*.*

The latter set of equalities is replaced by variational inequalities:

Find a vector θ^* *such that the set of variational inequalities*

$$\left\langle (r_L \circ \tilde{A}_{L,\theta_L^*}) \circ \ldots \circ (r_1 \circ \tilde{A}_{1,\theta_1^*})(x_i) - y_{L,i}, A_L^{\dagger}(\theta_L - \theta_L^*) \right\rangle \geq 0,\ \forall \theta,\ \ i \in \{1,\ldots,D\}. \tag{1.5}$$

hold.

Find a vector $\theta^* := (W^*, b^*) = (W_l^*, b_l^*)_{l\in\{1,2,\ldots,L\}}$ *such that*

$$\sum_{i=1}^{D} \omega_i' \left\langle (r_L \circ \tilde{A}_{L,\theta_L^*}) \circ \ldots \circ (r_1 \circ \tilde{A}_{1,\theta_1^*})(x_i) - y_{L,i}, A_L^{\dagger}(\theta_L - \theta_L^*) \right\rangle \geq 0,\ \forall \theta, \tag{1.6}$$

where $\omega_i' > 0$, $\sum_{i=1}^{D} \omega_i' = 1$*. Moreover, design an algorithm that approximates a* θ^**, a solution (if any) to (1.6).*

Due to the hierarchical structure of NN, the resulting squared norm of the operator $\hat{O}_L \circ \hat{O}_{L-1} \circ \ldots \circ \hat{O}_1$ is not necessarily convex in the weights W and biases b. The idea here is to exploit the hidden convexity within each layer to propose efficient algorithms for pre-training, training and for analyzing the outcomes of both shallow and deep learning.

Risk-awareness is vital in a generative machine intelligence because it helps identify and mitigate potential weaknesses and uncertainties within the system. It allows for the assessment of the confidence levels of individual subnetworks, enabling more reliable decision-making by weighting their outputs accordingly. Furthermore, being risk-aware is essential for detecting and defending against adversarial attacks, addressing ethical concerns related to bias and fairness, ensuring regulatory compliance, and making informed decisions about how to fuse the outputs of different subnetworks, ultimately enhancing the overall reliability and robustness of the network.

Problem 3 (Risk-aware design and training). *The expectile risk-aware training problem in generative machine intelligence is to find L and the operators $\hat{O}_1, \ldots, \hat{O}_L$:*

$$\arg\min_{L\geq 1} \arg\min_{\hat{O}_L,\ldots,\hat{O}_1} \mathcal{R}_\alpha[\|y_L - \hat{O}_L \circ (\hat{O}_{L-1} \circ \ldots \circ \hat{O}_1 x_0)\|^2], \tag{1.7}$$

where $\mathcal{R}_\alpha$ is a function generating a risk measure such as the α-expectile with $\frac{1}{2} < \alpha < 1$. The randomness here is the input-output pairs $(x_0, d_L = y_L)$. The expectile will be discussed in Definition 6.

Given the structure of the hierarchical information, we can invoke here the Stackelberg solution.

For $\alpha = \frac{1}{2}$ the α-expectile risk-aware design and training problem reduces to the risk-neutral case: find L and the operators $\hat{O}_1, \ldots, \hat{O}_L$:

$$\arg\min_{L\geq 1} \arg\min_{\hat{O}_L,\ldots,\hat{O}_1} \int_{x_0} \int_{y_L} [\|y_L - \hat{O}_L \circ (\hat{O}_{L-1} \circ \ldots \circ \hat{O}_1 x_0)\|^2] \mathbb{P}(dx_0 dy_L). \tag{1.8}$$

1.2 Averaged Non-Expansive Activation Functions

Averagedness is a property of interest for generative machine intelligence for the following reasons: (i) The averaged non-expansive operator $\hat{O}$ allows us to approximate $kernel(Id - \hat{O})$, which tackles the consistency problem. (ii) It provides the weak convergence of well-trained neural networks. (iii) It connects the fixed-points $\hat{O} = \hat{O}_L \circ \hat{O}_{L-1} \circ \ldots \circ \hat{O}_1$, i.e., $kernel(Id - \hat{O})$ with the intersection of fixed-points $\hat{O}_l$, i.e., $kernel(Id - \hat{O}_l)$.

Definition 2. *Let $\gamma \in (0, 1]$. An operator $\hat{O}: \mathcal{H}_0 \to \mathcal{H}_0$ is γ-averaged if $[Id + \frac{1}{\gamma}(\hat{O} - Id)]$ is 1-Lipschitz continuous.*

Some functions $R := (r_k)_k$ currently used in deep neural network architectures are displayed in Tables 1.1, 1.2 and 1.3 below.

Lemma 1. *For each of the entry-wise activation functions considered in Tables 1.1, 1.2 and 1.3, one has the following properties up to a certain scaling:*

(a) r_k *is a γ-averaged operator.*

TABLE 1.1
Activation functions that are γ-averaged, $\gamma \in (0, 1]$.

Activation function	Expression	γ-averagedness
Identity	$r_1(x) = Id(x) = x$	1
Linear	$r_2(x) = \langle \lambda, x \rangle + b$	$\frac{(1+\|\lambda\|)}{2}\mathbb{I}_{\{\|\lambda\| \leq 1\}}$
Rectified linear unit (ReLU)	$r_3(x) = \max(0, \langle \lambda, x \rangle + b)$	$\frac{(1+\|\lambda\|)}{2}\mathbb{I}_{\{\|\lambda\| \leq 1\}}$
Logistic (Sigmoid, Soft step)	$r_4(x) = \frac{1}{1+exp(-\langle \lambda, x \rangle - b)}$	$\frac{(4+\|\lambda\|)}{8}\mathbb{I}_{\{\|\lambda\| \leq 1\}}$
Sigmoid	$\sigma(x) = sigmoid(x) = \frac{1}{1+e^{-x}}$	$\frac{5}{8}$
Hyperbolic tangent	$r_5(x) = \lambda \frac{e^x - e^{-x}}{e^x + e^{-x}}$	$\frac{(1+\|\lambda\|)}{2}\mathbb{I}_{\{\|\lambda\| < 1\}}$
Softargmax	$r_6(x) = \frac{e^{\lambda x_i}}{\sum_{k=1}^K e^{\lambda x_k}}$	$\frac{(1+\|\lambda\|)}{2}\mathbb{I}_{\{\|\lambda\| < 1\}}$
Gaussian error linear unit (GELU2)	$r_7(x) = \lambda \frac{1}{2} x \left[1 + tanh\left(\sqrt{\frac{2}{\pi}}(x + 0.044715x^3)\right)\right]$	$\frac{18}{20}$
Gaussian error linear unit	$GELU(x) = \lambda(x)\mathbb{P}(X \leq x) = \lambda(x)\frac{1}{2}\left[1 + erf(\frac{x}{\sqrt{2}})\right], \quad X \sim \mathcal{N}(0, 1)$	$\frac{(1+\|\lambda\|)}{2}\mathbb{I}_{\{\|\lambda\| \leq 1\}}$
Softplus	$r_8(x) = \lambda \log(1 + e^x)$	$\frac{(1+\|\lambda\|)}{2}\mathbb{I}_{\{\|\lambda\| \leq 1\}}$
Softplus	$softplus(x) = \frac{1}{\lambda}\log(1 + e^{\lambda x})$	1
	$softplus(x) = \log(1 + \sum_{k=1}^d e^{x_k})$	1
Exponential linear unit (ELU)	$r_9(x) = \lambda(e^x - 1)\mathbb{I}_{\{x \leq 0\}} + x\mathbb{I}_{\{x > 0\}}$	1
Scaled exponential linear unit	$r_{10}(x) = \lambda[\alpha(e^x - 1)\mathbb{I}_{\{x<0\}} + \alpha x \mathbb{I}_{\{x \geq 0\}}], \quad (\alpha, \lambda) = (0.0507, 0.6733)$	$\lambda\alpha$
Leaky rectified linear unit	$r_{11}(x) = 0.01x\mathbb{I}_{\{x<0\}} + x\mathbb{I}_{\{x \geq 0\}}$	1
Parametric rectified linear unit	$r_{12}(x) = \lambda x \mathbb{I}_{\{x<0\}} + x\mathbb{I}_{\{x \geq 0\}}$	1

(b) *For all entry-wise activation function r_k in one-dimension that is γ-averaged with $\gamma < 1$, the operator r_k can be written as*

$$r_k = [Id + \partial f_k]^{-1} = \arg\min_y (f_k(y) + \frac{1}{2}\| \cdot - y\|^2) \tag{1.9}$$

for some proper lower semi-continuous convex function f_k, where $\partial f_k(x) := \{p, \ \forall y, \langle x - y, p\rangle + f_k(x) \leq f_k(y)\}$ denotes the subdifferential of f_k. Moreover, we have

$$r_k = \partial(\frac{1}{2}\| \cdot \|^2 + f_k)^*, \tag{1.10}$$

where $\psi^*(x) = \sup_y[\langle x, y\rangle - \psi(y)]$ is the Legendre-Fenchel conjugate of ψ.

(c) *$Fix(r_k) = \arg\min(f_k)$, i.e. the fixed-points of r_k are the minimizers of f_k.*

For a proof of the second equality in (b) see Proposition 6.a. in [147] (see also the equality

TABLE 1.2
Activation functions that are γ-averaged, $\gamma \in (0,1]$.(cont.)

Activation function	Expression	γ-averagedness
Sigmoid linear unit (Sigmoid shrinkage)	$r_{13}(\epsilon x) = \frac{\epsilon x}{1+e^{-\epsilon x}}$	1
Swish	$r_{14}(x) = \epsilon x\ sigmoid(\lambda x)$	$\frac{10+11\epsilon}{20}$
Gaussian	$r_{15}(x) = e^{-\langle x,x\rangle}$	$\frac{1+e^{-1}}{2}$
Maxout	$r_{16}(x) = \max_k x_k$	1
Approximate Heaviside / Binary step	$r_{17}(x) = \sigma(x/\epsilon)$	$\frac{(1+4\epsilon)}{8\epsilon}\mathbb{I}_{\{\epsilon\geq 1/4\}}$
Multiquadratics	$r_{18}(x) = \sqrt{(x-\alpha)^2+\lambda^2}$	1
Inverse multiquadratics	$r_{19}(x) = \frac{1}{\sqrt{(x-\alpha)^2+(1+\lambda)^2}}$	$\frac{2+\lambda}{2(1+\lambda)}$
Mish	$r_{20}(x) = x\ tanh(\ softplus(x))$	see Lemma 6
Metallic mean	$r_{21}(x) = \frac{x+\sqrt{x^2+4}}{2}$	$\frac{1}{2}$
Arc tangent	$r_{22}(x) = tan^{-1}(x)$	1
Softsign	$r_{23}(x) = \frac{x}{1+\lvert x\rvert}$	1
Inverse square root unit	$r_{24}(x) = \frac{x}{\sqrt{1+(1+\lambda)x^2}}$	$\frac{1+\sqrt{1+\lambda}}{2\sqrt{1+\lambda}}$
Inverse square root linear unit	$r_{25}(x) = \frac{x}{\sqrt{1+\lambda x^2}}\mathbb{I}_{\{x<0\}} + x\mathbb{I}_{\{x\geq 0\}}$	1
Square nonlinearity	$r_{25}(x) = -\mathbb{I}_{\{x<-2\}} + (x+\frac{x^2}{4})\mathbb{I}_{\{-2\leq x<0\}} + (x-\frac{x^2}{4})\mathbb{I}_{\{0\leq x\leq 2\}} + \mathbb{I}_{\{x>2\}}$	1
Bent identity	$r_{26}(x) = \frac{2}{3}\lambda(x+\frac{-1+\sqrt{1+x^2}}{2})$	λ

(16.31) in [35]). By Corollary 16.24 in [35], the equality (1.10) follows from

$$[\partial(\frac{1}{2}\|\cdot\|^2 + f_k)]^{-1} = \partial[(\frac{1}{2}\|\cdot\|^2 + f_k)^*],$$

where $\psi^*(x) = \sup_y[\langle x,y\rangle - \psi(y)]$ is the Legendre-Fenchel conjugate of ψ. For a proof of (c) see Proposition 12.28 in [35].

Lemma 2 (See Theorem 3.1 in [16] and Chap. 5 in [35]). *Let $L \geq 1$ be an integer and consider the sequence $0 < \gamma_l \leq 1,\ l \in \{1,\dots,L\}$.*

(i) Let $0 < \gamma \leq 1$. Then $[Id + \frac{1}{\gamma}(\hat{O} - Id)]$ is 1-Lipschitz continuous if and only if

$$\forall (x,y) \in \mathcal{H}_0^2,\ \|\hat{O}(x) - \hat{O}(y)\|^2 \leq \|x-y\|^2 - \frac{(1-\gamma)}{\gamma}\|x - \hat{O}(x) - y + \hat{O}(y)\|^2.$$

(ii) The composition $\hat{O}_L \circ \hat{O}_{L-1} \circ \dots \hat{O}_1$ of γ_l-averaged operators $\hat{O}_l : \mathcal{H}_0 \to \mathcal{H}_0$ is $\frac{1}{1+\frac{1}{\sum_{l=1}^{L}\frac{\gamma_l}{1-\gamma_l}}}$-averaged.

(iii) Assume each $\hat{O}_l$ is a γ_l-averaged operator. Let $\omega_l \geq 0$ such that $\sum_{l=1}^{L}\omega_l = 1$. Then the weighted sum operator $\sum_{l=1}^{L}\omega_l\hat{O}_l$ is $\sum_{l=1}^{L}\omega_l\gamma_l$-averaged.

TABLE 1.3
Activation functions that are γ-averaged, $\gamma \in (0,1]$ (cont.)

Softexponential	$r_{27}(x) = -\frac{\log(1-\lambda(x+\lambda))}{\lambda}\mathbb{I}_{\{\lambda<0\}} + x\mathbb{I}_{\{\lambda=0\}} + (\lambda + \frac{e^{\lambda x}-1}{\lambda})\mathbb{I}_{\{\lambda>0\}}$	1
Soft clipping	$r_{28}(x) = \frac{1}{\lambda}\log(\frac{1+e^{\lambda x}}{1+e^{\lambda(x-1)}})$	
Holonorm	$\tilde{r}_{28}(x) = (\frac{x_k}{1+\|x\|})_k$	see r_{23}
Sinusoid	$r_{29}(x) = sin(x)$	1
Sinc	$r_{29}(x) = \frac{sin(x)}{x}\mathbb{I}_{\{x\neq 0\}} + \mathbb{I}_{\{x=0\}}$	1
Piecewise linear	$r_{30}(x) = 0\mathbb{I}_{\{x\leq -\frac{1}{2}\}} + (x + \frac{1}{2})\mathbb{I}_{\{-\frac{1}{2}<x<\frac{1}{2}\}} + \mathbb{I}_{\{x>\frac{1}{2}\}}$	1
Sinu-sigmoidal Linear Unit	$r_{32}(x) = (x + \lambda\sin(\alpha x))\ sigmoid(x)$	see Lemma 6
Complementary Log-Log	$r_{33}(x) = 1 - e^{-e^{x}}$	3/4
Bipolar Sigmoid	$r_{34}(x) = \frac{1-e^{-x}}{1+e^{-x}}$	see r_5
Hard Tanh	$r_{35}(x) = \max(-1, min(1,x))$	1
Absolute value	$r_{36}(x) = \|x\|$	1
Logit	$r_{36}(x) = \frac{1}{10}\log(\frac{x}{1-x})\mathbb{I}_{[1/4,3/4]}(x)$	3/4
Softsign(Probit)	$r_{37}(x) = softsign(\Phi^{-1}(x)),\ \Phi(x) = \frac{1}{2}\left[1 + erf(\frac{x}{\sqrt{2}})\right]$	$\frac{9}{10}$
Linear Gaussian	$r_{38}(x) = xe^{-x^2}$	$\frac{2+\sqrt{e}}{4}$
Attention-based	$r_{39} = softargmax \circ r_0$	see Lemma 6
Attention-based	$r_{40} = r_{38} \circ softargmax \circ r_0$	see Lemma 6

(iv) If $\hat{O}$ is μ-Lipschitz continuous with $\mu < 1$. Then $\hat{O}$ is $\frac{(1+\mu)}{2}$-averaged.

Proof: We start by proving Lemma 6 (i). Let $0 < \gamma \leq 1$. The operator $[Id + \frac{1}{\gamma}(\hat{O} - Id)]$ is 1-Lipschitz if

$$\|(1-\frac{1}{\gamma})x + \frac{1}{\gamma}\hat{O}(x) - (1-\frac{1}{\gamma})y - \frac{1}{\gamma}\hat{O}(y)\| \leq \|x-y\|.$$

Consider the quantity

$$\begin{aligned} 0 &\leq \gamma[\|x-y\|^2 - \|(1-\tfrac{1}{\gamma})x + \tfrac{1}{\gamma}\hat{O}(x) - (1-\tfrac{1}{\gamma})y - \tfrac{1}{\gamma}\hat{O}(y)\|^2] \\ &= \gamma[\|x-y\|^2 - \tfrac{1}{\gamma^2}\| - (1-\gamma)(x-y) + \hat{O}(x) - \hat{O}(y)\|^2] \\ &= \gamma[\|x-y\|^2 - \tfrac{1}{\gamma^2}\Big((1-\gamma)^2\|x-y\|^2 + \|\hat{O}(x)-\hat{O}(y)\|^2 \\ &\quad -2(1-\gamma)\langle x-y, \hat{O}(x)-\hat{O}(y)\rangle\Big)]. \end{aligned} \tag{1.11}$$

Using the identity $-2\langle a,b\rangle = \|a-b\|^2 - \|a\|^2 - \|b\|^2$, we obtain

$$-2\langle x-y, \hat{O}(x)-\hat{O}(y)\rangle = \|x-\hat{O}(x)-(y-\hat{O}(y))\|^2 - \|x-y\|^2 - \|\hat{O}(x)-\hat{O}(y)\|^2.$$

It follows that

$$\begin{aligned}
0 &\leq \gamma[\|x-y\|^2 - \|(1-\tfrac{1}{\gamma})x + \tfrac{1}{\gamma}\hat{O}(x) - (1-\tfrac{1}{\gamma})y - \tfrac{1}{\gamma}\hat{O}(y)\|^2] \\
&= \gamma[\|x-y\|^2 - \tfrac{1}{\gamma^2}\Big((1-\gamma)^2\|x-y\|^2 + \|\hat{O}(x)-\hat{O}(y)\|^2 \\
&+(1-\gamma)\|x-\hat{O}(x)-(y-\hat{O}(y))\|^2 - (1-\gamma)\|x-y\|^2 \\
&-(1-\gamma)\|\hat{O}(x)-\hat{O}(y)\|^2\Big)] \\
&= \|x-y\|^2 - \|\hat{O}(x)-\hat{O}(y)\|^2 - \tfrac{(1-\gamma)}{\gamma}\|x-\hat{O}(x)-(y-\hat{O}(y))\|^2,
\end{aligned} \tag{1.12}$$

which completes the proof the γ-averaged operator inequality.

Lemma 6 (ii) is easily proved by induction. Start with the composition of two operators and apply the recursion.

We now prove (iii). We have

$$M_\gamma := Id + \frac{1}{\sum_{k=1}^{L}\omega_k\gamma_k}\left[\sum_{l=1}^{L}\omega_l\hat{O}_l - Id\right] = \sum_{l=1}^{L} p_l[Id + \tfrac{1}{\gamma_l}(\hat{O}_l - Id)] \tag{1.13}$$

where $p_l := \frac{\omega_l\gamma_l}{\sum_{k=1}^{L}\omega_k\gamma_k}$. Then $p_l > 0$ and $\sum_{l=1}^{L} p_l = 1$. Hence, M_γ is a 1-Lipschitz operator since it is a convex combination of 1-Lipschitz operators. This proves (iii).

We now focus on the statement (iv). We want to show that the operator $M_\mu = Id + \frac{2}{1+\mu}(\hat{O} - Id)$ 1-Lipschitz. We have $M_\mu = \frac{(1-\mu)}{1+\mu}(-Id) + \frac{2\mu}{1+\mu}(\frac{\hat{O}}{2})$, which is 1-Lipschitz as a convex combination of $-Id$ and $\frac{\hat{O}}{2}$, which are 1-Lipschitz operators. □

Lemma 3. *The following statements are equivalent: (i) The operator $\hat{O}$ is $\frac{1}{2}$-averaged. (ii)*

$$\|\hat{O}(x) - \hat{O}(y)\|^2 \leq \langle x-y, \hat{O}(x) - \hat{O}(y)\rangle, \ \forall (x,y).$$

(iii) $(1-\mu)Id + \mu\hat{O}$ is $\mu/2$-averaged for all $\mu \in (0,2)$.

Proof: $\hat{O}$ being $\frac{1}{2}$-averaged is equivalent to $\hat{O} = (Id+M)/2$ with M being 1−Lipschitz. This means that $M = 2\hat{O} - Id$. Hence, the operator $(1-\mu)Id + \mu\hat{O} = (1-\mu)Id + \mu(Id + M)/2 = (1-\mu/2)Id + (\mu/2)M$ is $\mu/2$-averaged for all $\mu \in (0,2)$.

Conversely, if $(1-\mu)Id + \mu\hat{O}$ is $\mu/2$-averaged for all $\mu \in (0,2)$, then by choosing $\mu = 1$ one obtains that O is $\frac{1}{2}$-averaged. □

1.2.1 From Averaged to Monotone Operators

Lemma 4. *$\hat{O}$ is $\frac{1}{2}$-averaged iff $O = \hat{O}^{-1} - Id$ is monotone.*

Proof: Let $\hat{O}$ be $\frac{1}{2}$-averaged. Then $\langle x - \hat{O}(x) - (y - \hat{O}(y)), \hat{O}(x) - \hat{O}(y)\rangle \geq 0, \ \forall (x,y)$. Let $\hat{O}(y_1 + x_1) = x_1$ and $\hat{O}(y_2 + x_2) = x_2$ with $y_i = Ox_i$. Set $z_i = y_i + x_i$. Let us compute $\langle y_1 - y_2, x_1 - x_2\rangle = \langle Ox_1 - Ox_2, x_1 - x_2\rangle = \langle z_1 - x_1 - (z_2 - x_2), \hat{O}(z_1) - \hat{O}(z_2)\rangle = \langle z_1 - \hat{O}(z_1) - (z_2 - \hat{O}(z_2)), \hat{O}(z_1) - \hat{O}(z_2)\rangle \geq 0$ hence $\langle Ox_1 - Ox_2, x_1 - x_2\rangle \geq 0$. This means that O is monotone.

Conversely, suppose that O is monotone. We choose $O(\hat{O}(x_i)) + \hat{O}(x_i) \ni x_i, \ i \in \{1,2\}$ in the graph. Then $O(\hat{O}(x_i)) \ni x_i - \hat{O}(x_i), \ i \in \{1,2\}$. By monotonicity $\langle O(\hat{O}(x_1)) - O(\hat{O}(x_2)), \hat{O}(x_1) - \hat{O}(x_2)\rangle \geq 0$ which leads to

$$\langle (x_1 - \hat{O}(x_1)) - (x_2 - \hat{O}(x_2)), \hat{O}(x_1) - \hat{O}(x_2)\rangle \geq 0,$$

which implies that $\hat{O}$ is $\frac{1}{2}$-averaged. □

The $\frac{1}{2}$-averagedness property is important because it is associated with monotone operators, which are crucial in variational inequalities.

Remark 1. *If $\hat{O}$ is $\frac{1}{2}$-averaged, then there is a monotone operator O such that $\hat{O} = (Id + O)^{-1}$. One can also use the operator $\hat{O}_\lambda = (Id + \lambda O)^{-1}$ approximation.*

1.2.2 Fixed-Points of Activation Functions

Lemma 5. *Let $(\rho_t, \mu_t, \epsilon_t)$ be nonnegative sequences such that $\sum_{t\geq 0} \epsilon_t < +\infty$ and $\rho_{t+1} \leq \rho_t - \mu_t + \epsilon_t$. Then, $\{\rho_t\}_t$ converges and $\sum_{t\geq 0} \mu_t < \infty$.*

The proof of Lemma 5 is straightforward.

The γ-averaged nonexpansive operators play key roles in the asymptotics of $(\hat{O} - Id)(x_t)$ even if the operator $\hat{O}$ has no fixed point. For example, the mapping $\hat{O}(x) = x + e$, with e non-zero, has no fixed-point. However, $\hat{O}(x_t) - x_t = e$, which converges to e. When $\hat{O}$ has a fixed-point, the difference $\hat{O}(x_t) - x_t$ goes to zero as t goes to infinity where x_t are the output of Algorithm in (1.2). These γ-averaged nonexpansive operators allow us to extend the well-known convergence results of the case of the contractive operators under Banach-Picard iterates.

Lemma 6 (See Theorem 3.1 in [16] and Chap. 5 in [35])**.** *Let $L \geq 1$ be an integer and consider the sequence $0 < \gamma_l \leq 1,\ l \in \{1, \ldots, L\}$.*

(v) Let Fixed($\hat{O}$) $= \{x \in \mathcal{H}_0 |\ x = \hat{O}(x)\}$ be the set of fixed-points of the operator $\hat{O}$. Assume that $\cap_{l=1}^{L}$ Fixed($\hat{O}_l$) $\neq \emptyset$. Then

$$\mathit{Fixed}(\sum_{l=1}^{L} \omega_l \hat{O}_l) = \mathit{Fixed}(\hat{O}_L \circ \hat{O}_{L-1} \circ \ldots \hat{O}_1) = \cap_{l=1}^{L} \mathit{Fixed}(\hat{O}_l).$$

(vi) (Propositions 5.14 and 5.15 in [35]) Let $0 < \gamma \leq 1$, and $\hat{O} : \mathcal{H}_0 \to \mathcal{H}_0$ be an γ-averaged operator such that Fixed($\hat{O}$) $\neq \emptyset$. Let $\{\lambda_t\}_{t\geq 1}$ be a sequence in $[0, \frac{1}{\gamma}]$ such that $\sum_{t\geq 1} \lambda_t(1 - \gamma \cdot \lambda_t) = +\infty$. Assume that $x_0 \in \mathcal{H}_0$ and set $x_{t+1} = x_t + \lambda_t(\hat{O}(x_t) - x_t)$. Then $\hat{O}(x_t) - x_t$ converges to 0 as t goes to infinity and x_t converges weakly to a point in Fixed($\hat{O}$).

Proof: The proof of (v) is elementary. We now focus on statement (vi). From the assumptions we note that $\epsilon_t = \gamma \cdot \lambda_t$ is a sequence in $[0, 1]$ and satisfies $\sum_{t\geq 1} \epsilon_t(1 - \epsilon_t) = +\infty$. Set $\widetilde{O} := (1 - \frac{1}{\gamma})Id + \frac{1}{\gamma}\hat{O}$. Then $x_{t+1} = x_t + \epsilon_t(\widetilde{O}(x_t) - x_t)$ and Fixed($\widetilde{O}$) = Fixed($\hat{O}$). Let $x^* \in$ Fixed($\hat{O}$) i.e., $\widetilde{O}(x^*) = \hat{O}(x^*) = x^*$. We evaluate $\|x_{t+1} - x^*\|$. Indeed,

$$\begin{aligned}
&\|x_{t+1} - x^*\|^2 = \|(1 - \epsilon_t)x_t + \epsilon_t\widetilde{O}(x_t) - \epsilon_t x^* - (1 - \epsilon_t)x^*\|^2 \\
&= \|(1 - \epsilon_t)(x_t - x^*) + \epsilon_t(\widetilde{O}(x_t) - O(x^*))\|^2 \\
&= (1 - \epsilon_t)^2\|x_t - x^*\|^2 + \epsilon_t^2\|\widetilde{O}(x_t) - \widetilde{O}(x^*)\|^2 \\
&+2\epsilon_t(1 - \epsilon_t)\langle x_t - x^*, \widetilde{O}(x_t) - \widetilde{O}(x^*)\rangle \\
&= (1 - \epsilon_t)^2\|x_t - x^*\|^2 + \epsilon_t^2\|\widetilde{O}(x_t) - \widetilde{O}(x^*)\|^2 \\
&-\epsilon_t(1 - \epsilon_t)[\|x_t - \widetilde{O}(x_t)\|^2 - \|x_t - x^*\|^2 - \|\widetilde{O}(x_t) - \widetilde{O}(x^*)\|^2] \\
&= [(1 - \epsilon_t)^2 + \epsilon_t(1 - \epsilon_t)]\|x_t - x^*\|^2 \\
&+[\epsilon_t^2 + \epsilon_t(1 - \epsilon_t)]\|\widetilde{O}(x_t) - \widetilde{O}(x^*)\|^2 - \epsilon_t(1 - \epsilon_t)\|x_t - \widetilde{O}(x_t)\|^2 \\
&= (1 - \epsilon_t)]\|x_t - x^*\|^2 + \epsilon_t\|\widetilde{O}(x_t) - \widetilde{O}(x^*)\|^2 \\
&-\epsilon_t(1 - \epsilon_t)\|x_t - \widetilde{O}(x_t)\|^2.
\end{aligned} \tag{1.14}$$

Therefore,

$$\begin{aligned}
&\|x_{t+1} - x^*\|^2 \\
&\quad = (1 - \epsilon_t)]\|x_t - x^*\|^2 + \epsilon_t\|\widetilde{O}(x_t) - \widetilde{O}(x^*)\|^2 - \epsilon_t(1 - \epsilon_t)\|x_t - \widetilde{O}(x_t)\|^2 \\
&\quad \leq \|x_t - x^*\|^2 - \epsilon_t(1 - \epsilon_t)\|x_t - \widetilde{O}(x_t)\|^2.
\end{aligned} \tag{1.15}$$

By a telescopic sum, we obtain

$$\begin{array}{ll} \sum_{t=0}^{T} \epsilon_t(1-\epsilon_t)\|x_t - \widetilde{O}(x_t)\|^2 & \leq \sum_{t=0}^{T}(\|x_t - x^*\|^2 - \|x_{t+1} - x^*\|^2) \\ & = \|x_0 - x^*\|^2 - \|x_{T+1} - x^*\|^2. \end{array} \tag{1.16}$$

As $\sum_{t=0}^{\infty} \epsilon_t(1-\epsilon_t)\|x_t - \widetilde{O}(x_t)\|^2 \leq \|x_0 - x^*\|^2 < +\infty$ and $\sum_{t=0}^{\infty} \epsilon_t(1-\epsilon_t) = +\infty$, it follows from Lemma 5 that the sequence $\|x_t - \widetilde{O}(x_t)\|^2 \to 0$ as $t \to \infty$. But, $\hat{O} - Id = \gamma(\widetilde{O} - Id)$. Thus, $\|x_t - \hat{O}(x_t)\|^2 \to 0$ as $t \to \infty$.

For a proof of the last statement that the sequence x_t converges weakly to a point in Fixed($\hat{O}$), we refer to the proof of Theorems 5.15 (iii) in [35] since it involves further preliminaries that we have chosen to not recall here. This completes the proof. □

Remark 2.

- *Lemma 6 (v) assumes the existence of a fixed-point because a* 1*-Lipschitz operator does not necessarily admit a fixed-point. It may need invariance and boundedness of the domain so that Brouwer-Schauder type fixed-point theorems can be applied. For instance, the mapping* $x \mapsto x + 2024$ *has no fixed-point in* $\mathbb{R}$ *but it is a* 1*-Lipschitz function. The domain in this particular case is unbounded.*
- *For a* γ*-averaged operator* $\hat{O}$*, the algorithm is given by*

$$\begin{cases} x_0 \in \mathcal{H}_0, \\ \textit{for } t \in \{1, 2, \ldots\} \\ x_{t+1} = x_t + \lambda_t(\hat{O}(x_t) - x_t), \end{cases} \tag{1.17}$$

where $\hat{O}$ *is not necessarily a contraction and we have more flexibility for the choice of* λ_t *up to* $\frac{1}{\gamma} \geq 1$.

1.3 Maximally Cyclically Monotone Activation Operators

We are interested in maximally cyclically monotone operators because they allow us to connect with anti-derivatives and proximal Bregman operators, which are useful for the training problem in generative machine intelligence. Anti-derivatives allow us to bypass gradient computation, and we will not need to use gradient descent in this case. It avoids all the unnecessary local extrema during the pre-training process.

Definition 3. *The operator* $\hat{O}$ *is* k*-cyclically monotone if every* $(x_1, \ldots, x_{k+1})$ *with* $x_{k+1} = x_1$ *and every* $(u_1, \ldots, u_k)$*,*

$$(x_i, u_i) \in graph(\hat{O}), 1 \leq i \leq k, x_{k+1} = x_1$$

implies that

$$\sum_{i=1}^{k} \langle x_{i+1} - x_i, u_i \rangle \leq 0.$$

$\hat{O}$ *is cyclically monotone if it is* k*-cyclically monotone for every integer* $k \geq 2$*. A maximally cyclically monotone* $\hat{O}$ *is a cyclically monotone such that there is no other cyclically monotone operator that properly contains the graph of* $\hat{O}$*.*

The next lemma shows that a subdifferential of a proper and convex function is cyclically monotone and maximal.

Lemma 7. *The subdifferential ∂h is maximally cyclically monotone for every proper and convex function h.*

Proof: To see it, we fix an integer $k \geq 2$. For every $1 \leq i \leq k$, consider (x_i, u_i) in the graph of the subdifferential of h. Set $x_{k+1} = x_1$. By definition of subdifferential, $\langle x_{i+1} - x_i, u_i \rangle \leq h(x_{i+1}) - h(x_i)$. Summing up these inequalities leads to

$$\sum_{i=1}^{k} \langle x_{i+1} - x_i, u_i \rangle \leq \sum_{i=1}^{k} h(x_{i+1}) - h(x_i) = h(x_{k+1}) - h(x_1) = h(x_1) - h(x_1) = 0.$$

□

Example 1. *Consider the softargmax is given by $\{softargmax(z)\} = (\partial\phi)(z)$ where*

$$\phi(z) = \epsilon \log \left(\sum_{j=1}^{k} e^{\frac{z_j}{\epsilon}} \right)$$

which is proper and convex. Hence, the softargmax activation is maximally cyclically monotone.

Example 2. *The rectified linear unit activation function has a convex anti-derivative h : $ReLU(x) = \partial_x h(x)$ with $h(x) = \frac{1}{2} x ReLU(x)$ and hence it is a maximally cyclically monotone activation operator.*

Example 3. *The sigmoid activation function has a convex anti-derivative $h : \sigma(x) = \partial_x h(x)$ with $h(x) = \log(1+e^x)$, and hence it is a maximally cyclically monotone activation operator.*

The next lemma shows that the converse is actually true. To show this, we construct explicitly a supremum of linear functions from the cyclically monotone inequalities.

Lemma 8. *Every maximally cyclically monotone operator $\hat{O}$ is a subdifferential for some proper convex function ϕ.*

Proof: Suppose that $\hat{O}$ is a maximally cyclically monotone operator. Then *graph*$(\hat{O})$ is non-empty. There is an element (x_0, u_0) in the graph of $\hat{O}$. Let us define the function:

$$\phi(x) = \sup_{k \geq 1} \sup_{(x_i, u_i) \in graph(\hat{O}), 1 \leq i \leq k} \langle x - x_k, u_k \rangle + \sum_{i=0}^{k-1} \langle x_{i+1} - x_i, u_i \rangle$$

By cyclic monotonicity, $\phi(x_0) = 0$ and ϕ is proper and convex as a supremum of linear maps. Take (x, u) in the graph of $\hat{O}$ and let $-\infty < \eta < \phi(x)$. There are finitely many points (x_i, u_i) in *graph*$(\hat{O})$, $0 \leq i \leq k$ such that $\langle x - x_k, u_k \rangle + \sum_{i=0}^{k-1} \langle x_{i+1} - x_i, u_i \rangle > \eta$.

We now set $(x_{k+1}, u_{k+1}) = (x, u)$. For every y, $\phi(y) \geq \langle y - x_{k+1}, u_{k+1} \rangle + \sum_{i=0}^{k} \langle x_{i+1} - x_i, u_i \rangle = \langle y - x, u \rangle + \langle x - x_k, u_k \rangle + \sum_{i=0}^{k-1} \langle x_{i+1} - x_i, u_i \rangle > \langle y - x, u \rangle + \eta$.

We deduce that $\phi(y) \geq \langle y - x, u \rangle + \eta$. Leting η go to $\phi(x)$ we obtain that $\phi(y) \geq \langle y - x, u \rangle + \phi(x)$. The latter inequality shows that $u \in \partial\phi$. This implies that the *graph*$(\hat{O})$ is a subset of *graph*$(\partial\phi)$. By maximality, we obtain $\hat{O} = \partial\phi$.

□

1.4 Deep Learning Outcomes as Games

Game theory studies strategic interactions between several decision-makers. In its basic form, each decision-maker has a choice space called action space. A decision-maker can build its strategy based on its type and available information to him/her. Each decision-maker has a preference structure that is often captured by a certain payoff function which depends on the choice of all the decision-makers, hence the interdependence between them. A decision-maker can have competitive, cooperative or co-opetitive behavior. Here, we are interested in basic sequential games and hierarchical games.

1.4.1 One-Shot Games

Definition 4. *The basic ingredients of a One-Shot Game are given by:*

- *The set of decision-makers $\mathcal{I}$, with cardinality $|\mathcal{I}| \geq 2$.*
- *For every decision-maker $i \in \mathcal{I}$, there is a set of actions $\mathcal{A}_i$ which is non-empty.*
- *Every decision-maker i has a preference structure that can be represented by an instant performance functional*

  $g_i : \prod_{j\in\mathcal{I}} \mathcal{A}_j \rightarrow \mathbb{R},\ a \mapsto g_i(a)$, *with* $a = (a_1, \ldots, a_I)$. *The expected performance functional of decision-maker i is given by*

$$E[g_i(a)] := \int_{b\in\prod_{j\in\mathcal{I}} \mathcal{A}_j} g_i(b) D_a(db).$$

The collection $\mathcal{G} = (\mathcal{I}, (\mathcal{A}_i, E[g_i])_{i\in\mathcal{I}})$ is called a game in strategic form.

1.4.2 Single Leader - Single Follower

We start with two decision-makers: 0 and 1. Decision-maker 0 (leader) moves first. Decision-maker 0 announces its pure strategy a_0 in $\mathcal{A}_0$. Then, decision-maker 1 (follower) reacts to 0's strategy a_0 by choosing a strategy $a_1 \in \mathcal{A}_1$. Each decision-maker $i \in \{0, 1\}$ has a performance functional $g_i(a)$ to be optimized.

To solve this hierarchical decision-making problem, we use a backward induction method. We start by solving the problem of decision-maker 1 given decision-maker 0's strategy. In the risk-neutral setting, decision-maker 1 solves

$$\inf_{a_1\in\mathcal{A}_1} \mathbb{E}[g_1(a_0, a_1)|a_0(.)].$$

This leads to a set-valued map called the reaction set or best response set of decision-maker 1 :

$$\text{rnBR}_1(a_0) = \arg\min_{v_1\in\mathcal{A}_1} \mathbb{E}[g_1(a_0, v_1)].$$

When the set $\text{rnBR}_1(a_0)$ has two or more elements, there is an ambiguity for decision-maker 0 on the possible response of decision-maker 1. In this case, several choices are possible: pessimistic/probabilistic/optimistic Stackelberg viewpoint have been proposed. Here, we provide sufficiency conditions under which the reaction set $\text{rnBR}_1(a_0)$ of decision-maker 1 is a singleton set. Then, decision-maker 0, who is the leader, solves a reverse problem to determine its best strategy. The leader 0 solves the following reverse problem:

$$\inf_{a_0\in\mathcal{A}_0} \{\mathbb{E}[g_0(a_0, a_1)],\ a_1 \in \text{rnBR}_1(a_0) \cap \mathcal{A}_1\}.$$

1.4.3 Other Notions of Stackelberg Solution

Here the generic spaces are X and Y. $\mathcal{B}_2(\mu, x) := \arg\max_{y\in Y} U_2(\mu, x, y)$

Weak Stackelberg solution

$x^* \in \arg\max_{x\in X} \min_{y\in\mathcal{B}_2(\mu,x)} U_1(\mu, x, y)$, $y^* \in \mathcal{B}_2(\mu, x^*)$, $(x^*, y^*) \in X \times Y$

Strong Stackelberg solution

$x^* \in \arg\max_{x\in X} \max_{y\in\mathcal{B}_2(\mu,x)} U_1(\mu, x, y)$, $y^* \in \mathcal{B}_2(\mu, x^*)$, $(x^*, y^*) \in X \times Y$

Inverse Stackelberg solution

$\mathcal{B}_1(\mu, \alpha) := \arg\max_{x\in X} U_1(\mu, x, \alpha(x))$

Weak Inverse Stackelberg solution

$\alpha^* \in \arg\max_{\alpha:X\to Y} \min_{x\in\mathcal{B}_1(\mu,\alpha)} U_2(\mu, x, \alpha(x))$, $x^* \in \mathcal{B}_1(\mu, \alpha^*)$, $y^* = \alpha^*(x^*)$, $(x^*, y^*) \in X \times Y$

Strong Inverse Stackelberg solution

$\alpha^* \in \arg\max_{\alpha:X\to Y} \max_{x\in\mathcal{B}_1(\mu,\alpha)} U_2(\mu, x, \alpha(x))$,
$x^* \in \mathcal{B}_1(\mu, \alpha^*)$, $y^* = \alpha^*(x^*)$, $(x^*, y^*) \in X \times Y$

In the unique response case, the leader, before announcing its $\alpha(.)$, can anticipate how the follower will react and tries to choose an

$$\alpha^* \in \arg\min_{\alpha:X\to Y} U_2(\mu, \mathcal{B}_1(\mu, \alpha(.)), \alpha(\mathcal{B}_1(\mu, \alpha(.))))$$

$$\mathcal{B}_1(\mu, \alpha^*(.)) \in \arg\max_{x\in X} U_1(\mu, x, \alpha^*(x))$$

Reverse Stackelberg solution

$\mathcal{B}_1(\mu, \beta) := \arg\max_{y\in Y} U_1(\mu, \beta(y), y)$

Weak Reverse Stackelberg solution

$\beta^* \in \arg\max_{\beta:Y\to X} \min_{y\in\mathcal{B}_1(\mu,\beta)} U_2(\mu, \beta(y), y)$, $y^* \in \mathcal{B}_1(\mu, \beta^*)$, $x^* = \beta^*(y^*)$, $(x^*, y^*) \in X \times Y$

Strong Reverse Stackelberg solution

$\beta^* \in \arg\max_{\beta:Y\to X} \max_{y\in\mathcal{B}_1(\mu,\beta)} U_2(\mu, \beta(y), y)$, $y^* \in \mathcal{B}_1(\mu, \beta^*)$, $x^* = \beta^*(y^*)$, $(x^*, y^*) \in X \times Y$

In the unique response case, the leader, before announcing its $\beta(.)$, can anticipate how the follower will react and tries to choose a

$$\beta^* \in \arg\min_{\beta:Y\to X} U_2(\mu, \beta(\mathcal{B}_1(\mu, \beta(.))), \mathcal{B}_1(\mu, \beta(.)))$$

$$\mathcal{B}_1(\mu, \beta^*(.)) \in \arg\max_{y\in Y} U_1(\mu, \beta^*(y), y)$$

Double Stackelberg solution

Fixed point condition: $x^* = \alpha(y^*), \quad y^* = \beta(x^*)$
Feasible set: $\text{Fix}(\alpha, \beta) := \{(x, y) \in X \times Y : x = \alpha(y),\ y = \beta(x)\}$

Weak Double Stackelberg solution

$(\alpha^*, \beta^*) \in \arg\max_{\alpha,\beta} \min_{(x,y)\in\text{Fix}(\alpha,\beta)}(U_1(\mu, x, y), U_2(\mu, x, y)),\ (x^*, y^*) \in \text{Fix}(\alpha^*, \beta^*)$

Strong Double Stackelberg solution

$(\alpha^*, \beta^*) \in \arg\max_{\alpha,\beta} \max_{(x,y)\in\text{Fix}(\alpha,\beta)}(U_1(\mu, x, y), U_2(\mu, x, y)),\ (x^*, y^*) \in \text{Fix}(\alpha^*, \beta^*)$

1.4.4 Multi-Layer Hierarchical Games

Definition 5. *The basic ingredients of a One-Shot Hierarchical Game with $L \geq 2$ hierarchy levels are given by:*

- *The set of decision-makers $\mathcal{I}$, with cardinality $|\mathcal{I}| \geq 2$ partitioned into L sets $\mathcal{I} = \cup_{l=1}^{L}\mathcal{I}_l$ where $\mathcal{I}_i \cap \mathcal{I}_j = \varnothing,\ i \neq j$.*
- *For every decision-maker $i \in \mathcal{I}$, at hierarchy level l, there is a set of actions $\mathcal{A}_{i,l}$ that is non-empty. The information available to decision-maker i contains the actions chosen by the decision-makers of all the preceding hierarchies 1 to $l-1$.*
- *Every decision-maker i has a preference structure that can be represented by an instant performance functional $g_i : \prod_{j\in\mathcal{I}}\mathcal{A}_j \to \mathbb{R},\ a \mapsto g_i(a)$, with $a = (a_1, \ldots, a_I)$. The expected performance functional of decision-maker i is given by*

$$E[g_i(a)] := \int_{b\in\prod_{j\in\mathcal{I}}\mathcal{A}_j} g_i(b)D_a(db).$$

The collection $\mathcal{G} = (\mathcal{I}, \mathcal{L}, (\mathcal{A}_i, E[g_i])_{i\in\mathcal{I}})$ is called a hierarchical game.

A fully hierarchical game is one where there is one decision-maker per layer, i.e., the number of layers is exactly the number of decision-makers. In this context, not only the set of decision-makers per level matters but also the number of hierarchical levels plays a key role in the global performance of the system. There are $|\mathcal{I}| = |\mathcal{L}|$ hierarchical levels. Each layer l, decision-maker $i = l$ chooses a strategy a_i knowing the strategy of the preceding decision-makers i.e., $\{i-1, \ldots, 1\}$. This becomes a sequential decision-making problem. We use a backward induction method to solve the hierarchical game problem.

1.4.5 Any Neural Network is a Hierarchical Game

A neural network is composed of multiple layers. At each layer, several operations are made forming a layer-operator. Any operator can be seen as a best-response correspondence. Best-response correspondence defines the reaction of a decision-maker in a (constrained) game. We then obtain a equivalent hierarchical constrained game corresponding to the neural network.

A important result is the connection between deep learning outcomes and Nash equilibria of a non-potential hierarchical game and this holds for general activation functions and arbitrary weights and biases. Table 1.4 displays connections between game theory and neural networks.

TABLE 1.4
Deep learning vs. game theory terminologies.

Game Theory	Deep Learning	Transformer
decision-maker	neuron unit	layer
action	weight/bias	query-key-value weight, bias
objective function	objective function	
sub-goal	feature	
measurement	output	
game design	architecture design	

Theorem 1. *The outcomes of any neural network given by the fixed-points (if any) of the operator $\hat{\mathcal{O}}_L \circ \ldots \circ \hat{\mathcal{O}}_1$ are exactly the constrained Nash equilibria (CNE) of the non-zero sum game*

$$\mathcal{G} = (\mathcal{L} = \{1,\ldots,L\}, (\mathcal{H}_l, 0, C_l(.))_{\mathcal{L}}),$$

where $C_l(y) = \hat{\mathcal{O}}_l(y^{(l-1)})$.

Proof: To prove it, observe that for each layer l of the deep neural network, the outcome can be written as $\hat{\mathcal{O}}_l(y^{(l-1)}) = y^{(l)}$ with $y^{(0)} = x_0$. It follows that

$$\left\{ \begin{array}{l} l \in \{1,\ldots,L\} \\ Inf_{\{y^{(l)} \in C_l(y)\}} 0 \end{array} \right. \Leftrightarrow \left\{ \begin{array}{l} y := (y^{(1)},\ldots,y^{(L)}), \\ y \in C(y) \end{array} \right. \Leftrightarrow \left\{ \begin{array}{l} \text{CNE of game:} \\ (\mathcal{L}, (\mathcal{H}_l, 0, C_l(.))_{\mathcal{L}}), \end{array} \right. \quad (1.18)$$

The latter set is the set of fixed-points of the operator C. □

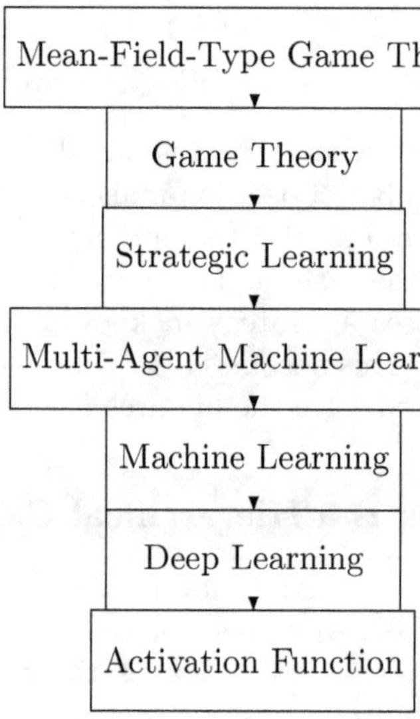

FIGURE 1.4
Connection between mean-field-type game theory and deep learning.

Next, we refine Theorem 1 with more explicit performance functions of the game. See Figure 1.4.

1.4.6 Hierarchical Games with Averaged NonExpansive Activations

$\frac{1}{2}$-averagedness property is connected to monotone operators. Games with monotone best-response correspondence have nice properties in terms of variational inequalities.

When all the activation functions are $\frac{1}{2}$-averaged one obtains a monotone game:

Proposition 1 (Averaged activations). *Consider a* $\mathbb{NN}$ *with activation functions* $r_1, \ldots r_L$ *that are* $\frac{1}{2}$*-averaged. Then the outcomes of* $\mathbb{NN}$ *(if any) are exactly the constrained Nash equilibria of the non-zero sum game*

$$\mathcal{G}_1 = (\mathcal{L} = \{1, \ldots, L\}, (\mathcal{H}_l, 0, C_l(.))_{\mathcal{L}}),$$

where $C_l(y) = (W_l y_{l-1} + b_l - O_l(y_l))$, O_l *is a monotone operator.*

Proof: For each layer l of the deep neural network, the outcome can be written as $r_l(W_l y_{l-1} + b_l) = y_l$ with $y_0 = x_0$. Since r_l is $\frac{1}{2}$-averaged, there is a monotone operator O such that $r_l = (Id + O_l)^{-1}$. It follows that $(W_l y_{l-1} + b_l - y_l) \in O_l(y_l)$ This means that $y_l \in (W_l y_{l-1} + b_l - O_l(y_l)) =: C_l(y)$.

$$\left\{\begin{array}{l} l \in \{1, \ldots, L\} \\ \\ Inf_{\{y_l \in C_l(y)\}} 0 \end{array}\right. \Longrightarrow \left\{\begin{array}{l} y := (y_1, \ldots, y_L), \\ y \in C(y) \end{array}\right. \Longrightarrow \left\{\begin{array}{l} \text{CNE of game:} \\ (\mathcal{L}, (\mathcal{H}_l, 0, C_l(.))_{\mathcal{L}}), \end{array}\right. \tag{1.19}$$

The latter set is the set of fixed-points of the operator C. □

1.4.7 Hierarchical Games with Maximally Cyclically Monotone Activations

We now connect neural networks to convex games using antiderivative of activations, proximal activations and maximally cyclically monotone activation operators.

When all the activation functions are maximally cyclically monotone one obtains an explicit component-wise convex game given by the anti-derivatives.

Theorem 2 (Maximally Cyclically Monotone Activations). *We have the following:*

- *The asymptotic outcomes of the deep neural network* $(x_0, R, W, b, L, \{\mathcal{H}_l\}_{1 \leq l \leq L})$ *with activation functions* $r_1, \ldots r_L$ *that are maximally cyclically monotone are exactly the Nash equilibria of the non-zero sum game* $\mathcal{G} = (\mathcal{L}, (\mathcal{H}_l,\ f_l(\cdot) + \frac{1}{2}\| \cdot - b_l - W_l x_{l-1}\|^2)_{l \in \mathcal{L}})$ *defined by*
  - *the decision-makers (players) are members of* $\mathcal{L}$
  - *the action space of the decision-maker* l *is* $\mathcal{H}_l$
  - *the objective function of decision-maker* l *output is* $z \mapsto f_l(z) + \frac{1}{2}\|z - b_l - W_l x_{l-1}\|^2$.
- *For three or more layers* ($L \geq 3$)*, there is a deep neural architecture such that the resulting game is not a potential game.*

Note that this also corresponds to the (multi-level) Stackelberg solution of the game $\mathcal{G}$ as a layer reacts to the previous layer design.

Proof: We start with the first statement of the theorem. The set of long-run outcomes of the neural network is the concatenation of all the outputs given by

$$x_1^* = \hat{O}_1 x_L^*,\ x_2^* = \hat{O}_2 \circ \hat{O}_1 x_L^*,\ \ \ldots\ \ x_{L-1}^* = \hat{O}_{L-1} \circ \hat{O}_{L-2} \ldots \hat{O}_2 \circ \hat{O}_1 x_L^*, \tag{1.20}$$

where $x_L^* \in \mathcal{H}_0$ satisfies $x_L^* = \hat{O}(x_L^*)$, which is rewritten as

$$\begin{cases} x^* = (x_1^*, x_2^*, \ldots, x_L^*) \in \prod_{l=1}^L \mathcal{H}_l, \\ x_1^* = r_1(W_1 x_L^* + b_1), \\ x_2^* = r_2(W_2 x_1^* + b_2), \\ \ldots \\ \ldots \\ x_l^* = r_l(W_l x_{l-1}^* + b_l) \\ \ldots \\ \ldots \\ x_{L-1}^* = r_{L-1}(W_{L-1} x_{L-2}^* + b_{L-1}), \\ x_L^* = r_L(W_L x_{L-1}^* + b_L), \end{cases} \tag{1.21}$$

or, by (1.9),

$$\begin{cases} x^* = (x_1^*, x_2^*, \ldots, x_L^*) \in \prod_{l=1}^L \mathcal{H}_l, \\ x_1^* = [Id + \partial f_1]^{-1}(W_1 x_L^* + b_1), \\ x_2^* = [Id + \partial f_2]^{-1}(W_2 x_1^* + b_2), \\ \ldots \\ \ldots \\ x_l^* = [Id + \partial f_l]^{-1}(W_l x_{l-1}^* + b_l), \\ \ldots \\ \ldots \\ x_L^* = [Id + \partial f_L]^{-1}(W_L x_{L-1}^* + b_L). \end{cases} \tag{1.22}$$

Again, by (1.9), this is equivalent to

$$\begin{cases} x^* = (x_1^*, x_2^*, \ldots, x_L^*) \in \prod_{l=1}^L \mathcal{H}_l, \\ x_1^* \in \arg\min_{y_1 \in \mathcal{H}_1} \frac{1}{2}\|y_1 - b_1 - W_1 x_L^*\|^2 + f_1(y_1), \\ x_2^* \in \arg\min_{y_2 \in \mathcal{H}_2} \frac{1}{2}\|y_2 - b_2 - W_2 x_1^*\|^2 + f_2(y_2), \\ \ldots \\ \ldots \\ x_l^* \in \arg\min_{y_l \in \mathcal{H}_l} \frac{1}{2}\|y_l - b_l - W_l x_{l-1}^*\|^2 + f_l(y_l), \\ \ldots \\ \ldots \\ x_L^* \in \arg\min_{y_L \in \mathcal{H}_0} \frac{1}{2}\|y_L - b_L - W_L x_{L-1}^*\|^2 + f_L(y_L). \end{cases} \tag{1.23}$$

By setting $x_0^* = x_L^*$ we can write

$$x_l^* \in \arg\min_{y_l \in \mathcal{H}_l} \frac{1}{2}\|y_l - b_l - W_l x_{l-1}^*\|^2 + f_l(y_l), \quad l \in \{1, \ldots, L\}. \tag{1.24}$$

In other words

$$(x_l^*)_{l \in \{1,\ldots,L\}} \text{ is a Nash equilibrium of the non-zero sum game } \mathcal{G}. \tag{1.25}$$

This completes the proof of the first statement. □

We now focus on the second statement with $L \geq 3$. Since there are at least three layers, due to the cycling behavior, i.e., rotating between these three sets, one can construct a cycle of mappings between these sets as illustrated next. Let $f_l = \mathbb{I}_{S_l}$ where $S_1 \neq S_2 \neq S_3$, S_l is not empty, convex and compact and $\cap_{l=1}^L S_l \neq \emptyset$. Set $W_l = Id$ and $b_l = 0$. Then, there is a fixed-point and the set of fixed-points is $\cap_{l=1}^L S_l$. However, there is no global best-response potential function associated with the game $\mathcal{G}$ defined above. Following [16], there is no global function $P : \prod_{l=1}^L \mathcal{H}_l \to \prod_{l=1}^L \mathcal{H}_l$ whose minimizers are in $\cap_{l=1}^L \text{Fixed}(\hat{O}_l)$. This means that the game $\mathcal{G}$ is not a best-response potential game. This completes the proof. □

Remark 3. *We have seen that the set of outcomes of neural networks in the long run is exactly the set of Nash equilibria of a constrained Nash game. The game is explicitly described in the case $\frac{1}{2}$-averaged activation operators as well as in the maximally cyclically monotone operators. These characterizations have several advantages. One of the advantages is the use of existing advanced strategic learning algorithms, specially in the case of operators that are not necessarily globally monotone in all the variables but component-wise (layer-wise) monotone.*

We highlight here the existence of fixed-point. For contraction operators (Lipschitzian with Lipschitz constant strictly less than 1) , the existence of a unique fixed-point follows from the Banach theorem. For non-expansive operators, those in which the Lipschitz constant is one, a fixed-point may not exist or there may be infinite number of fixed-points. For single-valued operators, one can use Brouwer-Schauder fixed-point theorem for continuous functions over convex, closed, bounded, non-empty set into itself. However, here the space is not bounded (the input signal can be arbitrarily high). Therefore, the Brouwer-Schauder fixed-point is not directly applicable. The closed unit ball is a compact set in finite dimension but it is not a compact set in Hilbert spaces (infinite dimensions). Another case of existence of a fixed-point is Tarski's fixed-point theorem for a monotonic operator over a partially ordered set.

Remark 4. *The following observations are in order.*

- *We do not need all the activation functions r_l to be contraction operators to have an overall composition $r_L \circ r_{L-1} \circ \ldots r_1$ to be a contraction operator.*
- *We do not need all the activation functions r_l to be averaged nonexpansive operators to have an overall composition $r_L \circ r_{L-1} \circ \ldots r_1$ to be an averaged operator.*
- *We do not need all the activation functions r_l to be maximally cyclically monotone operators to have an overall composition $r_L \circ r_{L-1} \circ \ldots r_1$ to be maximally cyclically monotone.*

1.5 Training in Deep Learning Architectures as Games

The outcomes of neural networks are analyzed above for well-trained neural networks. We are interested in pre-training and training the neural network. Given several input-output data points, we formulate the problem as variational inequality and progressively relax the formulation layer by layer and per data set variational inequality. The training problem is then seen as finding an element (if any) in an intersection of sets.

We now focus on the training problem given a data set. The equation $\hat{O}_L \circ \hat{O}_{L-1} \circ \ldots \hat{O}_1(x) = y_L$ can be written as

$$\begin{cases} \hat{O}_L y^{(L-1)} = y^{(L)}, \\ \hat{O}_{L-1} \circ \ldots \hat{O}_1(x) = y^{(L-1)}, \end{cases} \Leftrightarrow \begin{cases} \hat{O}_L y^{(L-1)} = y^{(L)}, \\ \hat{O}_{L-1} y^{(L-2)} = y^{(L-1)}, \\ \hat{O}_{L-2} y^{(L-3)} = y^{(L-2)}, \\ \vdots \\ \hat{O}_2 y^{(1)} = y^{(2)}, \\ \hat{O}_1 y^{(0)} = y^{(1)}, \\ y^{(0)} := x. \end{cases} \tag{1.26}$$

This means that the system of equations $\hat{O}_L \circ \hat{O}_{L-1} \circ \ldots \hat{O}_1(x_i) = y_{L,i}, \quad i \in \{1, \ldots, D\}$, can be written as

$$\begin{cases} i \in \{1, \ldots, D\} \\ \hat{O}_L y_i^{(L-1)} = y_i^{(L)}, \\ \hat{O}_{L-1} \circ \ldots \hat{O}_1(x_i) = y_i^{(L-1)} \end{cases} \Leftrightarrow \begin{cases} \hat{O}_L y_i^{(L-1)} = y_i^{(L)}, \ i \in \{1, \ldots, D\} \\ \hat{O}_{L-1} y_i^{(L-2)} = y_i^{(L-1)}, \ i \in \{1, \ldots, D\} \\ \hat{O}_{L-2} y_i^{(L-3)} = y_i^{(L-2)}, \ i \in \{1, \ldots, D\} \\ \vdots \\ \hat{O}_2 y_i^{(1)} = y_i^{(2)}, \ i \in \{1, \ldots, D\} \\ \hat{O}_1 y_i^{(0)} = y_i^{(1)}, \ i \in \{1, \ldots, D\} \\ y_i^{(0)} := x_i, \ i \in \{1, \ldots, D\}, \end{cases} \tag{1.27}$$

$$\Leftrightarrow \begin{cases} \theta = (\theta_1, \ldots, \theta_L) \\ \hat{O}_{L,\theta} y_i^{(L-1)} = y_i^{(L)}, \ i \in \{1, \ldots, D\} \\ \hat{O}_{L-1,\theta} y_i^{(L-2)} = y_i^{(L-1)}, \ i \in \{1, \ldots, D\} \\ \hat{O}_{L-2,\theta} y_i^{(L-3)} = y_i^{(L-2)}, \ i \in \{1, \ldots, D\} \\ \vdots \\ \hat{O}_{2,\theta} y_i^{(1)} = y_i^{(2)}, \ i \in \{1, \ldots, D\} \\ \hat{O}_{1,\theta} y_i^{(0)} = y_i^{(1)}, \ i \in \{1, \ldots, D\} \\ y_i^{(0)} := x_i, \ i \in \{1, \ldots, D\}, \end{cases} \Leftrightarrow \begin{cases} \theta = (\theta_1, \ldots, \theta_L) \\ l \in \{1, \ldots, L\} : \\ \hat{O}_{l,\theta_l} y_i^{(l-1)} = y_i^{(l)}, \\ i \in \{1, \ldots, D\} \end{cases} \tag{1.28}$$

This means that for each $l \in \{1, \ldots, L\}$, θ_l belongs to an intersection of sets. Given $(x_i, y_{L,i})_{i \in \{1,\ldots,D\}}$ the training problem (limited to weights and biases) is to find $\theta = (W, b) = (W_l, b_l)_l$ such that the neural network $(x_i, L, r, W, b, \{\mathcal{H}_l\}_{1 \leq l \leq L})$ produces an output that matches y_i at each iteration i. This means that

$$(r_L \circ A_{L,i,\theta_L}) \circ \ldots \circ (r_1 \circ A_{1,i,\theta_1})(x_i) = y_{L,i}, \quad i \in \{1, \ldots, D\},$$

where $\theta_l := (W_l, b_l)$ and $A_{l,i,\theta_l} = A_{l,i,(W_l,b_l)} : \ (W_l, b_l) \mapsto W_l y_{l-1,i} + b_l$ with $y_{l-1,i}$ being the output from layer $l - 1$.

Thus, given the data $(x_i, y_{L,i})_{i \in \{1,\ldots,D\}}$ the training problem is to find $\theta^* = (\theta_1^*, \ldots, \theta_L^*)$ such that

$$r_L \circ A_{L,i,\theta_L^*} \circ \ldots \circ (r_1 \circ A_{1,i,\theta_1^*})(x_i) - y_{L,i} = 0, \quad i \in \{1, \ldots, D\}. \tag{1.29}$$

In the sequel we will denote (for each i) $A_{l,i} := A_{l,i,\theta_l}$ and $A_{l,i}^{\dagger}$ is its adjoint operator.

Expectile training

Expectiles constitute a more generalized concept that allows for a non-linear partitioning of the data. Instead of dividing the data into two parts based on a fixed proportion, expectiles divide the data such that a certain proportion of the expected values of the data falls below the expectile. The expectile is law-invariant. For any $\frac{1}{2} \leq \alpha < 1$ the α-expectile is a coherent risk-measure. Expectiles are often used in risk engineering and financial modeling when dealing with asymmetric and non-normal distributions, where traditional quantiles may not provide an accurate representation of the data. They can be particularly useful in situations where one would like to focus on a specific range of expected outcomes rather than just looking at a fixed percentile.

Definition 6. *Let $(\Omega, \mathcal{F}, \mathbb{P})$ be an atomless probability space and $L^q(\Omega, \mathcal{F}, \mathbb{P})$ be the L^q space with $q \geq 1$. Let the α-expectile of an integrable random variable g, with $\alpha \in (0,1)$, be defined by*

$$\begin{cases} \rho_\alpha : \ L^2(\Omega, \mathcal{F}, \mathbb{P}) \to \mathbb{R}, \\ \rho_\alpha(g) \in \arg\min_{l' \in \mathbb{R}} \ E\{\alpha[(g-l')_+]^2 + (1-\alpha)[(l'-g)_+]^2\} \\ = \arg\min_{l' \in \mathbb{R}} \ E\{V_\alpha(g-l')\}, \\ V_\alpha : \ \mathbb{R} \to \mathbb{R}, \\ V_\alpha(x) = \alpha[x_+]^2 + (1-\alpha)[(-x)_+]^2 \\ = \alpha x^2 \mathbb{I}_{\{x \geq 0\}} + (1-\alpha) x^2 \mathbb{I}_{\{x<0\}} = |\alpha - \mathbb{I}_{\{x<0\}}| x^2 \end{cases} \tag{1.30}$$

For $\alpha = \frac{1}{2}$, the above definition yields $\arg\min_{l' \in \mathbb{R}} \ E[(g-l')^2]$. We know that $\{E[g]\} = \arg\min_{l' \in \mathbb{R}} \ E[(g-l')^2]$ which means that $\rho_{\frac{1}{2}}(g)$, the half-expectile of g, coincides with the expected value of g.

Lemma 9. *We have the following:*

(a) For each $\alpha \in (0,1)$, and a square integrable $g \in L^2$, the mapping $l' \mapsto E\{V_\alpha(g-l')\}$ is convex.

(b) For each $\alpha \in (0,1)$, and a square integrable $g \in L^2$, the mapping $l' \mapsto E\{V_\alpha(g-l')\}$ is differentiable, the derivative is well-defined in L^1, and it is a one-to-one mapping, given by $-2E\{U_\alpha(g-l')\}$ with

$$\begin{cases} U_\alpha : \ \mathbb{R} \to \mathbb{R}, \\ U_\alpha(x) = \alpha x \mathbb{I}_{\{x \geq 0\}} + (1-\alpha) x \mathbb{I}_{\{x<0\}} = |\alpha - \mathbb{I}_{\{x<0\}}| x. \end{cases} \tag{1.31}$$

(c) Let $\mathbb{U}_\alpha(l') = E\{U_\alpha(g-l')\}$. For each $\alpha \in (0,1)$, and a square integrable $g \in L^2$, the mapping $l' \mapsto E\{V_\alpha(g-l')\}$ has a unique minimizer and it is given by $\mathbb{U}_\alpha^{-1}(0) = \rho_\alpha(g)$. This means that the α-expectile of the loss g is the number l' uniquely determined by

$$\alpha E[(g-l')_+] = (1-\alpha) E[(l'-g)_+].$$

(d) For each $\frac{1}{2} \leq \alpha < 1$, $\rho_\alpha : L^2 \to \mathbb{R}$ is a coherent risk measure.

(e) For each $0 < \alpha < \frac{1}{2}$, $\tilde{\rho}_\alpha : \ L^2 \to \mathbb{R}$, defined by $\tilde{\rho}_\alpha(g) = -\rho_\alpha(-g)$ is a coherent risk measure.

(f) For each $0 < \alpha < 1$, $\rho_\alpha : L^1 \to \mathbb{R}$, is law-invariant i.e., $\mathcal{R}_\alpha(\mathbb{P}_{g_1}) = \mathcal{R}_\alpha(\mathbb{P}_{g_2})$ if $\mathbb{P}_{g_1} = \mathbb{P}_{g_2}$, where $\mathcal{R}_\alpha(\mathbb{P}_g) = \rho_\alpha(g) = \mathbb{U}_\alpha(g)^{-1}(0)$, and $\mathbb{U}_\alpha(g)(l') = E\{U_\alpha(g-l')\} = \int U_\alpha(y - l')\mathbb{P}_g(dy)$ for $l' \in \mathbb{R}$.

(g) $\mu \mapsto \mathcal{R}_\alpha(\mu)$ is Fréchet-differentiable

(h) The α-expectile of Gaussian random variable satisfies $\mathcal{R}_\alpha(\mathcal{N}(m, \sigma^2)) = m + (\alpha - \frac{1}{2})|\sigma|\sqrt{\frac{8}{\pi}} + O((\alpha - \frac{1}{2})^3)$. For $\frac{1}{2} < \alpha < 1$, set $\alpha - \frac{1}{2} = \frac{\sqrt{\beta}}{2}$, then

$$\mathcal{R}_{\frac{1+\sqrt{\beta}}{2}}(\mathcal{N}(m, \sigma^2)) = m + |\sigma|\sqrt{\frac{2\beta}{\pi}} + O(\frac{\beta\sqrt{\beta}}{8}).$$

Coherent Risk Quantification

The expectile satisfies the properties of coherent risk measures defined next.

Definition 7. *A risk measure of a random loss functional assigns a real value to it in order to determine its risk degree. A risk measure R is coherent if it satisfies the following properties:*

- *Translation principle:* $R(g_1 + \lambda) = R(g_1) + \lambda$ *for all* $\lambda \in \mathbb{R}$.
- *Subadditivity:* $R(g_1 + g_2) \leq R(g_1) + R(g_2)$ *for all* g_1, g_2
- *Monotonicity: If* $g_1 \leq g_2$, *then* $R(g_1) \leq R(g_2)$
- *Positive homogeneity:* $R(\lambda g) = \lambda R(g)$, *for all* $\lambda \geq 0$,

Example 4. *Non-coherent risk measures*

- *The expected value of g, denoted by $E[g]$, which is also the* arg min *of* $l' \mapsto E[(g - l')^2]$, *is a coherent risk measure. It is risk-neutral.*
- *The variance $var(g)$ is not coherent as $var(g + \lambda) = var(g)$ for all $\lambda \in \mathbb{R}$ and $g \in L^2$. Also, the variance is not positively homogeneous. Similarly, the standard deviation is not coherent.*
- *The mean-variance, mean-standard deviation, variance-to-mean ratio, mean-to-variance ratio, Sharpe ratio, and dispersion index, are not coherent risk measures.*
- *The quantile and the value-at-risk are not coherent risk measures.*
- *Similarly, the exponentiated return, and the entropic risk measure given by* $RS(g_l) = \frac{1}{\eta_l} \log \left(\frac{1}{\alpha_l} E_m[e^{\eta_l g_l(\theta, m)}] \right)$ *are not coherent as they do not satisfy positive homogeneity.*

Computing the Expectile as a Fixed-Point

Let $\alpha \in (0, 1)$ and g be an integrable real-valued random variable. Note that $g - l' = \max(g - l', 0) - \max(-(g - l'), 0)$ for all random variable g. We have that the expectile of g with level α is the unique solution of the equation:

$$\begin{array}{l} \rho_\alpha(g) = E[\max((1 + \epsilon)g - \epsilon \rho_\alpha(g), g)], \\ \epsilon := \frac{2\alpha - 1}{1 - \alpha}. \end{array} \tag{1.32}$$

By setting $\rho_\alpha(g) := x$ one has

$$x = E[\max((1 + \epsilon)g - \epsilon x, g)] =: F_{g,\epsilon}(x), \tag{1.33}$$

This means that a fixed-point equation $F_{g,\epsilon}(x) = x$ is to be solved.

1.5.1 Training under Averaged Nonexpansive Activations

We have the following lemma whose proof is immediate.

Lemma 10. *Let r_l be a $\frac{1}{2}$-averaged nonexpansive function. Then, for any z, we have*

$$r_l(A_l \theta_l) - z = 0 \Longleftrightarrow A_l \theta_l - (Id + O_l)z = 0,$$

where O_l is a maximally monotone operator.

Proposition 2. *Suppose that each r_l is an averaged nonexpansive activation. Then, the solutions (if any) of the training of weights and biases problem (1.29) are also solutions of the following variational inclusion: Given an input-output data set $\{(x^*_{0,i}, y^*_{L,i}),\ i \in \{1, 2, \dots, D\}$, find (W^*, b^*) such that*

$$0 \in A^{\dagger}_{l,i}[y^*_{l,i} + O_l(y^*_{l,i}) - (W^*_l y^*_{l-1,i} + b^*_l)], \quad i \in \{1, 2, \dots, D\}. \tag{1.34}$$

These are also Nash equilibria of the non-zero sum game $\mathcal{G}_n$ given by

$$\begin{cases} l \in \mathcal{L}, \\ (W^*_l, b^*_l) \in \arg\min_{W_l, b_l} \sum_{i=1}^{D} \omega_{l,i} \| y^*_{l,i} + O_l(y^*_{l,i}) - (W_l y^*_{l-1,i} + b_l)\|^2. \end{cases} \tag{1.35}$$

Proof: The proof uses Lemma 10 to connect the zeros to the minimization. □

Problem 2 is on sampled input-output data. One can also formulate the problem not on sampled data but on the exact distribution $\mathbb{P}(dx_0 dy_L)$.

Given $\mathbb{P}(dx_0 dy_L)$, find a vector $\theta^* := (W^*, b^*) = (W^*_l, b^*_l)_l$ such that

$$E\Big(\Big\langle (r_L \circ \tilde{A}_{L,\theta^*_L}) \circ \dots \circ (r_1 \circ \tilde{A}_{1,\theta^*_1})(x_0) - y_L, A^{\dagger}_L(\theta_L - \theta^*_L)\Big\rangle\Big) \geq 0, \ \forall \theta, \tag{1.36}$$

Moreover, design an algorithm that approximates a θ^*, a solution (if any) to (1.36).

1.5.2 Training under Maximally Cyclically Monotone Activations

Proposition 3. *Suppose that each r_l is a maximal cyclically monotone activation. Then, the solutions (if any) of the training of weights and biases problem (1.29) are also solutions of the following variational inclusion: Given an input-output data set $\{(x^*_{0,i}, y^*_{L,i}),\ i \in \{1, 2, \dots, D\}$, find (W^*, b^*) such that*

$$0 \in A^{\dagger}_{l,i}[y^*_{l,i} + \partial f_l(y^*_{l,i}) - (W^*_l y^*_{l-1,i} + b^*_l)], \quad i \in \{1, 2, \dots, D\}. \tag{1.37}$$

These are also Nash equilibria of the non-zero sum game $\mathcal{G}_3$ given by

$$\begin{cases} l \in \mathcal{L}, \\ (W^*_l, b^*_l) \in \arg\min_{W_l, b_l} \sum_{i=1}^{D} \omega_{l,i} \| y^*_{l,i} + \partial f_l(y^*_{l,i}) - (W_l y^*_{l-1,i} + b_l)\|^2. \end{cases} \tag{1.38}$$

Proof: Since each of the activation operators r_l is maximal cyclically monotone activation, $r_l = (Id + \partial f_l)^{-1}$ i.e., $O_l = \partial f_l$ for a proper, convex lower semi-continuous function f_l. the training problem is to find $\theta = (W, b) = (W_1, b_1, \dots, W_L, b_L)$ such that for each $1 \leq i \leq D$,

$$\begin{cases} (W_1 x^*_{0,t} + b_1) \in y^*_{1,i} + \partial f_1(y^*_{1,i}), \\ (W_2 x^*_{1,i} + b_2) \in y^*_{2,i} + \partial f_2(y^*_{2,i}), \\ \dots \\ \dots \\ (W_l x^*_{l-1,i} + b_l) \in y^*_{l,i} + \partial f_l(y^*_{l,i}), \\ \dots \\ \dots \\ (W_{L-1} x^*_{L-2,i} + b_{L-1}) \in y^*_{L-1,i} + \partial f_{L-1}(y^*_{L-1,i}), \\ (W_L x^*_{L-1,i} + b_L) \in y^*_{L,i} + \partial f_L(y^*_{L,i}), \end{cases} \tag{1.39}$$

which is rewritten as

$$\begin{cases} (W_1^*, b_1^*) \in \arg\min_{W_1,b_1} \|y_{1,i}^* + \partial f_1(y_{1,i}^*) - (W_1 x_{L,i}^* + b_1)\|^2, \\ (W_2^*, b_2^*) \in \arg\min_{W_2,b_2} \|y_{2,i}^* + \partial f_2(y_{2,i}^*) - (W_2 x_{1,i}^* + b_2)\|^2, \\ \cdots \\ \cdots \\ (W_l^*, b_l^*) \in \arg\min_{W_l,b_l} \|y_{l,i}^* + \partial f_l(y_{l,i}^*) - (W_l x_{l-1,i}^* + b_l)\|^2, \\ \cdots \\ \cdots \\ (W_L^*, b_L^*) \in \arg\min_{W_L,b_L} \|y_L^* + \partial f_L(y_{L,i}^*) - (W_L x_{L-1,i}^* + b_L)\|^2. \end{cases} \tag{1.40}$$

But, this set is exactly the set of Nash equilibria of the non-zero sum game

$$\mathcal{G}_3 = \Big(\mathcal{L}, \prod_{l=1}^{L} L_2(\mathcal{H}_{l-1}, \mathcal{H}_l) \times \mathcal{H}_l, \|y_l^* + \partial f_l(y_l^*) - b_l - W_l x_{l-1}^*\|^2\Big).$$

Moreover, by Fermat's principle, the last equation in (1.40) implies that

$$0 \in A_{l,i}^\dagger [y_{l,i}^* + \partial f_l(y_{l,i}^*) - (W_l^* x_{l-1,i}^* + b_l^*)],$$

which is the first part of the announced result.

For each $l \in \mathcal{L}$, let $\omega_{l,i} > 0$, $\sum_{i=1}^{D} \omega_{l,i} = 1$. Then, clearly the system of variational inequalities (1.40) yields

$$\langle (W_l, b_l) - (W_l^*, b_l^*), \sum_{i=1}^{D} \omega_{l,i} A_{l,i}^\dagger [y_{l,t}^* + \partial f_l(y_{l,i}^*) - (W_l^* y_{l-1,i}^* + b_l^*)] \rangle \geq 0$$

for all (W_l, b_l). Hence, the set of solutions of the training of weights and biases problem (1.29) is the set of Nash equilibria of the non-zero sum game

$$\begin{cases} l \in \mathcal{L}, \\ (W_l^*, b_l^*) \in \arg\min_{W_l,b_l} \sum_{i=1}^{D} \omega_{l,i} \|y_{l,i}^* + \partial f_l(y_{l,i}^*) - (W_l y_{l-1,i}^* + b_l)\|^2. \end{cases} \tag{1.41}$$

This completes the proof. □

We have the following lemma whose proof is immediate.

Lemma 11.

(i) *For any z, we have*

$$A_l^\dagger (r_l(A_l\theta) - z) = \partial_\theta [\phi_l(A_l\theta) - \langle A_l\theta, z\rangle],$$

where $\partial\phi_l = r_l$.

(ii) *If there exists θ^* solving $r_l(A_l\theta^*) = z$, then it coincides with the minimizer of $\theta \mapsto \phi_l(A_l\theta) - \langle A_l\theta, z\rangle$.*

Lemma 12. *Let L be an integer with $L \geq 1$ and, for every $l \in \{1, \ldots, L\}$ let $\hat{O}_l$ be γ-averaged and can be written as $\hat{O}_l = (Id + \partial f_l)^{-1}$ and let $0 < \epsilon < 1$. Suppose that there is a solution θ_l to the equation $z_l - \hat{O}_l(\theta_l) = 0$. Then, the algorithm $\theta_l^{p+1} = \theta_l^p + \epsilon(z_l - \hat{O}_l(\theta_l^p))$ given an initial point θ^0, converges to a solution to $\hat{O}_l(\theta_l) = z_l$ as p goes to infinity.*

Proof: By assumption, $Id - \hat{O}_l$ and $\hat{O}_l$ are both minimizers of proximal convex functions and are in particular 1-Lipschitz continuous. We can write $Id - \epsilon\hat{O}_l$ as an averaged nonexpansive operator as follows:

$$Id - \epsilon\hat{O}_l = (1-\epsilon)Id + \epsilon Id - \epsilon\hat{O}_l = (1-\epsilon)Id + \epsilon(Id - \hat{O}_l),$$

which is ϵ-averaged. This means that the operator $\tilde{O}_l = Id - \epsilon\hat{O}_l + \epsilon z_l$ is ϵ-averaged. Then, the result on the fixed-points of $\tilde{O}_l$ follows from Lemma 6. □

Lemma 13. *Let A_l be linear and non-zero and f_l be a proper convex function from $range(A_l)$ to $\mathbb{R}$. Let $h_l(x) = \inf_{z \in range(A_l)}(f_l^*(z) + \frac{1}{2}\|z - x\|^2)$. Then $x \in z + \partial f_l^*(z)$ i.e., $z \in (Id + \partial f_l^*)^{-1}(x)$.*
Then

$$\|A\|^2 \le 1 \implies A_l^\dagger \circ (Id + \partial f_l)^{-1} \circ A_l = (Id + \partial g_l)^{-1},$$

where $g_l(z) = (h_l \circ A_l)^*(z) - \frac{1}{2}\|z\|^2$.

Proof: From the Fenchel conjugate, one has

$$(Id + \partial f_l)^{-1} = [\partial(\frac{1}{2}\|.\|^2 + f_l)]^{-1} = \partial[(\frac{1}{2}\|.\|^2 + f_l)^*] = \partial p_l^*,$$

with $p_l = \frac{1}{2}\|.\|^2 + f_l$, i.e., $f_l = p_l - \frac{1}{2}\|.\|^2$.
Let $h_l(f_l^*, x) = \inf_z(f_l^*(z) + \frac{1}{2}\|z - x\|^2)$. Then

$$\inf_z(f_l^*(z) + \frac{1}{2}\|z - x\|^2) + \inf_z(f_l(z) + \frac{1}{2}\|z - x\|^2) = \frac{1}{2}\|x\|^2,$$

i.e.,
$h_l(f_l^*, x) + h_l(f_l, x) = \frac{1}{2}\|x\|^2$. By differentiating one obtains

$$\partial(h_l(f_l^*, x)) + \partial(h_l(f_l, x)) = x.$$

Hence,

$$A_l^\dagger \circ (Id + \partial f_l)^{-1} \circ A_l = A_l^\dagger \circ \partial p_l^* \circ A_l = \partial(p_l^* \circ A_l).$$

□

Proposition 4 (Activation functions that have antiderivatives). *Suppose the training problem has at least one solution. Then, the set of solutions of the training problem coincides with the set of Nash equilibria of the following layer by layer non-zero sum game $\mathcal{G}_4$:*

$$\arg\min_{\theta_l} \sum_{i=1}^{D} \omega_{l,i}[\phi_l(A_{l,i}\theta_l) - \langle A_{l,i}\theta_l, y_{l,i}\rangle] \quad l \in \mathcal{L}, \;\; \omega_{l,i} > 0, \; \sum_{i=1}^{D} \omega_{l,i} = 1. \tag{1.42}$$

Moreover, given θ_l^0, the algorithm defined for each i by $\theta_l^{p+1} = \theta_l^p - \frac{\gamma}{2\|A_{l,i}\|^2} A_{l,i}^\dagger[r_l(A_{l,i}\theta_l^p) - y_{l,i}]$, $0 < \gamma < 1$, converges to a minimizer θ_l of Problem (1.42), as $p \to \infty$.

Proof: Since, for each $i \in \{1, \ldots, D\}$, $r_l(A_{l,i}\theta) = y_{l,i}$ admits a solution $\theta_{l,i}$ then by Lemma 11 (i) we have

$$0 \in A_l^\dagger[r_l(A_{l,i}\theta_l) - y_{l,i}] = \partial_\theta[\phi_l(A_{l,i}\theta_l) - \langle A_{l,i}\theta_l, y_{l,i}\rangle], \quad i \in \{1, \ldots, D\},$$

with $\phi_l := (\frac{1}{2}\| \cdot \|^2 + f_l)^*$. Therefore, by Lemma 11 (ii), $\theta_{l,i}^*$ is a global minimizer of $\phi_l(A_{l,i}\theta_l) - \langle A_{l,i}\theta_l, y_{l,i}\rangle$. This in turn yields that θ_l^* minimizes the weighted function $\sum_{i=1}^{D} \omega_{l,i}[\phi_{l,i}(A_{l,t}\theta_l) - \langle A_{l,i}\theta_l, y_{l,i}\rangle]$.

Noting that the inclusion $0 \in A_{l,i}^{\dagger}[r_l(A_{l,i}\theta_l^*) - y_{l,i}]$ implies the inclusion $\theta_l^* \in \theta_l^* + A_{l,i}^{\dagger}[r_l(A_{l,i}\theta_l^*) - y_{l,i}]$, we examine the fixed-points of the operator $\theta \mapsto \theta + A_{l,i}^{\dagger}[r_l(A_{l,i}\theta) - y_{l,i}]$. Since $\frac{1}{2\|A_{l,i}\|^2}(A_{l,i}^{\dagger} \circ r_l \circ A_{l,i})$ is a γ-averaged operator for each i, given $\theta_{l,i}^0$, we can use Lemma 12: the iterates $\theta_l^{p+1} = \theta_l^p - \frac{\gamma}{2\|A_{l,i}\|^2} A_{l,i}^{\dagger}[r_l(A_{l,i}\theta_l^p) - y_{l,i}]$, $0 < \gamma < 1$, converge to $\theta_{l,i}$, as $p \to \infty$. The latter can also be retrieved from the convexity and Lipschitz gradient of $\theta \mapsto \phi_l(A_l\theta) - \langle A_l\theta, z\rangle$. This completes the proof. □

Remark 5. *In general, after a huge training phase, one would like to have weights and biases $\theta = (W, b), r_l$ be time-and-signal input invariant so that the architecture of deep neural network is fixed for any data set. This means that the operator $A_{l,i}$ which contains $x_{l,i} = y_{l-1,i}$ is signal input-dependent as the input changes. Thus, one would like to find $\theta_{l,i} = \theta_l$ at level l.*

Let us consider the formulation in (1.36) of the training problem based only on the distribution $\mathbb{P}(dx_0 dy_L)$. The algorithm is written in a stochastic approximation perspective with state-dependent noise:

$$\theta^{p+1} = \theta^p - \lambda_p B(\theta^p, x_0, y_L)$$

$$= \theta^p - \lambda_p\{E_{\mathbb{P}(dx_0' dy_L')}[B(\theta^p, x_0', y_L')] + B(\theta^p, x_0, y_L) - E_{\mathbb{P}(dx_0' dy_L')}[B(\theta^p, x_0', y_L')]\}$$

Here, the noise is $B(\theta_i, x_0, y_L) - E_{\mathbb{P}(dx_0' dy_L')}[B(\theta_i, x_0', y_L')]$.

One can also consider a mini-batch stochastic gradient with mini-batch size D :

$$\begin{cases} \theta^{p+1} = \theta^p - \lambda_p E_{\mathbb{P}(dx_0' dy_L')}[B(\theta^p, x_0', y_L')] \\ -\lambda_p\{\frac{1}{D}\sum_{i=1}^{D} B(\theta^p, x_{0,i}, y_{L,i}) - E_{\mathbb{P}(dx_0' dy_L')}[B(\theta^p, x_0', y_L')]\}. \end{cases} \tag{1.43}$$

Remark 6. *Why (standard) Gradient Descent Fails in Generative Adversarial Networks*

Consider $r(\theta_1, \theta_2) = \theta_1\theta_2$ defined over $[0, 10 \text{ trillion}]^2$. Then, the unconstrained descent algorithm where the Discriminator maximizes and the Generator minimizes yields:

$$\dot{\theta}_1(t) = \theta_2(t), \dot{\theta}_2(t) = -\theta_1(t), \quad \theta_1(0) = a, \quad \theta_2(0) = b,$$

which leads to

$$\ddot{x}(t) = -x(t), \quad x(0) = a, \quad \dot{x}(0) = b.$$

Let us start with an initial point (a, b) different from $(0, 0)$.

The particular solution, satisfying the initial conditions, is:

$$x(t) = a\cos(t) + b\sin(t), y(t) = b\cos(t) - a\sin(t)$$

Observe the solution satisfies:

$$\frac{(bx(t) - ay(t))^2}{a^2 + b^2} + \frac{(ax(t) + by(t))^2}{a^2 + b^2} = 1,$$

for any time t.

As the coordinates stay within the specified ellipse, they never reach zero. Consequently, gradient descent fails to converge in this simple minimax game.

The trajectory becomes periodical, exhibiting a cycling behavior.

Remark 7. • *Direct gradient descent algorithms and stochastic gradient algorithms have several limitations when the underlying cost function is weakly convex but not strongly convex, not η-smooth, unbounded gradient or non-Lipschitz gradient.*

- *Let ϕ_l be proper and convex, $x^* \in \arg\min \phi_l$ and given an initial point $x(0) = x_0$ run the gradient flow $\dot{x}(t) = -\nabla\phi_l(x(t))$. Then, the function $t \mapsto \phi_l(x(t))$ decreases. Indeed,*

$$\frac{d}{dt}\phi_l(x(t)) = \langle -\nabla\phi_l(x(t)), \nabla\phi_l(x(t))\rangle = -\|\nabla\phi_l(x(t))\|^2 \leq 0.$$

We deduce that $\phi_l(x(t)) \leq \phi_l(x(t'))$ for all $0 \leq t' < t$. Integrating over t', one obtains

$$\phi_l(x(t)) \leq \frac{1}{t}\int_0^t \phi_l(x(t'))dt'. \tag{1.44}$$

We now show that the average regret $\frac{1}{T}\int_0^T(\phi_l(x(t)) - \phi_l(x^*))dt$ vanishes. $\frac{d}{dt}(\frac{1}{2}\|x(t) - x^*\|^2) = -\langle\nabla\phi_l(x(t)), x(t) - x^*\rangle = \langle\nabla\phi_l(x(t)), x^* - x(t)\rangle + \phi_l(x(t)) - \phi_l(x(t)) \leq \phi_l(x^*) - \phi_l(x(t))$, where we have used the convexity of ϕ_l.

By integrating over time, $\frac{1}{2}\|x(t) - x^*\|^2 - \frac{1}{2}\|x(0) - x^*\|^2 \leq \int_0^t(\phi_l(x^*) - \phi_l(x(t')))dt'$. This implies that the average regret satisfies $\frac{1}{T}\int_0^T(\phi_l(x(t)) - \phi_l(x^*))dt \leq \frac{1}{2T}\|x(0) - x^*\|^2$ and by inequality (1.44),

$$0 \leq \phi_l(x(T)) - \phi_l(x^*) \leq \frac{1}{T}\int_0^T(\phi_l(x(t)) - \phi_l(x^*))dt \leq \frac{1}{2T}\|x(0) - x^*\|^2$$

for any $T > 0$.

- *In this case $\phi_l(x(t))$ converges, but $x(t)$ does not necessarily converge, and it may oscillate. To reduce oscillations, one can consider the time-average trajectory to break cyclic behavior and/or limit cycle. If the first-level time-average does not converge, one can use a second-level time-average, and so on.*

- *The time-average trajectory $\frac{1}{T}\int_0^T x(t)dt$ satisfies*

$$\phi_l\Big(\frac{1}{T}\int_0^T x(t)dt\Big) - \phi_l(x^*) \leq \frac{1}{T}\int_0^T(\phi_l(x(t)) - \phi_l(x^*))dt \leq \frac{1}{2T}\|x(0) - x^*\|^2. \tag{1.45}$$

- *For generative machine intelligence outputs, the trajectory of $x(t)$ is extremely important, as decisions may be made based on the output $x(T)$. For example, in the context of multimodal generative machine intelligence, if the output $x(T)$ changes, oscillates or has different meanings, it may not be reliable.*

- *We consider the implicit Euler method gradient descent written as $x_{t+1} - x_t = -\epsilon\nabla\phi_l(x_{t+1})$ i.e. $\frac{x_{t+1}-x_t}{\epsilon} + \nabla\phi_l(x_{t+1}) = 0$ which means that x_{t+1} is an extremum of*

$$y \longmapsto -\phi_l(y) - \frac{(y - x_t)^2}{2\epsilon}.$$

The methodology extends to metric space (with distance d) where the interpolation of the algorithm

$$x_{t+1} \in \arg\max_y[-\phi_l(y) - \frac{d(y, x_t)^2}{2\epsilon}]$$

converges to a solution of the gradient-flow:

$$\dot{x} = -\partial_x \phi_l(x),$$

as ϵ goes to zero.

These algorithms are improved in terms of speed in the Bregman learning section. The resultd do not assume strong convexity as they are not fulfilled by the layer operators.

1.6 Be Careful about the Minimization Formulation

Formulating the training problem as a variational inequality problem allows us to use an existing powerful algorithm for variational inequalities and for fixed-points. The advantage that game theory brings here is that most of these game-theoretic learning algorithms do not need gradient descent and therefore avoid all the unnecessary extrema. It uses instead antiderivatives or Bregman learning algorithms to tackle training problems. As the sufficient condition for convergence of gradient descent, Nesterov, Newton, Adam, Secant, Steffensen, etc. are not satisfied due to non-convexity of composition operator for neural network. We suggest exploiting the structure of the activations and check antiderivative, averagedness, proximal Bregman, or maximally cyclically monotonicity of activations.

1.6.1 Do Not Work With Gradient of Activation Functions

The training problem for large learning models is inherently non-convex in terms of the parameter space θ. Non-convex optimization landscapes are characterized by multiple local minima, saddle points, and flat regions, making it difficult to guarantee convergence to a global minimum. Currently, there is no general proof of convergence for gradient-based optimization methods like gradient descent, Adam, Nesterov, or even second-order methods like Newton's method when applied to non-convex functions. These algorithms were primarily designed for convex optimization problems, where convergence to a global minimum can be proven under certain conditions (such as strong convexity, coercivity, strong monotonicity, Lipschitz gradient, etc). In the non-convex setting, especially for deep learning models such as GAN, CNN, RNN, these algorithms may be trapped in suboptimal areas or get stuck in local minima. To fully exploit the internal operations of large learning models, such as the structure of layer normalization, projection to unit ball, attention mechanisms, multi-head attention, and layer-wise dependencies, reformulating the training problem could lead to more effective solutions. Formulating the training problem of large learning models as a variational inequality instead of a problem of finding zeros of the gradient of the squared error offers significant advantages, particularly in the context of the model's inherent non-convexity. In non-convex optimization, minimization techniques like gradient descent, Newton, Steffensen, Adam, or Nesterov often struggle to avoid local minima, saddle points, and flat regions, which can cause the model to be trapped around suboptimal solutions. By focusing on variational equilibrium rather than zeros of gradient of squared error, the original variational inequality formulation aims to find a point where no feasible perturbation in the parameters can provide a descent direction, making it more robust. This approach is particularly beneficial for large learning models, which exhibit rich internal structures such as multi-head self-attention and layer-wise dependencies. The original variational inequality formulation captures the interdependencies between different layers and components, providing a more holistic approach.

While gradient descent and its variants have been the cornerstone of the earlier deep learning optimization, there are limitations to its application, especially when dealing with non-convex architectures. Consider the variational inequality with $F(\theta) = \theta(1-\theta)$, i.e. find all θ^*

$$\langle F(\theta^*), (\theta - \theta^*)\rangle \geq 0, \ \forall \theta$$

which is the set $\{0, 1\}$. In this case there is a potential function ϕ such that $\nabla\phi = F$. It suffices to take $\phi = \frac{1}{2}\theta^2 - \frac{1}{3}\theta^3$. Next, we express the variational inequality for finding all zeros of $\nabla\phi$:

$$\langle \nabla\phi(\theta^*), (\theta - \theta^*)\rangle \geq 0, \ \forall \theta$$

which is the set $\{0, 1\}$.

Now consider the square error $\langle F, F\rangle$, in this particular case corresponds to F^2. We are attempting to find all values of θ that minimize F^2 and compare them with the values of θ that solve the original variational inequality problem, which involves finding the zeros of $\nabla\phi$. However it is not a good idea to work with the gradient of F^2 which in this case is $2(\nabla F)F = Hess(\phi)\nabla\phi$. Working with the gradient of the square error F^2 yields

$$\langle (\nabla(F^2))(\theta^*), (\theta - \theta^*)\rangle \geq 0, \ \forall \theta$$

which is

$$\langle (Hess(\phi)\nabla\phi)(\theta^*), (\theta - \theta^*)\rangle \geq 0, \ \forall \theta$$

which is the set $\{0, \frac{1}{2}, 1\}$. In other words, the zeros of the gradient includes the local extremum $\frac{1}{2}$. But $\frac{1}{2}$ is not solution to the original variational inequality problem. It is not a minimizer of F^2.

Through this simple example, we observe that the training problem is better formulated as a variational inequality in the output space, rather than as a finding of zeros of the gradient of the square error. Minimizing squared error norm differs from finding zeros of the gradient of the square error. This is why further stability analysis is required when employing algorithms for solving variational inequalities. It is important to exploit the hidden properties of the intermediary operators such as projection, proximal, normalization, ReLU which are all gradients of some other functions and use these properties in the training problem instead of limiting our analysis to finding zeros of the squared error. Note, however, that there is no global potential function for the generative machine intelligence architecture in the general setting. Layer by layer, all hidden potential and all hidden convexities will be useful in the sub-layer games involved.

1.6.2 Work With Anti-Derivative of Activations

We work with maximally cyclically monotone activation functions r_l. This allows us we exploit the hidden convexity of the model via anti-derivatives if even the activation function is not convex.

Lemma 14. *The zeros of maximally cyclically monotone activation r_l are exactly the minimizers of the proper convex function ϕ_l. The zeros of maximally cyclically monotone activation r_l are exactly the kernel $\partial\phi_l$.*

Proof: This follows immediately from Lemma 24. By working with the anti-derivative function ϕ_l we see that any vanishing gradient solves the training problem. While the global problem is non-convex, and sometimes the activation function is non-convex but the anti-derivative ϕ_l is convex for many of the activation operators implemented in the deep learning literature.

The activation functions that are defined entry-wise are special cases. In one dimension, the maximally cyclically monotone functions are exactly the complete non-decreasing curves. Hence, in one dimension, the set of functions from $\mathbb{R}$ to $\mathbb{R}$ which are 1-Lipschitzian, non-decreasing are exactly those which are the gradient of a function ϕ that is proper and convex. □

Armijo Gradient Flow

A continuous time analogue of the Armijo gradient flow, which is given by

$$\tfrac{d}{dt}\theta(t) = -g_{\theta\theta}^{-1}.\phi_\theta(\theta(t)), \tag{1.46}$$

where $\theta(0) = \theta_0$ is the initial value and g is a strictly convex function on θ. The advantage here is that it does not require the computation of the Hessian of ϕ as in the Newton scheme. The drawback is that it may not be sufficiently fast as established by the following result:

The advantage of the Armijo gradient flow algorithm is that it does not require the computation of the Hessian of ϕ as in the Newton scheme. The drawback is that it may not be sufficiently fast as established by the following result below.

Definition 8. *Let b be a strictly convex and twice continuously differentiable function $b : \Theta \to \mathbb{R}$. The Bregman discrepancy $d_b : \Theta \times reint(\Theta) \to \mathbb{R}$ is given by*

$$d_b(y, x) = b(y) - b(x) - \langle b_x(x), y - x\rangle,$$

with $reint(\Theta)$ being the relative interior of Θ. The Bregman discrepancy $d_b(y, x)$ is the difference between the value of b at the destination point y and the first order Taylor expansion of b around x. Since b is a convex function, $d_b(y, x) \geq 0$. Moreover, $d_b(y, x) > 0,\ \forall x \neq y$ (by strict convexity of b).

Proposition 5. *Let ϕ be a convex function over the weight space θ and $\theta(t)$ solution to (1.46). Then, for $t > 0$,*

$$0 \leq \phi(\theta(t)) - \phi(\theta^*) \leq \frac{d_b(\theta^*, \theta(0))}{t}, \tag{1.47}$$

where d_b denotes the Bregman divergence function.

As a corollary of Proposition 5 the average regret within $[t_0, T], t_0 > 0$ is bounded above by

$$regret_T = \frac{1}{T - t_0}\int_{t_0}^{T} \phi(\theta(t)) - \phi(\theta^*) \leq 2\rho d^2(\theta^*, \theta_0)\frac{\log \frac{T}{t_0}}{T - t_0}.$$

Proof of Proposition 5 Let

$$V(\theta(t)) = t(\phi(\theta(t)) - \phi(\theta^*)) + d_b(\theta^*, \theta(t)),$$

where θ is solution to (1.46). The function V is positive and $\frac{d}{dt}V = (\phi(\theta(t)) - \phi(\theta^*)) - t\langle\phi_\theta, b_{\theta\theta}^{-1}\phi_\theta(\theta(t))\rangle + \frac{d}{dt}d_b(\theta^*, \theta(t))$. By convexity of ϕ one has

$$\phi_\theta(\theta).(\theta^* - \theta) \leq [\phi(\theta^*) - \phi(\theta)],\ \forall\theta.$$

The above is equivalent to

$$[\phi(\theta(t)) - \phi(\theta^*)] \leq \langle\phi_\theta(\theta), \theta - \theta^*\rangle.$$

From

$$\partial_2 d_b(\theta^*, \theta) = -b_{\theta\theta}(\theta)(\theta^* - \theta),$$

one obtains

$$\frac{d}{dt}d_b(\theta^*,\theta(t)) = -b_{\theta\theta}^{-1}(\theta)\phi_\theta(\theta)\partial_2 d_b(\theta^*,\theta) = \langle \phi_\theta(\theta), \theta^* - \theta\rangle.$$

Hence,

$$\begin{aligned}\frac{d}{dt}V &\leq \langle \phi_\theta(\theta), \theta - \theta^*\rangle - t\langle \phi_\theta, b_{\theta\theta}^{-1}\phi_\theta(\theta(t))\rangle + \langle \phi_\theta(\theta), \theta^* - \theta\rangle \\ &= -t\langle \phi_\theta, b_{\theta\theta}^{-1}\phi_\theta(\theta(t))\rangle \leq 0.\end{aligned}$$

It follows that $\frac{d}{dt}V(\theta(t)) \leq 0$ along the path of the gradient dynamics. This decreasing property implies $0 \leq V(\theta(t)) \leq V(\theta(0)) = d_b(\theta^*, \theta(0))$. In particular, $0 \leq t(\phi(\theta(t)) - \phi(\theta^*)) \leq V(\theta(t)) \leq V(\theta(0)) < +\infty$. The error to the equilibrium loss value $\phi(\theta^*)$ is bounded by

$$0 \leq \phi(\theta(t)) - \phi(\theta^*) \leq \frac{V(\theta(0))}{t}, \ t > 0.$$

This completes the proof.

1.6.3 Beyond Gradient Descent of Anti-Derivatives: Bregman Training

To address these limitations, we introduce layer-by-layer Bregman learning, which extends beyond traditional gradient-based methods by incorporating Bregman proximal algorithms that exploit the hidden convexities. These algorithms operate in continuous time, providing a more flexible and potentially more powerful framework for optimization. Bregman learning allows us to define a broader class of discrepancy functions, which can be used to guide the optimization process more effectively than standard Euclidean distances. By leveraging Bregman divergences, we are able to capture more nuanced relationships between data points, leading to improved convergence properties and robustness in training. The continuous-time nature of Bregman proximal algorithms also opens up new avenues for incorporating second-order information into the optimization process, providing a richer understanding of the landscape and enabling more efficient navigation through complex, non-convex spaces.

Consider the optimal control problem $\inf_v \int_0^T \hat{\phi}(t,\theta,v)dt$ such that $\dot{\theta} = v$. The minimum principle is a necessary condition of optimality when the underlying function is sufficiently smooth. The adjoint variable is $\dot{p} = -H_\theta = -\hat{\phi}_\theta$. The optimal control minimizes the Hamiltonian

$$H(\theta, p) = \inf_v\{\hat{\phi} + pv\},$$

i.e., the Legendre-Fenchel transform of $-\hat{\phi}$ applied at the point $-p$. A closed-form expression of the optimal control can be obtained, and it is generically given by $v^* = H_p(\theta, p)$. A necessary condition for optimality says that $H_{v^*}(v^* - v) \geq 0$ for any v, where H_v denotes the sub-differential of H. This latter variational equation can be rewritten as

$$0 \leq H_{v^*}(v^* - v) = [\hat{\phi}_{v^*} + p](v^* - v), \ \forall v. \tag{1.48}$$

In particular, an interior solution v^* should solve $p = -\hat{\phi}_{v^*}$ and the adjoint equation becomes $\dot{p} = \frac{d}{dt}(-\hat{\phi}_{v^*}) = -\hat{\phi}_\theta(\theta, v^*)$, which implies that

$$\frac{d}{dt}\hat{\phi}_{\dot{\theta}} = \hat{\phi}_\theta(\theta, \dot{\theta}). \tag{1.49}$$

The latter equation is also called the Euler-Lagrange equation in the field of calculus of variations. Since the minimization is among all possible curves, this minimum principle may exhibit features that allow investigation of faster time curves. We investigate Equation (1.49) for a class of quantity-of-interest $\hat{\phi}$. Let the family of Bregman-based Lagrangians be

$$\hat{\phi}(\theta, v) = e^{\alpha+\gamma}[d_b(\theta + e^{-\alpha}v, \theta) - e^{\beta}\phi(\theta)],$$

for a certain smooth and convex function g. If $\dot{\gamma} = e^{\alpha}$ then the Euler-Lagrange equation (1.49) reduces to the following second-order ordinary differential system

$$\frac{d}{dt}[b_\theta(\theta + e^{-\alpha}\dot{\theta})] = -e^{\alpha+\beta}\phi_\theta(\theta). \tag{1.50}$$

Note that a discrete-time Euler-Lagrange can be obtained by considering the optimization of $\sum_{k=1}^{K} \hat{\phi}(\theta_k, u_k)$ subject to $\theta_{k+1} = \theta_k + \delta u_k,\ \delta > 0$. The Bellman optimality equation yields $V_k(\theta_k) = \min_{u_k}\{\hat{\phi} + V_{k+1}(\theta_{k+1})\}$ which means that the discrete-time Euler-Lagrange equation yields

$$\frac{\hat{\phi}_{u_k} - \hat{\phi}_{u_{k-1}}}{\delta} = \hat{\phi}_{\theta_k}.$$

In view of (1.50), the Bregman algorithms for static optimization problems $\inf_{\theta\in\Theta}\phi(\theta)$ yield

$$\begin{array}{l} \frac{d}{dt}[b_\theta(y(t))] = -e^{\alpha(t)+\beta(t)}\phi_\theta(\theta(t)), \\ y(t) = \theta(t) + e^{-\alpha(t)}v(t), \\ v(t) = \dot{\theta}(t) \end{array} \tag{1.51}$$

where the initial values are $\theta(0), v(0)$.

Proposition 6. *Let θ be a Hilbert space and ϕ a convex function over θ and $\dot{\beta}(t) \leq e^{\alpha(t)}$. Then, the Bregman learning algorithm (1.51) generates an error as*

$$0 \leq \phi(\theta(t)) - \phi(\theta^*) \leq e^{-\beta(t)}c_0, \tag{1.52}$$

where $c_0 := d_b(\theta^*, \theta(0) + e^{-\alpha(0)}v(0)) + e^{\beta(0)}[-\phi(\theta^*) + \phi(\theta(0))] > 0.$

Proof of Proposition 6: Let us define function V as follows.

$$V(\theta, v, \theta^*) = d_b(\theta^*, \theta + e^{-\alpha}v) + e^{\beta}[-\phi(\theta^*) + \phi(\theta)]. \tag{1.53}$$

Then $V(\theta(t), v(t), \theta^*)$ is positive. Let us compute the time derivative of V over the path $\theta(t), v(t)$ generated by the Bregman algorithm.

$$\frac{d}{dt}[b_\theta(\theta + e^{-\alpha}\dot{\theta})] = -e^{\alpha+\beta}\phi_\theta(\theta).$$

We also have that

$$\begin{array}{l} \frac{d}{dt}V(\theta(t), v(t), \theta^*) = \\ -\frac{d}{dt}[\theta + e^{-\alpha}v]b_{\theta\theta}(\theta + e^{-\alpha}v).(\theta^* - \theta - e^{-\alpha}v) \\ +\dot{\beta}e^{\beta}[\phi(\theta) - \phi(\theta^*)] + e^{\beta}\phi_\theta(\theta)\dot{\theta}. \end{array} \tag{1.54}$$

By adding and subtracting the same term $\phi_\theta(\theta^* - \theta)$ we have

$$\begin{array}{l} \frac{d}{dt}V(\theta(t), v(t), \theta^*) = \\ -\frac{d}{dt}[\theta + e^{-\alpha}v]g_{\theta\theta}(\theta + e^{-\alpha}v).(\theta^* - \theta - e^{-\alpha}v) \\ +\dot{\beta}e^{\beta}[\phi(\theta) - \phi(\theta^*) + \phi_\theta(\theta^* - \theta) - \phi_\theta(\theta^* - \theta)] \\ +e^{\beta}\phi_\theta(\theta)\dot{\theta} = -\frac{d}{dt}[b_\theta(\theta + e^{-\alpha}v)].(\theta^* - \theta - e^{-\alpha}v) \\ +\dot{\beta}e^{\beta}[\phi(\theta) - \phi(\theta^*) + \phi_\theta(\theta^* - \theta)] \\ -\dot{\beta}e^{\beta}\phi_\theta(\theta^* - \theta) + e^{\beta}\phi_\theta(\theta)\dot{\theta}. \end{array} \tag{1.55}$$

By further expansion of the above expression we obtain the following equations

$$\begin{aligned}
&\tfrac{d}{dt}V(\theta(t), v(t), \theta^*) = e^{\alpha+\beta}\phi_\theta(\theta)(\theta^* - \theta - e^{-\alpha}v) + \\
&\dot{\beta}e^{\beta}[\phi(\theta) - \phi(\theta^*) + \phi_\theta(\theta^* - \theta)] \\
&-\dot{\beta}e^{\beta}\phi_\theta(\theta^* - \theta) + e^{\beta}\phi_\theta(\theta)\dot{\theta} \\
&= e^{\alpha+\beta}\phi_\theta(\theta)(\theta^* - \theta) + e^{\beta}\phi_\theta(\theta)(\dot{\theta} - v) \\
&-\dot{\beta}e^{\beta}\phi_\theta(\theta^* - \theta) + \dot{\beta}e^{\beta}[\phi(\theta) - \phi(\theta^*) - \phi_\theta(\theta - \theta^*)] \\
&= e^{\beta}(e^{\alpha} - \dot{\beta})\phi_\theta(\theta^* - \theta) \\
&+\dot{\beta}e^{\beta}[\phi(\theta) - \phi(\theta^*) - \phi_\theta(\theta - \theta^*)].
\end{aligned} \tag{1.56}$$

By convexity of the function ϕ, $[\phi(\theta) - \phi(\theta^*) - \phi_\theta(\theta - \theta^*)] \leq 0$ and $\phi_\theta(\theta^* - \theta) \leq 0$. If $e^{\alpha} - \dot{\beta} \geq 0$ then

$$\begin{aligned}
&\tfrac{d}{dt}V(\theta(t), v(t), \theta^*) = e^{\beta}(e^{\alpha} - \dot{\beta})\phi_\theta(\theta^* - \theta) \\
&+\dot{\beta}e^{\beta}[\phi(\theta) - \phi(\theta^*) - \phi_\theta(\theta - \theta^*)] \leq 0.
\end{aligned} \tag{1.57}$$

Thus, $\frac{d}{dt}V(\theta(t), v(t), \theta^*) \leq 0$ for $\dot{\beta} \leq e^{\alpha}$. Then the function V is decreasing over the path of the Bregman algorithm. It follows that

$$e^{\beta}[\phi(\theta) - \phi(\theta^*)] \leq V(\theta, v, \theta^*) \leq V(\theta_0, v_0, \theta^*).$$

Then, the global error is

$$0 \leq \phi(\theta) - \phi(\theta^*) \leq e^{-\beta}V(\theta_0, v_0, \theta^*),$$

with $\dot{\beta} \leq e^{\alpha}$, which shows a quick convergence to θ^*. This completes the proof. □

Definition 9 (Convergence time). *Let $\delta > 0$ and $\theta(t)$ be the trajectory generated by Bregman algorithm starting from θ_0 at time t_0. The convergence time to be within a ball $B_\delta(\phi(\theta^*))$ of radius $\delta > 0$ from the center $\phi(\theta^*)$ is given by*

$$T_\delta = \inf\{t \mid \phi(\theta(t)) - \phi(\theta^*) \leq \delta, \ t > t_0\}.$$

Proposition 7. *Under the assumptions above, the error generated by the algorithm is at most (1.52) which means that it takes at most $T_\delta = \beta^{-1}[\log \frac{c_0}{\delta}]$ time units for the algorithm to be within a ball $B_\delta(\phi(a^*))$ of radius $\delta > 0$ from the center $\phi(a^*)$.*

The proof is immediate. For $\delta > 0$ the average regret bound of Proposition 9,

$$\text{regret}_T \leq \frac{c_0}{T - t_0}\int_{t_0}^{T} e^{-\beta(s)}ds \leq \delta, \tag{1.58}$$

provides the announced convergence time bound. This completes the proof. □

Under the same assumption as in Proposition 6, it takes at most $T_\delta = \beta^{-1}[\log \frac{c_0}{\delta}]$ time units to the Bregman algorithms to be within a ball $B_\delta(\phi(\theta^*))$ of radius $\delta > 0$ from the center $\phi(\theta^*)$. See Table 1.5 for detailed parametric functions on the bound T_δ.

Note that Proposition 6 does not require the strong convexity property often used in the proof of convergence in gradient dynamics and Newton-based gradient methods. This is because the Bregman divergence is carefully designed to compensate for that part as a regularizer or a penalty function. Tables 1.5 and 1.6 summarize the theoretical speedup advantages of Bregman algorithms over the state-of-the-art algorithms.

TABLE 1.5
Bregman convergence rate under different set of functions

Convergence	Error Bound	Time-to-Reach
Triple exponential $\alpha(t) = t + e^t,\ \beta(t) = e^{e^t}$	$e^{-e^{e^t}} c_0$	$\log[\log(\log \frac{c_0}{\delta})]$
Double exponential rate $\alpha(t) = t,\ \beta(t) = e^t$	$e^{-e^t} c_0$	$\log(\log \frac{c_0}{\delta})$
Exponential rate $\alpha(t) = 0,\ \beta(t) = t$	$e^{-t} c_0$	$\log \frac{c_0}{\delta}$
Polynomial order k $\alpha(t) = \log k - \log t,\ \beta(t) = k \log t$	$\frac{c_0}{t^k}$	$\frac{c_0^{1/k}}{\delta^{1/k}}$

TABLE 1.6
Performance of the proposed Bregman algorithm compared to the classical ones with a precision error within $\delta > 0$.

	Accuracy	Time-to-Reach
Bregman	$O(e^{-\beta(t)})$	$O(\beta^{-1}(\log(\frac{1}{\delta})))$
Bregman 2	$O(e^{-e^{e^t}})$	$O(\log\log(\log(\frac{1}{\delta})))$
Bregman 1	$O(e^{-e^t})$	$O(\log(\log(\frac{1}{\delta})))$
Ishikawa-Nesterov Newton	$O(\frac{1}{t^2})$	$O(\frac{1}{\sqrt{\delta}})$
Conjugate/proximal gradient	$O(\frac{1}{t})$	$O(\frac{1}{\delta})$
Gradient ascent	$O(\frac{1}{t})$	$O(\frac{1}{\delta})$
Regret-min	$O(\frac{\log t}{t})$	-
Standard black-box	$O(\frac{1}{\sqrt{t}} + .)$	$O(\frac{1}{\delta^2})$

Risk-Aware Bregman dynamics

In order to reduce the oscillatory phase, we introduce a risk-aware Bregman dynamics which is a speed-up-and-average version of Equation (1.51) called *mean dynamics* m of ω given by

$$\begin{array}{l} \dddot{m} = -\frac{3}{t}\ddot{m} - (e^{\alpha} - \dot{\alpha})(\ddot{m} + \frac{2}{t}\dot{m}) \\ -\frac{e^{2\alpha+\beta}}{t} g_{mm}^{-1}(m + [t + 2e^{-\alpha}]\dot{m} + te^{-\alpha}\ddot{m})\phi_m(t\dot{m} + m), \end{array} \tag{1.59}$$

with starting vector $m(0) = \omega(0), \dot{m}(0), \ddot{m}(0)$.

Proposition 8. *The time-average trajectory of the Bregman learning algorithm (1.51) generates the mean dynamics (1.59)*

Proof: We use the average relation $m(t) = \frac{1}{t}\int_0^t x(s)\, ds$ where x solves Equation (1.51). From the definition of m, and by L'Hôpital's rule, $m(0) = x(0)$. Moreover, $m(t)$ and $x(t)$ share the following equations:

$$\begin{array}{l} x(t) = m(t) + t\dot{m}(t), \\ \dot{x}(t) = 2\dot{m}(t) + t\ddot{m}(t), \\ \ddot{x}(t) = 3\ddot{m}(t) + t\dddot{m}(t). \end{array} \tag{1.60}$$

Substituting these values in Equation (1.51) yields the mean dynamics (1.59). This completes the proof. □

The mean dynamics generates a less oscillatory trajectory. The next result provides an accuracy bound for (1.59).

Proposition 9. *The mean dynamics (1.59) satisfies*

$$0 \leq \phi(m(t)) - \phi(x^*) \leq \frac{c_0}{t} \int_0^t e^{-\beta(s)} ds. \tag{1.61}$$

Proof: Let $m(t) = \frac{1}{t} \int_0^t x(s) ds$. Then, $m(t) = \int_{\mathbb{R}} x(s) \left(\frac{1}{t}\mathbb{I}_{[0,t]}(s)\right) \, ds$. Thus, $m(t) = \mathbb{E}_{\mu(t)} x$ where $\mu(t)$ is the measure with density $d\mu(t)[s] = \frac{1}{t}\mathbb{I}_{[0,t]}(ds)$. By convexity of ϕ we apply the Jensen's inequality:

$$l\left(\frac{1}{t}\int_0^t x(s)ds\right) = \phi(m(t)) = \phi(\mathbb{E}_{\mu(t)} x)$$

$$\leq \mathbb{E}_{\mu(t)}\phi(x) = \frac{1}{t}\int_0^t \phi(x(s))ds.$$

In view of (1.52) one has

$$0 \leq \phi\left(\frac{1}{t}\int_0^t x(s)ds\right) - \phi(x^*) \leq \frac{1}{t}\int_0^t [\phi(x(s)) - \phi(x^*)]ds$$

$$\leq c_0 \frac{1}{t}\int_0^t e^{-\beta(s)} ds,$$

$$0 \leq \phi(m(t)) - \phi(x^*) \leq \frac{c_0}{t}\int_0^t e^{-\beta(s)} ds.$$

In particular, for $\beta(s) = -s + e^s$, one obtains an error bound to the minimum value as

$$\frac{c_0}{t}\int_0^t e^{-\beta(s)} ds = \frac{c_0}{t}\int_0^t e^s e^{-e^s} ds = \frac{c_0(\frac{1}{e} - e^{-e^t})}{t},$$

and for $\beta(s) = s$,

$$\frac{c_0}{t}\int_0^t e^{-\beta(s)} ds = \frac{c_0(1 - e^{-t})}{t}.$$

This completes the proof. □

The mean dynamics (1.59) satisfies

$$0 \leq \phi(m(t)) - \phi(\theta^*) \leq \frac{c_0}{t}\int_0^t e^{-\beta(s)} ds.$$

In particular, for $\beta(s) = -s + e^s$, one obtains an error bound to the minimum value as

$$\frac{c_0}{t}\int_0^t e^{-\beta(s)} ds = \frac{c_0}{t}\int_0^t e^s e^{-e^s} ds = \frac{c_0(\frac{1}{e} - e^{-e^t})}{t},$$

and for $\beta(s) = s$,

$$\frac{c_0}{t}\int_0^t e^{-\beta(s)} ds = \frac{c_0(1 - e^{-t})}{t}.$$

1.6.3.1 Training when the True Distribution is UNknown.

The quality, distribution, diversity and integrity of the data $\mathcal{D} = \{(x_{0,o}, y_{L,o}) \in \mathcal{H}_0 \times \mathcal{H}_L, 1 \leq o \leq D\}$, for an integer $D \geq 1$, used in the pre-training and training phase are crucial in maintaining model's output reliability. In many cases, the true distribution of the underlying sample data $\mathcal{D}$ over $\Omega = \mathcal{H}_0 \times \mathcal{H}_L$ is not available or is unknown to the decision-maker. We formulate the training problem from the variations $\tilde{m}$ around the data-driven measured or observed distribution m.

We introduce next the notion of $f-$divergence. Let m and $\tilde{m}$ be two probability measures over a space Ω such that m is absolutely continuous with respect to $\tilde{m}$. Then, for a convex and proper function f, the f-divergence of $\tilde{m}$ from m is defined as

$$D_f(m \parallel \tilde{m}) \equiv \int_\Omega f\left(\frac{dm}{d\tilde{m}}\right) d\tilde{m} - f(1),$$

where $\frac{dm}{d\tilde{m}}$ is the Radon-Nikodym derivative of the measure m with respect to the measure $\tilde{m}$. The term $f(1)$ is used for controlling the sign. The f-divergence can be used to capture discrepancies between two probability measures.

By Jensen's inequality,

$$\begin{array}{l} D_f(m \parallel \tilde{m}) = \int_\Omega f\left(\frac{dm}{d\tilde{m}}\right) d\tilde{m} - f(1) \\ \geq f\left(\int_\Omega \frac{dm}{d\tilde{m}} d\tilde{m}\right) - f(1) \\ = f\left(\int_\Omega dm\right) - f(1) = f(1) - f(1) = 0. \end{array} \tag{1.62}$$

Thus, $D_f(m \parallel \tilde{m}) \geq 0$ for any convex function f. Note however that the $f-$divergence $D_f(m \parallel \tilde{m})$ is not a distance (for example, it does not satisfy the symmetry property). For $\rho_l > 0$, the distributional uncertainty set driven by f-divergence ball is

$$B_{\rho_l}(m) = \{\tilde{m} | \int_\Omega \tilde{m}(d\omega) = 1, \; D_f(\tilde{m}||m) \leq \rho_l\}.$$

Let $(\Omega, \mathcal{F}, \tilde{m})$ be a measure space, f a measurable function and ν an absolutely continuous measure, i.e., there exists $\rho(.) > 0$ such that $\nu(dx) = \rho(x)\tilde{m}(dx)$. Let g be a function such that e^g is $\tilde{m}$-integrable. The Legendre-Fenchel transform of the relative entropy yields:

$$\sup_\nu \left\{ \int g(x)\nu(dx) - \int \log(\rho(x))\nu(dx) \right\} = \log\left(\int_\Omega e^{g(\omega)} \tilde{m}(d\omega) \right).$$

From the above relation we deduce that

$$\begin{array}{l} \sup_\nu \left\{ \int g_l(x)\nu(dx) - \frac{1}{\epsilon_l} \int \log(\rho(x))\nu(dx) \right\} \\ = \sup_\nu \left\{ \int g_l(x)\nu(dx) - \frac{1}{\epsilon_l} \int \log(\frac{\nu(dx)}{\tilde{m}(dx)})\nu(dx) \right\} = \frac{1}{\epsilon_l} \log\left(\int_\Omega e^{\epsilon_l g_l(\omega)} \tilde{m}(d\omega) \right), \end{array} \tag{1.63}$$

which is the log of the expected value of exponentiated loss. For a very small ϵ_l, one can connect the latter expression to mean-variance through

$$\frac{1}{\epsilon_l} \log\left(\int_\Omega e^{\epsilon_l g_l(\omega)} \tilde{m}(d\omega) \right) = E_{\tilde{m}} g_l + \frac{\epsilon_l}{2} \mathrm{var}_{\tilde{m}}[g_l] + O(\epsilon_l^2).$$

It follows from the previous observation that minimizing the log of the expected value of exponentiated loss can be seen as a distributionally robust optimization:

$$\begin{array}{l} \inf_{\theta_l \in \mathcal{H}_l} \frac{1}{\epsilon_l} \log\left(\int_\Omega e^{\epsilon_l g_l(\theta,\omega)} \tilde{m}(d\omega) \right) \\ = \inf_{\theta_l \in \mathcal{H}_l} \sup_\nu \left\{ \int g_l(\theta,\omega)\nu(dx) - \frac{1}{\epsilon_l} \int \log(\frac{\nu(dx)}{\tilde{m}(dx)})\nu(dx) \right\}, \end{array} \tag{1.64}$$

with the distributional constraint that the measure μ is absolutely continuous with respect to $\tilde{m}$.

We work with the sample game

$$\mathcal{G} = (\mathcal{L} = \{1, \ldots, L\}, \Omega = (\mathcal{H}_0 \times \mathcal{H}_L), (\Theta_l, g_l)_{\mathcal{L}}),$$

with a given $\omega = (x, y)$, and where $g_l(\theta, x, y) = g_l(\theta, \omega) = \phi_l(A_l\theta) - \langle A_l\theta, y_l \rangle$.

Definition 10. *Distributionally robust risk-neutral game:*

Consider a set $\mathcal{L}$ of decision-makers with $|\mathcal{L}| = L \geq 2$. Each decision-maker $l \in \mathcal{L}$ has its own loss functional $E_{\tilde{m}} g_l$ with $g_l : \prod_{l\in\mathcal{L}} \Theta_l \times \Omega \to \mathbb{R}$ expressed as $g_l(\theta, \omega)$. Further, the distributionally robust risk-neutral game is given by

$$G_{dr} = (\mathcal{L}, (\Theta_l \times B_{\rho_l}(m), E_{\tilde{m}} g_l)_{l\in\mathcal{L}}, (\rho_l)_{l\in\mathcal{L}}, m)\,.$$

The distributionally robust equilibrium problem of G_{dr} is formulated as follows:

$$\begin{array}{l} l \in \mathcal{L}, \\ \inf_{\theta_l \in \Theta_l} \sup_{\tilde{m} \in B_{\rho_l}(m)} E_{\tilde{m}} g_l(\theta, \omega). \end{array} \tag{1.65}$$

Example 5. *As machine intelligence evolves, it is crucial to recognize and address the unique challenges faced by diverse communities. In many countries in Africa, where audio literacy plays a central role in communication and cultural heritage, the integration of machine intelligence technologies presents both unprecedented opportunities and obstacles. The significance of ensuring that the benefits of machine intelligence reach audio-literate populations cannot be overstated. Despite the growing recognition of machine intelligence's potential in Africa, there exists a notable research gap in tailoring risk assessment frameworks to the specific needs of audio-literate communities. Conventional risk measures often fall short in capturing the nuances of oral traditions and audio-centric communication prevalent in many parts of the continent. This type of analysis with G_{dr} bridges this gap by proposing a data-driven, distributionally robust, coherent risk assessment framework explicitly designed for machine intelligence applications in Africa. m is the audio data sampled in 7000 dialects in the continent and 2000 local languages in Africa. The true distribution of these local languages are unknown to current machine intelligence systems. The imperative lies not only in acknowledging the existence of this gap but also in actively addressing it. By doing so, we pave the way for responsible and culturally sensitive integration of machine intelligence technologies, ensuring that they empower, rather than marginalize, the diverse and vibrant audio-literate communities across the continent.*

> We know what duality theory is in optimization, game theory, economics as well as in physics. Duality inequalities are extremely useful when tackling adversarial capacity bounds and robustness approaches. For distributional robustness, it involves an infinite dimensional optimization of a set of measures. Using the Legendre-Fenchel transform and $f-$-divergence, we map the optimization over measure as inf sup or sup inf which in turns leads to three terms inf sup inf or sup inf sup. The study then involves three terms and there is a need for triality theory to understand the outcomes.

Proposition 10 (Triality theory). *Let ϕ_3 be a real-valued function $(\theta_1, \theta_2, \theta_3) \mapsto \phi_3(\theta_1, \theta_2, \theta_3)$ defined on $\prod_{k=1}^{3} \Theta_k$. Then, the following inequalities hold:*

$$\begin{array}{l} \sup_{\theta_2 \in \Theta_2} \inf_{\theta_1 \in \Theta_1, \theta_3 \in \Theta_3} \phi_3(\theta_1, \theta_2, \theta_3) \leq \\ \inf_{\theta_3 \in \Theta_3} \sup_{\theta_2 \in \Theta_2} \inf_{\theta_1 \in \Theta_1} \phi_3(\theta_1, \theta_2, \theta_3) \leq \\ \inf_{\theta_1 \in \Theta_1, \theta_3 \in \Theta_3} \sup_{\theta_2 \in \Theta_2} \phi_3(\theta_1, \theta_2, \theta_3), \end{array} \tag{1.66}$$

and similarly

$$\begin{aligned}&\sup_{\theta_1\in\Theta_1,\theta_3\in\Theta_3}\inf_{\theta_2\in\Theta_2}\phi_3(\theta_1,\theta_2,\theta_3)\le\\&\sup_{\theta_3\in\Theta_3}\inf_{\theta_2\in\Theta_2}\sup_{\theta_1\in\Theta_1}\phi_3(\theta_1,\theta_2,\theta_3)\le\\&\inf_{\theta_2\in\Theta_2}\sup_{\theta_1\in\Theta_1,\theta_3\in\Theta_3}\phi_3(\theta_1,\theta_2,\theta_3).\end{aligned}\tag{1.67}$$

□

Proof: A proof of Proposition 10 can be obtained from the classical duality inequality. First we shall prove the sup inf inequality. Define

$$g(\theta_2,\theta_3)=\inf_{\theta_1\in\Theta_1}\phi_3(\theta_1,\theta_2,\theta_3).$$

Thus, for all $\theta_2,\theta_3,$ one has $g(\theta_2,\theta_3)\le\phi_3(\theta_1,\theta_2,\theta_3)$. It follows that, for any $\theta_1,\theta_3,$

$$\sup_{\theta_2\in\Theta_2} g(\theta_2,\theta_3)\le\sup_{\theta_2\in\Theta_2}\phi_3(\theta_1,\theta_2,\theta_3).$$

Using the definition of g, one obtains

$$\sup_{\theta_2\in\Theta_2}\inf_{\theta_1\in\Theta_1}\phi_3(\theta_1,\theta_2,\theta_3)\le\sup_{\theta_2\in\Theta_2}\phi_3(\theta_1,\theta_2,\theta_3),\ \forall\theta_1,\theta_3.$$

Taking the infimum in θ_1 yields:

$$\sup_{\theta_2\in\Theta_2}\inf_{\theta_1\in\Theta_1}\phi_3(\theta_1,\theta_2,\theta_3)\le\inf_{\theta_1\in\Theta_1}\sup_{\theta_2\in\Theta_2}\phi_3(\theta_1,\theta_2,\theta_3),\ \forall\theta_3\tag{1.68}$$

Now, for the variable in θ_3 we use two operations:

- Taking the infimum in inequality (1.68) in θ_3 yields

$$\begin{aligned}\inf_{\theta_3\in\Theta_3}\sup_{\theta_2\in\Theta_2}\inf_{\theta_1\in\Theta_1}\phi_3(\theta_1,\theta_2,\theta_3)&\le\inf_{\theta_3\in\Theta_3}\inf_{\theta_1\in\Theta_1}\sup_{\theta_2\in\Theta_2}\phi_3(\theta_1,\theta_2,\theta_3)\\&=\inf_{(\theta_1,\theta_3)\in\Theta_1\times\Theta_3}\sup_{\theta_2\in\Theta_2}\phi_3(\theta_1,\theta_2,\theta_3),\end{aligned}$$

  which proves the second part of the inequalities (1.66). The first part of the inequalities (1.66) follows immediately from (1.68).

- Taking the supremum in inequality (1.68) in θ_3 yields

$$\sup_{(\theta_2,\theta_3)\in\Theta_2\times\Theta_3}\inf_{\theta_1\in\Theta_1}\phi_3(\theta_1,\theta_2,\theta_3)\le\sup_{\theta_3\in\Theta_3}\inf_{\theta_1\in\Theta_1}\sup_{\theta_2\in\Theta_2}\phi_3(\theta_1,\theta_2,\theta_3),$$

  which proves the first part of the inequalities (1.67). The second part of the inequalities (1.67) follows immediately from (1.68). Thus the set of inequalities in (1.67) are proved.

This completes the proof. □

We transform the distributionally robust term as follows.

Lemma 15. *The distributionally robust term is given by*

$$\sup_{\tilde m\in B_{\rho_l}(m)}E_{\tilde m}\phi_l(\theta,\omega)=\inf_{\lambda_l\ge0,\mu}E_m\left\{\lambda_l(\rho_l+f(1))+\mu+\lambda_l f^*\left(\frac{\phi_l(\theta,\omega)-\mu}{\lambda_l}\right)\right\}.\tag{1.69}$$

Proof: We prove the equivalence formula under f-divergence. The (distributionally) robust best-response problem of player i under f-divergence is equivalent to

$$\left\{ \inf_{\theta_i} \sup_{G\in G_\rho(m)} \mathbb{E}_m[\phi_i G]; \right. \tag{1.70}$$

where $G(\tilde{\omega}) = \frac{d\tilde{m}}{dm}(\tilde{\omega})$ is the likelihood and set $G_\rho(m)$ is

$$G_\rho(m) = \{G \mid \int_{\tilde{\omega}} f(G(\tilde{\omega}))dm - f(1) \leq \rho, \int_{\tilde{\omega}} G(\tilde{\omega})dm = 1\}.$$

We introduce the Lagrangian as

$$\begin{aligned}\tilde{\phi}_i(\theta, G, \lambda, \mu) = & \int_{\tilde{\omega}} \phi_i(a, \tilde{\omega})G(\tilde{\omega})dm + \lambda(\rho + f(1) - \int_{\tilde{\omega}} f(G(\tilde{\omega}))dm) \\ & +\mu(1 - \int_{\tilde{\omega}} G(\tilde{\omega})dm(\tilde{\omega})),\end{aligned}$$

where $\lambda \geq 0$ and $\mu \in \mathbb{R}$. The problem solved by player i is

$$(\tilde{P}_i^*) \left\{ \inf_{\theta_i} \sup_{G\in G_\rho(m)} \inf_{\lambda\geq 0, \mu\in\mathbb{R}} \tilde{\phi}_i(\theta, G, \lambda, \mu). \right. \tag{1.71}$$

A full understanding of the problem $(\tilde{P}_i^*)$ requires a triality theory (not a duality theory). The underlying idea is that one can use a transformation of the last two terms to derive a finite-dimensional optimization problem. The Lagrangian $\tilde{\phi}_i$ of player i is clearly concave in G and convex in λ, μ and is semi-continuous jointly. By the triality theory above, $\tilde{\phi}_i : (\theta, G, \lambda, \mu) \mapsto \tilde{\phi}_i(\theta, G, \lambda, \mu)$ satisfies the sup inf inequality, the inequalities (1.66), (1.67) and one has the following:

$$\inf_{\theta_i} \sup_{G\in G_\rho(m)} \inf_{\lambda\geq 0, \mu\in\mathbb{R}} \tilde{\phi}_i(.) \leq \inf_{\theta_i} \inf_{\lambda\geq 0, \mu\in\mathbb{R}} \sup_{G\in G_\rho(m)} \tilde{\phi}_i(.).$$

In this case $\tilde{\phi}_i$ is concave in G and convex in λ, μ and is semi-continuous jointly, thus there is no gap in the second part of the optimization and the following equality holds:

$$\inf_{\theta_i} \sup_{G\in G_\rho(m)} \inf_{\lambda\geq 0, \mu\in\mathbb{R}} \tilde{\phi}_i(.) = \inf_{\theta_i} \inf_{\lambda\geq 0, \mu\in\mathbb{R}} \sup_{G\in G_\rho(m)} \tilde{\phi}_i(.).$$

The latter problem can be rewritten as

$$(\tilde{P}_i^*) \left\{ \inf_{\theta_i\in\Omega_i, \lambda\geq 0, \mu\in\mathbb{R}} [\sup_{G\in G_\rho(m)} \tilde{\phi}_i(\theta, G, \lambda, \mu)]. \right. \tag{1.72}$$

The Lagrangian function takes the form as $\tilde{\phi}_i = \lambda(\rho + f(1)) + \mu + \int \{G[\phi_i - \mu] - \lambda f(G)\}dm$. It follows that

$$\sup_{G\in G_\rho(m)} \tilde{\phi}_i(\theta, G, \lambda, \mu) = \lambda(\rho + f(1)) + \mu + \sup_G \int \{G[\phi_i - \mu] - \lambda f(G)\}dm.$$

Introducing the Fenchel-Legendre transform on G and exchanging sup and $\int$, one gets

$$\sup_{G\in G_\rho(m)} \tilde{\phi}_i(.) = \lambda(\rho + f(1)) + \mu + \int \lambda f^*(\frac{\phi_i - \mu}{\lambda})dm.$$

Since the dimension of the problem is reduced to the dimension of $\Omega_i \times \mathbb{R}_+ \times \mathbb{R}$ per player i, it follows that the curse of dimensionality of the robust best-response problem of player i is significantly reduced to an equivalent of the finite-dimensional stochastic optimization problem. This completes the proof.

□

Based on the above, the distributionally robust game can be transformed into the following game $\hat{G}_{rdr}$

Proposition 11. *Reduced distributionally robust risk-neutral game*

Consider a set $\mathcal{L}$ of decision-makers with $|\mathcal{L}| \geq 2$. Each decision-maker $l \in \mathcal{L}$ has its own loss functional $E_m \hat{g}_l$ with $\hat{g}_l : \prod_{j\in\mathcal{L}} \Theta_j \times \mathbb{R}_+ \times \mathbb{R} \times \Omega \to \mathbb{R}$ expressed as $\hat{g}_l(\theta, \lambda_l, \mu, \omega)$. The game G_{dr} is equivalent to the reduced distributionally robust risk-neutral game given by

$$\hat{G}_{rdr} = (\mathcal{L}, (\Theta_l \times \mathbb{R}_+ \times \mathbb{R}, E_m \hat{g}_l)_{l\in\mathcal{L}}, \rho_l, m)\,.$$

Further, the reduced distributionally robust equilibrium problem of $\hat{G}_{rdr}$ is formulated as follows:

$$\begin{array}{l} l \in \mathcal{L}, \\ \inf_{(\theta_l,\lambda_l,\mu)\in\Theta_l\times\mathbb{R}_+\times\mathbb{R}} E_m \hat{g}_l(\theta, \lambda_l, \mu, \omega) \end{array} \tag{1.73}$$

The advantage of the reduced distributionally robust game $\hat{G}_{rdr}$ is that the action space $\Theta_l \times B_{\rho_l}(m)$ is reduced to the space $\Theta_l \times \mathbb{R}_+ \times \mathbb{R}$. See Figure 1.5.

Proof: We now use triality theory (Proposition 10) to derive the following expression:

$$\begin{array}{l} \inf_{\theta_l\in\Theta_l} \sup_{\tilde{m}\in B_{\rho_l}(m)} E_{\tilde{m}} g_l(\theta, \omega) \\ = \inf_{\theta_l\in\Theta_l} \inf_{\lambda_l\geq 0,\mu} E_m \left\{ \lambda_l(\rho_l + f(1)) + \mu + \lambda_l f^* \left(\frac{g_l(\theta,\omega)-\mu}{\lambda_l} \right) \right\} \\ = \inf_{(\theta_l,\lambda_l,\mu)\in\Theta_l\times\mathbb{R}_+\times\mathbb{R}} E_m \hat{g}_l(\theta, \lambda_l, \mu, \omega), \end{array} \tag{1.74}$$

where

$$\hat{g}_l(\theta, \lambda_l, \mu, \omega) = \lambda_l(\rho_l + f(1)) + \mu + \lambda_l f^* \left(\frac{g_l(\theta, \omega) - \mu}{\lambda_l} \right).$$

□

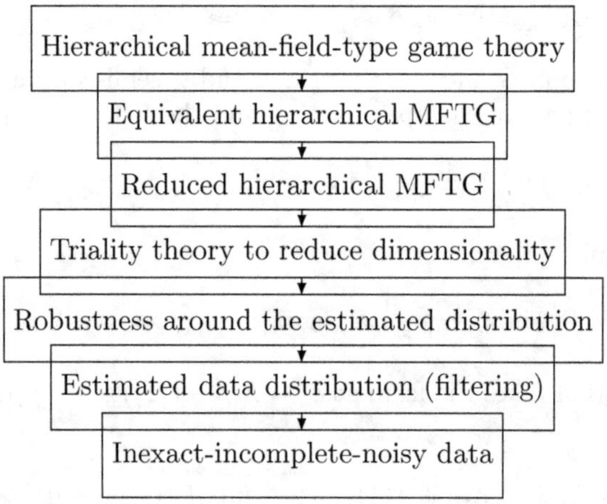

FIGURE 1.5
Connection between distributional-robust mean-field-type game theory and transformers with distributional robust training cost.

1.6.3.2 Mean-Variance Loss

We now focus on a game with mean-variance loss function. This class of games is widely used in risk quantification and belongs to the class of mean-field-type games (MFTGs). The basic ingredients of a One-Shot MFTG are as follows:

Definition 11. *MFTG*

- *The set of decision-makers $\mathcal{L}$, with cardinality $|\mathcal{L}| = L \geq 2$.*

- *For every decision-maker $l \in \mathcal{L}$, there is a set of actions $\mathcal{A}_l$ that is non-empty.*
- *Every decision-maker l has a preference structure that can be represented by an instant performance functional*

  $P_l : \prod_{j\in\mathcal{L}} \mathcal{A}_j \times \mathcal{P}\left(\prod_{j\in\mathcal{L}} \mathcal{A}_j\right) \to \mathbb{R},\ (a, D_a) \mapsto P_l(a, D_a)$, *with* $a = (a_1, \ldots, a_L)$. *The expected performance functional of decision-maker l is given by* $E[P_l(a, D_a)] := \int_{b\in\prod_{j\in\mathcal{L}} \mathcal{A}_j} P_l(b, D_a)D_a(db)$.

  The collection $\mathcal{G} = (\mathcal{L}, (\mathcal{A}_l, E[P_l])_{l\in\mathcal{L}})$ is called a one-shot MFTG in strategic form.

Note that $D \mapsto E[P_l(a, D)]$ is not necessarily linear.

Definition 12. *Mean-Field-Type Game with Mean-Variance Loss.*
Consider a set $\mathcal{L}$ of decision-makers with $|\mathcal{L}| \geq 2$. Each decision-maker $l \in \mathcal{L}$ has its own loss functional $mv_l : \prod_{j\in\mathcal{L}} \Theta_j \times \mathbb{P}(\Omega) \to \mathbb{R}$ expressed as

$$mv_l(\theta, \tilde{m}) = E_{\tilde{m}}[g_l(\theta, \omega)] + \hat{r}_l \cdot var_{\tilde{m}}[g_l(\theta, \omega)].$$

Here $\hat{r}_l = \frac{c_l}{\epsilon + |g_l(x^0, \omega^0)|}$, $c_l > 0, \epsilon > 0$ with c_l being the risk-sensitivity index of the decision-maker l.

The mean-variance game is given by $G_{mv} = (\mathcal{L}, (\Theta_l, mv_l)_{l\in\mathcal{L}}, \tilde{m})$. Further, the mean-variance equilibrium problem of G_{mv} is formulated as follows:

$$\left\{ \begin{array}{l} l \in \mathcal{L}, \\ \theta_l \in \arg\min_{\theta_l' \in \Theta_l} mv_l(\theta_l', \theta_{-l}, \tilde{m}) \end{array} \right. \tag{1.75}$$

Note that $\tilde{m} \mapsto \mathrm{mv}_l(\theta, \tilde{m})$ is non-linear. This non-linearity in the measure $\tilde{m}$ creates an extra challenge in the search for the best-response strategy.

The Mean-Variance as the Best Case of an Expected Loss

Lemma 16. *Let mv be a mean-variance trade-off of g, $\hat{\lambda} > 0$. We have:*

$$\begin{array}{ll} mv & = E[g] + \hat{\lambda} \cdot var[g] \\ & = \inf_{z\in\mathbb{R}} E\left[\left\{(1 - 2\hat{\lambda}z)g + \hat{\lambda}z^2\right\} + \hat{\lambda}g^2\right] = \inf_{z\in\mathbb{R}} E[\tilde{g}(z)], \end{array} \tag{1.76}$$

where $\tilde{g}(z) = (1 - 2z\hat{\lambda})g + \hat{\lambda}z^2 + \hat{\lambda}g^2$.

Proof:

$$\begin{array}{ll} mv & = E[g] + \hat{\lambda} \cdot \mathrm{var}[g] = E[g] + \hat{\lambda}(E[g^2]) - \hat{\lambda}(E[g])^2 \\ & = E[g] + \hat{\lambda}E[g^2] - \hat{\lambda} \sup_{z\in\mathbb{R}} \left\{2\langle z, E[g]\rangle - z^2\right\} \\ & = E[g] + \hat{\lambda}E[g^2] + \hat{\lambda} \inf_{z\in\mathbb{R}} \left\{-2\langle z, E[g]\rangle + z^2\right\} \\ & = \inf_{z\in\mathbb{R}} \left[E[g] + \hat{\lambda}E[g^2] + \hat{\lambda}\left\{-2\langle z, E[g]\rangle + z^2\right\}\right] \\ & = \inf_{z\in\mathbb{R}} \left\{(1 - 2\hat{\lambda}z)E[g] + \hat{\lambda}z^2\right\} + \hat{\lambda}E[g^2] \\ & = \inf_{z\in\mathbb{R}} E\left[\left\{(1 - 2\hat{\lambda}z)g + \hat{\lambda}z^2\right\} + \hat{\lambda}g^2\right] = \inf_{z\in\mathbb{R}} E[\tilde{g}(z)], \end{array} \tag{1.77}$$

where $\tilde{g}(z) = (1 - 2z\hat{\lambda})g + \hat{\lambda}z^2 + \hat{\lambda}g^2$. □

Proposition 12. *MFTG with Measure-Linearized Mean-Variance Loss*
Consider a set $\mathcal{L}$ of decision-makers with $|\mathcal{L}| \geq 2$. Each decision-maker $l \in \mathcal{L}$ has their own loss functional $E_{\tilde{m}}\tilde{g}_l(\theta, z_l, \omega)$ with $\tilde{g}_l : \prod_{j\in\mathcal{L}} \Theta_j \times \mathbb{R} \times \Omega \to \mathbb{R}$ expressed as $\tilde{g}_l(\theta, z_l, \omega)$.

The mean-variance game G_{mv} is equivalent to the reduced mean-variance game given by $\tilde{G}_{rmv} = (\mathcal{L}, (\Theta_l \times \mathbb{R}, E_{\tilde{m}}\tilde{g}_l)_{l\in\mathcal{L}}, \tilde{m})$. The reduced mean-variance equilibrium problem of $\tilde{G}_{rmv}$ is formulated as follows:

$$\begin{array}{l} l \in \mathcal{L}, \\ \inf_{(\theta_l, z_l)\in\Theta_l\times\mathbb{R}} E_{\tilde{m}}\tilde{g}_l(\theta, z_l, \omega) \end{array} \tag{1.78}$$

Proof: Based on Lemma 16, we transform the mean-variance game G_{mv} as follows:

$$\inf_{\theta_l\in\in\Theta_l} E_{\tilde{m}}[g_l(\theta,\omega)] + \hat{r}_l \cdot var_{\tilde{m}}[g_l(\theta,\omega)] = \inf_{(\theta_l,z_l)\in\Theta_l\times\mathbb{R}} E_{\tilde{m}}\tilde{g}_l(\theta, z_l, \omega), \tag{1.79}$$

where

$$\tilde{g}_l(\theta, z_l, \omega) = (1 - 2z_l\hat{r}_l)g_l + \hat{r}_l z_l^2 + \hat{r}_l g_l^2.$$

□

The advantage of $\tilde{G}_{rmv}$ formulation is that $\tilde{m} \mapsto E_{\tilde{m}}\tilde{g}_l(\theta, z_l, \omega)$ is linear. The price to pay is that the action space of decision-maker l is now augmented from Θ_l to $\Theta_l \times \mathbb{R}$.

1.6.3.3 Distributional Robust Mean-Variance of a Loss Functional

The exact probability measure $\tilde{m}$ over Ω is known to the decision-makers. Given a data set with size D, a certain data-driven distribution m is estimated. The decision-makers consider the distributional robustness of their estimation m.

Definition 13. *Distributionally robust mean-variance MFTG*

Consider a set $\mathcal{L}$ of decision-makers with $|\mathcal{L}| \geq 2$. Each decision-maker $l \in \mathcal{L}$ has own loss functional $drmv_l : (\prod_{j\in\mathcal{L}} \Theta_j) \times \mathbb{P}(\Omega) \to \mathbb{R}$ expressed as $drmv_l(\theta, \tilde{m}) = \sup_{\tilde{m}\in B_{\rho_l}(m)} mv_l(\theta, \tilde{m})$. The distributionally robust mean-variance game is given by

$$G_{drmv} = (\mathcal{L}, (\Theta_l \times B_{\rho_l}(m), drmv_l)_{l\in\mathcal{L}}, \rho_l, m).$$

The distributionally robust mean-variance equilibrium problem of G_{drmv} is formulated as follows:

$$\begin{array}{l} l \in \mathcal{L}, \\ \inf_{\theta_l\in\Theta_l} \sup_{\tilde{m}\in B_{\rho_l}(m)} mv_l(\theta, \tilde{m}) \end{array} \tag{1.80}$$

The challenge of G_{drmv} is that the function $\tilde{m} \mapsto \text{mv}_l(\theta, \tilde{m})$ is non-linear (in the measure) and the distributional uncertainty set $B_{\rho_l}(m)$ is infinite dimensional (in general).

Proposition 13. *Distributionally robust mean-variance MFTG: linearized measure and f-divergence*

Consider a set $\mathcal{L}$ of decision-makers with $|\mathcal{L}| \geq 2$. Each decision-maker $l \in \mathcal{L}$ has its own loss functional $E_m\, \hat{mv}_l(\theta, z_l, \lambda_l, \mu, \omega)$ with

$$\hat{mv}_l : (\prod_{j\in\mathcal{L}} \Theta_j) \times \mathbb{R} \times \mathbb{R}_+ \times \mathbb{R} \times \Omega \to \mathbb{R}.$$

The game G_{drmv} is equivalent to the reduced distributionally robust mean-variance game

$$\hat{G}_{drmv} = (\mathcal{L}, (\Theta_l \times \mathbb{R} \times \mathbb{R}_+ \times \mathbb{R}, E_m\hat{mv}_l)_{l\in\mathcal{L}}, \rho_l, m),$$

whose solution is obtained through

$$\begin{array}{l} l \in \mathcal{L}, \\ \inf_{(\theta_l, z_l, \lambda_l, \mu)\in\Theta_l\times\mathbb{R}\times\mathbb{R}_+\times\mathbb{R}} E_m\hat{mv}_l(\theta, z_l, \lambda_l, \mu, \omega). \end{array} \tag{1.81}$$

Proof:

$$\begin{aligned}
&\inf_{\theta_l\in\Theta_l}\sup_{\tilde{m}\in B_{\rho_l}(m)} mv(g_l(\theta,\omega);\tilde{m}||m)\\
&=\inf_{\theta_l\in\Theta_l}\sup_{\tilde{m}\in B_{\rho_l}(m)} E_{\tilde{m}}g_l(\theta,\omega)+\hat{r}_l\text{var}_{\tilde{m}}[g_l(\theta,\omega)]\\
&=\inf_{\theta_l\in\Theta_l}\ \sup_{\tilde{m}\in B_{\rho_l}(m)}\inf_{z_l\in\mathbb{R}} E_{\tilde{m}}\left[(1-2z_l\hat{r}_l)g_l(\theta,\omega)+\hat{r}_l z_l^2+\hat{r}_l g_l^2(\theta,\omega)\right]\\
&=\inf_{(\theta_l,z_l)\in\Theta_l\times\mathbb{R}}\ \sup_{\tilde{m}\in B_{\rho_l}(m)} E_{\tilde{m}}\tilde{g}_l(\theta,z_l,\omega)\\
&=\inf_{(\theta_l,z_l)\in\Theta_l\times\mathbb{R}}\ \inf_{\lambda_l\geq 0,\mu} E_m\left\{\lambda_l(\rho_l+f(1))+\mu+\lambda_l f^*\left(\frac{\tilde{g}_l(\theta,z_l,\omega)-\mu}{\lambda_l}\right)\right\}\\
&=\inf_{(\theta_l,z_l,\lambda_l,\mu)\in\Theta_l\times\mathbb{R}\times\mathbb{R}_+\times\mathbb{R}} E_m\hat{\text{mv}}_l(\theta,z_l,\lambda_l,\mu,\omega),
\end{aligned}\tag{1.82}$$

where

$$\hat{\text{mv}}_l(\theta,z_l,\lambda_l,\mu,\omega)=\lambda_l(\rho_l+f(1))+\mu+\lambda_l f^*\left(\frac{\tilde{g}_l(\theta,z_l,\omega)-\mu}{\lambda_l}\right).$$

$$\tilde{g}_l(\theta,z_l,\omega)=(1-2z_l\hat{r}_l)g_l+\hat{r}_l z_l^2+\hat{r}_l g_l^2,$$

and where we have used (1.74). □

Thanks to the previous transformation, the function $\text{mv}_l(\theta,\tilde{m})$ is transformed into infimization over a linearized function in the measure $\tilde{m}$ resulting in an augmented action space. As a consequence of this step, the action space $\Theta_l\times B_{\rho_l}(m)$ is augmented to be $\Theta_l\times\mathbb{R}\times B_{\rho_l}(m)$ for decision-maker l. As a second step, the action space $\Theta_l\times\mathbb{R}\times B_{\rho_l}(m)$ is reduced to $\Theta_l\times\mathbb{R}\times\mathbb{R}_+\times\mathbb{R}$ thanks to the Fenchel transformation above (See Figure 1.6).

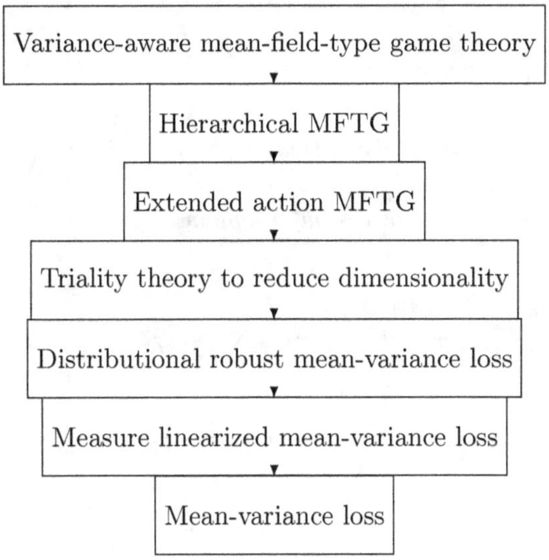

FIGURE 1.6
Connection between variance-aware mean-field-type game theory and transformers with variance-aware training cost.

1.6.3.4 Variance-Aware Bregman Learning

Let

$$\hat{\text{mv}}_l(x,y,\lambda,\mu)=E_m\lambda(\rho+f(1))+\mu+\lambda f^*\left(\frac{\tilde{g}(x,y,\omega)-\mu}{\lambda}\right),$$

where

$$\tilde{g}_l(x,y,\omega) = (1-2y\hat{r}_l)g_l(x,y,\omega) + \hat{r}_l y^2 + \hat{r}_l g_l^2(x,y,\omega).$$

Proposition 14 (Regret bound). *Let $s=(x,y,\lambda,\mu)$ and consider the following dynamics*

$$\begin{aligned} s(0) &= s_{00},\ \dot{s}(0)=s_{01},\ \ddot{s}(0)=s_{02}, \\ \dddot{s} &= -\frac{3}{t}\ddot{s} - (e^{\alpha}-\dot{\alpha})(\ddot{s}+\frac{2}{t}\dot{s}) \\ & -\frac{e^{2\alpha+\beta}}{t} b_{ss}^{-1}(s+[t+2e^{-\alpha}]\dot{s}+te^{-\alpha}\ddot{s})\hat{mv}_{l,s}(t\dot{s}+s), \end{aligned} \tag{1.83}$$

Then, we have the following bound:

$$0 \le \hat{mv}(s(t)) - \hat{mv}(s^*) \le \frac{c_0}{t}\int_0^t e^{-\beta(t')}dt',$$

where $c_0 = e^{\beta(0)}[\hat{mv}(s(0)) - \hat{mv}(s^*)] + d_b(s^*, s_0 + e^{-\alpha(0)}\dot{s}_0) > 0$ is the initial error plus the initial Bregman discrepancy.

The proof follows from Proposition 6. □

Ensemble Variance-Aware Bregman Learning

We now focus on the accuracy bounds over the sample data $\{\tilde{\omega}_k \sim m, k\in\{1,\dots,D\}\}$ with $D\ge 1$ points.

Proposition 15 (Sample Regret bound). *Let*

$$\hat{mv}(x,y,\lambda,\mu,\omega) = \lambda(\rho+f(1)) + \mu + \lambda f^*\left(\frac{\tilde{g}(x,y,\omega)-\mu}{\lambda}\right),$$

and particle swarm-based stochastic Bregman dynamics given by

$$\begin{array}{l} z(0)=z_{00},\ \dot{z}(0)=z_{01} \\ \frac{d}{dt}[b_z(z+e^{-\alpha}\dot{z})] = -e^{\alpha+\beta}\frac{1}{D}\sum_{k=1}^D \hat{mv}_z(x,y,\lambda,\mu,\tilde{\omega}_k), \\ \tilde{\omega}_k \sim m, k\in\{1,\dots,D\}. \end{array} \tag{1.84}$$

For each sample $(\tilde{\omega}_k)_{k\in\{1,\dots,D\}}\in\Omega^D$, and convex function $\tilde{g}$ one has

$$0 \le \frac{1}{D}\sum_{k=1}^D \hat{mv}(z(t),\tilde{\omega}_k) - \hat{mv}(z_D^*,\tilde{\omega}_k) \le \frac{c_{0,D}}{t}\int_0^t e^{-\beta(s)}ds,$$

where $c_{0,D} = e^{\beta(0)}[\hat{mv}(z_{0,D}) - \hat{mv}(z_D^*)] + d_b(z_D^*, z_{0,D} + e^{-\alpha(0)}\dot{z}_0) > 0$ is the initial error and $z_D^* \in \arg\min_z \frac{1}{D}\sum_{k=1}^D \hat{mv}(z,\tilde{\omega}_k)$.

Proof. The proof follows from the inequalities in Proposition 14. □

Training when $\hat{O}_L \circ \hat{O}_{L-1} \circ \dots \circ \hat{O}_1(x) = y$ has no solution. The above analysis was conducted under the assumption of the existence of a solution. Sometimes the output equation may not have an exact solution. For example, the equation $-1 = ReLU(x)$ has no solution. But one can look for an x that will make the difference $ReLU(x)+1$ as close as possible. We formulate a criterion to reduce the gap when the output data y is not in the range of $\hat{O}_L \circ \hat{O}_{L-1} \circ \dots \circ \hat{O}_1$. One can, of course, consider $\|\hat{O}_L \circ \hat{O}_{L-1} \circ \dots \circ \hat{O}_1(x) - y\|^2$ and optimize through the parameters inside $\hat{O}_l$.

1.7 Deep Neural Network Examples

Residual neural networks

We consider a residual neural network (Figure 1.7) with $L \geq 1$ layers. Let $\{\mathcal{H}_l\}_{0 \leq l \leq L}$ be non-zero Hilbert spaces. For every $l \in \{1, \ldots, L\}$, let $W_l : \mathcal{H}_{l-1} \to \mathcal{H}_l$ be a bounded linear operator and consider the family of activation operators $R_l : \mathcal{H}_l \to \mathcal{H}_l$. The deep neural network is defined by the composition of operators

$$(Id + R_L \circ (W_L \cdot + b_L)) \circ (Id + R_{L-1} \circ (W_{L-1} \cdot + b_{L-1})) \circ \ldots \circ (Id + R_1(W_1 \cdot + b_1)), \quad (1.85)$$

where W_l is the weight operator and b_l captures the bias parameter. Let $x_0 \in \mathcal{H}_0$, $\{\lambda_t\}_{t \geq 0}$ be a non-negative sequence, and consider the map defined by

$$\hat{O}_l : \left| \begin{array}{l} \mathcal{H}_{l-1} \to \mathcal{H}_l \\ x \mapsto x + R_l(W_l x + b_l). \end{array} \right.$$

Iterate layer by layer and by timestep the following:

$$\text{for } t \in \{0, 1, 2, \ldots\} \left| \begin{array}{l} y_{1,t} = \hat{O}_1(x_t), \\ \text{for } l \in \{2, \ldots, L\}, \; y_{l,t} = \hat{O}_l(y_{l-1,t}), \\ x_{t+1} = x_t + \lambda_t(y_{L,t} - x_t). \end{array} \right. \quad (1.86)$$

which means that

$$x_{t+1} = x_t + \lambda_t(\hat{O}_L \circ \hat{O}_{L-1} \circ \ldots \circ \hat{O}_1(x_t) - x_t),$$

starting from x_0. The sequence $(\lambda_t)_t$, with $\lambda_t \geq 0$, is the step-size (learning rate). For consistency of the operation, the output of $\hat{O}_L$ should be in the same space as the input i.e., $\mathcal{H}_L = \mathcal{H}_0$. As we want to examine situations in which time goes to infinity, the cumulative step-size $\sum_{t=1}^{T} \lambda_t \to +\infty$ as T goes to infinity.

Definition 14. *The residual neural network specification is given by the family*

$$res\mathbb{NN} := (x_0, L, \{R_l, W_l, b_l, \mathcal{H}_l\}_{1 \leq l \leq L}, \lambda).$$

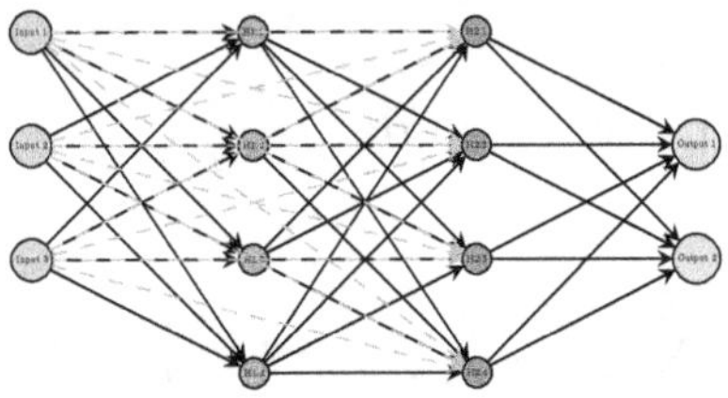

FIGURE 1.7
An example of a residual neural network: The input layer nodes are colored in orange. The first hidden layer nodes are colored in blue. The second hidden layer nodes are colored in purple. The output layer nodes are colored in green. The residual connections are represented by dashed lines in pink.

Deep ReLU Networks

A deep ReLU network is a neural network (Figure 1.2) defined by (1.1) with the Rectified Linear Unit (ReLU) activation function $r_l(x) = ReLU(x) = \frac{x+|x|}{2}$. The ReLU function defined on real numbers is extended entrywise to elements of the Hilbert space. $R_l(x) = \sum_{k=1}^{\infty} ReLU(\langle x, e_k \rangle) e_k$ where $(e_k)_{k\geq 0}$ is an orthonormal basis of the Hilbert space.

Remark 8. *On the composition of ReLU activations: Let x be the input signal and y be the last layer output. The following mappings are not convex and not differentiable.*

- $x \mapsto ReLU(x)$ *is convex.*
- $x \mapsto ReLU(x)$ *is* $1-$*Lipschitz (non-expansive).*
- $x \mapsto ReLU(x)$ *is not differentiable at* 0 *but has a subdifferential set at zero.*
- $x \mapsto ReLU(x)$ *is* $\frac{1}{2}$*-averaged. The extension to Hilbert space is done entry-wise. Thus,* $X \mapsto ReLU(X)$ *is* $\frac{1}{2}$*-averaged.*
- $x \mapsto ReLU(x)$ *has a convex antiderivative* h*:* $ReLU(x) = \partial_x h(x)$ *with* $h(x) = \frac{1}{2} x ReLU(x) = \frac{1}{2} ReLU^2(x)$.
- $x \mapsto ReLU(x)$ *is a proximal, i.e., there is a function* f *such that* $ReLU(x) = (Id + \partial f)^{-1}(x)$*. It suffices to choose* $f(x) = \infty \mathbb{I}_{x<0} + 0\mathbb{I}_{x\geq 0}$.
- $x \mapsto x - ReLU(x)$ *is* $1-$*Lipschitz (non-expansive), not differentiable at* 0*,* $\frac{1}{2}$*-averaged, has a convex antiderivative and is a proximal.*

Remark 9. *On the composition of ReLU activations: Let x be the input signal and y be the last layer output. Then, the following mappings are not convex and not differentiable.*

- $$(\omega_1, b_1, \omega_2, b_2) \mapsto ReLU(\omega_2(ReLU(\omega_1 x + b_1)) + b_2).$$
- *(x random)*
$$(\omega_1, b_1, \omega_2, b_2) \mapsto \mathbb{E}_x ReLU(\omega_2(ReLU(\omega_1 x + b_1)) + b_2).$$
- $$(\omega_1, b_1, \omega_2, b_2) \mapsto \|ReLU(\omega_2(ReLU(\omega_1 x + b_1)) + b_2) - y\|^2.$$
- *(x and y random):*
$$(\omega_1, b_1, \omega_2, b_2) \mapsto \mathbb{E}_{x,y}\|ReLU(\omega_2(ReLU(\omega_1 x + b_1)) + b_2) - y\|^2.$$
- $$(\omega, b) \mapsto \frac{1}{D}\sum_{i=1}^{D} \|ReLU(\omega_2(ReLU(\omega_1 x_i + b_1)) + b_2) - y_i\|^2,$$
given a set of input signals and layer outputs $\{(x_i, y_i), i \in \{1, \ldots, D\}\}$, $D \geq 1$.
- $$(\omega_1, b_1, \omega_2, b_2) \mapsto \omega_2(ReLU(\omega_1 x + b_1)) + b_2.$$
- *(x random)*
$$(\omega_1, b_1, \omega_2, b_2) \mapsto \mathbb{E}_x \omega_2(ReLU(\omega_1 x + b_1)) + b_2.$$

- $$(\omega_1, b_1, \omega_2, b_2) \mapsto \|\omega_2(ReLU(\omega_1 x + b_1)) + b_2 - y\|^2.$$

- *(x and y random):*
$$(\omega_1, b_1, \omega_2, b_2) \mapsto \mathbb{E}_{x,y}\|\omega_2(ReLU(\omega_1 x + b_1)) + b_2 - y\|^2.$$

- $$(\omega, b) \mapsto \frac{1}{D}\sum_{i=1}^{D} \|\omega_2(ReLU(\omega_1 x_i + b_1)) + b_2 - y_i\|^2,$$
given a set of input signals and layer outputs $\{(x_i, y_i), i \in \{1, \ldots, D\}\}$, $D \geq 1$.

- *The composition* $x \mapsto a_1 x + b \mapsto \max(0, a_1 x + b_1) \mapsto a_2 \max(0, a_1 x + b_1) + b_2 \mapsto \max(0, a_2 \max(0, a_1 x + b_1) + b_2)$ *is not convex in* (a, b). *For training we consider the mapping*
$$check_3 : ((a_1, b_1), (a_2, b_2); (x, y_2)) \mapsto |y_2 - \max(0, a_2 \max(0, a_1 x + b_1) + b_2)|^2$$
for the input-output (x, y_2) *which reduces to* $(a_1, a_2) \mapsto |1 - \max(0, a_2 \max(0, a_1))|^2$ *for* $(x, y_2) = (1, 1), (b_1, b_2) = (0, 0)$. *The function* $check_3((a_1, 0), (a_2, 0); (1, 1))$ *evaluated at* $(a_1, a_2) = (1, 1)$ *gives* 0 *and at* $(-1, -1)$ *gives* 1 *meaning that* $check_3((1, 0), (1, 0); (1, 1)) = 0$, $check_3((-1, 0), (-1, 0); (1, 1)) = 1$,
$$\begin{aligned} check_3((0,0),(0,0);(1,1)) = 1 \ &> \tfrac{1}{2}(0+1) \\ &= \tfrac{1}{2} check_3((1,0),(1,0);(1,1)) \\ &+ \tfrac{1}{2} check_3((-1,0),(-1,0);(1,1)). \end{aligned}$$
Thus, $check_3$ *is not convex in* (a_1, a_2). *This means that even if ReLU is convex, the output function is not convex in the weight parameters. Since we optimize the weights when training the neural networks, it leads to non-convex optimization. We will see that there are still some interesting features of the model even if the operator* $O_2 \circ O_1$ *is non-convex in* (a_1, b_1, a_2, b_2) *for a given pair* (x, y_2).

Deep Sigmoid Networks

The activation operators r_k are not all convex.

Example 6. • *The sigmoid, the S-shaped function, which is widely used in the architecture of generative machine intelligence, is given by* $\sigma(x) = \frac{1}{1+e^{-x}} - \frac{1}{2}$. *Then* $\sigma'(x) = \frac{e^{-x}}{(1+e^{-x})^2}$ *and* $\sigma''(x) = \frac{-e^{-x}(1+e^{-x})^2 - e^{-x}2(-e^{-x})(1+e^{-x})}{(1+e^{-x})^4} = \frac{1-e^{x}}{e^{2x}(1+e^{-x})^2}$ *The second derivative is* $\sigma''(x) > 0$ *for* $x < 0$, $\sigma''(x) = 0$ *for* $x = 0$ *and* $\sigma''(x) < 0$ *for* $x > 0$. *This means that on the positive axis branch, the function is not convex: we shift by* $\frac{1}{2}$ *to obtain* $\sigma(\log 2) = \frac{2}{3}$, $\sigma(\log 8) = \frac{8}{9}$
$$\begin{aligned} &\tfrac{4}{5} = \sigma(\tfrac{1}{2}\log 2 + \tfrac{1}{2}\log 8) = \sigma(\log\ 4) \\ &> \tfrac{1}{2}[\tfrac{2}{3} + \tfrac{8}{9}] = \tfrac{14}{18} = \tfrac{1}{2}[\sigma(\log 2) + \sigma(\log 8)]. \end{aligned}$$

- *The sigmoid activation function* σ *is* $\frac{5}{8}$*-averaged. The extension to Hilbert space is done entry-by-entry. Thus,* $X \mapsto \sigma(X)$ *is* $\frac{5}{8}$*-averaged.*

- *The one-dimensional sigmoid function has antiderivative function* h *given by* $\log(1 + e^x) - \frac{1}{2}x$.

- *The one-dimensional sigmoid function is a proximal of f_l i.e.*

$$\frac{1}{1+e^{-x}} - \frac{1}{2} = [Id + \partial f_l]^{-1}(x) = \arg\min_y (f_l(y) + \frac{1}{2}\|x-y\|^2)$$

where

$$f_l(x) = \begin{cases} \frac{1+2x}{2}\log(\frac{1+2x}{2}) + \frac{1-2x}{2}\log(\frac{1-2x}{2}) - \frac{4x^2+1}{8} & \text{if } |x| < \frac{1}{2}, \\ -\frac{1}{4} & \text{if } |x| = \frac{1}{2}, \\ +\infty & \text{if } |x| > \frac{1}{2}. \end{cases} \quad (1.87)$$

Deep SiLU Networks

Example 7. *The scaled sigmoid linear unit (SiLU) activation function, which is widely used in the architecture of generative machine intelligence, is given by $\epsilon x \sigma(\epsilon x)$, with $\epsilon < \frac{1}{2}$, is $\frac{1+2\epsilon}{2}$-averaged. The extension to Hilbert space is done entry-wise. Thus, $X \mapsto \sigma(X)$ is $\frac{1+2\epsilon}{2}$-averaged.*

Deep Softargmax Networks

Example 8. *The softargmax, which is the Boltzmann-Gibbs distribution, is widely used in the architecture of generative AI. Let $k \geq 2, \epsilon > 0, j \in \{1, 2, \ldots, k\}$,*

$$softargmax: \mathbb{R}^k \ni z \mapsto \left(\frac{e^{z_1}}{\sum_{j=1}^k e^{z_j}}, \ldots, \frac{e^{z_k}}{\sum_{j=1}^k e^{z_j}} \right) \in \mathbb{R}^k_+.$$

The scaled softargmax

$$\epsilon \; softargmax(\frac{z}{\epsilon}) = \epsilon \left(\frac{e^{\frac{z_1}{\epsilon}}}{\sum_{j=1}^k e^{\frac{z_j}{\epsilon}}}, \ldots, \frac{e^{\frac{z_k}{\epsilon}}}{\sum_{j=1}^k e^{\frac{z_j}{\epsilon}}} \right)$$

is 1−averaged. The extension to real-valued matrix is done row by row or column by column. Thus, $X \mapsto softargmax(X)$ is 1-averaged.

Deep Kolmogorov-Arnold Networks

$$\begin{aligned} y_i^{(0)} &= x_i \\ h^{(l)} &= W^{(l)} y^{(l-1)} + b^{(l)}, \quad 1 \leq l \leq L \\ y_i^{(l)} &= r_i(\textstyle\sum_{j=1}^{n_{l-1}} \phi_{ij}^{(l)}(h_j^l)), \quad 1 \leq l \leq L \end{aligned} \quad (1.88)$$

where L is the number of layers in the network, n_l is the number of nodes in layer l, $y_i^{(l)}$ is the output of the i-th node in layer l, $\phi_{(ij)}^{(l)}$: Learnable activation function on the edge connecting the j-th node in layer $l-1$ to the i-th node in layer l, r_i is an activation function.

2

Mathematics of Transformers

Machine intelligence is transforming industries from blockchained retail to manufacturing, with the ability to analyze data, generate insights, and automate decisions at unprecedented scales. At the forefront of this movement is the transformer, a cutting-edge model that has rapidly become the backbone of systems used for everything from text processing to image processing to autonomous driving. But despite its widespread adoption, the transformer architecture remains a complex mathematical block of libraries, requiring a deeper understanding to unlock its full potential. We pull back the curtain on the transformer, its mathematical structure and offer a fresh perspective on how it operates through the lens of game theory. The result is a comprehensive guide to how machine intelligence models can be optimized and understood using strategic interactions at their core. At the heart of the transformer is its "attention mechanism" a feature that allows the model to focus on the most relevant parts of its input, much like how human attention works. But this chapter goes beyond the basics and examines the properties of the transformer's architecture, specifically, how its layered operators and normalization processes can be thought of as players in a noncooperative-coopetitive-cooperative-coalitional game. As we will see, *the initial architecture of attention is not all we need* and there are many other variations and architectures to be built. The chapter introduces the Boltzmann-Gibbs transformer, a variant that traditional transformer models by addressing key issues around layer normalization, a common bottleneck in improving model performance. By framing normalization and attention mechanisms as game-theoretic components, we offer a new pathway for improving efficiency and performance in large-scale machine intelligence models. But perhaps the most striking contribution of this chapter is its exploration of transformers as tensor-graph neural networks. We present transformers with Boltzmann-Gibbs, sigmoid or difference self-attention with 2-layered Rectified Linear Unit (ReLU), Sigmoid Linear Unit (SiLU), Swish or Mixture of Experts feedforward networks. We show that these models are not just mathematical objects but strategic systems, where each layer interacts with the others to optimize an overarching objective. The strategic interplay between layers can be seen as a multi-agent game, where every component has its own role in achieving the model's ultimate goal. We walk through practical concerns, such as the limitations of fixed-size transformers and the introduction of simpler normalization schemes, all while keeping their focus on the strategic dynamics that govern model behavior. In addition to technical refinements, the chapter touches on the emerging field of mixture-of-experts transformers' models that, much like a boardroom of specialists, delegate different tasks to different sub-networks to optimize efficiency. Again, the game-theory lens shines here and it offers new ways to think about how these expert sub-networks can be coordinated for maximum effect. As machine intelligence continues to grow in scope and complexity, understanding the transformer's mathematical foundation and its game-like internal interactions will be key to pushing the boundaries of what these systems can achieve. For industries relying on machine intelligence, the message is clear: mastering the transformer architecture is not just about scaling up data or computation. It is about understanding how every part of the system plays its role in the larger game of optimization and decision-making. Beyond the neural network itself, we also show that the training of transformers can be seen as an aggregative game, where the agents (layers) work together to produce a coherent result.

Ingredients of a transformer

The transformer architecture, which has been widely used in various tasks, consists of several key ingredients: Transformers employ an encoder-decoder structure, where the encoder processes the input sequence and the decoder generates the output sequence. This structure allows the model to capture and generate complex dependencies between words or data points. The self-attention mechanism is a crucial component of the transformer architecture (see Figure 2). It allows the model to weigh the importance of different words in the input sequence when generating a representation for each word. Self-attention enables the model to consider global context and capture long-range dependencies effectively. Transformers utilize multi-head attention to capture different types of information and attend to multiple parts of the input sequence simultaneously. Each attention head focuses on different aspects of the input, enabling the model to capture diverse and complementary information. Since transformers do not inherently encode the order or position of words in the input sequence, positional encoding is introduced. Positional encoding adds position-specific information to the word embeddings, providing the model with knowledge of word order and position within the sequence. Transformers incorporate feed-forward neural networks, typically with a multi-layer perceptron structure, to process and transform the representations learned through self-attention. The feed-forward networks add non-linearity to the model and facilitate learning more complex patterns and representations. Residual connections are connections that bypass certain layers and add the original input to the output of subsequent layers. Layer normalization is applied after each layer to stabilize the learning process and improve the model's ability to handle different input sequences. In certain tasks, the decoder needs to generate output tokens one at a time while attending only to the previously generated tokens. Masking is used to prevent the decoder from seeing future tokens during training and ensure an autoregressive generation process. These ingredients collectively enable transformers to model complex language patterns, capture long-range dependencies, and generate coherent and contextually relevant responses. By leveraging self-attention and parallelization, transformers have been highly successful in many tasks.

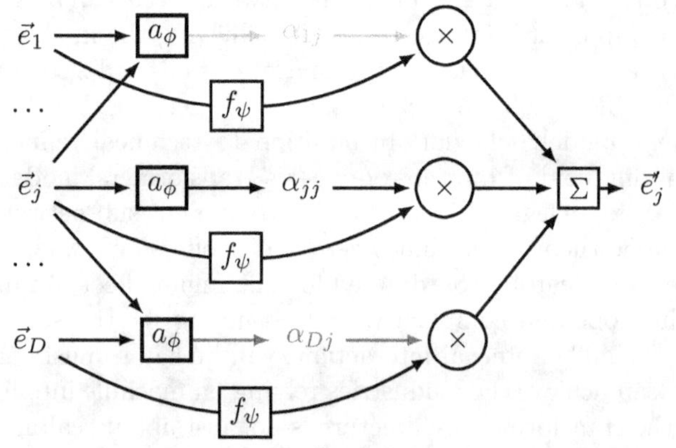

FIGURE 2.1
Scheme of the attention mechanism.

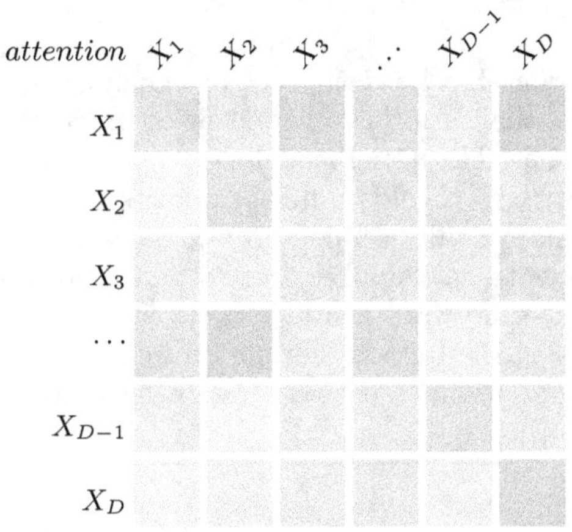

FIGURE 2.2
Pairwise interaction in the attention mechanism at each layer

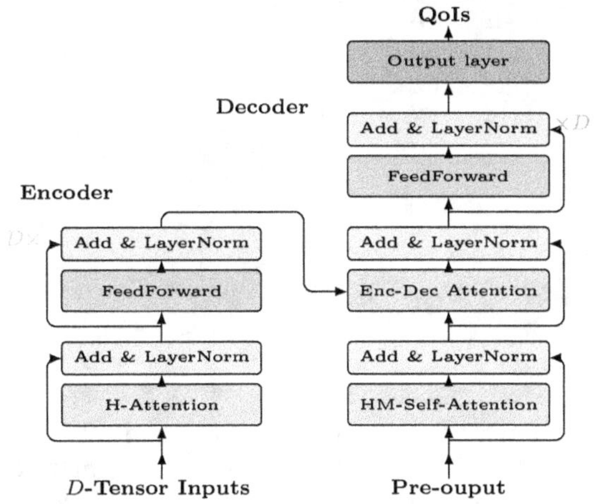

FIGURE 2.3
Overall architecture of the transformer, which follows the encoder-decoder paradigm. H is the number of heads of the attention. HM is the masked version of the multi-head attention. Enc-Dec denotes encoder-decoder. QoIs stands for quantities-of-interest.

2.1 Transformer

A generative pretrained transformer is composed of input preprocessing, positional encoding, embedding, and formatting followed by several transformer blocks and an output layer. A transformer block is composed of signal normalization which maps the signal to a ball. The normalization is followed by an interaction component between data points, masked or not. The initial model is a pairwise interaction called attention. In the auto-regressive setting, each data point interacts with the preceding data points leading to self-attention. We will examine both masked and unmasked self-attention. To the attention output it again applies an add and a norm operator. The latter is followed by a two-layer feedforward neural network. The process is repeated until the final output layer, making the full transformer. We describe and analyze this process using operator theory, variational inequality theory, fixed-point theory, control theory, optimization and game theory.

$$
\begin{array}{l}
\text{Parameters } : L, D, d, k, k', q, H \in \mathbb{N} \\
\quad 1 \leq l \leq L : \\
\quad\quad 1 \leq h \leq H : Q_{lh}, K_{lh}, \in \mathcal{L}(\mathcal{H}^d, \mathcal{H}^k), V_{lh} \in \mathcal{L}(\mathcal{H}^d, \mathcal{H}^{k'}), W_{lh} \in \mathcal{L}(\mathcal{H}^{k'}, \mathcal{H}^d), \\
\quad\quad W_{1l} \in \mathcal{L}(\mathcal{H}^d, \mathcal{H}^q), b_{1l} \in \mathcal{H}^q, W_{2l} \in \mathcal{L}(\mathcal{H}^q, \mathcal{H}^d), b_{2l} \in \mathcal{H}^d, \\
\text{Input } : y_0 = x_0 \in \mathcal{H}^{Dd}, \\
1 \leq l \leq L : \\
\quad \text{Input at } l : \ x_l = y_{l-1} \in \mathcal{H}^{Dd}, \\
\\
\quad 1 \leq i \leq D : \ u_{li} = \text{normalized}(x_{li}), \\
\quad 1 \leq i \leq D : \ \tilde{u}_{li} = \frac{1}{\sqrt{H}} \sum_{h=1}^{H} W_{lh} \sum_{j=1}^{i} \text{attention}_{ij}(Q_{lh}, K_{lh}, u_l) V_{lh} u_{lj}, \\
\quad 1 \leq i \leq D : \ \hat{u}_{li} = x_{li} + \tilde{u}_{li}, \\
\\
\quad 1 \leq i \leq D : \ \hat{y}_{li} = \text{normalized}(\hat{u}_{li}), \\
\quad 1 \leq i \leq D : \ \tilde{y}_{li} = W_{2l} r_l (W_{1l} \hat{y}_{li} + b_{1l}) + b_{2l}, \\
\quad 1 \leq i \leq D : \ y_{li} = \hat{u}_{li} + \tilde{y}_{li}, \\
\\
\quad \text{Output at } l : y_l, \\
\text{Return } y_L \in \mathcal{H}^{Dd}
\end{array}
\tag{2.1}
$$

where

- normalized $: \mathcal{O}_{l,nn} : \mathcal{H}^d \to \mathcal{H}^d$
- for $1 \leq i, j \leq D$, the interaction between entry i and the other entries is captured by $\text{attention}_{ij} : \mathcal{L}(\mathcal{H}^d, \mathcal{H}^k)^2 \times \mathcal{H}^{Dd} \to [0, 1]$. The map $\mathcal{O}_{l,att}$ is the $H-$head (masked) attention mechanism with $1 \leq i \leq D$, and $\mathcal{O}_{l,att,i} : \mathcal{H}^{Dd} \ni x \mapsto \frac{1}{\sqrt{H}} \sum_{h=1}^{H} W_{lh} \sum_{j=1}^{i} \text{attention}_{ij}(Q_{lh}, K_{lh}, x) V_{lh} x_j \in \mathcal{H}^d$.
- activation $r_l : \mathcal{H}_l \to \mathcal{H}_l$. The map $\mathcal{O}_{l,ff} : \mathcal{H}^d \ni x \mapsto W_{2l} r_l (W_{1l} \hat{y}_{li} + b_{1l}) + b_{2l} \in \mathcal{H}^d$ is a 2-layered feedforward operator.

The output of the full transformer is $y_L = \hat{\mathcal{O}}_L \circ \hat{\mathcal{O}}_{L-1} \circ \ldots \circ \hat{\mathcal{O}}_1(x_0)$ with the full operator at layer l given by

$$
\hat{\mathcal{O}}_l = (Id + \mathcal{O}_{l,ff} \circ \mathcal{O}_{l,nn}) \circ (Id + \mathcal{O}_{l,att} \circ \mathcal{O}_{l,nn}),
$$

where $Id(x) = x$ is the identity operator. The internal steps can be summarized as follows:

$$
\begin{aligned}
&Parameters : L, D, d, k, k', q, H \in \mathbb{N} \\
&\quad 1 \leq l \leq L : \\
&\quad\quad 1 \leq h \leq H : Q_{lh}, K_{lh}, \in \mathcal{L}(\mathcal{H}^d, \mathcal{H}^k), V_{lh} \in \mathcal{L}(\mathcal{H}^d, \mathcal{H}^{k'}), W_{lh} \in \mathcal{L}(\mathcal{H}^{k'}, \mathcal{H}^d), \\
&\quad\quad W_{1l} \in \mathcal{L}(\mathcal{H}^d, \mathcal{H}^q), b_{1l} \in \mathcal{H}^q, W_{2l} \in \mathcal{L}(\mathcal{H}^q, \mathcal{H}^d), b_{2l} \in \mathcal{H}^d, \\
&\text{Input } : y_0 = x_0 \in \mathcal{H}^{Dd}, \\
&\text{for } l \in \{1, \ldots, L\} : \\
&\quad y_l = (Id + \mathcal{O}_{l,ff} \circ \mathcal{O}_{l,nn}) \circ (Id + \mathcal{O}_{l,att} \circ \mathcal{O}_{l,nn})(y_{l-1}), \\
&\text{Return } y_L \in \mathcal{H}^{Dd}
\end{aligned}
\tag{2.2}
$$

Note that there is a small modification in the 2-layered feedforward network to be studied for transformers. This modification is due to the normalization operator that appears after the add-and-norm operator. The modified 2-layer neural network specification is given by the family

$$\mathbb{NN}_2 := (x_0, L, \{r_l, W_{1l}, b_{1l}, W_{2l}, b_{2l}, \mathcal{H}_l\}_{1 \leq l \leq L}, \lambda).$$

$$
\hat{O}_l : \left| \begin{array}{l} \mathcal{H}_{l-1} \to \mathcal{H}_l \\ x \mapsto x + W_{2l} r_l (W_l \frac{x}{1+\|x\|} + b_{1l}) + b_{2l}. \end{array} \right.
$$

$$
\text{for } t \in \{0, 1, 2, \ldots\} \left| \begin{array}{l} y_{1,t} = \hat{O}_1(x_t), \\ \text{for } l \in \{2, \ldots, L\}, \ y_{l,t} = \hat{O}_l(y_{l-1,t}), \\ x_{t+1} = x_t + \lambda_t (y_{L,t} - x_t). \end{array} \right.
\tag{2.3}
$$

2.2 Boltzmann-Gibbs Transformer

We start with the model used in Generative Pre-Trained Transformer (GPT) which is based on Boltzmann-Gibbs attention mechanism.

2.2.1 Layer Operator of the Generative Pre-Trained Transformer

The full operator at layer l is given by

$$\hat{\mathcal{O}}_l = (Id + \mathcal{O}_{l,ff} \circ \mathcal{O}_{l,nn}) \circ (Id + \mathcal{O}_{l,att} \circ \mathcal{O}_{l,nn}),$$

where

Parameters $: x_0, l, D, d, k, \gamma_{1l,i}, \beta_{1l,i}, H, Q_{lh}, K_{lh}, V_{lh}, W_{lh}, \gamma_{2l,i}, \beta_{2l,i},$
$W_{1li}, W_{2li}, b_{1li}, b_{2li}, \epsilon_l > 0,$
Input at $l :\ x_l \in \mathcal{H}^{Dd},$

$$
\begin{aligned}
u_{l,i} &= \gamma_{1l,i}\frac{\sqrt{d-1}(x_{l,i}-\frac{\sum_{k=1}^{d} x_{li,k}}{d}\mathbb{1})}{\sqrt{\epsilon_l+\langle(x_{l,i}-\frac{\sum_{k=1}^{d} x_{lik}}{d}\mathbb{1}),(x_{l,i}-\frac{\sum_{k=1}^{d} x_{li,k}}{d}\mathbb{1})\rangle}}+\beta_{1l,i},\\
\tilde{u}_{l,i} &= \frac{1}{\sqrt{H}}\sum_{h=1}^{H} W_{lh}\sum_{j=1}^{i}\frac{e^{\frac{1}{\sqrt{k}}\langle Q_{lh}u_i,K_{lh}u_j\rangle}}{\sum_{j'}^{i} e^{\frac{1}{\sqrt{k}}\langle Q_{lh}u_i,K_{lh}u_{j'}\rangle}}V_{lh}u_j,\\
\hat{u}_{l,i} &= x_{l,i}+\tilde{u}_{l,i}\\
\hat{y}_{l,i} &= \gamma_{2l,i}\frac{\sqrt{d-1}(\hat{u}_{li}-\frac{\sum_{k=1}^{d}\hat{u}_{lik}}{d}\mathbb{1})}{\sqrt{\epsilon_l+\langle(\hat{u}_{li}-\frac{\sum_{k=1}^{d}\hat{u}_{lik}}{d}\mathbb{1}),(\hat{u}_{li}-\frac{\sum_{k=1}^{d}\hat{u}_{lik}}{d}\mathbb{1})\rangle}}+\beta_{2l,i},
\end{aligned}
\tag{2.4}
$$

$$
\begin{aligned}
\tilde{y}_{li} &= W_{2li}r_l(W_{1li}\hat{y}_{li}+b_{1li})+b_{2li},\\
y_{li} &= \hat{u}_{li}+\tilde{y}_{li},
\end{aligned}
$$

$$
\begin{aligned}
&\text{Output at } l: y_l\\
&\text{Input at } l+1: x_{l+1}=y_l
\end{aligned}
$$

and this holds for $1 \leq l \leq L,\ L \geq 2$.

Sum of Composition of the Operators

Let us consider the operator $E_1 = (Id + \mathcal{O}_{l,att} \circ \mathcal{O}_{l,nn})$. As the normalization occurs before taking the self-attention, $\mathcal{O}_{l,att}$ restricted at the unit ball is of course Lipschitz, as it is a continuously differentiable function, and the closed unit ball is a convex, closed, bounded and non-empty set but not compact (in Hilbert spaces). Hence the Jacobian has a bounded norm in the closed unit ball. We can scale with V to get an averaged operator on the unit ball. This holds not only for the standard normalization $\mathcal{O}_{l,nn} = A_{1l,nn} \circ \text{proj}_{C_2} \circ \text{proj}_{C_1}$ but also for the modified normalization $\mathcal{O}_{l,nn} = A_l \circ \hat{\sigma}$. The second operator $E_2 = (Id + \mathcal{O}_{l,ff} \circ \mathcal{O}_{l,nn})$ with $\mathcal{O}_{l,ff} = A_{2l,ff} \circ r_l \circ A_{1l,ff}$ can be made averaged by designing the linear operators involved. However, for training, we do not need to do such restriction on weights and biases if the activation function of the feedforward is a gradient. For consistency and coherence check, however, we need a well-behaved fixed-point operation of the overall network. The full operator at layer l is

$$\hat{\mathcal{O}}_l = E_2 \circ E_1$$

$$= (Id + A_{2l,ff} \circ r_l \circ A_{1l,ff} \circ A_{2l,nn} \circ \text{proj}_{C_2} \circ \text{proj}_{C_1}) \circ (Id + \mathcal{O}_{l,att} \circ A_{1l,nn} \circ \text{proj}_{C_2} \circ \text{proj}_{C_1}).$$

or

$$\hat{\mathcal{O}}_l = (Id + A_{2l,ff} \circ r_l \circ A_{1l,ff} \circ A_{2l,nn} \circ \hat{\sigma}_{2,nn}) \circ (Id + \mathcal{O}_{l,att} \circ A_{1l,nn} \circ \hat{\sigma}_{1,nn}).$$

In both cases, each layer operation can be easily decomposed as a unique Nash equilibrium of a properly designed hierarchical game.

Outcome of the Full Transformer

The output of the full GPT is $\hat{\mathcal{O}}_L \circ \hat{\mathcal{O}}_{L-1} \circ \ldots \circ \hat{\mathcal{O}}_1$. Iterate layer by layer and by timestep the following:

$$
\left|
\begin{aligned}
&\textit{Parameters}\\
&\textit{Input}: x_0\\
&\text{for } t \in \{0,1,2,\ldots\}:\\
&\quad y_{1,t} = (Id + \mathcal{O}_{1,ff} \circ \mathcal{O}_{1,nn}) \circ (Id + \mathcal{O}_{1,att} \circ \mathcal{O}_{1,nn})(x_t),\\
&\quad \text{for } l \in \{2,\ldots,L\},\ y_{l,t} = (Id + \mathcal{O}_{l,ff} \circ \mathcal{O}_{l,nn}) \circ (Id + \mathcal{O}_{l,att} \circ \mathcal{O}_{l,nn})(y_{l-1,t}),\\
&\quad x_{t+1} = x_t + \lambda_t(y_{L,t} - x_t).
\end{aligned}
\right.
\tag{2.5}
$$

which means that

$$x_{t+1} = x_t + \lambda_t(\hat{O}_L \circ \ldots \circ \hat{O}_1(x_t) - x_t),$$

starting from x_0. For consistency of the operation, the output of $\hat{O}_L$ should be in the same space as the input i.e., $\mathcal{H}_L = \mathcal{H}_0$. See Figure 2.4.

Definition 15. *The GPT is given by the family* $x_0, L, D, d, k,$ $(r_l, \gamma_{1l,i}, \beta_{1l,i}, H, Q_{lh}, K_{lh}, V_{lh}, W_{lh}, \gamma_{2l,i}, \beta_{2l,i}, W_{1li}, W_{2li}, b_{1li}, b_{2li})_{1\leq l\leq L, 1\leq i\leq D}, \lambda)$.

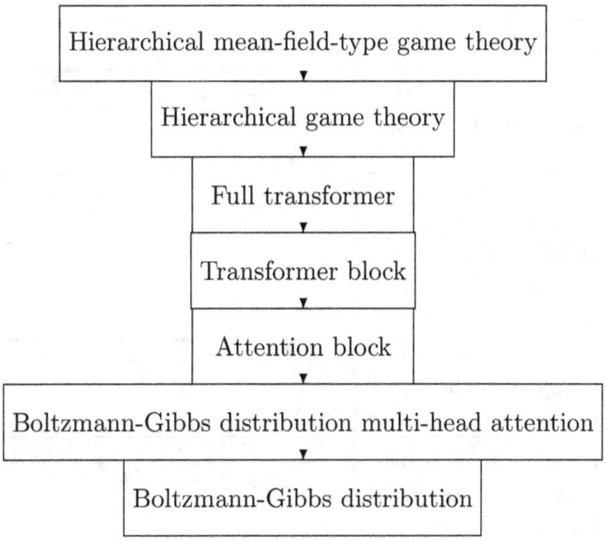

FIGURE 2.4
Connection between mean-field-type game theory and transformers with Boltzmann-Gibbs attention mechanism.

2.2.2 Layer Normalization Issue

> The normalization operator is used twice in a transformer block. Sometimes the basic normalization is followed by a linear bounded operator. The normalization operator allows us to apply fixed-point theorems as it maps the unbounded into a bounded set. With composition operators that are continuous over non-empty, convex, closed, and bounded sets, one can examine the set of fixed-points (if any) of the transformer.

Table 2.1 displays some normalization functions proposed in the generative machine intelligence literature.

Layer Normalization as weighted double projection

Consider $x = (x_1, \ldots x_D)'$, $D \geq 2$ elements of $\mathcal{H}^d$, $d \geq 2$. The layer normalization is a concatenation row by row:

$$\mathcal{O}_{l,nn}(x_1, \ldots, x_D) = (\mathcal{O}_{l,nn,1}(x_1), \ldots \mathcal{O}_{l,nn,D}(x_D))'$$

with

$$\mathcal{O}_{l,nn,i}(x_i) = \gamma_{l,i} \frac{\sqrt{d-1}(x_i - \frac{\sum_{k=1}^d x_{ik}}{d}\mathbb{1})}{\sqrt{\langle (x_i - \frac{\sum_{k=1}^d x_{ik}}{d}\mathbb{1}), (x_i - \frac{\sum_{k=1}^d x_{ik}}{d}\mathbb{1})\rangle}} + \beta_{l,i}.$$

TABLE 2.1
Normalization requires a design in itself

Type	Properties	Formula
center/std	limited in Hilbert spaces. constant issue	$\gamma_{li}\frac{\sqrt{d-1}(x_{li}-\frac{\sum_{k=1}^{d} x_{lik}}{d}\mathbb{1})}{\|(x_{li}-\frac{\sum_{k=1}^{d} x_{lik}}{d}\mathbb{1})\|_2}+\beta_{l,i}$
RMSnorm	not invertible. zero issue. Maps into the unit sphere	$\frac{x_{li}}{\|x_{li}\|}\mathbb{I}_{x_{li}\neq 0}$
RMSenorm	design $\epsilon_l > 0$. Maps into the the unit ball	$\frac{x_{li}}{\sqrt{\epsilon_l+\langle x_{li},x_{li}\rangle}} \Leftrightarrow y_{li}\sqrt{\frac{\epsilon_l}{1-\|y_{li}\|^2}}$
ϵRMSnorm	design $\epsilon_l > 0$. Maps into the unit ball.	$\frac{x_{li}}{\epsilon_l+\sqrt{\langle x_{li},x_{li}\rangle}} \Leftrightarrow \epsilon_l\frac{\|y_{li}\|}{1-\|y_{li}\|}\frac{y_{li}}{\|y_{li}\|}$
Tanh	element-wise. extension to Hilbert space is restrictive	
$id/(1+norm)$	one-to-one map. Maps into the unit ball	$\frac{x_{li}}{1+\|x_{li}\|} \Leftrightarrow \frac{y_{li}}{1-\|y_{li}\|}$
CapsuleNorm	maps the entire space into the unit ball	$\frac{\|x_{li}\|^2}{1+\|x_{li}\|^2}\frac{x_{li}}{\|x_{li}\|} \Leftrightarrow \frac{y_{li}}{\|y_{li}\|}\sqrt{\frac{\|y_{li}\|}{1-\|y_{li}\|}}$

We describe the normalization as a composition of three operators: two projection operators followed by an affine operator.

Lemma 17. *Let $d \geq 2$. Let $C = \{w \in \mathcal{H}^d,\ \langle w, w\rangle \leq d-1, \langle w, \mathbb{1}\rangle = 0\}$ which is the intersection of the ball with radius $\sqrt{d-1}$ and the hyperplane directed by vector $\mathbb{1}$ with element 1 at each entry. The hyperplane is the orthogonal to $\mathbb{1}$. Note that these two sets are convex and closed, and their intersection is non-empty. Denote the two sets by $C_1 = \{w \in \mathbb{R}^k,\ \langle w, \mathbb{1}\rangle = 0\}$ and $C_2 = \{w \in \mathbb{R}^k,\ \langle w, w\rangle \leq d-1\}$.*

- *The projection x into C_1 is given by*

$$Proj_{C_1}(x) = x - \frac{\langle x, \mathbb{1}\rangle}{\langle \mathbb{1}, \mathbb{1}\rangle}\mathbb{1}.$$

- *The projection of x into C_2 is*

$$Proj_{C_2}(x) = x\mathbb{I}_{x\in C_2} + \sqrt{d-1}\frac{x}{\sqrt{\langle x, x\rangle}}\mathbb{I}_{x\notin C_2}.$$

In particular for the sphere with radius $\sqrt{d-1}$, the projection can be obtained any non-zero element and is given by

$$\sqrt{d-1}\frac{x}{\sqrt{\langle x, x\rangle}}\mathbb{I}_{x\neq 0}.$$

- *The i-th component of the normalization operation at layer l is given by*

$$\mathcal{O}_{l,nn,i}(x_i) = \gamma_{l,i}(Proj_{C_2} \circ Proj_{C_1})(x_i) + \beta_{l,i} = (A_{l,i} \circ Proj_{C_2} \circ Proj_{C_1})(x_i),$$

where $A_{l,i} : w_i \mapsto \gamma_{l,i}w_i + \beta_{l,i}$.

Proof: Let $e \in \mathcal{H}, e \neq 0$, $\delta \in \mathbb{R}$ and $C_3 = \{w \in \mathcal{H},\ \langle w, e\rangle = \delta\}$. Then the projection x to C_3 is explicitly given by

$$Proj_{C_1}(x) = x + \frac{\delta - \langle x, e\rangle}{\langle e, e\rangle} e.$$

We apply this result for $\delta = 0$ and $e = \mathbb{1}$, which non-zero. The second and third assertions are immediate. □

In terms of practical implementation of this normalization layer in transformer-based foundational large learning models, there are several issues. The first issue is with zero in the denominator, which needs to be handled. Another issue is on the geometric aspect. In the finite dimension, the hyperplane is of dimension $d-1$ which could lead to a loss of data. It is reasonable to ask whether this projection to the hyperplane is really useful for the quality of the output. For example, the input matrix is

$$X_0 = \begin{pmatrix} c_1 & c_1 & \dots & \dots & c_1 & c_1 \\ c_2 & c_2 & \dots & \dots & c_2 & c_2 \\ \dots & \dots & \dots & \dots & \dots & \dots \\ c_{D-1} & c_{D-1} & \dots & \dots & c_{D-1} & c_{D-1} \\ c_D & c_D & \dots & \dots & c_D & c_D \end{pmatrix} \tag{2.6}$$

with $c_i \neq c_j, i \neq j$. The first projection is applied row by row. Note the projection of $c_i\mathbb{1}$ to the orthogonal of $\mathbb{1}$ yielding the null vector. There is no chance to reconstruct something similar to X_0 after this operator. Some of the recent implementations of generative machine intelligence have decided to bypass the first projection.

Layer Normalization Operator is not Invertible

The function

$$z \mapsto \frac{z}{\sqrt{\langle z, z\rangle}} \mathbb{I}_{z\neq 0}$$

is not invertible. As a consequence, the second projection is not necessarily a good idea as one cannot retrieve the input signals that are outside the unit sphere.

Layer Normalization is a Weighted Gradient Flow

The modified normalization operator is given by $\mathcal{O}_{l,nn}(x_1, \dots, x_D) = (y_1, \dots, y_D)$ where

$$y_i = \gamma_{l,i}\sqrt{d-1}\frac{(x_i - \frac{\sum_{k=1}^d x_{ik}}{d}\mathbb{1})}{\sqrt{\epsilon_l + \langle (x_i - \frac{\sum_{k=1}^d x_{ik}}{d}), (x_i - \frac{\sum_{k=1}^d x_{ik}}{d}\mathbb{1})\rangle}} + \beta_{l,i},$$

where $\beta_{l,i}, \gamma_{l,i}$ are weights to be determined and $\epsilon_l > 0$ is fixed. $\beta_{l,i}$ is a vector in $\mathcal{H}^d$ and $\gamma_{l,i}$ is a diagonal matrix with d diagonal elements in $\mathcal{H}^d$. Here we divide the denominator by $d-1$ instant of d to an unbiased empirical variance.

Lemma 18.
- *Let $\epsilon_l > 0$. The mapping $z \mapsto \sqrt{\epsilon_l + \langle z, z\rangle}$ is differentiable with gradient $\frac{z}{\sqrt{\epsilon_l + \langle z, z\rangle}}$.*
- *The mapping $\phi_i :\ x_i \mapsto \sqrt{\frac{\epsilon_l + \langle (x_i - \frac{\sum_{k=1}^d x_{ik}}{d}\mathbb{1}), (x_i - \frac{\sum_{k=1}^d x_{ik}}{d}\mathbb{1})\rangle}{d-1}}$ is differentiable with gradient*

$$\frac{\sqrt{d-1}(x_i - \frac{\sum_{k=1}^d x_{ik}}{d}\mathbb{1})}{\sqrt{\epsilon_l + \langle (x_i - \frac{\sum_{k=1}^d x_{ik}}{d}\mathbb{1}), (x_i - \frac{\sum_{k=1}^d x_{ik}}{d}\mathbb{1})\rangle}}$$

- *Layer Normalization is a weighted gradient flow: We have* $\mathcal{O}_{l,nn}(x_1,\ldots,x_D) = (\mathcal{O}_{l,nn,1}(x_1),\ldots \mathcal{O}_{l,nn,D}(x_D))$ *with* $\mathcal{O}_{l,nn,i}(x_i) = (A_{l,i} \circ (\nabla\phi_i))(x_i)$.

The proof is immediate following from the standard differentiation. □

Add & Norm differs from Norm & Add

The Norm & Add operator is $u_i + \mathcal{O}_{l,nn,i}(x_i)$ differs from the Add & Norm operator $\mathcal{O}_{l,nn,i}(x_i + u_i)$.

Alternative Normalization Operators to z-Score-like Layer Normalization

One of the key inconveniences of z-Score-like LayerNorm used in the earlier transformer architectures, such as GPT, arises from its underlying mechanism, which involves the composition of two projections. The first projection is onto a hyperplane that is orthogonal to the vector where all components are equal to 1. Essentially, z-score LayerNorm centers the input by subtracting the mean, ensuring that the data lies on a hyperplane of reduced dimensionality. This projection reduces the degrees of freedom of the data, as it essentially forces the input to conform to a lower-dimensional subspace. This dimensionality reduction can lead to a loss of information and structure within the data. The second projection is onto a sphere by scaling the centered data to have unit variance. Together, these two operations, first a projection onto a hyperplane and then onto a sphere, can cause the input data to lose crucial scaling, nuances, and relationships, potentially affecting its quality. If these two operations are repeated across multiple layers, as they are in deep transformers, the compounded loss of structure and data quality can result in what is known as *model collapse*, where all input data may converge to a similar form, reducing the model's ability to distinguish between different inputs. Particularly, all constant vectors will be normalized to the same output, while zero vectors are excluded from the normalization process, further contributing to potential collapse.

The second significant issue with z-score-like LayerNorm is its non-invertibility. Unlike transformations such as linear layers, which have clearly defined inverses, z-score LayerNorm irreversibly alters the data by normalizing it, meaning that the original input cannot be reconstructed from the output. This introduces a risk of inconsistency between the input and the output, as the transformation process distorts or compresses the input data in a way that is not recoverable. For example, in a transformer architecture, the output embeddings of each layer depend heavily on the normalized representations from z-score LayerNorm. Since z-score LayerNorm discards absolute magnitude information by normalizing the variance, inputs that were originally distinct may end up having similar normalized representations, causing ambiguities. This can be particularly problematic in tasks that require fine-grained distinctions between token embeddings, such as in autoregressive generation tasks like in GPT, where every token influences the next one. If subtle variations between tokens are smoothed out due to z-score LayerNorm, the model might generate inconsistent or incoherent outputs. The inability to reverse this transformation means that errors or distortions introduced by z-score LayerNorm can propagate through the network, potentially leading to degraded performance and reduced ability to capture interesting patterns in the input data.

The normalization, defined as $\text{HoloNorm}(z) = \frac{z}{1+\|z\|}$, offers a promising alternative to traditional HoloNorm by addressing several key issues that arise in transformer architectures. First and foremost, this new normalization avoids the two consecutive projections that can degrade data quality in standard HoloNorm. In traditional HoloNorm, data is projected onto a lower-dimensional hyperplane (by removing the mean) and then projected onto a sphere by rescaling based on variance. These projections can cause a loss of critical information and lead to the risk of model collapse, where input vectors of constant values get mapped to the same output, and zero vectors are excluded entirely. In contrast, the proposed normalization directly scales the input vector z by dividing it by $1 + \|z\|$, where $\|z\|$ is the norm (or magnitude) of the input. This approach ensures that the magnitude of the input is smoothly adjusted, without the need for projecting onto a hyperplane or sphere. As a result, the original structure and relationships within the data are better preserved, minimizing the risk of information loss or distortions.

Another key advantage is that this new normalization is less prone to cause model collapse. Since it scales the input by its 1+ norm, it avoids the issue where all constant vectors map to the same output. By incorporating the norm into the scaling factor, it introduces a natural way to maintain variability in the input, even when the components are identical. Additionally, unlike traditional HoloNorm, this method does not exclude the null vector. When z is zero, the normalization simply returns zero, making the operation well-defined even for inputs of zero magnitude, which improves the model's robustness.

Moreover, this form of normalization retains crucial information about the input's original magnitude. In traditional z-score LayerNorm, scaling by the variance effectively removes information about the input's magnitude, which can lead to inconsistencies between the input and output, especially in scenarios where magnitude plays an important role in the decision-making process of the model. By dividing by $1 + \|z\|$, the magnitude of the input is controlled without completely eliminating it, allowing the model to retain some notion of the input's relative scale. This can be particularly beneficial in transformer models like GPT, where token embeddings may have significant variation in magnitude, and preserving this information can lead to more precise representations of the input. An additional major advantage of the normalization

$$\text{HoloNorm}(z) = \frac{z}{1 + \|z\|},$$

is that it is fully invertible and represents a one-to-one mapping. This is a significant improvement over traditional z-score LayerNorm, which lacks invertibility due to its reliance on projection steps (onto a hyperplane and then a sphere). With traditional z-score LayerNorm, once the data has been transformed, it is impossible to reconstruct the original input, as some information about the input's absolute magnitude and individual components is lost. This non-invertibility can lead to inconsistencies between the inputs and outputs, making it harder for the model to maintain internal coherence, particularly in deep networks like transformers. In contrast, the proposed normalization allows for a clear and direct inverse function. Given the output HoloNorm(z) the original input z can be recovered by simply inverting the normalization process. Specifically, for a given normalized output w, we can recover z as follows:

$$z = \text{HoloNorm}^{-1}(w) = \frac{w}{1 - \sqrt{\langle w, w \rangle}}.$$

This property ensures that no information is irretrievably lost during the normalization process. Since the transformation is smooth and bijective (one-to-one), each unique input corresponds to a unique output, and vice versa. This bijective property guarantees that all the structure of the input is preserved in a reversible manner. In contexts like transformer models, where maintaining precise relationships between inputs and outputs across layers is crucial for tasks like autoregressive language modeling, having a normalization method that is invertible is highly beneficial.

Moreover, this invertibility addresses a core limitation of traditional z-score LayerNorm, where normalization can lead to ambiguities or inconsistencies because different inputs may produce the same normalized output (constant vectors being mapped to the same output). In the new normalization HoloNorm, no such ambiguities arise. Each distinct input vector is mapped to a distinct output, preserving the input's uniqueness, which is critical for models that rely on subtle differences in embeddings to make accurate predictions. Therefore, this HoloNorm not only preserves the magnitude and structure of the input in a way that traditional LayerNorm does not but also ensures full invertibility and a one-to-one mapping, making it a much more robust choice for architectures like GPT that demand high fidelity in internal data representations.

The derivative of the proposed layer normalization is $\frac{x}{(1+|x|)^2}$, which is positive, and its anti-derivative $|x| + \ln(1 + |x|) + C$. Figures 2.5 and 2.6 provide sample examples. *Note that the HoloNorm differs from the softsign in 2D, 3D and beyond.* In Hibert space, they are two different functions. There is no sign of a vector. The softsign function applied element-wise to the coefficients of an element in an Hilbert space is a totally different function.

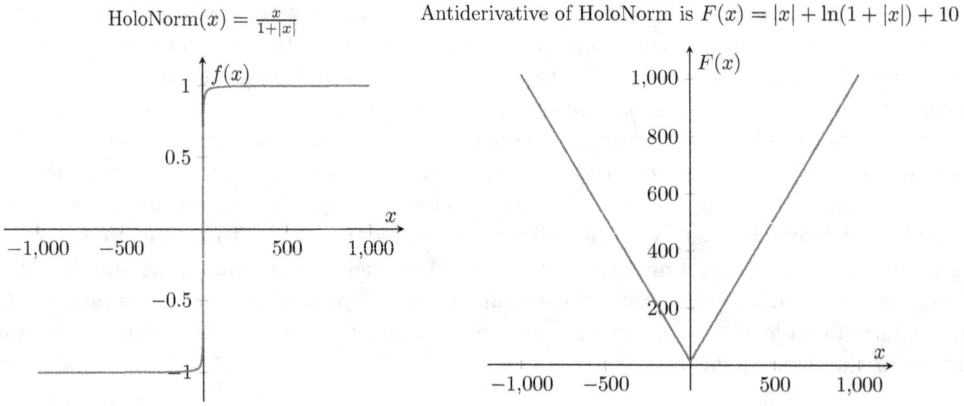

FIGURE 2.5
Holonorm and its anti-derivative.

Another normalization is the capsule normalization defined as CapsuleNorm$(z) = \frac{\|z\|^2}{1+\|z\|^2}\frac{z}{\|z\|}$, which also maps the entire space to the unit ball.

For y in the unit ball, we can solve $\frac{\|z\|^2}{1+\|z\|^2}\frac{z}{\|z\|} = y$.

By taking the norm of both sides, we arrive at $\frac{\|z\|^2}{1+\|z\|^2} = \|y\|$, i.e., $\|z\| = \sqrt{\frac{\|y\|}{1-\|y\|}}$. This means that $\frac{z}{\|z\|} = \frac{y}{\|y\|}$. Thus,

$$z = \frac{y}{\|y\|}\sqrt{\frac{\|y\|}{1 - \|y\|}}.$$

In one-dimension, one can also use hyperbolic tangent $tanh(x)$ as its structure is similar to scaled version of softsign $\frac{x}{1+|x|}$. However, $\frac{x}{1+|x|}$ is the advantage provided above for a simple formula for signal inversion and signal recovery.

2.3 Transformer-based Tensor-Graph Neural Networks

We consider a transformer-based tensor-graph neural network with higher-order tensor data and $L \geq 1$ layers. Let $\{\mathcal{H}_l\}_{0 \leq l \leq L}$ be non-zero Hilbert spaces, equipped with an inner product $\langle ., . \rangle$ and the induced norm given by $\|X_o\| := \sqrt{\langle X_o, X_o \rangle}$.

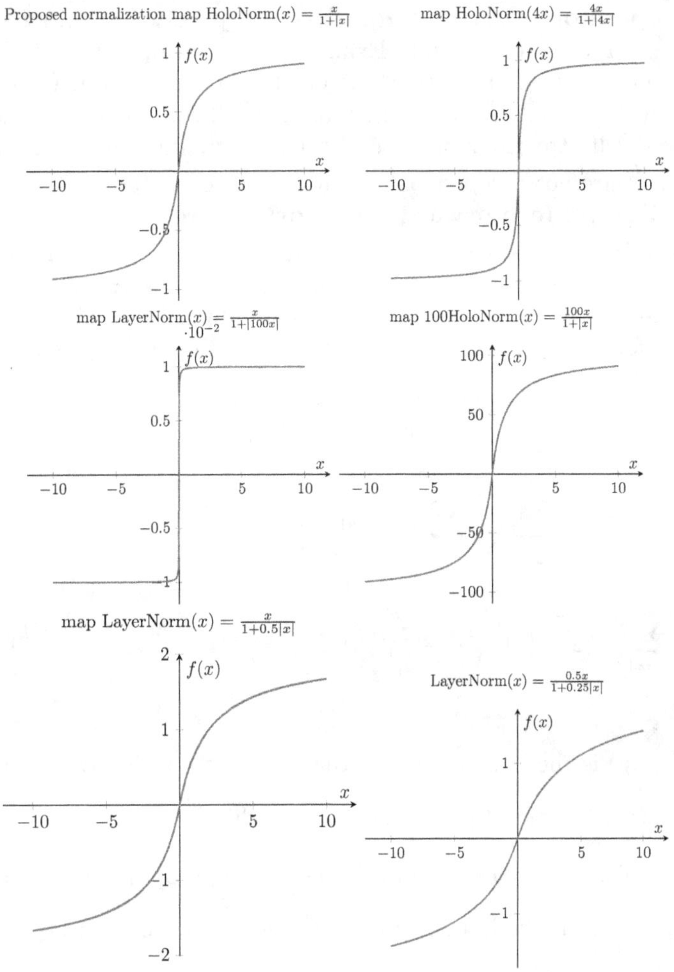

FIGURE 2.6
Scaled HoloNorms.

The tensor-graph neural network is defined by the composition of operators

$$\hat{\mathcal{O}}_L \circ \hat{\mathcal{O}}_{L-1} \circ \ldots \circ \hat{\mathcal{O}}_1, \tag{2.7}$$

where $\hat{\mathcal{O}}_l : \mathcal{H}_{l-1} \to \mathcal{H}_l$ given by

$$\hat{\mathcal{O}}_l = (Id + \mathcal{O}_{l,ff} \circ \mathcal{O}_{l,nn}) \circ (Id + \mathcal{O}_{l,att} \circ \mathcal{O}_{l,nn}).$$

Input Layer: Let $X^{(0)} = (X_1, \ldots, X_D)$ be the input as a sequence of D tensors, each $X_o \in \mathcal{H}^{(d_1 \times d_2 \times \ldots \times d_k)} =: \mathcal{H}_0$, $o \in \{1, \ldots, D\}$. The input space of the transformer is $\mathcal{H}_0^D$.

For layer $1 \le l \le L$, we calculate new tensor representations by applying a tensor operation:

Normalization operator: $\mathcal{O}_{l,nn} : \mathcal{H}_{l-1} \to \mathcal{B}_1 \subset \mathcal{H}_{l-1}$ is applied to each X_o separately, where $1 \le o \le D$ and $\mathcal{B}_1$ is the unit ball of $\mathcal{H}_0$, i.e., $\mathcal{B}_1 = \{X \in \mathcal{H}_0, \langle X, X \rangle \le 1\}$. Examples include projective normalization, HoloNorm (hn), double projection (to the hyperplane followed by the unit ball), one-to-one mapping from unbounded space to a bounded one such as $\hat{\sigma}(X_1, \ldots, X_D) = (\hat{\sigma}(X_1), \hat{\sigma}(X_2), \ldots, \hat{\sigma}(X_D))$ with $\hat{\sigma}(X_o) = \frac{X_o}{1+\sqrt{\langle X_o, X_o \rangle}}$.

Attention operator: $\mathcal{O}_{l,att} : \mathcal{H}_{l-1} \to \mathcal{H}_{l-1}$ between an aggregative interaction between X_o for $o \in \{1,\dots,D\}$. Examples of such operators include Boltzmann-Gibbs masked/unmasked self-attention, exponentially weighted masked/unmasked self-attention, entry-wise normalized masked/unmasked self-attention, and entry-wise sigmoid masked/unmasked self-attention. Because of the self-attention interaction, the tensor processes $Y_1^{(l)},\dots,Y_D^{(l)}$ are not independent. They are all correlated.

Two-layered graph-feedforward operator: For every $l \in \{1,\dots,L\}$, we introduce the operator $\mathcal{O}_{l,ff} : \mathcal{H}_{l-1} \to \mathcal{H}_l$. Let $W_p^{(l)} : \mathcal{H}_{l-1} \to \mathcal{H}_l$ be a bounded linear operator and consider the family of activation operators $r_l : \mathcal{H}_l \to \mathcal{H}_l$. An activation function $r_l : \mathcal{H}_l \to \mathcal{H}_l$ and two pairs of weights and biases are applied to each X_o separately.

For each position $(i_1, i_2, \dots, i_k)$, the new representation $Y^{(l)}(i_1, i_2, \dots, i_k)$ is computed as follows:

$$\tilde{Y}^{(l)}(i_1, i_2, \dots, i_k) =$$

$$\sum_{i_1'=1}^{d_1}\sum_{i_2'=1}^{d_2}\cdots\sum_{i_k'=1}^{d_k} W_2^{(l)}(i_1, i_2, \dots, i_k, i_1', i_2', \dots, i_k')\cdot$$

$$r_l\left(\sum_{j_1=1}^{d_1}\sum_{j_2=1}^{d_2}\cdots\sum_{j_k=1}^{d_k} W_1^{(l)}(i_1', i_2', \dots, i_k', j_1, j_2, \dots, j_k)\cdot Y^{(l-1)}(j) + b_1^{(l)}\right) + b_2^{(l)},$$

where:

- $\tilde{Y}^{(l)}(i_1, i_2, \dots, i_k)$ is the hidden state of the tensor at position $(i_1, i_2, \dots, i_k)$ in layer l.
- r_l is an activation function (e.g., ReLU, GELU, SiLU).
- $W_1^{(l)}, W_2^{(l)}$ are learnable linear bounded tensors of appropriate dimensions.
- $Y^{(l-1)}(j_1, j_2, \dots, j_k)$ is the hidden state of the tensor at position $(j_1, j_2, \dots, j_k)$ in layer $l-1$.
- $b_1^{(l)}, b_2^{(l)}$ are learnable bias tensors.

The kernels $W_p^{(l)}(i_1, i_2, \dots, i_k, i_1', i_2', \dots, i_k')$ are designed over a graph (V', E') with

$$W_p^{(l)}(i, i') = \tilde{W}_p^{(l)}(i, i')\mathbb{I}_{\{i' \in \mathcal{N}_i\}},\ p \in \{1, 2\ \}$$

where $\mathcal{N}_i = \{i', (i, i') \in E\}$ is the set of neighbors of node i. The operation is followed by a tensor-graph aggregation procedure:

$$Y^{(l)}(i_1, i_2, \dots, i_k) = \frac{\sum_{i_1'=1}^{d_1}\sum_{i_2'=1}^{d_2}\cdots\sum_{i_k'=1}^{d_k}\tilde{Y}^{(l)}(i')\mathbb{I}_{\{i' \in \mathcal{N}_i\}}}{\sum_{j_1'=1}^{d_1}\sum_{j_2'=1}^{d_2}\cdots\sum_{j_k'=1}^{d_k}\mathbb{I}_{\{j' \in \mathcal{N}_i\}}},$$

where the denominator is the number of neighbors of $i = (i_1, i_2, \dots, i_k)$. The output of the network is $Y^{(L)} = \hat{\mathcal{O}}_L \circ \hat{\mathcal{O}}_{L-1} \circ \dots \circ \hat{\mathcal{O}}_1 X^{(0)}$. This is an essential tool for engineering tasks involving multi-dimensional data, such as images, videos, and sensor arrays. It extends traditional neural networks to handle tensors of higher orders. It operates on tensor data, where tensors represent multi-dimensional arrays. It uses tensor convolution and pooling operations. The layers process data in a hierarchical and localized manner. It is well-suited for structured data with multiple dimensions. It can capture spatial patterns and hierarchies in data.

Next we introduce a blocktree and a blockchain. A blocktree is a 2-tuple $(\mathcal{B}, \mathcal{E})$ where $\mathcal{B} \subset \mathbb{B}$, is a non-empty set of valid blocks, including at least one element $\tilde{b}_0$ (the genesis block) and $\mathcal{E} \subset \mathcal{B}^2$. There exists a unique path from the initial block $\tilde{b}_0$ to any other block. A total order $\succ$ over $\mathcal{B}$, preserving $\mathcal{E}$, i.e., $(\tilde{b}, \tilde{b}') \in \mathcal{E} \implies \tilde{b}' \succ \tilde{b}$. The blockchain is the unique path from the genesis block $\tilde{b}_0$ to the last block $\max_{\succ} \mathcal{B}$. A blockchain is built by a network of processes applying sequentially the expansion operation starting from the initial blocktree $\{\tilde{b}_0, \{\}\}$. In the distributed case, every processing node refers to its own view of the blocktree and strives to build a consensus with each other on the shared blockchain while increasing its depth.

Regular nodes try to build valid blocks. When such a valid block is found by a node, it expands its local blocktree from its last block (i.e. the last block of its own view of the blockchain) and broadcasts it to the others. When a regular node receives a new block from another one, it updates its local blocktree expanding it with the new block. During this updating phase, forks may actually be observed due to blocks broadcasting delay. Building a new valid block is an operation that may take many forms depending on the blockchain protocols. Several types of nodes are considered depending on the behaviors: empathy-altruism, regular, spiteful and malicious behaviors.

The network integrates a blockchain step in the Tensor-Graph Neural Network, which involves maintaining a decentralized ledger of transactions. Each node i in the tensor-graph may represent a participant in the blockchain network. The blockchain step involves validating and processing transactions from neighboring nodes, updating the local tensor-graph features accordingly, propagating validated transactions to neighboring nodes for consensus. Incorporating a blockchain step into a Tensor-Graph Neural Network enhances the model's ability to handle decentralized and distributed data. The blockchain layer enables secure and tamper-resistant data exchange among participants while leveraging the power of tensor and graph representations provided by this architecture.

The blockchain records all transactions. It allows us to incentivize all participant in the development of machine intelligence through blockchain tokens or any other form of reward. It also allows us to make all facts and legislative entries in a subnetwork prevent the machine intelligence from falsifying the facts. This will reduce error in historical well-known events. It can also help in fact-checking specifically in the area of medical statements in online media.

The algorithm is as follows. Let $X^{(0)} \in \mathcal{H}_0^D$ be the input of the transformer-based tensor-graph network, $\{\lambda_t\}_{t \geq 0}$ be a non-negative sequence, and consider the following map defined iteratively layer by layer and by timestep:

$$\left|\begin{array}{l} \text{Parameters} \\ \text{Input} : X^{(0)} \\ \text{for } t \in \{0,1,2,\ldots\}, \\ \quad Y_t^{(1)} = (Id + \mathcal{O}_{1,ff} \circ \mathcal{O}_{1,nn}) \circ (Id + \mathcal{O}_{1,att} \circ \mathcal{O}_{1,nn})(X_t), \\ \quad \text{for } l \in \{2,\ldots,L\}, \\ \qquad Y_t^{(l)} = (Id + \mathcal{O}_{l,ff} \circ \mathcal{O}_{l,nn}) \circ (Id + \mathcal{O}_{l,att} \circ \mathcal{O}_{l,nn})(Y_t^{(l-1)}), \\ \quad X_{t+1} = X_t + \lambda_t (Y_t^{(L)} - X_t). \end{array}\right. \tag{2.8}$$

which means that

$$X_{t+1} = X_t + \lambda_t (\hat{O}_L \circ \ldots \circ \hat{O}_1(X_t) - X_t),$$

starting from $X_0 := X^{(0)}$. For consistency of the operation, the output of $\hat{O}_L$ should be in the same space as the input, i.e., $\mathcal{H}_L = \mathcal{H}_0$. The sequence $(\lambda_t)_t$, with $\lambda_t \geq 0$, is the step-size (learning rate). As we want to examine situations in which time goes to infinity, the cumulative step-size $\sum_{t=1}^{T} \lambda_t \to +\infty$ as T goes to infinity.

Definition 16. *The transformer-based tensor-graph neural network* $\mathbb{TGN}$ *specification is given by the family* $(L, D, \mathcal{H}_0^D, (V', E'), (\mathcal{H}_l, \mathcal{O}_{l,ff}, \mathcal{O}_{l,nn}, \mathcal{O}_{l,att}, \theta_l)_{1\leq l\leq L}, \lambda)$, *where* θ_l *denotes the collection of all parameters involved the operators at layer* $1 \leq l \leq L$.

Example of tensors used in practice are videos, blockchained medical records (of patients across hospitals, sectors, departments, countries), genomics data, historical spatio-temporal blockchained data, etc.

2.3.1 Properties of the Normalization

The normalization operator applies to each X_o separately. A typical normalization operator we work with is the mapping $\hat{\sigma} : X_o \mapsto \frac{X_o}{1+\sqrt{\langle X_o, X_o\rangle}}$. The layer Normalization operator is now $\mathcal{O}_{l,nn}(X_1, \ldots, X_D) = (\mathcal{O}_{l,nn}(X_1), \ldots \mathcal{O}_{l,nn}(X_D))$ with $\mathcal{O}_{l,nn}(X_o) = \hat{\sigma}(X_o)$. The advantage of this normalization is that it does not reduce the input dimensionality. Moreover it is invertible in the sense that if $\hat{\sigma}(X_o) = Z_o$ then $\frac{X_o}{1+\sqrt{\langle X_o, X_o\rangle}} = Z_o$.

Taking the norm on both sides yields

$$\frac{\sqrt{\langle X_o, X_o\rangle}}{1+\sqrt{\langle X_o, X_o\rangle}} = \sqrt{\langle Z_o, Z_o\rangle} = 1 - \frac{1}{1+\sqrt{\langle X_o, X_o\rangle}}.$$

Thus, $X_o = \frac{Z_o}{1-\sqrt{\langle Z_o, Z_o\rangle}} = \hat{\sigma}_o^{-1}(Z_o)$ for $\langle Z_o, Z_o\rangle < 1$. This normalization operator is a one-to-one mapping from the input space to the open unit ball.

Let X_1 and X_2 be two elements of $\mathcal{H}$. Then

$$\begin{aligned} &\|\hat{\sigma}(X_1) - \hat{\sigma}(X_2)\| \leq 2, \\ &\|\hat{\sigma}(X_o)\| \leq \min(1, \|X_o\|). \end{aligned} \tag{2.9}$$

The normalization can be rewritten as $\hat{\sigma}(X_o) = \frac{\|X_o\|}{1+\|X_o\|}\frac{X_o}{\|X_o\|}$ which, for $X_o \neq 0$, is weighted differential $\hat{\sigma}(X_o) = (\frac{1}{1+\|X_o\|})\nabla(\frac{1}{2}\|X_o\|^2) = (\frac{\|X_o\|}{1+\|X_o\|})\nabla(\|X_o\|)$.

The Jacobian is $\nabla(\hat{\sigma}(X_o)) = J(X_o) = \frac{\mathbb{I}}{1+\|X_o\|} - \frac{X_o^{\dagger}X_o}{(1+\|X_o\|)^2\|X_o\|}$. Since the maximum eigenvalue $\frac{1}{1+\|X_o\|} \leq 1$ is achieved at $X_o = 0$, we deduce that $\hat{\sigma}$ is 1-Lipschitz.

2.3.2 Properties of the Attention Operator

SoftArgMax is a gradient

The log-sum up to $o \geq 2$, mapping $\phi_o(z) = \epsilon \log\left(\sum_{j=1}^{o} e^{\frac{z_j}{\epsilon}}\right)$ is differentiable and its gradient the

$$\text{softargmax}_o : \mathbb{R}^o \ni z \mapsto \left(\frac{e^{\frac{z_1}{\epsilon}}}{\sum_{j=1}^{o} e^{\frac{z_j}{\epsilon}}}, \ldots, \frac{e^{\frac{z_o}{\epsilon}}}{\sum_{j=1}^{o} e^{\frac{z_j}{\epsilon}}}\right) \in \mathbb{R}^o_+.$$

softargmax_o is a probability distribution over $\{1, \ldots, o\}$. The vector softargmax_o is extended to $\mathbb{R}^D$ by completing with 0, i.e., $\text{softargmax}^*_{oj} = \text{softargmax}_o \mathbb{I}_{j\leq o}$. The softargmax, which is the Boltzmann-Gibbs distribution, is widely used in the architecture of generative machine intelligence.

SoftArgMax as ArgMax

Consider the entropy function $p = (p_1, \ldots, p_o) \mapsto \sum_{j=1}^{o} p_j \log p_j$, defined over $\sum_{j=1}^{o} p_j = 1,\ p_j > 0,\ j \in \{1, \ldots, o\}$ the $(o-1)$-simplex of $\mathbb{R}^o$. We use the convention $0 \log 0 = 0$

as $\lim_{x\to 0_+} x\log x = 0$. This entropy is a convex function. Consider the penalized function $z \mapsto \sup_p\{\langle z,p\rangle - \epsilon\sum_{j=1}^{o} p_j \log p_j\}$, then p is the argmax which, for a very small ϵ can be seen as an argument of the approximate max of the components of z. Hence, the name "soft" argument maximum or softargmax. The value associated with it is the Fenchel-Legendre conjugate of the entropy.

SoftArgMax is not Necessarily an Averaged Operator

We introduce the notion of averaged nonexpansive operator.

Definition 17. *Let $0 < \gamma \leq 1$. An operator $\hat{O}: \mathcal{H}_0^D \to \mathcal{H}_0^D$ is γ-averaged if $[Id + \frac{1}{\gamma}(\hat{O} - Id)]$ is 1-Lipschitz continuous.*

The γ-averaged nonexpansive operators play key roles in the asymptotics of $(\hat{O} - Id)(X_t)$ even if the operator $\hat{O}$ has no fixed point. For example, the mapping $\hat{O}(X) = X + e$, with e non-zero element, has no fixed-point. However, $\hat{O}(X_t) - X_t = e$ which converges to e. When $\hat{O}$ has a fixed-point, the difference $\hat{O}(X_t) - X_t$ goes to zero as t goes to infinity where X_t are the output of Algorithm in (2.8). These γ-averaged nonexpansive operators allow us to extend the well-known convergence results of the case of the contractive operators under Banach-Picard iterates.

The gradient of softargmax is the matrix $\epsilon\nabla_z \text{softargmax}_D(\frac{z}{\epsilon}) = \text{diag}(p) - p^\dagger p$ where $p = \text{softargmax}_D(\frac{z}{\epsilon})$. Each row sums up to zero. To be a contraction, the parameter ϵ must greater than 1, and to be closer to the softargmax, ϵ needs to be small enough (closer to zero). These two extremes are not consistent. The scaled softargmax

$$\epsilon \text{ softargmax}(\frac{z}{\epsilon}) = \epsilon\left(\frac{e^{\frac{z_1}{\epsilon}}}{\sum_{j=1}^k e^{\frac{z_j}{\epsilon}}}, \ldots, \frac{e^{\frac{z_k}{\epsilon}}}{\sum_{j=1}^k e^{\frac{z_j}{\epsilon}}}\right)$$

is 1−averaged.

Lemma 19. *The operator $\hat{O}$ is $\frac{1}{2}$-averaged iff $\|\hat{O}(X) - \hat{O}(Y)\|^2 \leq \langle X - Y, \hat{O}(X) - \hat{O}(Y)\rangle$, $\forall (X,Y)$ iff $(1-\mu)Id + \mu\hat{O}$ is $\mu/2$-averaged for all $\mu \in (0,2)$.*

Proof: $\hat{O}$ is $\frac{1}{2}$-averaged and is equivalent to $\hat{O} = (Id + M)/2$ with M being 1−Lipschitz. This means that $M = 2\hat{O} - Id$. Hence, the operator $(1-\mu)Id + \mu\hat{O} = (1-\mu)Id + \mu(Id + M)/2 = (1-\mu/2)Id + (\mu/2)M$ is $\mu/2$-averaged for all $\mu \in (0,2)$.

Conversely, if $(1-\mu)Id + \mu\hat{O}$ is $\mu/2$-averaged for all $\mu \in (0,2)$, then by choosing $\mu = 1$ one obtains that O is $\frac{1}{2}$-averaged. □

Lemma 20. *$\hat{O}$ is $\frac{1}{2}$-averaged iff $O = \hat{O}^{-1} - Id$ is monotone.*

Proof: Let $\hat{O}$ be $\frac{1}{2}$-averaged. Then $\langle X - \hat{O}(X) - (Y - \hat{O}(Y)), \hat{O}(X) - \hat{O}(Y)\rangle \geq 0$, $\forall (X,Y)$. Let $\hat{O}(Y_1 + X_1) = X_1$ and $\hat{O}(Y_2 + X_2) = X_2$ with $Y_o = OX_o$. Set $Z_o = Y_o + X_o$. Let us compute $\langle Y_1 - Y_2, X_1 - X_2\rangle = \langle OX_1 - OX_2, X_1 - X_2\rangle = \langle Z_1 - X_1 - (Z_2 - X_2), \hat{O}(Z_1) - \hat{O}(Z_2)\rangle = \langle Z_1 - \hat{O}(Z_1) - (Z_2 - \hat{O}(Z_2)), \hat{O}(Z_1) - \hat{O}(Z_2)\rangle \geq 0$ hence $\langle OX_1 - OX_2, X_1 - X_2\rangle \geq 0$. This means that O is monotone.

Conversely, suppose that O is monotone. We choose $O(\hat{O}(X_o)) + \hat{O}(X_o) \ni X_o$, $o \in \{1,2\}$ in the graph. Then $O(\hat{O}(X_o)) \ni X_o - \hat{O}(X_o)$, $o \in \{1,2\}$. By monotonicity $\langle O(\hat{O}(X_1)) - O(\hat{O}(X_2)), \hat{O}(X_1) - \hat{O}(X_2)\rangle \geq 0$, which is

$$\langle (X_1 - \hat{O}(X_1)) - (X_2 - \hat{O}(X_2)), \hat{O}(X_1) - \hat{O}(X_2)\rangle \geq 0,$$

which implies that $\hat{O}$ is $\frac{1}{2}$-averaged. □

The $\frac{1}{2}$-averagedness property is important because it is associated with monotone operators which are crucial in variational inequalities.

Remark 10. *If $\hat{O}$ is $\frac{1}{2}$-averaged, then there is a monotone operator O such that $\hat{O} = (Id + O)^{-1}$. One can also use the operator $\hat{O}_\lambda = (Id + \lambda O)^{-1}$ approximation.*

Our interest in averaged operators is due to the weak convergence properties of their Banach-Picard or Mann iterates to their fixed-points (if any).

Proposition 16. *Let $0 < \gamma \leq 1$, and $\hat{O} = \hat{O}_L \circ \hat{O}_{L-1} \circ \ldots \hat{O}_1 : \mathcal{H}_0^D \to \mathcal{H}_0^D$ be an γ-averaged operator such that $\hat{O}$ has at least one fixed-point.*

Let $\{\lambda_t\}_{t\geq 1}$ be a sequence in $[0, \frac{1}{\gamma}]$ such that $\sum_{t\geq 1} \lambda_t(1 - \gamma \cdot \lambda_t) = +\infty$. Assume that $Y_0 \in \mathcal{H}_0^D$ and set $Y_{t+1} = Y_t + \lambda_t(\hat{O}(Y_t) - Y_t)$. Then $\hat{O}(Y_t) - Y_t$ converges to 0 as t goes to infinity and Y_t converges weakly to a point in Fixed($\hat{O}$) $= \{X \in \mathcal{H}_0^D, \hat{O}(X) = X\} = ker(Id - \hat{O})$.

Proof: From the assumptions we note that $\epsilon_t = \gamma \cdot \lambda_t$ is a sequence in $[0, 1]$ and satisfies $\sum_{t\geq 1} \epsilon_t(1 - \epsilon_t) = +\infty$. Set $\widetilde{O} := (1 - \frac{1}{\gamma})Id + \frac{1}{\gamma}\hat{O}$. Then $X_{t+1} = X_t + \epsilon_t(\widetilde{O}(X_t) - X_t)$ and Fixed($\widetilde{O}$) = Fixed($\hat{O}$). Let $X^* \in$ Fixed($\hat{O}$) i.e., $\widetilde{O}(X^*) = \hat{O}(X^*) = X^*$. We evaluate $\|X_{t+1} - X^*\|$. Indeed,

$$
\begin{array}{l}
\|X_{t+1} - X^*\|^2 = \|(1 - \epsilon_t)X_t + \epsilon_t\widetilde{O}(X_t) - \epsilon_t X^* - (1 - \epsilon_t)X^*\|^2 \\
= \|(1 - \epsilon_t)(X_t - X^*) + \epsilon_t(\widetilde{O}(X_t) - \hat{O}(X^*))\|^2 \\
= (1 - \epsilon_t)^2\|X_t - X^*\|^2 + \epsilon_t^2\|\widetilde{O}(X_t) - \widetilde{O}(X^*)\|^2 \\
+2\epsilon_t(1 - \epsilon_t)\langle X_t - X^*, \widetilde{O}(X_t) - \widetilde{O}(X^*)\rangle \\
= (1 - \epsilon_t)^2\|X_t - X^*\|^2 + \epsilon_t^2\|\widetilde{O}(X_t) - \widetilde{O}(X^*)\|^2 \\
-\epsilon_t(1 - \epsilon_t)[\|X_t - \widetilde{O}(X_t)\|^2 - \|X_t - X^*\|^2 - \|\widetilde{O}(X_t) - \widetilde{O}(X^*)\|^2] \\
= [(1 - \epsilon_t)^2 + \epsilon_t(1 - \epsilon_t)]\|X_t - X^*\|^2 \\
+[\epsilon_t^2 + \epsilon_t(1 - \epsilon_t)]\|\widetilde{O}(X_t) - \widetilde{O}(X^*)\|^2 - \epsilon_t(1 - \epsilon_t)\|X_t - \widetilde{O}(X_t)\|^2 \\
= (1 - \epsilon_t)]\|X_t - X^*\|^2 + \epsilon_t\|\widetilde{O}(X_t) - \widetilde{O}(X^*)\|^2 \\
-\epsilon_t(1 - \epsilon_t)\|X_t - \widetilde{O}(X_t)\|^2.
\end{array}
\tag{2.10}
$$

Therefore,

$$
\begin{array}{l}
\|X_{t+1} - X^*\|^2 \\
\quad = (1 - \epsilon_t)]\|X_t - X^*\|^2 + \epsilon_t\|\widetilde{O}(X_t) - \widetilde{O}(X^*)\|^2 - \epsilon_t(1 - \epsilon_t)\|X_t - \widetilde{O}(X_t)\|^2 \\
\quad \leq \|X_t - X^*\|^2 - \epsilon_t(1 - \epsilon_t)\|X_t - \widetilde{O}(X_t)\|^2.
\end{array}
\tag{2.11}
$$

By a telescopic sum, we obtain

$$
\begin{array}{ll}
\sum_{t=0}^T \epsilon_t(1 - \epsilon_t)\|X_t - \widetilde{O}(X_t)\|^2 & \leq \sum_{t=0}^T(\|X_t - X^*\|^2 - \|X_{t+1} - X^*\|^2) \\
& = \|X_0 - X^*\|^2 - \|X_{T+1} - X^*\|^2.
\end{array}
\tag{2.12}
$$

As $\sum_{t=0}^\infty \epsilon_t(1 - \epsilon_t)\|X_t - \widetilde{O}(X_t)\|^2 \leq \|X_0 - X^*\|^2 < +\infty$ and $\sum_{t=0}^\infty \epsilon_t(1 - \epsilon_t) = +\infty$, it follows that the sequence $\|X_t - \widetilde{O}(X_t)\|^2 \to 0$ as $t \to \infty$. But, $\hat{O} - Id = \gamma(\widetilde{O} - Id)$. Thus, $\|X_t - \hat{O}(X_t)\|^2 \to 0$ as $t \to \infty$. □

The attention operator is not globally Lipschitz

Lemma 21. *Let $D \geq 2$, $z_{oj} = \frac{1}{\sqrt{k}}\langle QX_o, KX_j\rangle$ for $Q, K \in \mathcal{L}(\mathcal{H}^d, \mathcal{H}^k)$ and $z_o = \{z_{oj}\}_{1\leq j\leq o}$. For every $i \geq 2$, the mapping $\mathcal{H}_{l-1} \to \mathcal{H}_{l-1}$ given by*

$$
x \mapsto \sum_{l=1}^D softargmax_{il}(\{\frac{1}{\sqrt{k}}\langle QX_o, KX_j\rangle\}_{1\leq j\leq o})X_l \mathbb{I}_{l\leq o},
$$

is a not Lipschitz.

Proof: The assertion follows from the derivative of the composition and the fact the quadratic term is not Lipschitz in X.

It suffices to show that $X \mapsto \frac{\sum_{j=1}^{o} e^{\frac{1}{\sqrt{k}}\langle QX_o, KX_j\rangle} X_j}{\sum_{l=1}^{o} e^{\frac{1}{\sqrt{k}}\langle QX_o, KX_l\rangle}}$ is not Lipschitz.

We examine

$$\frac{e^{\frac{1}{\sqrt{k}}\langle QX_o, KX_o\rangle} X_o + \sum_{j=1}^{o-1} e^{\frac{1}{\sqrt{k}}\langle QX_o, KX_j\rangle} X_j}{e^{\frac{1}{\sqrt{k}}\langle QX_o, KX_o\rangle} + \sum_{l=1}^{o-1} e^{\frac{1}{\sqrt{k}}\langle QX_o, KX_l\rangle}} = p_{oo} X_o + \sum_{j=1}^{o-1} p_{oj} X_j$$

The gradient in X_o leads to the following matrix

$$(\frac{1}{\sqrt{k}}(QK^\dagger + KQ^\dagger)X_o^\dagger X_o + \mathbb{I}_{dd})p_{oo} + \sum_{j=1}^{o-1} p_{oj} \frac{1}{\sqrt{k}} Q^\dagger K X_j^\dagger X_j$$

$$-(\frac{1}{\sqrt{k}}(QK^\dagger + KQ^\dagger)X_o^\dagger p_{oo} + \sum_{j=1}^{o-1} p_{oj} \frac{1}{\sqrt{k}} Q^\dagger K X_j^\dagger)(p_{oo} X_o + \sum_{j=1}^{o-1} p_{oj} X_j)$$

When $X_o = \frac{1}{\epsilon}\mathbb{1}_d = X_j$ the probabilities are uniform (on the left).

$$(\frac{1}{\sqrt{k}}(QK^\dagger + KQ^\dagger)X_o^\dagger X_o + \mathbb{I}_{dd} + \frac{o-1}{\sqrt{k}} Q^\dagger K X_o^\dagger X_o)\frac{1}{o}$$

$$-(\frac{1}{\sqrt{k}}(QK^\dagger + KQ^\dagger)X_o^\dagger + \frac{o-1}{\sqrt{k}} Q^\dagger K X_o^\dagger)\frac{X_o}{o}$$

can be written as $\frac{1}{o}\mathbb{I}_{dd} + BX_o^\dagger X_o$, $B \neq 0$, whose norm goes to infinity as ϵ goes to zero. This means that the (matrix) Jacobian is unbounded. □

The Boltzmann-Gibbs attention operator is not an averaged operator

As a corollary of Lemma 21, we have:

Corollary 1. *The Boltzman-Gibbs masked self-attention operator $\mathcal{O}_{att,o} : \mathcal{H}_{l-1} \to \mathcal{H}_{l-1}$ given by $X \mapsto W(\sum_{j=1}^{o} softargmax_{oj}(z_o)VX_j)$ with non-zero linear operators $V \in \mathcal{L}(\mathcal{H}^d, \mathcal{H}^k)$ and $W \in \mathcal{L}(\mathcal{H}^k, \mathcal{H}^d)$, is not an averaged operator.*

Note all these (masked and unmasked) self-attentions are not Lipschitz on the entire domain, but they are Lipschitz when restricted to any non-empty, bounded, closed domain. The unit ball of Hilbert space is not compact. In any finite dimension however, the (closed) unit ball is not compact. In Hilbert spaces (infinite dimensions) the closed unit ball is weakly compact. Thus, the following result holds:

Corollary 2. *The restricted operator $\mathcal{O}_{att,o} : \mathcal{B}_1 \to \mathcal{H}_{l-1}$ is Lipschitz.*

The single-head self-attention operator

$$\mathcal{O}_{att}(X) = (WV)\text{softargmax}(X(\frac{1}{\sqrt{k}}QK^\dagger)X^\dagger)X$$

$$= (M_1)\text{softargmax}(X(M_2)X^\dagger)X$$

where $M_1 = WV, M_2 = \frac{1}{\sqrt{k}}QK^\dagger$.

Definition 18. *The multi-head self-attention is*

$$(M_{1h}, M_{2h})_{1\leq h\leq H} \mapsto \frac{1}{\sqrt{H}} \sum_{h=1}^{H} M_{1h} \, softargmax(X(M_{2h})X^{\dagger})X$$

As a finite sum of Lipschitz continuous operators, the restricted multi-head self-attention operator $\mathcal{O}_{att} : \mathcal{B}_1 \to \mathcal{H}_{l-1}$ is Lipschitz.

2.3.3 Properties of the Tensor-Graph Feedforward

Definition 19. *The entry-wise activation function in an Hilbert space is given by*

$$R_l(X) = \sum_{k=0}^{\infty} r_l(\langle X, e_k\rangle)e_k,$$

where $(e_k)_k$ is an orthonormal basis of the Hilbert space.

As a particular case, for Euclidean spaces, this basis is finite.

Lemma 22. *If each entry k, the map is $X_k \mapsto r_l(X_k)$ is γ_l-averaged then*

$$X = (X_k)_k \mapsto R_l(X) = \sum_{k=0}^{\infty} r_l(\langle X, e_k\rangle)e_k$$

is also is γ_l-averaged.

Proof:

$$\begin{aligned} R_l(X) &= \textstyle\sum_{k=0}^{\infty} r_l(\langle X, e_k\rangle)e_k = \sum_{k=0}^{\infty}((1-\gamma_l)\langle X, e_k\rangle + \gamma_l \tilde{r}_l(\langle X, e_k\rangle))e_k \\ &= (1-\gamma_l)\textstyle\sum_{k=0}^{\infty}\langle X, e_k\rangle e_k + \gamma_l \sum_{k=0}^{\infty} \tilde{r}_l(\langle X, e_k\rangle)e_k \\ &= ((1-\gamma_l)Id + \gamma_l\tilde{r}_l)(X) \end{aligned} \tag{2.13}$$

with $\tilde{r}_l(X) := \sum_{k=0}^{\infty} \tilde{r}_l(\langle X, e_k\rangle)e_k$. One has $\tilde{r}_l(X) - \tilde{r}_l(Y) = \sum_{k=0}^{\infty}(\tilde{r}_l(\langle X, e_k\rangle) - \tilde{r}_l(\langle Y, e_k\rangle))e_k$.

Then, $\|\tilde{r}_l(X) - \tilde{r}_l(Y)\|^2 = \sum_{k=0}^{\infty} \|\tilde{r}_l(\langle X, e_k\rangle) - \tilde{r}_l(\langle Y, e_k\rangle)\|^2 \leq \sum_{k=0}^{\infty} \|\langle X, e_k\rangle - \langle Y, e_k\rangle\|^2$.

As $\langle X, e_k\rangle \mapsto \tilde{r}_l(\langle X, e_k\rangle)$ is 1-Lipschitz, we have $\|\tilde{r}_l(X) - \tilde{r}_l(Y)\|^2 \leq \sum_{k=0}^{\infty} \|\langle X - Y, e_k\rangle\|^2 \leq \|X-Y\|^2$. This implies that, $\tilde{r}_l(x)$ is 1-Lipschitz. Hence $r_l = (1-\gamma_l)Id + \gamma_l\tilde{r}_l$ is an γ_l-averaged operator. □

Example 9. *The rectified linear unit (ReLU) activation function, which is widely used in the architecture of generative machine intelligence, is given by $ReLU(x) = \max(0, x) = \frac{x+|x|}{2}$ is $\frac{1}{2}$-averaged. The extension to Hilbert space is done entry-wise. Thus, $X \mapsto ReLU(X)$ is $\frac{1}{2}$-averaged.*

Example 10. *The sigmoid activation function, which is widely used in the architecture of generative machine intelligence, is given by $\sigma(x) = \frac{1}{1+e^{-x}}$ is $\frac{5}{8}$-averaged. The extension to Hilbert space is done entry-wise. Thus, $X \mapsto \sigma(X)$ is $\frac{5}{8}$-averaged.*

Example 11. *The scaled sigmoid linear unit (SiLU) activation function, which is widely used in the architecture of generative machine intelligence, is given by $\epsilon x\sigma(\epsilon x)$, with $\epsilon < \frac{1}{2}$, is $\frac{1+2\epsilon}{2}$-averaged. The extension to Hilbert space is done entry-wise. Thus, $X \mapsto \sigma(X)$ is $\frac{1+2\epsilon}{2}$-averaged.*

Example 12. *The basic softargmax activation function for a real-valued vector, which is widely used in the output layer for selection and classification problems in the architecture of generative machine intelligence is* 1*-averaged. The extension to the real-valued matrix is done row by row or column by column. Thus,* $X \mapsto softargmax(X)$ *is* 1*-averaged.*

Example 13. *The unit (open) ball normalization function (HoloNorm)* $X \mapsto \hat{\sigma}(X) = \frac{X}{1+\|X\|}$ *defined on Hilbert space is* 1*-averaged.*

$\mathbb{TGN}$ Outcomes under Maximally Cyclically Monotone Activation Operators

The operator $\hat{O}$ is k-cyclically monotone if every $(X_1, \dots, X_{k+1})$ with $X_{k+1} = X_1$ and every $(u_1, \dots, u_k)$, $(X_o, u_o) \in graph(\hat{O}), 1 \leq o \leq k, X_{k+1} = X_1$ implies that $\sum_{o=1}^{k} \langle X_{o+1} - X_o, u_o \rangle \leq 0$.

$\hat{O}$ is cyclically monotone if it is k-cyclically monotone for every integer $k \geq 2$.

We next introduce the notion of maximal cyclic monotonicity. A maximally cyclically monotone $\hat{O}$ is a cyclically monotone such that there is no other cyclically monotone operator that properly contains the graph of $\hat{O}$.

The next lemma shows that a subdifferential of a proper and convex function is cyclically monotone and it is maximal.

Lemma 23. *The subdifferential* ∂h *is maximally cyclically monotone for every proper and convex function* h*.*

Proof: To see it, we fix an integer $k \geq 2$. For every $1 \leq o \leq k$, consider (X_o, u_o) in the graph of the subdifferential of h. Set $X_{k+1} = X_1$. By definition of subdifferential, $\langle X_{o+1} - X_o, u_o \rangle \leq h(X_{o+1}) - h(X_o)$. Summing up these inequalities leads to

$$\sum_{o=1}^{k} \langle X_{o+1} - X_o, u_o \rangle \leq \sum_{o=1}^{k} h(X_{o+1}) - h(X_o) = h(X_{k+1}) - h(X_1) = 0.$$

□

Example 14. *Consider the softargmax. One has* $\{softargmax(z)\} = (\partial\phi)(z)$ *where*

$$\phi(z) = \epsilon \log \left(\sum_{j=1}^{k} e^{\frac{z_j}{\epsilon}} \right)$$

which is a proper and convex function. Hence, the softargmax activation is maximally cyclically monotone.

Example 15. *The rectified linear unit activation function has a convex anti-derivative* ϕ *:* $ReLU(x) = \partial_x \phi(x)$ *with* $\phi(x) = \frac{1}{2} x ReLU(x)$ *and hence it is a maximally cyclically monotone activation operator.*

Example 16. *The sigmoid activation function has a convex anti-derivative* $\phi : \sigma(x) = \partial_x \phi(x)$ *with* $\phi(x) = \log(1 + e^x)$ *and hence it is a maximally cyclically monotone activation operator.*

The next lemma shows that the converse is actually true. To show it we construct explicitly a supremum of linear functions from the cyclically monotone inequalities.

Lemma 24. *Every maximally cyclically monotone operator* $\hat{O}$ *is a subdifferential for some proper convex function* ϕ*.*

Proof: Suppose that $\hat{O}$ is a maximally cyclically monotone operator. Then $graph(\hat{O})$ is non-empty. There is an element (X_0, u_0) in the graph of $\hat{O}$. Let us define the function:

$$\phi(X) = \sup_{k\geq 1} \sup_{(X_o,u_o)\in graph(\hat{O}), 1\leq o\leq k} \langle X - X_k, u_k\rangle + \sum_{o=0}^{k-1} \langle X_{o+1} - X_o, u_o\rangle.$$

By cyclic monotonicity, $\phi(X_0) = 0$ and ϕ is proper and convex as a supremum of linear maps. Take (X, u) in the graph of $\hat{O}$ and let $-\infty < \eta < \phi(X)$. There are finitely many points (X_o, u_o) in $graph(\hat{O})$, $0 \leq o \leq k$ such that $\langle X - X_k, u_k\rangle + \sum_{o=0}^{k-1} \langle X_{o+1} - X_o, u_o\rangle > \eta$.

We now set $(X_{k+1}, u_{k+1}) = (X, u)$. For every Y, $\phi(Y) \geq \langle Y - X_{k+1}, u_{k+1}\rangle + \sum_{o=0}^{k} \langle X_{o+1} - X_o, u_o\rangle = \langle Y - X, u\rangle + \langle X - X_k, u_k\rangle + \sum_{o=0}^{k-1} \langle X_{o+1} - X_o, u_o\rangle > \langle Y - X, u\rangle + \eta$.

We deduce that $\phi(Y) \geq \langle Y - X, u\rangle + \eta$. Letting η go to $\phi(X)$ we obtain that $\phi(Y) \geq \langle Y - X, u\rangle + \phi(X)$. The latter inequality shows that $u \in \partial\phi$. This implies that the $graph(\hat{O})$ is a subset of $graph(\partial\phi)$. By maximality, we obtain $\hat{O} = \partial\phi$

□

From the relation $r_l = [Id + \partial f_l]^{-1} = [\partial(\frac{1}{2}\|\cdot\|^2 + f_l)]^{-1} = \partial(\frac{1}{2}\|\cdot\|^2 + f_l)^*$, we have that $\phi_l = (\frac{1}{2}\|\cdot\|^2 + f_l)^*$ and $r_l = \partial\phi_l$, where, as usual, $\psi^*(X) = \sup_Y[\langle X, Y\rangle - \psi(Y)]$ is the Legendre-Fenchel conjugate of ψ.

Lemma 25. *The normed two-layer feedforward map* $(W_1, b_1, W_2, b_2) \mapsto \|W_2 r_l(W_1 X + b_1) + b_2\|^2$ *is not necessarily convex even if* r_l *is convex.*

Proof: The non-convexity is because of the multiplicative factor. Take $r_l = ReLU$ (which is a convex function) as an illustration. Then, the feedforward operator $\|\mathcal{O}_{ff}\|$ is not convex in the parameter (W_1, b_1, W_2, b_2). □

Observe that the two-layered feedforward operator $\mathcal{O}_{l,ff}$ can be replaced by group-rational polynomial two-layered Kolmogorov-Arnold networks with activation functions $r_l(x) = \frac{P_*(x)}{1+|Q_*(x)|}$ where P_* is a polynomial of degree n_p and Q_* is a polynomial of degree n_q.

2.3.4 Attention as a Partial Anti-Derivative

Let $\phi_{o,l}(X) = \frac{1}{\sqrt{H}}\sqrt{k}\log\left(\prod_{h=1}^{H}\sum_{j=1}^{o} e^{\frac{1}{\sqrt{k}}\langle Q_{lh}X_o, K_{lh}X_j\rangle}\right)$.

Its derivative in X_o yields

$$X \mapsto \frac{1}{\sqrt{H}}\sum_{h=1}^{H}\frac{\sum_{j=1}^{o} e^{\frac{1}{\sqrt{k}}\langle Q_{lh}X_o, K_{lh}X_j\rangle} Q_{lh}^* K_{lh} X_j}{\sum_{l=1}^{o} e^{\frac{1}{\sqrt{k}}\langle Q_{lh}X_o, K_{lh}X_l\rangle}}.$$

2.3.5 Properties of the Tensor-Graph Aggregation

The mapping $aggr: \tilde{Y}^{(l)} \mapsto Y^{(l)}$ given by

$$Y^{(l)}(i_1, i_2, \ldots, i_k) = \frac{\sum_{i'_1=1}^{d_1}\sum_{i'_2=1}^{d_2}\cdots\sum_{i'_k=1}^{d_k}\tilde{Y}^{(l)}(i')\mathbb{I}_{\{i'\in\mathcal{N}_i\}}}{\sum_{j'_1=1}^{d_1}\sum_{j'_2=1}^{d_2}\cdots\sum_{j'_k=1}^{d_k}\mathbb{I}_{\{j'\in\mathcal{N}_i\}}},$$

is $\frac{1}{1+\delta}$-Lipschitz, where δ is the minimum degree of the graph vertices. The aggregation operator $aggr$ is $\frac{2+\delta}{2(1+\delta)}$-averaged.

2.4 Analysis of the Fixed Size Transformer

2.4.1 Transformer-based Tensor-Graph Outcomes

When dealing with finite, sampled datasets, neural networks must look for consistency in their outcomes. Achieving this consistency requires iterating the outcomes of the network repeatedly, a process that can be conceptualized as a feedback loop.

Problem 4. *What does a well-trained transformer-based Deep Neural Network do? To answer this question, we aim to find and characterize the behavior of the neural network for a large class of architectures,* $\mathbb{TGN}$*. To this end, we examine the possible limit (if any) of* $(X_t, Y_t^{(1)}, \ldots, Y_t^{(L)})$ *as* t *goes to infinity.*

$$\begin{aligned}
&l = 1,\\
&Y^{(1)} = \hat{\mathcal{O}}_1(X_*),\\
&\ldots\\
&l \in \{2, \ldots, L-1\}: \; Y^{(l)} = \hat{\mathcal{O}}_l(Y^{(l-1)}),\\
&\ldots\\
&X^* = Y^{(L)} = \hat{\mathcal{O}}_L(Y^{(L-1)}),
\end{aligned} \tag{2.14}$$

which is rewritten in a two-step operation per layer as follows:

$$\begin{aligned}
&l = 1,\\
&Y^{(\frac{1}{2})} = (Id + \mathcal{O}_{1,att} \circ \mathcal{O}_{1,nn})(X^*),\\
&Y^{(1)} = (Id + \mathcal{O}_{1,ff} \circ \mathcal{O}_{1,nn})(Y^{(\frac{1}{2})}),\\
&\ldots\\
&l \in \{2, \ldots, L-1\}:\\
&Y^{(l-\frac{1}{2})} = (Id + \mathcal{O}_{l,att} \circ \mathcal{O}_{l,nn})(Y^{(l-1)}),\\
&Y^{(l)} = (Id + \mathcal{O}_{l,ff} \circ \mathcal{O}_{l,nn})(Y^{(l-\frac{1}{2})}),\\
&\ldots\\
&Y^{(L-\frac{1}{2})} = (Id + \mathcal{O}_{L,att} \circ \mathcal{O}_{L,nn})(Y^{(L-1)}),\\
&X_* = Y^{(L)} = (Id + \mathcal{O}_{L,ff} \circ \mathcal{O}_{L,nn})(Y^{(L-\frac{1}{2})}),
\end{aligned} \tag{2.15}$$

which means that $(Y^{(\frac{1}{2})}, Y^{(1)}, Y^{(\frac{3}{2})}, Y^{(2)}, \ldots, Y^{(L-\frac{1}{2})}, Y^{(L)})$ is a fixed-point of the operator

$$(\hat{\mathcal{F}}_{\frac{1}{2}}, \hat{\mathcal{F}}_1, \ldots, \hat{\mathcal{F}}_{L-\frac{1}{2}}, \hat{\mathcal{F}}_L)$$

where

$$
\begin{aligned}
& l = 1, \\
& Y^{(\frac{1}{2})} = \hat{\mathcal{F}}_{\frac{1}{2}}(X_*), \\
& Y^{(1)} = \hat{\mathcal{F}}_1(Y^{\frac{1}{2}}), \\
& \dots \\
& l \in \{2, \dots, L-1\} : \\
& Y^{(l-\frac{1}{2})} = \hat{\mathcal{F}}_{l-\frac{1}{2}}(Y^{(l-1)}), \\
& Y^{(l)} = \hat{\mathcal{F}}_l(Y^{(l-\frac{1}{2})}), \\
& \dots \\
& Y^{(L-\frac{1}{2})} = \hat{\mathcal{F}}_{L-\frac{1}{2}}(Y^{(L-1)}), \\
& X_* = Y^{(L)} = \hat{\mathcal{F}}_L(Y^{(L-\frac{1}{2})}).
\end{aligned}
\tag{2.16}
$$

From Proposition 16, we know that if the operator $\hat{\mathcal{O}} := \hat{\mathcal{O}}_L \circ \ldots \circ \hat{\mathcal{O}}_1 : \mathcal{H}_0^D \to \mathcal{H}_0^D$ is γ-averaged operator, $0 < \gamma \leq 1$, such that $\hat{\mathcal{O}}$ has at least one fixed-point, then the algorithm with $X_0 \in \mathcal{H}_0^D$ and $X_{t+1} = X_t + \lambda_t(\hat{\mathcal{O}}(X_t) - X_t)$ generates a sequence $\hat{\mathcal{O}}(X_t) - X_t$ that converges to 0 as t goes to infinity and X_t converges weakly to a point in Fixed$(\hat{\mathcal{O}})$ as long as $\{\lambda_t\}_{t\geq 1}$ is a chosen sequence in $[0, \frac{1}{\gamma}]$ such that $\sum_{t\geq 1} \lambda_t(1 - \gamma \cdot \lambda_t) = +\infty$.

The above result assumes that there exists a fixed-point.

We now discuss the existence of a fixed-point. For contraction operators (Lipschitzian with Lipschitz constant strictly less than 1), the existence of a unique fixed-point follows from Banach's contraction theorem. For non-expansive operators (those with a Lipschitz constant equals one), a fixed-point may not exist or there may be infinite number of fixed-points. For single-valued operators, one can use the Brouwer-Schauder fixed-point theorem for continuous functions over convex, compact, non-empty set into itself. However, here the space is not compact (the input signal can be arbitrary high). Therefore, the Brouwer-Schauder fixed-point is not directly applicable. Another case of existence of a fixed-point is the Tarski's fixed-point theorem for monotonic operator over partially ordered set.

As a corollary, $\mathcal{O}_{L,nn} \circ \hat{\mathcal{O}}$ is at least one fixed-point in the (closed) unit ball $\mathcal{B}_1$. The proof follows from the Brouwer-Schauder fixed-point theorem applied to the composition $\mathcal{O}_{L,nn} \circ \hat{\mathcal{O}}$ which is Lipschitz continuous (hence continuous) and which now preserves the unit ball because of $\mathcal{O}_{L,nn}$ applied at the final output.

Next we connect the transformer-based Tensor-Graph outcomes to game theory.

2.4.2 Transformer-based Tensor-Graph as an Hierarchical Aggregative Game

The problem of machine intelligence consistency can be reformulated as a hierarchical game, where each layer in the network relies on information from the previous layer to make decisions about its outputs. This hierarchical information structure reflects how layers in a deep network process data, with each layer transforming the input it receives from the preceding layer.

The consistency game can be described as follows: the layers of the neural network act as the decision-makers, and their decision space is defined by the output at each respective layer. The information structure is based on the input from the preceding layer, which corresponds to the action taken by the previous decision-maker. This forms a fully hierarchical game, specifically a fully Stackelberg game, where each layer operates as a leader, making decisions based on the actions of the preceding layer.

Normalization operator as a best-response in a one-shot game

$$\left\{\begin{array}{l} \text{for } o \in \{1,2,\ldots,D\}, \\ X'_o \in \arg\min_{Y_o} \frac{1}{2}\|\hat{\sigma}(X_o) - Y_o\|^2 = \{\hat{\sigma}(X_o)\}. \end{array}\right. \tag{2.17}$$

At each layer l, each normalization step, the system (2.18) defines a D-decision-maker non-zero-sum game played independently.

Softargmax operator as a best-response in a one-shot game

From the collection of tensors X from the precedent layer l, each player i minimizes

$$a_o = (a_{oj})_{1\leq j\leq D} \mapsto g_o(X, a_o) = -\sum_{j=1}^{o} z_{oj}a_{oj} + \sum_{j=1}^{o} a_{oj}\log a_{oj} + \sum_{j=o+1}^{D} a_{oj}^2,$$

defined over

$$\Delta^*_{o-1} = \{a_o \in \mathbb{R}^D, \sum_{j=1}^{o} a_{oj} = 1, a_{oj} > 0, j \in \{1,\ldots,o\}, \text{ and } a_{oj} = 0, o < j \leq D\}.$$

The dependence of the preceding layer decision-makers is via $z_{oj} = \frac{1}{\sqrt{k}}\langle QX_o, KX_j\rangle$ for $Q, K \in \mathcal{L}(\mathcal{H}^d, \mathcal{H}^k)$.

Given X, consider the D-decision-maker non-sum game

$$\hat{G}_{\text{softargmin}} = \left(\mathcal{I} = \{1,\ldots,D\}, (\Delta^*_{o-1}, g_o)_{o\in\mathcal{I}}, X\right).$$

The best-response of this game for $o \in \mathcal{I}$ is $\{a_o\} = \text{argmin}_{a'_o} g_o(X, a'_o) = \{\text{softargmax}^*_o(z_o)\}$ which is a probability measure.

Lemma 26. *Given the collection X, the D-decision-maker non-sum game*

$$\hat{G}_{\mathit{softargmin}} = \left(\mathcal{I}, (\Delta^*_{o-1}, g_o)_{o\in\mathcal{I}}, X\right)$$

has a unique Nash equilibrium and it is given by $(\mathit{softargmax}^*_o(z_o))_{o\in\mathcal{I}}$.

Averaging as a Game

Let $\sum_{j=1}^{D} p_{oj} = 1, p_{oj} \geq 0$, and consider the convex function $Y_o \mapsto \sum_{j=1}^{D} p_{oj}\frac{1}{2}\|Y_o - X_j\|^2$. This function is minimized at $Y^*_o = \sum_{j=1}^{D} p_{oj}X_j$ which is the convex combination of the elements $X_1,\ldots,X_D$.

Multi-head Self-Attention as a best-response in a one-shot game

From the previous observation we see that $Y^*_o = \sum_{j=1}^{D} \text{softargmax}^*_{oj}X_j$ is the minimizer of the convex function $Y_o \mapsto \sum_{j=1}^{D} \text{softargmax}^*_{oj}\frac{1}{2}\|X_j - Y_o\|^2$.

Lemma 27. *Let h_o be the cost function*

$$h_o : Y_o \mapsto \sum_{j=1}^{D} \mathit{softargmax}^*_{oj}\frac{1}{2}\|(WV)X_j - Y_o\|^2.$$

Given the collection of tensors X, the D-decision-maker non-sum game

$$\hat{G}_{att} = (\mathcal{I}, (\mathcal{H}_0, h_o)_{o\in\mathcal{I}}, X)$$

has a unique Nash equilibrium and it is given by $(\mathcal{O}_{att,o}(X))_{o\in\mathcal{I}}$ *with*

$$\mathcal{O}_{att,o}(X) = (WV)\sum_{j=1}^{D} softargmax^*_{oj}X_j.$$

Lemma 28. *Let* $\hat{g}_o$ *be the cost function*

$$\hat{g}_o : Y_o \mapsto \sum_{j=1}^{D}\sum_{h=1}^{H} softargmax^*_{oj,h}\frac{1}{2H}\|\sqrt{H}(W_hV_h)X_j - Y_o\|^2,$$

with $softargmax^*_{oj,h} = softargmax^*_{oj,h}(Q_h, K_h)$. *Given the tensor X, the D-decision-maker non-sum game*

$$\hat{G}_{multihead} = (\mathcal{I}, (\mathcal{H}_0, \hat{g}_o)_{o\in\mathcal{I}}, X)$$

has a unique Nash equilibrium and it is given by $(\mathcal{O}_{att,o}(X))_{o\in\mathcal{I}}$ *with*

$$\mathcal{O}_{att,o}(X) = \frac{1}{\sqrt{H}}\sum_{h=1}^{H}(W_hV_h)\sum_{j=1}^{D} softargmax^*_{oj,h}X_j.$$

Corollary 3. *Let* $\tilde{g}_o$ *be the cost function*

$$\tilde{g}_o : Y_o \mapsto \sum_{j=1}^{D}\sum_{h=1}^{H} softargmax^*_{oj,h}\frac{1}{2H}\|\sqrt{H}(W_hV_h)X_j + X_o - Y_o\|^2.$$

Given the collection of tensors X, the D-decision-maker non-sum game

$$\hat{G}_{Idplusmultihead} = (\mathcal{I}, (\mathcal{H}_0, \tilde{g}_o)_{o\in\mathcal{I}}, X)$$

has a unique Nash equilibrium and it is given by $(Id + \mathcal{O}_{att,o}(X))_{o\in\mathcal{I}}$ *with* $(Id + \mathcal{O}_{Id+att,o})(X) = X_o + \frac{1}{\sqrt{H}}\sum_{h=1}^{H}(W_hV_h)\sum_{j=1}^{D} softargmax^*_{oj,h}X_j$.

Remark 11.

- *As the parameter* $\frac{1}{\sqrt{k}}$ *is not too large, the softargmax operation is not really an approximate maximizer. This is only probabilistic weights assigned to the vectors. The denominator which is a sum of exponentials creates some difficulties in practice. One may consider instead a logarithmic scale: for* $j \le i$, $\chi_{ij} = \log p_{ij} = z_{ij} - \log(\sum_{l=1}^{i} e^{z_{il}})$ *i.e.,*

$$\chi_{ij} = \frac{1}{\sqrt{k}}\langle Qx_i, Kx_j\rangle - \log\left(\sum_{l=1}^{i} e^{\frac{1}{\sqrt{k}}\langle Qx_i, Kx_l\rangle}\right).$$

- *The softargmax in the attention is not needed at this stage of the architecture. We can use simpler scaling factor such as sigmoid attention. For each head h, compute the weight matrix*

$$S_{ij,h} = \frac{1}{\sqrt{k}}\langle Q_hx_i, K_hx_j\rangle.$$

One can apply a sigmoid function to each entry of S_h *weight* $S'_{ij,h} = \sigma(S_{ij,h}) = \frac{e^{S_{ij,h}}}{D+e^{S_{ij,h}}}$. *The rows are not normalized to probabilities in this case. However* $x_i + \frac{1}{\sqrt{H}}\sum_{h=1}^{H}(W_hV_h)\sum_{j=1}^{D} S'_{ij,h}x_j$ *can still be seen as a game.*

- *Another attention mechanism is obtained with the function*

$$S'_{ij,h} = e^{-\|Q_h^\dagger K_h(x_i - x_j)\|^2}.$$

- *Yet, another attention mechanism is obtained with the function*

$$S'_{ij,h} = \frac{\|Q_h^\dagger K_h(x_i - x_j)\|}{D + \|Q_h^\dagger K_h(x_i - x_j)\|}.$$

Graph aggregation as a best-response in a one-shot game

$$\left\{ \begin{array}{l} \text{for } o \in \{1, 2, \ldots, D\}, \\ X'_o(i) \in \arg\min_y \sum_{i' \in \mathcal{N}_i} \frac{1}{2}\|X_o(i') - y\|^2 = \{\frac{\sum_{i' \in \mathcal{N}_i} X_o(i')}{\sum_{i'} \mathbb{I}_{i' \in \mathcal{N}_i}}\}. \end{array} \right. \tag{2.18}$$

Combining all the layer-by-layer interaction between inputs and outputs, we arrive at the following first main result.

Proposition 17. *The outcomes of the transformer-based tensor-graph neural network* $\mathbb{TGN}$ *are exactly the Nash equilibria of a hierarchical game.*

Proof: We have shown above that each operator $\hat{\mathcal{F}}_{l-\frac{1}{2}}$ and $\hat{\mathcal{F}}_l$ is a result of a game. By collecting all these games, we arrive at a hierarchical game where each hierarchical level l is a multi-step sequential game. At the hierarchical level l, each decision-maker has a layer input layer, which is the action output of the preceding layer decision-makers. If the $\mathbb{TGN}$ has an outcome tensor X_*, it means that it is fixed-point of $\hat{\mathcal{F}}$. This leads to a fixed-point to $\hat{\mathcal{O}}$. At this fixed-point X_*, each decision-maker is playing its unique best-response to preceding input and preceding decision-maker at the same sub-layer. □

Lemma 29. *(Existence) Let* $F: \mathcal{H}_0^D \to \mathcal{H}_0^D$ *be a continuous operator and* $\mathcal{B}_\rho^D$ *the closed ball of* $\mathcal{H}_0$ *with radius* $\rho > 0$. *The restricted variational inequality problem: Find* $X_* \in \mathcal{B}_\rho^D$ *such that* $\langle F(X_*), X - X_* \rangle \geq 0$, $\forall X \in \mathcal{B}_\rho^D$ *has at least one solution. In particular, the restriction of* F *to the closed ball* $\mathcal{B}_\rho^D$ *admits is a Nash equilibrium.*

This existence result is extremely useful in generative machine intelligence as the normalization operation brings an element to a certain closed product-ball $\mathcal{B}_\rho^D$, which is a convex, closed, bounded and non-empty set of $\mathcal{H}_0^D$ and the operators involved are continuous (modified normalization, projection, attention, averaging, ReLU, softargmax, affine map are all continuous).

Proof: The variational inequality restricted to a convex, bounded, closed and non-empty set is equivalent to $X = \text{proj}_{\mathcal{B}_\rho}(X - \gamma F(X)), \gamma > 0$, which admits at least one solution according to Brouwer-Schauder fixed-point theorem. □

This result provides an existence of a consistent outcome up to a normalization or scaling. If $\|X_*\| < 1$ then one can re-scale or up-scale using the mapping $\hat{\sigma}^{-1}$.

Remark 12. *The outcomes of any neural network given by the fixed-points (if any) of the operator* $\hat{\mathcal{O}}_L \circ \ldots \circ \hat{\mathcal{O}}_1$ *are exactly the constrained Nash equilibria (CNE) of the non-zero sum game*

$$\mathcal{G} = (\mathcal{L} = \{1, \ldots, L\}, (\mathcal{H}_l, 0, C_l(.))_{\mathcal{L}}),$$

where $C_l(Y) = \hat{\mathcal{O}}_l(Y)$,

To prove it, observe that for each layer l of the deep neural network, the outcome can be written as $\hat{O}_l(Y^{(l-1)}) = Y^{(l)}$ with $Y^{(0)} = X_0$. It follows that

$$\begin{cases} l \in \{1, \ldots, L\} \\ Inf_{\{Y^{(l)} \in C_l(Y)\}} 0 \end{cases} \Leftrightarrow \begin{cases} Y := (Y^{(1)}, \ldots, Y^{(L)}), \\ Y \in C(Y) \end{cases} \Leftrightarrow \begin{cases} \text{CNE of game:} \\ (\mathcal{L}, (\mathcal{H}_l, 0, C_l(.))_{\mathcal{L}}). \end{cases} \tag{2.19}$$

The latter set is the set of fixed-points of the operator C.

2.4.3 Training Problem

A strong formulation of the design problem is as follows:

Find L, H, D, d and the operators $\hat{O}_1, \ldots, \hat{O}_L$:

$$Y_o = \hat{O}_L \circ \ldots \circ \hat{O}_1(X_o), \ o \in \{1, \ldots, D\}. \tag{2.20}$$

We can turn it into a variational inequality as follows

$$\langle (Y_o - \hat{O}_L \circ \ldots \circ \hat{O}_1(X_o)), X'_o - X_o \rangle \geq 0, \ \forall X'_o \in \mathcal{H}_0, o \in \{1, \ldots, D\}. \tag{2.21}$$

This is a very strong requirement as the operators are fixed in advance and should be satisfied with the any data. This is no reason for such equality to hold in general.

We formulate another input-output data distribution-dependent design problem as follows:

Problem 5. *Design Problem: Find L, H, D, d and the operators $\hat{O}_1, \ldots, \hat{O}_L$:*

$$\arg\min_{L,D \geq 1} \arg\min_{\hat{O}_L, \ldots, \hat{O}_1} \int_{X_0} \int_{Y^{(L)}} [\Phi(Y^{(L)}, \hat{O}_L \circ \ldots \circ \hat{O}_1 X_0)] \mathbb{P}(dX_0 dY^{(L)}). \tag{2.22}$$

where Φ is a real-valued risk-aware cost function.

In this formulation the input-data is turned into $\mathbb{P}(dX_0 dY^{(L)})$.

The minimization over functional space, and transport map are more involved. As an illustration, let us consider the Monge problem: Given two measures m_0, ν_1, Design: Find $\hat{O}_1 : \mathcal{H}_0^D \to \mathcal{H}_0^D$

$$\arg\min_{\hat{O}_1} \{ \int_{X_0} [\Phi(X_0, \hat{O}_1 X_0)] m_0(dX_0), \ \hat{O}_1 \# m_0 = \nu_1 \}, \tag{2.23}$$

where $\hat{O}_1 \# m_0$ is the push forward operator induced by $\hat{O}_1$, i.e., $\nu_1(E) = m_0(\hat{O}_1^{-1}(E))$ for any measurable subset.

Some special cases of the Monge problem can be solved for initial absolutely continuous measures m_0 through the Kantorovich relaxation:

Given two measures m_0, ν_1, find a joint probability measure $\mathbb{P}(dX_0 dY^{(1)})$ such that

$$\arg\min_{\mathbb{P}} \{ \int_{X_0} \int_{Y^{(1)}} [\Phi(Y^{(1)}, \hat{O}_1 X_0)] \mathbb{P}(dX_0 dY^{(1)}), \ m_0 = \mathbb{P}_X, \nu_1 = \mathbb{P}_Y \}, \tag{2.24}$$

where $\mathbb{P}_X = m_0$ is the first marginal and $\mathbb{P}_Y = \nu_1$ is the second marginal and m_0 is assumed to be absolutely continuous measure.

Remark 13. *A number of observations are in order. First, the absolute continuity for the initial measure means the initial data input to have density. The data input is currently large but still finite. Thus, it is not density-based data. This means that one needs to get an estimated density from the training data. Second, using mean-field-type control, the optimal* *map $\hat{\mathcal{O}}_1$ depends on the measure (training data) which is problematic for the architectures of generative machine intelligence. Third, the optimal transport map $\hat{\mathcal{O}}_1$ does not have the shape of ReLU, softargmax, masked/unmasked self-attention used at the layers. Can we instead design activation functions that all collectively approximate the optimal transport map? This question is related to approximation capability of transformer-based neural networks.*

We therefore focus on a sub-goal, which is the minimization over parameters

$$\theta = (Q_{lh}, K_{lh}, V_{lh}, W_{lh}, W_{1l}, W_{2l}, b_{1l}, b_{2l})_{1\leq l\leq L}$$

the variational inequality becomes

$$\langle (Y^{(L)} - \hat{\mathcal{O}}_L \circ \ldots \circ \hat{\mathcal{O}}_1(X_o))(\theta), A_L^{\dagger}(\theta' - \theta)\rangle \geq 0, \ \forall \theta', o \in \{1, \ldots, D\}, \tag{2.25}$$

where $A_L^{\dagger}$ is the adjoint operator which returns in the space as $Y^{(L)}$.

An alternative formulation to consider is a cost functional formulation. However $\theta \mapsto \int_{X_0}\int_{Y^{(L)}}[\Phi(Y^{(L)}, \hat{\mathcal{O}}_L \circ \ldots \circ \hat{\mathcal{O}}_1 X_0)]\mathbb{P}(dX_0 dY^{(L)})$ is non-convex, i.e., the risk-aware minimization over θ is non-convex even if Φ is a convex function. This means that some hidden convexity structures in the operator $(Id + \mathcal{O}_{l,ff} \circ \mathcal{O}_{l,nn}) \circ (Id + \mathcal{O}_{l,att} \circ \mathcal{O}_{l,nn})$ need to be exploited.

Given $\mathcal{D} = \{(\hat{X}_o, \hat{Y}_o) \in \mathcal{H}_0 \times \mathcal{H}_L, o \in \{1, 2, \ldots, D\}\}$, the mini-batch training problem (limited to weights and biases) is to find a time-and state- independent control action θ such that

$$\begin{cases} \mathcal{D} = \{(\hat{X}_o, \hat{Y}_o) \in \mathcal{H}_0 \times \mathcal{H}_L, o \in \{1, 2, \ldots, D\}\}, \\ \inf_{\theta\in\Theta} \frac{1}{D}\sum_{o=1}^{D} \Phi(Y_{o,T}, \hat{Y}_o), \\ \text{such that} \\ Y_0 = \hat{X}, \\ Y_{t+1} = Y_t + \lambda_t(\hat{\mathcal{O}}_L \circ \ldots \circ \hat{\mathcal{O}}_1(Y_t) - Y_t). \end{cases} \tag{2.26}$$

2.5 Small Learning Rate Regime of Finite Sequence Transformer

We now examine the dynamics of transformer-based neural network when λ_t is very small. One can relate this case to a pseudo-trajectory of an ordinary differential system or stochastic differential system. Here D is finite. The pseudo-trajectory of the algorithm in (2.8) is captured by the following ODE:

$$\begin{cases} Y(0) = X_0 \in \mathcal{H}_0^D, \\ \dot{Y} = (\hat{\mathcal{O}}_L \circ \ldots \circ \hat{\mathcal{O}}_1)(Y) - Y, \ t > 0. \end{cases} \tag{2.27}$$

Even if the operator is not globally Lipschitz, it does not create a difficulty in numerous situations. When restricted to any not empty and compact set of $\mathcal{H}_0$, the operator $\hat{\mathcal{O}}_L \circ \ldots \circ \hat{\mathcal{O}}_1$ is locally Lipschitz on compact sets and linear maps involved are all assumed to be bounded. As locally Lipschitz drifts define a unique solution to the ODE for any initial condition, we conclude that there is a unique trajectory for such cases. Note, however, that some of the interaction terms such as the Coulomb interaction require a more careful analysis to establish existence due to singularities at zero. As expected, the steady states (if any) of this ODE are the fixed-points of the operator $\hat{\mathcal{O}}_L \circ \ldots \circ \hat{\mathcal{O}}_1$.

The ODE (2.27) can be obtained for a fixed and finite number of layers L and finite sequence length D by letting the learning rate λ_t go to zero. This ODE approximation methodology is similar to the Euler scheme or stochastic approximation method. As a consequence we do not need to have an infinite number of layers to have an ODE or continuous time model. The infinite number of layers approach is not practical and is out of range of the current computational capabilities. A nice property about this ODE is that any rest point provides a Nash equilibrium of the hierarchical game between the D entries of the L layers. This property is also called Nash stationarity in evolutionary game theory, when the rest-points of the dynamics are exactly the Nash equilibria. Not all rest points are stable.

Nash equilibria differs from Stable Nash equilibria: When $(\hat{O}_L \circ \ldots \circ \hat{O}_1)$ is a strict contraction, its unique fixed-point which is an element of the kernel of $Id - (\hat{O}_L \circ \ldots \circ \hat{O}_1)$ is a stable rest point of (2.27) and there is a global convergence to that rest-point starting from an initial signal input of sequence size $D \geq 1$. However, for general $(\hat{O}_L \circ \ldots \circ \hat{O}_1)$, some rest points may not be locally asymptotically stable. If the initial point does not coincide with the rest point, the system may move to another a point in the long run. The locally asymptotically stable rest points have nice features in the asymptotic outcomes of $\mathbb{TGN}$.

The training problem is to find a time-and state-independent control action θ such that

$$\begin{cases} \mathcal{D} = \{(\hat{X}_o, \hat{Y}_o) \in \mathcal{H}_0 \times \mathcal{H}_L, o \in \{1, 2, \ldots, D\}\}, \\ \inf_{\theta \in \Theta} \frac{1}{D} \sum_{o=1}^{D} \Phi(Y_o(T), \hat{Y}_o), \\ \text{such that} \\ Y(0) = \hat{X}, \\ \dot{Y} = (\hat{O}_L \circ \ldots \circ \hat{O}_1)(Y) - Y, \ T > t > 0. \end{cases} \tag{2.28}$$

Because the control action is restricted the time-and state-independent control Θ is because of a data-independent architecture at the end of the training process. We will see later how to implement state-feedback weight/bias in transformers.

2.6 Transformer with a Simpler Normalization

2.6.1 Boltzmann-Gibbs Transformer with a Simpler Normalization

Here we change the mean-variance normalization with a simpler normalization to the unit ball. It involves less parameters and extend to Hilbert spaces. Note that for Hilbert spaces, using the mean does not necessarily make sense as the infinite sum may be infinite.

$$\begin{aligned} &\text{Parameters } : L, D, d, k, k', q, H \in \mathbb{N}, \\ &\quad 1 \leq l \leq L : \\ &\quad 1 \leq h \leq H : Q_{lh}, K_{lh} \in \mathcal{L}(\mathcal{H}^d, \mathcal{H}^k), V_{lh} \in \mathcal{L}(\mathcal{H}^d, \mathcal{H}^{k'}), W_{lh} \in \mathcal{L}(\mathcal{H}^{k'}, \mathcal{H}^d), \\ &\quad W_{1l} \in \mathcal{L}(\mathcal{H}^d, \mathcal{H}^q), b_{1l} \in \mathcal{H}^q, W_{2l} \in \mathcal{L}(\mathcal{H}^q, \mathcal{H}^d), b_{2l} \in \mathcal{H}^d, \\ &\text{Input } : y_0 = x_0 \in \mathcal{H}^{Dd}, \\ &1 \leq l \leq L : \\ &\quad \text{Input at } l : \ x_l = y_{l-1} \in \mathcal{H}^{Dd}, \end{aligned}$$

$$
\begin{aligned}
&1 \le i \le D: \ u_{li} = \tfrac{x_{li}}{1+\|x_{li}\|}, \\
&1 \le i \le D: \ \tilde{u}_{li} = \tfrac{1}{\sqrt{H}} \textstyle\sum_{h=1}^{H} W_{lh} \sum_{j=1}^{i} \frac{e^{\frac{1}{\sqrt{k}}\langle Q_{lh}u_{li}, K_{lh}u_{lj}\rangle}}{\sum_{j'}^{i} e^{\frac{1}{\sqrt{k}}\langle Q_{lh}u_{li}, K_{lh}u_{lj'}\rangle}} V_{lh}u_{lj}, \\
&1 \le i \le D: \ \hat{u}_{li} = x_{li} + \tilde{u}_{li}, \\
&1 \le i \le D: \ \hat{y}_{li} = \tfrac{\hat{u}_{li}}{1+\|\hat{u}_{li}\|}, \\
&1 \le i \le D: \ \tilde{y}_{li} = W_{2l} r_l (W_{1l}\hat{y}_{li} + b_{1l}) + b_{2l}, \\
&1 \le i \le D: \ y_{li} = \hat{u}_{li} + \tilde{y}_{li},
\end{aligned}
\tag{2.29}
$$

Output at $l : y_l$

Return $y_L \in \mathcal{H}^{Dd}$

2.6.2 Sigmoid Transformer with a Simpler Normalization

Here we replace Boltzmann-Gibbs by Sigmoid self-attention (Figure 2.7).

Parameters $: L, D, d, k, k', q, H \in \mathbb{N}$,

$1 \le l \le L:$

$1 \le h \le H : Q_{lh}, K_{lh} \in \mathcal{L}(\mathcal{H}^d, \mathcal{H}^k), V_{lh} \in \mathcal{L}(\mathcal{H}^d, \mathcal{H}^{k'}), W_{lh} \in \mathcal{L}(\mathcal{H}^{k'}, \mathcal{H}^d)$,

$W_{1l} \in \mathcal{L}(\mathcal{H}^d, \mathcal{H}^q), b_{1l} \in \mathcal{H}^q, W_{2l} \in \mathcal{L}(\mathcal{H}^q, \mathcal{H}^d), b_{2l} \in \mathcal{H}^d$,

Input $: y_0 = x_0 \in \mathcal{H}^{Dd}$,

$1 \le l \le L:$

Input at $l: \ x_l = y_{l-1} \in \mathcal{H}^{Dd}$,

$$
\begin{aligned}
&1 \le i \le D: \ u_{li} = \tfrac{x_{li}}{1+\|x_{li}\|}, \\
&1 \le i \le D: \ \tilde{u}_{li} = \tfrac{1}{\sqrt{H}} \textstyle\sum_{h=1}^{H} W_{lh} \sum_{j=1}^{i} \frac{e^{\frac{1}{\sqrt{k}}\langle Q_{lh}u_{li}, K_{lh}u_{lj}\rangle}}{D+e^{\frac{1}{\sqrt{k}}\langle Q_{lh}u_{li}, K_{lh}u_{lj}\rangle}} V_{lh}u_{j}, \\
&1 \le i \le D: \ \hat{u}_{li} = x_{li} + \tilde{u}_{li}, \\
&1 \le i \le D: \ \hat{y}_{li} = \tfrac{\hat{u}_{li}}{1+\|\hat{u}_{li}\|}, \\
&1 \le i \le D: \ \tilde{y}_{li} = W_{2li} r_l (W_{1li}\hat{y}_{li} + b_{1li}) + b_{2li}, \\
&1 \le i \le D: \ y_{li} = \hat{u}_{li} + \tilde{y}_{li},
\end{aligned}
\tag{2.30}
$$

Output at $l : y_l$

Return $y_L \in \mathcal{H}^{Dd}$

2.7 Mixture-of-Experts Transformers

In a Mixture of Experts model (Figure 2.8), an expert refers to a sub-model or specialized neural network that is part of a larger collection of experts. Each expert focuses on handling specific types of input data or tasks, making it a modular component within the larger architecture. In MoE, instead of using all experts for every input, a gating mechanism dynamically selects a subset of experts (often just one or a few) based on the characteristics of the input. The experts can be thought of as independent units trained to specialize in certain parts of the data distribution, allowing the model to allocate computational resources more efficiently. This selective activation of experts allows MoE models to scale to larger

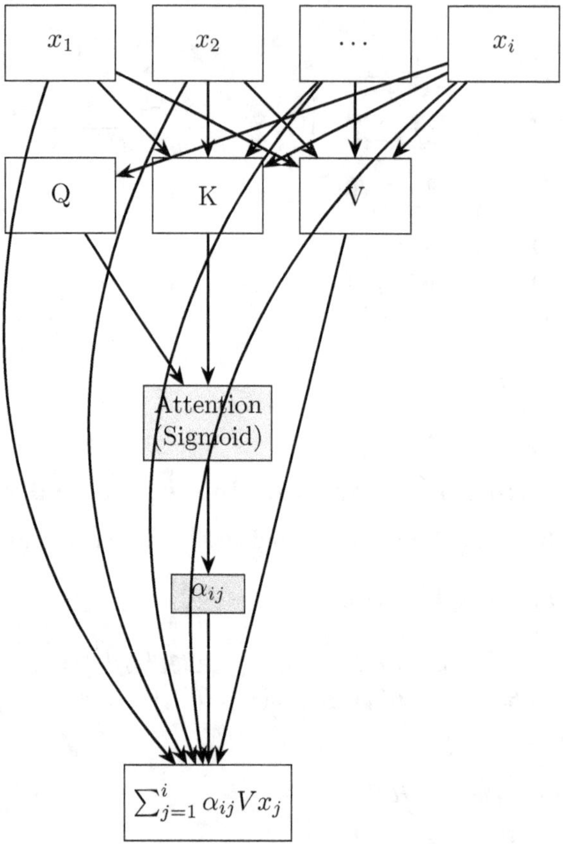

FIGURE 2.7
α_{ij} = D-scaled sigmoid($\langle Qx_i, Kx_j \rangle$). The figure shows the calculation for a single x_i. The full autoregressive computation sums over all j up to i.

sizes while keeping computational costs relatively low, as only a few experts are used per input, rather than the entire model being activated.

The Mixture of Experts model incorporates dynamic sparsity in its architecture to increase its computational efficiency without sacrificing performance. This is done by selectively activating a subset of the model's parameters (the experts) during each forward pass, rather than utilizing all parameters at once. The basic concept behind MoE is to have multiple "experts" (sub-models) within a network, but for each input, only a few experts are used, enabling scaling up model size while keeping computational costs low. The distinction between MoE Transformers and standard transformers begins with how the feed-forward network is modified. MoE Layers replace or augment some of the feed-forward layers in the transformer architecture. This layer introduces multiple "expert" sub-networks, and for each input, only a subset of these experts is chosen. Each expert is a separate feed-forward network, typically with a similar structure to that of a standard Transformer feed-forward network but with varying weights across experts. If the model contains experts, each expert will process the input in parallel, but only a few experts will actually participate in the final computation for a given input.

The gating mechanism determines which experts should be used for a given input. The gate is typically a learned function that takes in the token embeddings and produces a probability distribution over all the experts. For each input token, a sparse selection of

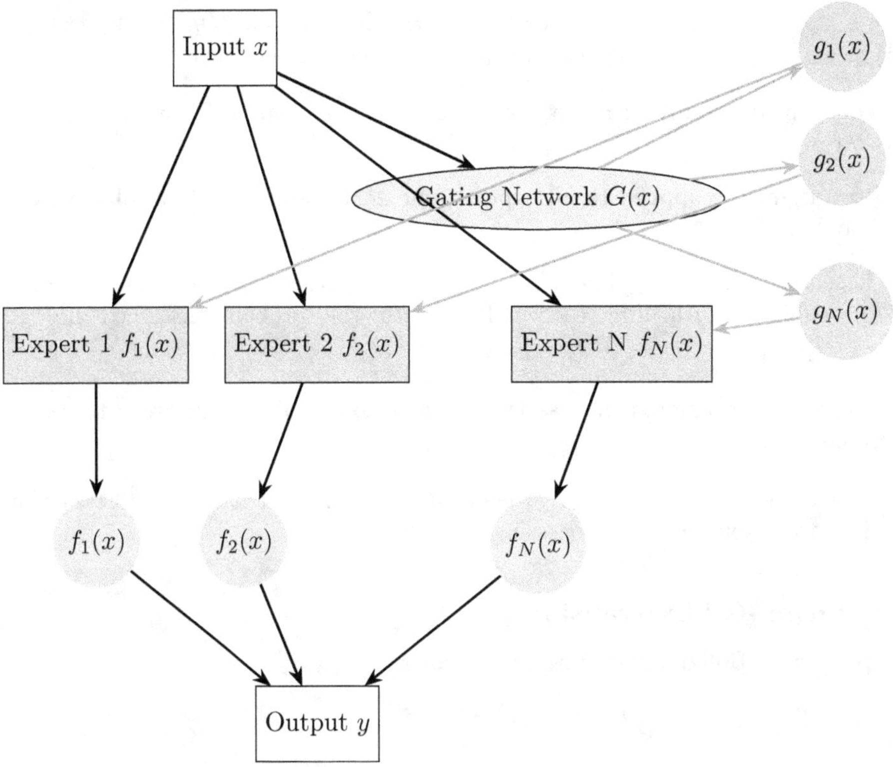

FIGURE 2.8
$y = \sum_{i=1}^{N} g_i(x) f_i(x)$. The gating network $G(x)$ outputs weights $g_i(x)$ that determine the contribution of each expert $f_i(x)$ to the final output y.

experts (usually 1 or 2) is chosen based on the highest probability values produced by the gate. This mechanism introduces sparsity: although the model has access to many experts, only a small fraction is used for each token, drastically reducing computational cost. The selected experts are then routed the input token, which they process, and their outputs are combined (typically via a weighted sum) to form the final output for that token. This routing process can introduce load-balancing challenges, as some experts may be chosen far more often than others. Methods like regularization and routing constraints can be applied to avoid overloading specific experts.

Like in the standard transformer model, the input to the MoE layer is added to the output, helping preserve signals and allowing for deeper networks. The layer normalization is applied before or after the MoE layer to stabilize training. After passing through multiple MoE-enhanced transformer blocks, the final output is fed through a linear transformation to generate logits, which are then used for the desired task.

We keep the attention block as in the previous section. The feedforward network of the transformer involves several learnable parameters that already exceed several trillion. We examine the activation of some of these operations by splitting into shared experts, activated non-shared experts, and non-activated non-shared experts. There are $E+1$ experts in total. One expert, say, $E+1$ is designated as the pool of shared experts. The rest of the experts is denoted by $\mathcal{E} = \{1, \ldots, E\}$.

$$O_{ff}(x_i) = \eta_{i,E+1} O_{ff,E+1}(x_i) + \sum_{e \in \mathcal{E}} \eta_{ie} O_{ff,e}(x_i),$$

$\eta_{ie} = s_{ie}\mathbb{I}_{\{s_{ie}\in Topk(\{s_{ie'},e'\in\mathcal{E}\},\mathcal{E}_2)\}}$. $W_{\mathcal{E}l} \in \mathcal{L}(\mathcal{H}^d, \mathbb{R}^E), b_{\mathcal{E}l} \in \mathbb{R}^E$, $s_{ie} = softmax_e(W_{\mathcal{E}l}x_i + b_{\mathcal{E}})$, $\eta_{i,E+1} = \sigma(W_{E+1}x_i + b_{E+1})$. The experts are sparsely activated per token.

- For each x_i, the projection from Hilbert to lower dimensional space $\mathbb{R}^E$ allows selection of experts for x_i.
- If the full softargmax is applied to all experts, one enables all experts available (with positive probability).
- The Top-k selection works as follows. We select only the top k experts in terms of these probabilities. All are re-normalized into the probabilities, i.e., $\eta_{ie} := \frac{e^{z_{li,e}}}{\sum_{e'\in Topk} e^{z_{li,e'}}}\mathbb{I}_{\{e\in Topk\}}$, $z_{li,e} = (W_{\mathcal{E}l}x_i + b_{l\mathcal{E}})_e$.
- Unselected experts are ignored, i.e., assigned a probability of zero in order to reduce computational cost.

This process, often referred to as $Gate(i, e)$ determines how much trust or importance the model the assigns to an expert e for processing the token i.

MoE - Boltzmann-Gibbs Transformer

Mixture of experts with Boltzmann-Gibbs Transformer.

$$
\begin{aligned}
&\text{Parameters } : \mathcal{E} = \{1, \ldots, E\}, L, D, d, k, k', q, H \in \mathbb{N},\\
&\quad 1 \le l \le L :\\
&\quad\quad 1 \le h \le H : Q_{lh}, K_{lh} \in \mathcal{L}(\mathcal{H}^d, \mathcal{H}^k), V_{lh} \in \mathcal{L}(\mathcal{H}^d, \mathcal{H}^{k'}), W_{lh} \in \mathcal{L}(\mathcal{H}^{k'}, \mathcal{H}^d),\\
&\quad\quad W_{1l,e} \in \mathcal{L}(\mathcal{H}^d, \mathcal{H}^q), b_{1l,e} \in \mathcal{H}^q, W_{2l,e} \in \mathcal{L}(\mathcal{H}^q, \mathcal{H}^d), b_{2l,e} \in \mathcal{H}^d, e \in \mathcal{E}\\
&\quad\quad W_{3l} \in \mathcal{L}(\mathcal{H}^d, \mathbb{R}^E), b_{3l} \in \mathbb{R}^E,\\
&\text{Input } : y_0 = x_0 \in \mathcal{H}^{Dd},\\
&1 \le l \le L :\\
&\quad \text{Input at } l : \ x_l = y_{l-1} \in \mathcal{H}^{Dd},\\
&\quad\quad 1 \le i \le D : \ u_{li} = \tfrac{x_{li}}{1+\|x_{li}\|},\\
&\quad\quad 1 \le i \le D : \ \tilde{u}_{li} = \tfrac{1}{\sqrt{H}} \textstyle\sum_{h=1}^{H} W_{lh} \sum_{j=1}^{i} \frac{e^{\frac{1}{\sqrt{k}}\langle Q_{lh}u_{li}, K_{lh}u_{lj}\rangle}}{\sum_{j'}^{i} e^{\frac{1}{\sqrt{k}}\langle Q_{lh}u_{li}, K_{lh}u_{lj'}\rangle}} V_{lh}u_{lj},\\
&\quad\quad 1 \le i \le D : \ \hat{u}_{li} = x_{li} + \tilde{u}_{li}, \qquad\qquad (2.31)\\
&\quad\quad 1 \le i \le D : \ \hat{y}_{li} = \tfrac{\hat{u}_{li}}{1+\|\hat{u}_{li}\|},\\
&\quad\quad 1 \le i \le D : \text{Choose } \mathcal{E}_i,\\
&\quad\quad\quad \eta_{lie} = \frac{e^{(\tilde{W}_{3l}\hat{y}_{li}+\tilde{b}_{3l})_e}}{\sum_{e'\in\mathcal{E}_i} e^{(\tilde{W}_{3l}\hat{y}_{li}+\tilde{b}_{3l})_{e'}}}\mathbb{I}_{e\in\mathcal{E}_i},\\
&\quad\quad\quad e = E+1 : \tilde{y}_{li,e} = W_{2l,e}r_l(W_{1l,e}\hat{y}_{li,e} + b_{1l,e}) + b_{2l,e},\\
&\quad\quad\quad 1 \le e \le E_i : \ \tilde{y}_{li,e} = W_{2l,e}r_l(W_{1l,e}\hat{y}_{li,e} + b_{1l,e}) + b_{2l,e},\\
&\quad\quad\quad \tilde{y}_{li} = \eta_{li,E+1}\tilde{y}_{li,E+1} + \textstyle\sum_{e\in E_i} \eta_{lie}\tilde{y}_{li,e},\\
&\quad\quad 1 \le i \le D : \ y_{li} = \hat{u}_{li} + \tilde{y}_{li},\\
&\quad \text{Output at } l : y_l\\
&\text{Return } y_L \in \mathcal{H}^{Dd}
\end{aligned}
$$

One typical example is to choose $\mathcal{E}_i = Topk$, the top k experts with respect to the output $(\tilde{W}_{3l}\hat{y}_{li} + \tilde{b}_{3l})$. In the sum over e, only the selected top k are then active. All others available are on sleep mode.

MoE - Sigmoid Transformer

Mixture of experts with Sigmoid Transformer.

$$
\begin{aligned}
&\text{Parameters } : \mathcal{E} = \{1, \dots, E\}, L, D, d, k, k', q, H \in \mathbb{N},\\
&\quad 1 \le l \le L :\\
&\quad\quad 1 \le h \le H : Q_{lh}, K_{lh} \in \mathcal{L}(\mathcal{H}^d, \mathcal{H}^k), V_{lh} \in \mathcal{L}(\mathcal{H}^d, \mathcal{H}^{k'}), W_{lh} \in \mathcal{L}(\mathcal{H}^{k'}, \mathcal{H}^d),\\
&\quad\quad 1 \le e \le E : W_{1l,e} \in \mathcal{L}(\mathcal{H}^d, \mathcal{H}^q), b_{1l,e} \in \mathcal{H}^q, W_{2l,e} \in \mathcal{L}(\mathcal{H}^q, \mathcal{H}^d), b_{2l,e} \in \mathcal{H}^d,\\
&\quad\quad W_{3l} \in \mathcal{L}(\mathcal{H}^d, \mathbb{R}^E), b_{3l} \in \mathbb{R}^E,\\
&\text{Input } : y_0 = x_0 \in \mathcal{H}^{Dd},\\
&1 \le l \le L :\\
&\quad \text{Input at } l : \; x_l = y_{l-1} \in \mathcal{H}^{Dd},\\
&\\
&\quad\quad 1 \le i \le D : \; u_{li} = \tfrac{x_{li}}{1+\|x_{li}\|},\\
&\quad\quad 1 \le i \le D : \; \tilde{u}_{li} = \tfrac{1}{\sqrt{H}} \textstyle\sum_{h=1}^{H} W_{lh} \sum_{j=1}^{i} \frac{e^{\frac{1}{\sqrt{k}}\langle Q_{lh}u_{li}, K_{lh}u_{lj}\rangle}}{D + e^{\frac{1}{\sqrt{k}}\langle Q_{lh}u_{li}, K_{lh}u_{lj}\rangle}} V_{lh} u_j,\\
&\quad\quad 1 \le i \le D : \; \hat{u}_{li} = x_{li} + \tilde{u}_{li},\\
&\\
&\quad\quad 1 \le i \le D : \; \hat{y}_{li} = \tfrac{\hat{u}_{li}}{1+\|\hat{u}_{li}\|},\\
&\quad\quad 1 \le i \le D : \text{Choose } \mathcal{E}_i,\\
&\quad\quad\quad \eta_{lie} = \frac{e^{(\tilde{W}_{3l}\hat{y}_{li} + \tilde{b}_{3l})_e}}{\sum_{e' \in \mathcal{E}_i} e^{(\tilde{W}_{3l}\hat{y}_{li} + \tilde{b}_{3l})_{e'}}} \mathbb{I}_{e \in \mathcal{E}_i},\\
&\quad\quad\quad e = E+1 : \tilde{y}_{li,e} = W_{2l,e} r_l (W_{1l,e} \hat{y}_{li,e} + b_{1l,e}) + b_{2l,e},\\
&\quad\quad\quad 1 \le e \le E_i : \; \tilde{y}_{li,e} = W_{2l,e} r_l (W_{1l,e} \hat{y}_{li,e} + b_{1l,e}) + b_{2l,e},\\
&\quad\quad\quad \tilde{y}_{li} = \eta_{li,E+1} \tilde{y}_{li,E+1} + \textstyle\sum_{e \in E_i} \eta_{lie} \tilde{y}_{li,e},\\
&\quad\quad 1 \le i \le D : \; y_{li} = \hat{u}_{li} + \tilde{y}_{li},\\
&\\
&\quad \text{Output at } l : y_l\\
&\text{Return } y_L \in \mathcal{H}^{Dd}
\end{aligned}
\tag{2.32}
$$

2.8 Difference Transformer

$$
\begin{aligned}
&\text{Parameters } : L, D, d, k, k', q, H \in \mathbb{N}, \chi \in \mathbb{R}\\
&\quad 1 \le l \le L :\\
&\quad\quad 1 \le h \le H : Q_{lh}^{(1)}, K_{lh}^{(1)}, \in \mathcal{L}(\mathcal{H}^d, \mathcal{H}^k), Q_{lh}^{(2)}, K_{lh}^{(2)}, \in \mathcal{L}(\mathcal{H}^d, \mathcal{H}^k),\\
&\quad\quad\quad\quad V_{lh} \in \mathcal{L}(\mathcal{H}^d, \mathcal{H}^{k'}), W_{lh} \in \mathcal{L}(\mathcal{H}^{k'}, \mathcal{H}^d),\\
&\quad\quad W_{1l} \in \mathcal{L}(\mathcal{H}^d, \mathcal{H}^q), b_{1l} \in \mathcal{H}^q, W_{2l} \in \mathcal{L}(\mathcal{H}^q, \mathcal{H}^d), b_{2l} \in \mathcal{H}^d,\\
&\text{Input } : y_0 = x_0 \in \mathcal{H}^{Dd},\\
&1 \le l \le L :
\end{aligned}
$$

$$
\begin{aligned}
&\text{Input at } l: \ x_l = y_{l-1} \in \mathcal{H}^{Dd},\\
&\quad 1 \le i \le D: \ u_{li} = \text{normalized}(x_{li}),\\
&\quad 1 \le i \le D: \ \tilde{u}_{li} = \tfrac{1}{\sqrt{H}} \textstyle\sum_{h=1}^{H} W_{lh} \sum_{j=1}^{i} (\text{attention}_{ij}(Q_{lh}^{(1)}, K_{lh}^{(1)}, u_l)\\
&\qquad\qquad -\chi \text{attention}_{ij}(Q_{lh}^{(2)}, K_{lh}^{(2)}, u_l)) V_{lh} u_{lj},\\
&\quad 1 \le i \le D: \ \hat{u}_{li} = x_{li} + \tilde{u}_{li},\\
&\quad 1 \le i \le D: \ \hat{y}_{li} = \text{normalized}(\hat{u}_{li}),\\
&\quad 1 \le i \le D: \ \tilde{y}_{li} = W_{2l} r_l (W_{1l} \hat{y}_{li} + b_{1l}) + b_{2l},\\
&\quad 1 \le i \le D: \ y_{li} = \hat{u}_{li} + \tilde{y}_{li},\\
&\text{Output at } l: y_l,\\
&\text{Return } y_L \in \mathcal{H}^{Dd}
\end{aligned}
\tag{2.33}
$$

where

- normalized $: \mathcal{O}_{l,nn} : \mathcal{H}^d \to \mathcal{H}^d$
- for $1 \le i, j \le D$, the interaction between entry i and the other entries is captured by $\text{attention}_{ij} : \mathcal{L}(\mathcal{H}^d, \mathcal{H}^k)^2 \times \mathcal{H}^{Dd} \to [0,1]$. The map $\mathcal{O}_{l,datt}$ is the $H-$head (masked) difference attention mechanism with $1 \le i \le D$, and $\mathcal{O}_{l,datt,i} : \mathcal{H}^{Dd} \ni x_l \mapsto \frac{1}{\sqrt{H}} \sum_{h=1}^{H} W_{lh} \sum_{j=1}^{i} (\text{attention}_{ij}(Q_{lh}^{(1)}, K_{lh}^{(1)}, x_l) - \chi \text{attention}_{ij}(Q_{lh}^{(2)}, K_{lh}^{(2)}, x_l)) V_{lh} x_{lj} \in \mathcal{H}^d$.
- activation $r_l : \mathcal{H}_l \to \mathcal{H}_l$. The map $\mathcal{O}_{l,ff} : \mathcal{H}^d \ni x \mapsto W_{2l} r_l (W_{1l} \hat{y}_{li} + b_{1l}) + b_{2l} \in \mathcal{H}^d$ is a 2-layered feedforward operator.

The output of the full transformer with difference attention is $y_L = \hat{\mathcal{O}}_L \circ \hat{\mathcal{O}}_{L-1} \circ \ldots \circ \hat{\mathcal{O}}_1(x_0)$ with the full operator at layer l given by

$$\hat{\mathcal{O}}_l = (Id + \mathcal{O}_{l,ff} \circ \mathcal{O}_{l,nn}) \circ (Id + \mathcal{O}_{l,att} \circ \mathcal{O}_{l,nn}).$$

where $Id(x) = x$ is the identity operator. The internal steps can be summarized as follows:

$$
\begin{aligned}
&Parameters: L, D, d, k, k', q, H \in \mathbb{N}\\
&\quad 1 \le l \le L:\\
&\qquad 1 \le h \le H: Q_{lh}, K_{lh}, \in \mathcal{L}(\mathcal{H}^d, \mathcal{H}^k), V_{lh} \in \mathcal{L}(\mathcal{H}^d, \mathcal{H}^{k'}), W_{lh} \in \mathcal{L}(\mathcal{H}^{k'}, \mathcal{H}^d),\\
&\qquad W_{1l} \in \mathcal{L}(\mathcal{H}^d, \mathcal{H}^q), b_{1l} \in \mathcal{H}^q, W_{2l} \in \mathcal{L}(\mathcal{H}^q, \mathcal{H}^d), b_{2l} \in \mathcal{H}^d,\\
&\text{Input } : y_0 = x_0 \in \mathcal{H}^{Dd},\\
&\text{for } l \in \{1, \ldots, L\}:\\
&\quad y_l = (Id + \mathcal{O}_{l,ff} \circ \mathcal{O}_{l,nn}) \circ (Id + \mathcal{O}_{l,datt} \circ \mathcal{O}_{l,nn})(y_{l-1}),\\
&\text{Return } y_L \in \mathcal{H}^{Dd}
\end{aligned}
\tag{2.34}
$$

2.8.1 Difference BG Transformer

The difference transformer modifies the Boltzmann-Gibbs self-attention into a difference of Boltzmann-Gibbs with two different pairs (Q_1, K_1) (Q_2, K_2) keeping the values and the weights matrices. This difference of Boltzmann-Gibbs is a weighted difference by a real number χ.

$$
\begin{aligned}
&\text{Parameters } : L, D, d, k, k', q, H \in \mathbb{N}, \chi \in \mathbb{R} \\
&\quad 1 \leq l \leq L : \\
&\quad\quad 1 \leq h \leq H : Q_{lh}^{(1)}, K_{lh}^{(1)}, Q_{lh}^{(2)}, K_{lh}^{(2)} \in \mathcal{L}(\mathcal{H}^d, \mathcal{H}^k), \\
&\quad\quad\quad V_{lh} \in \mathcal{L}(\mathcal{H}^d, \mathcal{H}^{k'}), W_{lh} \in \mathcal{L}(\mathcal{H}^{k'}, \mathcal{H}^d), \\
&\quad\quad W_{1l} \in \mathcal{L}(\mathcal{H}^d, \mathcal{H}^q), b_{1l} \in \mathcal{H}^q, W_{2l} \in \mathcal{L}(\mathcal{H}^q, \mathcal{H}^d), b_{2l} \in \mathcal{H}^d, \\
&\text{Input } : y_0 = x_0 \in \mathcal{H}^{Dd}, \\
&1 \leq l \leq L : \\
&\quad \text{Input at } l : \; x_l = y_{l-1} \in \mathcal{H}^{Dd}, \\
&\quad 1 \leq i \leq D : \; u_{li} = \tfrac{x_{li}}{1 + \|x_{li}\|}, \\
&\quad 1 \leq i \leq D : \; O_{i,lh} = \left(\frac{e^{\frac{1}{\sqrt{k}} \langle Q_{lh}^{(1)} u_{li}, K_{lh}^{(1)} u_{lj} \rangle}}{\sum_{j'}^{i} e^{\frac{1}{\sqrt{k}} \langle Q_{lh}^{(1)} u_{li}, K_{lh}^{(1)} u_{lj'} \rangle}} - \chi \frac{e^{\frac{1}{\sqrt{k}} \langle Q_{lh}^{(2)} u_{li}, K_{lh}^{(2)} u_{lj} \rangle}}{\sum_{j'}^{i} e^{\frac{1}{\sqrt{k}} \langle Q_{lh}^{(2)} u_{li}, K_{lh}^{(2)} u_{lj'} \rangle}} \right), \\
&\quad\quad \tilde{u}_{li} = \tfrac{1}{\sqrt{H}} \textstyle\sum_{h=1}^{H} W_{lh} \sum_{j=1}^{i} O_{i,lh} V_{lh} u_{lj}, \\
&\quad 1 \leq i \leq D : \; \hat{u}_{li} = x_{li} + \tilde{u}_{li}, \\
&\quad 1 \leq i \leq D : \; \hat{y}_{li} = \tfrac{\hat{u}_{li}}{1 + \|\hat{u}_{li}\|}, \\
&\quad 1 \leq i \leq D : \; \tilde{y}_{li} = W_{2l} r_l (W_{1l} \hat{y}_{li} + b_{1l}) + b_{2l}, \\
&\quad 1 \leq i \leq D : \; y_{li} = \hat{u}_{li} + \tilde{y}_{li}, \\
&\quad \text{Output at } l : y_l, \\
&\text{Return } y_L \in \mathcal{H}^{Dd}
\end{aligned}
\tag{2.35}
$$

2.8.2 Difference Sigmoid Transformer

The difference Sigmoid transformer modifies the Sigmoid self-attention into a difference of Sigmoid with two different pairs (Q_1, K_1), (Q_2, K_2) keeping the values and the weights matrices.

$$
\begin{aligned}
&\text{Parameters } : L, D, d, k, k', q, H \in \mathbb{N}, \chi \in \mathbb{R} \\
&\quad 1 \leq l \leq L : \\
&\quad\quad 1 \leq h \leq H : Q_{lh}^{(1)}, K_{lh}^{(1)}, Q_{lh}^{(2)}, K_{lh}^{(2)} \in \mathcal{L}(\mathcal{H}^d, \mathcal{H}^k), \\
&\quad\quad\quad V_{lh} \in \mathcal{L}(\mathcal{H}^d, \mathcal{H}^{k'}), W_{lh} \in \mathcal{L}(\mathcal{H}^{k'}, \mathcal{H}^d), \\
&\quad\quad W_{1l} \in \mathcal{L}(\mathcal{H}^d, \mathcal{H}^q), b_{1l} \in \mathcal{H}^q, W_{2l} \in \mathcal{L}(\mathcal{H}^q, \mathcal{H}^d), b_{2l} \in \mathcal{H}^d, \\
&\text{Input } : y_0 = x_0 \in \mathcal{H}^{Dd}, \\
&1 \leq l \leq L :
\end{aligned}
$$

$$
\begin{aligned}
&\text{Input at } l: \ x_l = y_{l-1} \in \mathcal{H}^{Dd}, \\
&\quad 1 \le i \le D: \ u_{li} = \tfrac{x_{li}}{1+\|x_{li}\|}, \\
&\quad 1 \le i \le D: \ \ O_{i,lh} = \left(\frac{e^{\frac{1}{\sqrt{k}}\langle Q^{(1)}_{lh}u_{li}, K^{(1)}_{lh}u_{lj}\rangle}}{D+e^{\frac{1}{\sqrt{k}}\langle Q^{(1)}_{lh}u_{li}, K^{(1)}_{lh}u_{lj}\rangle}} - \chi \frac{e^{\frac{1}{\sqrt{k}}\langle Q^{(2)}_{lh}u_{li}, K^{(2)}_{lh}u_{lj}\rangle}}{D+e^{\frac{1}{\sqrt{k}}\langle Q^{(2)}_{lh}u_{li}, K^{(2)}_{lh}u_{lj'}\rangle}} \right) \\
&\qquad \tilde{u}_{li} = \tfrac{1}{\sqrt{H}} \textstyle\sum_{h=1}^{H} W_{lh} \sum_{j=1}^{i} O_{i,lh} V_{lh} u_{lj}, \\
&\quad 1 \le i \le D: \ \hat{u}_{li} = x_{li} + \tilde{u}_{li}, \\
&\quad 1 \le i \le D: \ \hat{y}_{li} = \tfrac{\hat{u}_{li}}{1+\|\hat{u}_{li}\|}, \\
&\quad 1 \le i \le D: \ \tilde{y}_{li} = W_{2l} r_l (W_{1l}\hat{y}_{li} + b_{1l}) + b_{2l}, \\
&\quad 1 \le i \le D: \ y_{li} = \hat{u}_{li} + \tilde{y}_{li}, \\
&\text{Output at } l: y_l, \\
&\text{Return } y_L \in \mathcal{H}^{Dd}.
\end{aligned}
\tag{2.36}
$$

2.8.3 Sigmoid Transformer with Mixture-of-Heads and Mixture-of-Experts

For convenience we provide below the full sigmoid Transformer with mixture-of-heads and mixture-of-experts that operates one a lengthy sequence $(x_i, 1 \le i \le D)$.

$$
\begin{aligned}
&\text{Parameters } : \mathcal{E} = \{1, \ldots, E\}, D, L, d, k, k', q, H \in \mathbb{N}, \\
&\quad 1 \le l \le L: \\
&\qquad 1 \le h \le H: Q_{lh}, K_{lh} \in \mathcal{L}(\mathcal{H}^d, \mathcal{H}^k), V_{lh} \in \mathcal{L}(\mathcal{H}^d, \mathcal{H}^{k'}), W_{lh} \in \mathcal{L}(\mathcal{H}^{k'}, \mathcal{H}^d), \\
&\qquad W_{1l} \in \mathcal{L}(\mathcal{H}^d, \mathcal{H}^q), b_{1l} \in \mathcal{H}^q, W_{2l} \in \mathcal{L}(\mathcal{H}^q, \mathcal{H}^d), b_{2l} \in \mathcal{H}^d, p, \lambda \\
&\qquad W_{3l} \in \mathcal{L}(\mathcal{H}^d, \mathbb{R}^E), b_{3l} \in \mathbb{R}^E, \\
&\text{Input } : 1 \le i \le D : y_{i0} = x_{i0} = y_{i0,6} \in \mathcal{H}^d, \\
&t \in \{0, \ldots, T-1\}: \\
&\quad 1 \le l \le L: \\
&\qquad \text{Input at } l: \ 1 \le i \le D: y_{i,l-1,6} \in \mathcal{H}^d, \\
&\qquad\quad 1 \le i \le D: y_{il,1} = \tfrac{y_{i,l-1,6}}{1+\|y_{i,l-1,6}\|} =: \text{Normalized}(y_{i,l-1,6}), \\
&\qquad\quad \text{SelectH} \subseteq \{1, \ldots, H\}, p_l \in \mathbb{R}^H_+ \\
&\qquad\quad 1 \le i \le D: O_{i,lh} = \textstyle\sum_{j=1}^{i} \frac{e^{\frac{1}{\sqrt{k}}\langle Q_{lh}y_{il,1}, K_{lh}y_{jl,1}\rangle}}{D+e^{\frac{1}{\sqrt{k}}\langle Q_{lh}y_{il,1}, K_{lh}y_{jl,1}\rangle}} \\
&\qquad\quad 1 \le i \le D: y_{il,2} = \tfrac{1}{\sqrt{H}} \textstyle\sum_{h=1}^{H} p_{lh} \mathbb{I}_{h \in SelectH} W_{lh} O_{i,lh} V_{lh} y_{jl,1} \\
&\qquad\quad 1 \le i \le D: y_{il,3} = Add(y_{i,l-1,6}, y_{il,2}) = y_{i,l-1,6} + y_{il,2}, \\
&\qquad\quad 1 \le i \le D: y_{il,4} = \tfrac{y_{il,3}}{1+\|y_{il,3}\|} = \text{Normalized}(y_{il,3}),
\end{aligned}
$$

$$\begin{aligned}
&1 \leq i \leq D : \text{Choose } \mathcal{E}_i,\\
&\quad 1 \leq i \leq D : \eta_{le} = \frac{e^{(\tilde{W}_{3l} y_{il,4} + \tilde{b}_{3l})_e}}{\sum_{e' \in \mathcal{E}_i} e^{(\tilde{W}_{3l} y_{il,4} + \tilde{b}_{3l})_{e'}}} \mathbb{I}_{e \in \mathcal{E}_i},\\
&\quad 1 \leq i \leq D : e = E+1 :\ \tilde{y}_{il,e} = W_{2l,e} r_l (W_{1l,e} y_{il,4} + b_{1l,e}) + b_{2l,e},\\
&\quad 1 \leq i \leq D : 1 \leq e \leq E :\ \tilde{y}_{il,e} = W_{2l,e} r_l (W_{1l,e} y_{il,4} + b_{1l,e}) + b_{2l,e},\\
&\quad 1 \leq i \leq D : y_{il,5} = \eta_{l,E+1} \tilde{y}_{il,E+1} + \textstyle\sum_{e \in \mathcal{E}_i} \eta_{le} \tilde{y}_{il,e},\\
&1 \leq i \leq D : y_{il,6} = \text{Add}(y_{il,5}, y_{il,3}) = y_{il,5} + y_{il,3},
\end{aligned} \tag{2.37}$$

Output at l : $y_{l,6} \in \mathcal{H}^{Dd}$

Return: $y_{L,6} \in \mathcal{H}^{Dd}$,

$x_{t+1} = [(1 - \lambda_t) x_t + \lambda_t y_{L,6}]$,

Return: $x_T \in \mathcal{H}^{Dd}$

2.9 There is No Best Generative Machine Intelligence

Transformers, diffusions, and other large learning models can be used to analyze blockchain tokens, cryptocurrencies and central bank digital currencies. These generative foundational machine intelligence models can look at past data (backcasting), current data (nowcasting) and estimate future trends (forecasting) of the financial technologies (FinTech) market. By doing this, they can help spot investment opportunities, track market movements and provide valuable insights. The results can be delivered through an easy-to-use dashboard, making it accessible for investors, traders, and Fintech professionals to make informed decisions. This dashboard could highlight key trends, risks, and opportunities in almost real-time, helping users stay ahead in the market. The precision of the output from generative machine intelligence displayed in a dashboard is crucial because it directly affects the quality of the decisions users make. Inaccurate or imprecise insights can lead to poor investment choices, misinterpreting market trends, or overlooking key threads and opportunities. For investors, traders, and Fintech professionals, even small errors in machine intelligence-generated forecasts or recommendations can result in significant financial losses. Therefore, ensuring high precision not only builds trust in the machine intelligence system but also helps users confidently rely on it for accurate, actionable insights that support smarter decision-making.

Let us start with the case of backcasting. FinTech has generated vast amounts of online data, particularly in the blockchain space. The dashboard should be tested on this historical data. If the model does not perform well on past data generated by FinTech, it is unlikely that it will be trusted for identifying future opportunities. That is why backcasting is a critical first step. It is also worth noting that this is easily verifiable because the data already exists, and there is no need for a prediction window to assess its performance ex-post.

Reviewing the literature on transformer-based deep neural networks, such as masked multi-head Boltzmann-Gibbs self-attention with 2-layer feed-forward networks using ReLU, SiLU, or GeLU, or weighted Kolmogorov-Arnold networks, or diffusion-transformers (transfusion), all tend to introduce statistical bias after optimizing trillions of parameters. Transformers, with their large number of parameters, are trained on massive datasets that may contain imbalances or patterns that lead the model to favor certain predictions. Additionally, the training process involves minimizing a loss function or a variational inequality across many parameters, which can lead to overfitting or underfitting, especially when the data distribution is skewed or lacks diversity. Moreover, techniques like masked self-attention, which transformers rely on, can unintentionally emphasize certain parts of the input data, further contributing to (statistically) biased outputs. Having such a statistical bias is not a good indicator of learning accuracy. In the business world, sometimes being able to foresee a trend in advance, even without full market details, can make a big difference. However, in some areas, errors in trend prediction can lead to significant financial losses in the FinTech market, which can have global consequences. The same explanation holds for nowcasting, forecasting, forward-looking, estimation, approximation, etc.

This means that after the post-design, post-training, post-optimization, post-fine-tuning and post-self consistency check of these class of learning algorithms, most of these would lead to:

$$E[\hat{O}_L \circ \hat{O}_{L-1} \circ \ldots \circ \hat{O}_1(x_{t-1}) - (a_{t+0}, \ldots, a_{t+h-1})] \neq 0$$

where $x_{t-1} = (a_1, \ldots, a_{t-1})$ is the historical data up to $t-1$.

The connection with the mean-field-type game theory here is the variance reduction component of the algorithms.

2.9.1 Point Forecasting

We consider $A \geq 2$ interactive assets distributed across a platform of platforms blockchain companies, crypto-tokens and Central Bank Digital Currencies. In practice, the number of assets of top investors is around $A = 30$. For each asset i we display a dashboard on trends and indicators. An example of price dynamics for asset o is given by

$$\begin{cases} p_{o,0} > 0, \ o \in \{1, 2, \ldots, A\}, \\ p_{o,t+1} = p_{o,t}(1 + f_{o,t}), \ o \in \{1, 2, \ldots, A\}, \\ f_{o,t} = (d_{o,t}(r'_t, n_{o,t}) - s_{o,t}(r'_t, n_{o,t}))\delta + \eta_{o,t+1} \end{cases} \tag{2.38}$$

where d_o is the demand (usage) of o in the market, s_o is the supply, δ is a time-scale and $\eta_{o,t+1}$ is a combination of noise-differences times $\sqrt{\delta}$. The interaction between the A assets is through the demand-supply which depends on the amount invested by the active decision-makers. We choose d_o, s_o to be functions of the total number of active decision-makers $n_{o,t} := \sum_{j=1}^{I} \mathbb{I}_{a_{j,o,t}>0}$, where r'_t is a regime switching process over a finite set S. The action $a_{j,o,t}$ is the number of unit tokens o held by decision-maker j at time t. $a_{j,o,t} > 0$ means that j is active on token o at time t.

The evolution of return is given by

$$\begin{cases} w_{i,0} > 0, \ i \in \{1, 2, \ldots, I\}, \\ w_{i,t+1} = w_{i,t} \\ +(d_{1,t} - s_{1,t})\delta(w_{i,t} - \sum_{o=2}^{A} p_{o,t}a_o) + (w_{i,t} - \sum_{o=2}^{A} p_{o,t}a_o)\eta_{1,t+1} \\ +\sum_{o=2}^{A} p_{o,t}(d_{o,t} - s_{o,t})\delta a_o + \sum_{o=2}^{A} p_{o,t}a_o\eta_{o,t+1}. \end{cases} \tag{2.39}$$

The returns are correlated via the prices, which in turn are correlated via demand-supply.

Consider the system (2.39) and associated with it a vector of quantities-of-interest which are for example,

- the wealth, $w_{i,T}$ the variance $var(w_{i,T})$, the trends such as the (weekly, monthly, quarterly, yearly) moving averages

$$\frac{1}{M}\sum_{k=1}^{M} w_{i,T-k}, \ M \in \{7, 30, 90, 365\}$$

and the (de-)growth rates $\frac{w_{i,T}}{w_{i,T-1}}, \frac{w_{i,T}}{w_{i,T-730}}, \frac{w_{i,T}}{w_{i,T-365}}, \frac{w_{i,T-365}}{w_{i,T-730}}$ for each decision-maker i.

- The o-th asset wealth $w_{i,o,T}$, the variance $var(w_{i,o,T})$, the trends such as the (weekly, monthly, quarterly, yearly) moving averages

$$\frac{1}{M}\sum_{k=1}^{M} w_{i,o,T-k}, \quad M \in \{7, 30, 90, 365\}$$

and the (de-)growth rates $\frac{w_{i,o,T}}{w_{i,o,T-1}}, \frac{w_{i,o,T}}{w_{i,o,T-730}}, \frac{w_{i,o,T}}{w_{i,o,T-365}}, \frac{w_{i,o,T-365}}{w_{i,o,T-730}}$ per asset o for decision-maker i.

- The market price $p_{o,T}$, the variance price $var(p_{o,T})$, the (weekly, monthly, quarterly, yearly) moving averages

$$\frac{1}{M}\sum_{k=1}^{M} p_{o,T-k}, \quad M \in \{7, 30, 90, 365\}$$

and the (de-)growth rates

$$\log(\frac{p_{o,T}}{p_{o,T-1}}), \log(\frac{p_{o,T}}{p_{o,T-730}}), \log(\frac{p_{o,T}}{p_{o,T-365}}), \log(\frac{p_{o,T-365}}{p_{o,T-730}})$$

for each asset o.

We collect all these quantities-of-interest and obtain an element of $\mathbb{R}^D$ where D, the total of number of these quantities-of-interest to be displayed in the dashboard of the decision-makers. These quantities of interest are quantized into a set V which is a non-empty and finite set. We then consider A historical sequences $(r_{1,k}, \ldots, r_{A,k})_{1 \le k \le t-1} \in \mathbb{R}^{(t-1)AD}$ for $t \ge 2$. The realized result $r_{o,t}$ of token o at time t takes values in a non-empty and finite set V of $\mathbb{R}^D$. An example of a component of V could be, for example,

- increase by 10+ %,
- decrease by 2 % or
- no change from the previous time.

A well-trained financial technology generative artificial intelligence is used to forecast the next $h \ge 1$ trends of the market: h-ahead forecast of the vector $(r_{1,t+j}, \ldots, r_{A,t+j})_{0 \le j \le h-1}$ given $(r_{1,k}, \ldots, r_{A,k})_{k \le t-1}$ the past realization up to $t-1$. The generative machine intelligence can be a top deep learning, transformer-based, diffusion-based, generative adversarial network or any other considered as state-of-the-art algorithm for processing sequences, time series, tabular, sheets, etc. The output of the h-ahead forecast by generative machine intelligence is denoted by

$$(f_{1,t+j}, \ldots, f_{A,t+j})_{0 \le j \le h-1}.$$

The transformer-based generated output is

$$\hat{\mathcal{O}}_L \circ \hat{\mathcal{O}}_{L-1} \circ \ldots \circ \hat{\mathcal{O}}_1(h_{t-1})$$

with the operator at layer l given by

$$\hat{\mathcal{O}}_l = (Id + \mathcal{O}_{l,ff} \circ \mathcal{O}_{l,nn}) \circ (Id + \mathcal{O}_{l,att} \circ \mathcal{O}_{l,nn}),$$

with

- L is the total number of layers. $1 \le l \le L$ denotes the l-th layer of the neural network.

- Input space: $\mathcal{H}_0 := \mathbb{R}^{(t-1)AD}$ where $(t-1)$ is the size of the historical data window length. The input is a three-entry tensor with time index, asset index and trend index.
- $\mathcal{O}_{l,nn}$: is the normalization operator followed by a linear operator. The normalization part could be $\frac{z}{1+\|z\|}$ or $proj_{C_2} \circ proj_{C_1}(z)$ where $C_1 = \{w \in \mathbb{R}^k,\ \langle w, \mathbb{1}\rangle = 0\}$ and $C_2 = \{w \in \mathbb{R}^k,\ \langle w, w\rangle \leq \sqrt{D-1}\}$.
- $\mathcal{O}_{l,att}$ is the masked self-attention mechanism. It is the part where interactions between the historical data of the token occurs. An example of masked self-attentions include Boltzmann-Gibbs (applied to each row of each same token), sigmoid (entry-wise), etc.
- $\mathcal{O}_{l,ff}$ is a (two-layered) feedforward neural network such as $W_{2l} r_l(W_{1l}x + b_{1l}) + b_{2l}$ with r_l being $ReLU, GeLU, SiLU$ etc. And $(W_{2l}, b_{2l}, W_{1l}, b_{1l})$ are bounded linear weights and bias. One can also replace it by a properly designed weighted Kolmogorov-Arnold network with rational polynomial basis.
- The output at layer L is extracted to be in the form

$$(f_{1,t+j'}, \ldots, f_{A,t+j'})_{0 \leq j' \leq h-1} \in \mathbb{R}^{hAD} =: \mathcal{H}_L$$

  where h is the forecasting window. When $h = 0$ it corresponds to nowcasting. The change of dimension can be obtained from a linear bounded operator $W_* : \mathbb{R}^{(t-1)AD} \to \mathbb{R}^{hAD}$. It could also a final masked softmax operator applied to the vector.

Definition 20. *The generative machine intelligence is given by*

$$(L, t, A, D, \hat{\mathcal{O}}_L \circ \hat{\mathcal{O}}_{L-1} \circ \ldots \circ \hat{\mathcal{O}}_1, \mathbb{R}^{(t-1)AD}, \mathbb{R}^{hAD}).$$

We define the squared error of the generative AI-based (back/now/fore)-casting evaluated ex-post is as

$$Err_{h,t}(f|r) = \frac{1}{A} \sum_{1 \leq o \leq A} \frac{1}{t_o - h} \sum_{1 \leq t' \leq t_o - h} \frac{1}{hD} \sum_{0 \leq h' \leq h-1} \mathbb{E} d^2(f_{o,t'+h'}, r_{o,t'+h'}),$$

where $d = d_2$ is the distance induced by the 2−norm of $\mathbb{R}^D$.

This h-ahead averaged forecasting error is not scaled by data magnitude but normalized by the time horizon length, forecasting horizon length, number of assets and number of forecasting trend indicators. $Err_{h,t}(f|r)$ will be used as the performance evaluation criteria of the algorithm. The forecasting f is considered as good when r is not too small but $Err_{h,t}(f|r)$ is small enough. We have chosen a different updating time. Each asset has its own clock t_o which is updated at time of the day/night. We work with the available data (the number of times it has been updated but not the number of seconds since the beginning). To build f we are allowed to convexify the domain V which arises naturally when the outcomes are seen and one uses averaging or frequency of occurences.

Problem 6. *Consider a well-trained generative machine intelligence which takes an historical data as input* $(r_{1,k}, \ldots, r_{A,k})_{1 \leq k \leq t-1} \in \mathbb{R}^{(t-1)AD}$*, operates over* L *layers and provides an output*

$$(f_{1,t+j'}, \ldots, f_{A,t+j'})_{0 \leq j' \leq h-1} \in \mathbb{R}^{hAD}.$$

Define the now/fore-casting error as

$$Err_{h,t}((f_{1,t+j'}, \ldots, f_{A,t+j'})_{0 \leq j' \leq h-1} | (r_{1,k}, \ldots, r_{A,k})_{1 \leq k \leq t-1}).$$

Can we design a new learning algorithm based on the input, the output and that does better than f *in terms of error?*

We show that the answer to this question is positive. This is an important result as it shows the limitation of current generative machine intelligence architectures in pointwise back/now/fore-casting. The forecasting here is point-wise related. And it is clear that point-wise forecasting is too much asking to an approximation method such as generative machine intelligence foundational learning models. This is one of the reasons why one has attempted to focus on weaker version such as probability-based, cumulative function-based, interval-based, localized or aggregated quantities-of-interest. Typically, given the historical data, forecasting a mean (expected value) is easier than forecasting a (point) value that will be realized. The limitation is also coming from the fact that the approximation tool used during the training yields an optimal weight and bias that are dependent on input which is not a constant. A constant weight and bias obtained at the end of the training process. This constant weight and bias is therefore unrelated to the specific historical data in hand. The conditional expectation however depends strongly on the nature of the historical data.

Next we explain in an explicit way how we can do better than the best generative machine intelligence.

2.9.2 Doing Better Than the Existing Best Generative Intelligence

As $\mathbb{E}X^2 = var(X) + (\mathbb{E}X)^2$ one can decompose $Err_{h,t}(f|r)$ into two parts. To do so we introduce the frequency of realization at entry z of asset i. $\nu_{r,t_o}(z) = \frac{1}{t_o}\sum_{1\leq t\leq t_o} \mathbb{I}_{r_{o,t}=z}$. One can write the time-average realization up to t_o as $\bar{r}_{o,t_o} = \frac{1}{t_o}\sum_{1\leq t\leq t_o} r_{o,t} = \sum_{z\in V} z(\frac{1}{t_o}\sum_{1\leq t\leq t_o}\mathbb{I}_{r_{o,t}=z}) = \mathbb{E}_{\nu_{r,t_o}} r_o$. We do this operation for all the A assets available on the platform.

The following real-valued random variables

$$d_2(f_{o,t}, \bar{f}_{o,t}), d_2(\bar{f}_{o,t}, \bar{r}_{o,t}), d_2(\bar{r}_{o,t}, r_{o,t})$$

are all within two times the diameter of V.

$$\begin{aligned} Err_{h,t}(f|r) &= \tfrac{1}{D}\mathbb{E}\sum_{(o,t',h')} \lambda_{o,t',h'} d^2(f_{o,t'+h'}, r_{o,t'+h'}) \\ &= \tfrac{1}{D}\mathbb{E}\sum_{(o,t',h')}\sum_{z\in V} \lambda_{o,t',h'} d^2(z, r_{o,t'+h'})\mathbb{I}_{f_{o,t'+h'}=z}, \end{aligned} \tag{2.40}$$

where $\lambda_{o,t',h'} = \frac{1}{A(t_o-h)h} > 0$.

The idea here is to write this expression in terms of the second moment of a conditional random variable. We write

$$\mathbb{E}(X^2) = \mathbb{E}(\mathbb{E}(X^2|f)) = \mathbb{E}(\text{var}(X|f)) + \mathbb{E}\{(\mathbb{E}(X|f))^2\}.$$

The idea now is to build another mechanism g that does better than f for the performance $Err_{h,t}(f|r)$. Let $\mathcal{F}_t$ be the filtration generated by the union of events $\{r_{t'}, t' \leq t\}$. The function $g \mapsto \mathbb{E}\sum_{1\leq o\leq A, 1\leq t'\leq t} d^2(g_o, r_{o,t'})$ is minimized at the conditional expectation $g_{o,t} = \mathbb{E}(r_{o,t}|\ \mathcal{F}_{t-1})$. The next h steps ahead forecast is $\mathbb{E}((r_{o,t}, \ldots, r_{o,t+h-1})|\ \mathcal{F}_{t-1})$. This estimation is unbiased. However, the output of the generative machine intelligence is an estimated version of $(g_{o,t}, \ldots, g_{o,t+h-1})$ and not the exact conditional expectation. This leads to an estimation error or approximation for the well-trained generative machine intelligence f.

Let us replace the conditional expectation by a very simple mechanism which averages of the previous actions.

$$g_{o,t}(z) = \frac{\sum_{1\leq t'\leq t} \lambda_{o,t'} r_{o,t'} \mathbb{I}_{f_{o,t'}=z}}{\sum_{1\leq t''\leq t} \lambda_{o,t''} \mathbb{I}_{f_{o,t''}=z}}.$$

Proposition 18. *We have* $Err_{h,t}(g|r) \leq Err_{h,t}(f|r) - e_t^2$ *where* e_t *goes zero as* t *goes to infinity.*

This means in particular that for every generative machine intelligence output, there is another simple algorithm, one that performs strictly better in terms of accuracy as long as the term $\mathbb{E}(X|f) \neq 0$. This answers positively to the question of Problem 6. The strategy g_t uses $\{(r_{t'}, f_{t'}), t' \leq t-h\}$ but does not use the past $\{g_{t'}, t' \leq t-2\}$. One can also build a sequence that does not use f and that does better than f. It suffices to replace $\{f_{t'}, t' \leq t-h\}$ by $\{r_{t'}, t' \leq t-h\}$ in the definition of g. This also says that generative machine intelligence combined with reinforcement learning can do better than generative machine intelligence alone. We prove Proposition 18 step-by-step. The algorithm induced by g tracks the entire average. We write it an iterative version so that the next average is obtained by a simple convex combination between the new data and the previous average. Indeed, the weighted average

$$m_t = \frac{\sum_{1\leq t'\leq t} \lambda_{t'} r_{t'} \mathbb{I}_{f_{t'}=z}}{\sum_{1\leq t'\leq t} \lambda_{t'} \mathbb{I}_{f_{t'}=z}}$$

with $\sum_{1\leq t'\leq t} \lambda_{t'} \mathbb{I}_{f_{t'}=z} > 0$ and $\lambda_{t'} \geq 0$ for every t', can be rewritten as

$$m_t = m_{t-1} + \frac{\lambda_t \mathbb{I}_{f_t=z}}{\sum_{1\leq t'\leq t} \lambda_{t'} \mathbb{I}_{f_{t'}=z}} (r_t - m_{t-1}).$$

This recursive equation is crucial in online update. As we can see, the latter $m_t(z)$ is $g_t(z)$. The recursive equation for m is made as an online update as the next average is obtained from the previous update and the new data. We ask whether a similar online update exist for the error induced by g. Let

$$\begin{aligned}
\tilde{m}_{2,t} &= \textstyle\sum_{1\leq t'\leq t} \lambda_{t'} \mathbb{I}_{f_{t'}=z} d^2(r_{t'}, m_t(z)) \\
&= \textstyle\sum_{1\leq t'\leq t} \lambda_{t'} \mathbb{I}_{f_{t'}=z} d^2(r_{t'}, m_{t-1} + \frac{\lambda_t \mathbb{I}_{f_t=z}}{\sum_{1\leq t'\leq t} \lambda_{t'} \mathbb{I}_{f_{t'}=z}} (r_t - m_{t-1})) \\
&= \textstyle\sum_{1\leq t'\leq t} \lambda_{t'} \mathbb{I}_{f_{t'}=z} d^2(r_{t'}, m_{t-1}) \\
&\quad + \textstyle\sum_{1\leq t'\leq t} \lambda_{t'} \mathbb{I}_{f_{t'}=z} d^2(0, \frac{\lambda_t \mathbb{I}_{f_t=z}}{\sum_{1\leq t'\leq t} \lambda_{t'} \mathbb{I}_{f_{t'}=z}} (r_t - m_{t-1})) \\
&\quad -2\langle \textstyle\sum_{1\leq t'\leq t} \lambda_{t'} \mathbb{I}_{f_{t'}=z} r_{t'}, \frac{\lambda_t \mathbb{I}_{f_t=z}}{\sum_{1\leq t''\leq t} \lambda_{t''} \mathbb{I}_{f_{t''}=z}} (r_t - m_{t-1})) \rangle \\
&= \tilde{m}_{2,t-1} + \lambda_t \mathbb{I}_{f_t=z} d^2(r_t, m_{t-1}) \\
&\quad + \textstyle\sum_{1\leq t'\leq t} \lambda_{t'} \mathbb{I}_{f_{t'}=z} d^2(0, \frac{\lambda_t \mathbb{I}_{f_t=z}}{\sum_{1\leq t''\leq t} \lambda_{t''} \mathbb{I}_{f_{t''}=z}} (r_t - m_{t-1})) \\
&\quad -2\langle m_{t-1} + \frac{\lambda_t \mathbb{I}_{f_t=z}}{\sum_{1\leq t'\leq t} \lambda_{t'} \mathbb{I}_{f_{t'}=z}} (r_t - m_{t-1}), \lambda_t \mathbb{I}_{f_t=z} (r_t - m_{t-1}) \rangle \\
&= \tilde{m}_{2,t-1} + \lambda_t \mathbb{I}_{f_t=z} (1 - \frac{\lambda_t \mathbb{I}_{f_t=z}}{\sum_{1\leq t'\leq t} \lambda_{t'} \mathbb{I}_{f_{t'}=z}}) d^2(r_t, m_{t-1}) \\
&\quad -2\lambda_t \mathbb{I}_{f_t=z} \langle m_{t-1}, (r_t - m_{t-1}) \rangle.
\end{aligned} \tag{2.41}$$

Hence,

$$\begin{aligned}
\tilde{m}_{2,t} &= \textstyle\sum_{1\leq t'\leq t} \lambda_{t'} \mathbb{I}_{f_{t'}=z} (1 - \frac{\lambda_{t'} \mathbb{I}_{f_{t'}=z}}{\sum_{1\leq t''\leq t'} \lambda_{t''} \mathbb{I}_{f_{t''}=z}}) d^2(r_{t'}, m_{t'-1}) \\
&\quad -2 \textstyle\sum_{1\leq t'\leq t} \lambda_{t'} \mathbb{I}_{f_{t'}=z} \langle m_{t'-1}, r_{t'} \rangle + 2 \sum_{1\leq t'\leq t} \lambda_{t'} \mathbb{I}_{f_{t'}=z} d^2(m_{t'-1}, 0).
\end{aligned} \tag{2.42}$$

Thus,

$$\begin{aligned}
m_{2,t} &= \frac{\tilde{m}_{2,t}}{\sum_{1\le t''\le t}\lambda_{t''}\mathbb{I}_{f_{t''}=z}} \\
&= \frac{\sum_{1\le t'\le t}\lambda_{t'}\mathbb{I}_{f_{t'}=z}d^2(r_{t'},m_{t'-1})}{\sum_{1\le t''\le t}\lambda_{t''}\mathbb{I}_{f_{t''}=z}} \\
&= -\frac{\sum_{1\le t'\le t}\frac{(\lambda_{t'}\mathbb{I}_{f_{t'}=z})^2}{\sum_{1\le t''\le t'}\lambda_{t''}\mathbb{I}_{f_{t''}=z}}d^2(r_{t'},m_{t'-1})}{\sum_{1\le t''\le t}\lambda_{t''}\mathbb{I}_{f_{t''}=z}} \\
&+\frac{(-2\sum_{1\le t'\le t}\lambda_{t'}\mathbb{I}_{f_{t'}=z}\langle m_{t'-1},r_{t'}\rangle+2\sum_{1\le t'\le t}\lambda_{t'}\mathbb{I}_{f_{t'}=z}d^2(m_{t'-1},0))}{\sum_{1\le t''\le t}\lambda_{t''}\mathbb{I}_{f_{t''}=z}}.
\end{aligned} \tag{2.43}$$

The extra error made by working with online computation and the error is given by

$$\begin{aligned}
&\frac{\sum_{1\le t'\le t}\lambda_{t'}\mathbb{I}_{f_{t'}=z}d^2(r_{t'},m_{t'-1})}{\sum_{1\le t''\le t}\lambda_{t''}\mathbb{I}_{f_{t''}=z}} - \frac{\sum_{1\le t'\le t}\lambda_{t'}\mathbb{I}_{f_{t'}=z}d^2(r_{t'},m_{t})}{\sum_{1\le t''\le t}\lambda_{t''}\mathbb{I}_{f_{t''}=z}} \\
&= \frac{\sum_{1\le t'\le t}\frac{(\lambda_{t'}\mathbb{I}_{f_{t'}=z})^2}{\sum_{1\le t''\le t'}\lambda_{t''}\mathbb{I}_{f_{t''}=z}}d^2(r_{t'},m_{t'-1})}{\sum_{1\le t''\le t}\lambda_{t''}\mathbb{I}_{f_{t''}=z}} \\
&+\frac{(2\sum_{1\le t'\le t}\lambda_{t'}\mathbb{I}_{f_{t'}=z}\langle m_{t'-1},r_{t'}\rangle-2\sum_{1\le t'\le t}\lambda_{t'}\mathbb{I}_{f_{t'}=z}d^2(m_{t'-1},0))}{\sum_{1\le t''\le t}\lambda_{t''}\mathbb{I}_{f_{t''}=z}}.
\end{aligned} \tag{2.44}$$

Using the diameter and the convex combination in V we arrive at

$$\begin{aligned}
&\frac{\sum_{1\le t'\le t}\lambda_{t'}\mathbb{I}_{f_{t'}=z}d^2(r_{t'},m_{t'-1})}{\sum_{1\le t''\le t}\lambda_{t''}\mathbb{I}_{f_{t''}=z}} - \frac{\sum_{1\le t'\le t}\lambda_{t'}\mathbb{I}_{f_{t'}=z}d^2(r_{t'},m_{t})}{\sum_{1\le t''\le t}\lambda_{t''}\mathbb{I}_{f_{t''}=z}} \\
&\le \frac{1+\log(\sum_{1\le t'\le t}\lambda_{t'}\mathbb{I}_{f_{t'}=z})}{\sum_{1\le t''\le t}\lambda_{t''}\mathbb{I}_{f_{t''}=z}}4\ \textit{diameter}^2(V) \\
&+\frac{(2\sum_{1\le t'\le t}\lambda_{t'}\mathbb{I}_{f_{t'}=z}\langle m_{t'-1},r_{t'}\rangle-2\sum_{1\le t'\le t}\lambda_{t'}\mathbb{I}_{f_{t'}=z}d^2(m_{t'-1},0))}{\sum_{1\le t''\le t}\lambda_{t''}\mathbb{I}_{f_{t''}=z}}.
\end{aligned} \tag{2.45}$$

Based on the above calculations it follows that

$$\begin{aligned}
&0 \le Err_{h,t}(f|r) - Err_{h,t}(g|r) \\
&= \tfrac{1}{D}\mathbb{E}\sum_{(o,t',h')}\lambda_{o,t',h'}(d^2(f_{o,t'+h'},r_{i,t'+h'}) - d^2(g_{o,t'+h'},r_{o,t'+h'})) \\
&\le \mathbb{E}\frac{1+\log(\sum_{1\le t'\le t}\lambda_{t'}\mathbb{I}_{f_{t'}=z})}{\sum_{1\le t''\le t}\lambda_{t''}\mathbb{I}_{f_{t''}=z}}4\text{diameter}^2(V).
\end{aligned} \tag{2.46}$$

□

Why is Generative Machine Intelligence alone very limited for pointwise forecasting?

In order to understand the limitation of generative machine intelligence, let us consider a very simple and well-known example which is the linear regression of $y_j = Wx_j + b + n_j$ where (W, b) are fixed and $\mathbb{E}n_j = 0$. We sample $(t-1)$ times and obtain a sequence (x_j, y_j) from that equation for $1 \le j \le t-1$. This sequence is sent to a top generative machine intelligence. The goal is to reduce the global error. The coefficient b is easily obtained as $b = \mathbb{E}y - W\mathbb{E}x$ which clearly depends on the distribution via the expected values. From the relation $\mathbb{E}(y - \mathbb{E}y)(x - \mathbb{E}x)^\dagger = W\mathbb{E}(x - \mathbb{E}x)(x - \mathbb{E}x)^\dagger$ we obtain that W is also a function of the distribution of (x, y) via the covariance and the variance term. This means the optimal weight/bias that minimizes the error is dependent on the moments. They are not distribution-independent. The fixed weight and bias in the architecture of generative machine intelligence can approximate but are not optimal even for a basic linear regression problem:

- Pointwise matching: Based on this reasoning, we can see that the equation $W_L x_L + b_L = y_L + f_L(y_L)$ cannot be satisfied with all sequences (x_L, y_L) with a constant (W_L, b_L), where f_L is such that the activation function at L is $r_L = (Id + \partial f_L)^{-1}$, where ∂f_L is the sub-differential operator of f_L. The optimal choice (W_L, b_L) to match $W_L x_L + b_L = y_L + f_L(y_L)$ is possible only when we allow (W_L, b_L) to be dependent on (x_L, y_L).

- Data set based average matching: If we work with a given sequence, we get

$$b_L = \bar{y}_L + \bar{f}_L(y_L) - W_L \bar{x}_L,$$

and

$$W_L(x_L - \bar{x}_L)(x_L - \bar{x}_L)^\dagger$$
$$= (y_L - \bar{y}_L)(x_L - \bar{x}_L)^\dagger + (f_L(y_L) - \bar{f}_L(y_L))(x_L - \bar{x}_L)^\dagger,$$

where $\bar{a} = \frac{1}{t}\sum_{t'=1}^t a_{t'}$. If x_L is zero, there is no W_L needed and b_L should match $y_L + f_L(y_L)$. Clearly, the optimal weight/bias are sequence-dependent via the empirical average and the empirical covariance and variance.

- And in the stochastic case, given the distribution of (y_L, x_L), the optimal bias is

$$b_L = \mathbb{E}y_L + \mathbb{E}f_L(y_L) - W_L \mathbb{E}x_L$$

and

$$W_L \mathbb{E}(x_L - \mathbb{E}x_L)(x_L - \mathbb{E}x_L)^\dagger$$
$$= \mathbb{E}(y_L - \mathbb{E}y_L)(x_L - \mathbb{E}x_L)^\dagger + \mathbb{E}(f_L(y_L) - \mathbb{E}f_L(y_L))(x_L - \mathbb{E}x_L)^\dagger.$$

By taking the inverse of the matrix $\mathbb{E}(x_L - \mathbb{E}x_L)(x_L - \mathbb{E}x_L)^\dagger$ we obtain the optimal weight which again involves the second moment of x_L, the covariance (y_L, x_L) and the covariance $(f_L(y_L), x_L)$. We conclude the optimal training weight/bias are data-dependent in the finite regime and moment/covariance-dependent in the long-run. Restricting the design to constant (data-independent, distribution-independent) coefficients lead therefore to suboptimality. In practice however, we may not need the optimal weight/bias to perform relatively well in some specific tasks.

Observe that the time-average performs better than any other approximation tool such as the current golden standard generative machine intelligence is because the average $\frac{1}{t}\sum_{t'=1}^t a_{t'}$ is the argmin of the cumulative pointwise squared error $x \mapsto \sum_{t'=1}^t d_2^2(x, a_{t'})$. The fact that the weighted time-average performs better than the current golden standard generative machine intelligence in pointwise forecasting is because the average $\frac{\sum_{t'=1}^t \lambda_{t'} a_{t'}}{\sum_{t'=1}^t \lambda_{t'}}$ is the argmin of the cumulative pointwise squared error $x \mapsto \sum_{t'=1}^t \lambda_{t'} d_2^2(x, a_{t'})$. Note that if we change d_2 or the 2-norm to another one such as p-norm, $p \geq 1, p \neq 2$, the argmin is changed, time-average is not the argmin anymore.

We have examined the use of generative machine intelligence in pointwise back/now/forecasting algorithms. We have shown that all algorithms in this category are limited in terms of statistical bias and approximation bias. This bias creates in turn a strong limitation for reducing the global accuracy. As a consequence, one can always find another simple averaging algorithm that does better than generative machine intelligence in terms of accuracy. Of course, accuracy is not the only key performance metric in generative machine intelligence and pointwise forecasting is not the only golden standard approach. However, in the FinTech area where important decisions are made by central banks, regulators, investors, insurers, traders and users based on generative machine intelligence output displayed on dashboards, the accuracy of these algorithms needs to be checked and improved with knowledge such as the one proposed here.

2.10 Constant Weight/Bias are Suboptimal

While current transformer models with fixed parameters is promising with respect to the state-of-the-art performance in many domains, this constant parameterization poses challenges for generalization, particularly when dealing with evolving, diverse, or unseen data. Addressing these issues by making parameters more dynamic and adaptive to the input or underlying data distribution could lead to more flexible, robust models better equipped for real-world applications.

Let us recall what a well-trained transformer with fixed parameters looks like. In the standard implementation of transformers, the key components such as query Q, key K, value V, weights W, and bias b, are typically fixed after training. These parameters function within the transformer architecture as follows:

Query, Key, and Value

For each token in the input sequence, a linear transformation is applied to compute the query Q, key K, and value V matrices. These matrices represent different projections of the input data. These are fixed learned parameters that stay constant during inference, computed during the training process. Once the training is complete, they no longer adapt based on new input.

Weights W

The weights of the linear layers in the Transformer are learned during training and then fixed. These weights help transform the input embeddings and hidden layers into meaningful projections and are crucial for the model's ability to represent regular functions.

Query Bias b

Bias terms, which adjust the linear transformations slightly to account for shifts in the data, are also learned during training and then held constant during inference.

Attention Mechanism

The attention mechanism uses the Q, K, and V projections to compute attention scores. These scores are then used to weight the value vectors, capturing the importance of different parts of the input. Importantly, the way Q, K, and V interact with the input is fixed once training is completed, meaning they are no longer able to adapt to changes in the underlying input distribution. After the model has been trained, all parameters (Q, K, V, W, b) become constant, meaning that the relationships these parameters capture during training are "frozen" and applied to any future input without further adaptation.

Why Constant Parameters Are Problematic for Generalizability

While the transformer architecture is promising in tasks such as text, this constant parameterization where Q, K, V, W, and b remain fixed after training, poses several challenges for generalizability. Below we explain why this approach can be problematic: Fixed parameters do not adjust to shift in the data distribution. Once the transformer is trained, it assumes the input distribution during inference is similar to what it saw during training. However, real-world data can change over time (e.g., domain shifts), and static parameters struggle to generalize effectively when the input data distribution differs significantly from the training data. For instance, a transformer trained on one language domain may not generalize well to another domain where the syntax or vocabulary differs slightly, simply because its Q, K, V, W, and b cannot adjust to these differences. Transformers rely heavily on attention mechanisms to determine which tokens or inputs are most relevant. However, with fixed Q, K, and V, the model's attention scores are calculated based on a static transformation of inputs. This can limit the model's ability to dynamically adjust its focus depending on the specific context of the input, especially for data outside the training set. When facing out-of-distribution data, the fixed parameters might miss key contextual variations that were not present during training, leading to degraded performance in tasks like machine translation or text summarization. Constant parameters inherently lack the flexibility to adjust to new patterns or unseen data that were not well-represented in the training set. Even with techniques like transfer learning, the parameter values of Q, K, V, W, and b remain static during the fine-tuning process, and the model may struggle with generalization to entirely new tasks or domains. Dynamic input-specific parameter adjustments could allow the model to recalibrate and interpret novel or unseen inputs more accurately, but this is not possible with fixed parameters. In large-scale tasks, especially those involving highly diverse datasets, different data points can have significantly different structures. A single, static set of parameters may fail to capture the variability in such datasets, leading to suboptimal generalization. For example, a transformer model trained on a large dataset with multiple languages or contexts may fail to effectively generalize across these different contexts due to the lack of input-adaptive parameters. A more flexible parameterization would allow the model to adjust based on the specific characteristics of each data subset. In many real-world tasks, data can evolve over time. This is particularly true in tasks like recommendation systems, where user behavior may change, or in time-series forecasting. Fixed parameters in the Transformer model are not well-suited to handle such non-stationary data, where the input distribution shifts over time. Without the ability to adapt to these changes, the model's performance can degrade quickly, as it is no longer optimized for the evolving nature of the data.

On the Feasibility of the implementation of state-feedback loop

Current implementations of residual neural networks incorporate a signal input state-feedback loop in the architecture, but that state is not used in the design of weights and biases. The add-and-Norm operator, used either before or after the self-attention mechanism, includes the state feedback, though it is not exploited in weight design. The same applies to the feedforward block in transformers. Introducing dynamic parameters that adjust based on the current input or the overall data distribution can help overcome the limitations of fixed parameters. Training the model to learn how to adjust its parameters on-the-fly or in response to new data points can improve its adaptability and generalization capacity. Neural Architecture Search (NAS) techniques can be used to discover architectures that are capable of dynamic parameter adjustments based on specific characteristics of the input data, improving the generalization of the model.

On the Suboptimality of State-Independent Parameters

In neural networks, especially in architectures such as Transformers, the question of optimal parameter dependence on input signals becomes crucial for achieving high-performance outcomes. Optimal Control Theory and Dynamic Programming provide useful frameworks to analyze how parameterization strategies affect learning and performance. From the viewpoint of Optimal Control Theory, the control law (or policy) that governs the system's behavior must be adaptable to the state of the system to minimize a cost function over time. Analogously, in neural networks, parameters such as the query, key, value, weight, and bias act as control variables that influence the network's behavior in transforming input signals to output predictions. Optimal control theory suggests that these parameters should not be fixed or state-independent. Instead, they must dynamically adapt based on the current state (in the neural network's case, the input data). State-independent parameters, where parameters such as weights and biases are set independently of the nature of the data, are inherently suboptimal. This stems from the fact that they fail to adjust to varying input signals or changing data distributions, which can lead to inefficient transformations and higher error rates. Optimal control theory emphasizes that control variables (parameters) should be state-dependent for the best outcome, which directly aligns with the need for input-adaptive parameters in neural networks. Dynamic Programming (DP) extends this principle further by focusing on the recursive decomposition of decision problems. In neural networks, training a model involves finding optimal parameters that minimize a loss function across all layers and data points. The Bellman Principle of Optimality states that any optimal policy must consider not just the current state but also the future states, which in the context of neural networks, translates into parameters that are responsive to data distributions rather than being static. Therefore, state-independent parameters fail to capture the structure of data, leading to suboptimality when handling heterogeneous datasets. For large datasets, the parameters should ideally depend not just on individual input signals but also on the distribution of the data itself. A state-independent parameterization assumes a uniformity of data structure that rarely exists in real-world datasets, leading to inefficient learning. By contrast, optimal parameterization should vary according to both the input signal and the broader statistical properties of the dataset, ensuring more effective learning and generalization. The suboptimality of state-independent parameters in neural networks, particularly in transformer-like architectures, is well-explained by principles from optimal control theory and dynamic programming. These frameworks highlight the importance of dynamic, state- and data-dependent parameterization for efficient, adaptive learning in high-dimensional data environments.

In this chapter, we investigated Blockchainized Tensor-Graph Neural Networks and their underlying mathematical structures. The aim is to highlight to the key the interactions between tensor-based architectures and blockchain technologies using non-zero-sum game theory.

2.11 Notes

Transformer-based architectures have been widely examined as candidates for time series analysis, backcasting, nowcasting, forecasting and risk quantification in finance [73, 81, 171, 220, 47, 130, 225, 8, 216, 74, 137, 128, 213, 153, 133, 76, 65, 221, 166, 135, 217, 69, 144, 222, 95, 90, 134, 210, 227, 214, 1, 116, 226, 113, 151, 115, 39, 136, 45, 160, 94, 172, 156, 152, 168, 215, 125, 46, 92, 91, 219, 100, 56, 34, 228, 132, 155, 111, 82, 138, 114]. The connection between deep learning and game theory has been examined in several recent papers. The work [191] surveys the interplay between the two fields. On the one hand, Game theory has developed different solution concepts. Many of them involve fixed-point arguments and variational inequalities. In a game theoretic setting, the objective functions do not have to be convex (in the global variable) and there are several strategic learning algorithms for solving game-theoretic problems beyond the class of potential convex games. On the other hand, deep neural networks input-output outcomes [157, 211, 37] are fixed-points of a composition of activation, attention, and weight operators. The composition of multiple operators may not be convex in the weight parameters which need to be designed during the training phase of the neural network. Several *convex optimization tools* have been used to address the training problem which is a *non-convex optimization* problem in the weight parameters. As a consequence, most of the error bounds, stability and accuracy analysis provided in the literature are not exploitable for the training problem. The convexity in the input signal variable does not necessarily mean or imply convexity in the weight parameters which also differs from the convexity of the functionals involved in another problem at hand. The works in [50, 51] have established properties of activation functions used in neural networks and used averaged-operators to prove convergence in the weak sense. The large learning model (LLM) properties based on these references to establish connections with the Stackelberg solution and Nash equilibria in [64]. Acceleration techniques are provided in [87, 36, 89, 122, 84, 85, 120].

The training process can also be framed as a game, where the layers serve as the decision-makers. In this case, the actions correspond to the adjustment of weights and biases. The information structure is determined by the input from the previous layer, leading to a scenario of partial hierarchical information. However, unlike a classical Stackelberg game, the current decision-maker (layer) does not have direct knowledge of the precise action taken by the preceding layer, as it can only observe the resulting input. This creates a unique variation of the Stackelberg structure in the training process. Within this context, the training process of neural networks can be viewed as a game played over the sampled data, where each layer acts based on the information provided by the layer below it. The challenge is to ensure that this iterative process leads to stable, consistent outcomes. By examining the consistency problem as a hierarchical game, we gain new insights into how to stabilize the training process and ensure that the network performs reliably across different samples. This approach also opens up possibilities for more effective game theory techniques that can exploit the structure of the game to improve convergence and performance on real-world data.

3

Extremely Large Transformers

In the relentless pursuit of advancing machine intelligence, one thing has become clear: size matters when it is strategically exploited. Qwen 2.5 (18 trillion parameters), DeepSeek(130 billion parameters), GPT-4 (OpenAI, 175 billion parameters), Gemini 1.5 (Google DeepMind, over 100 billion parameters), Claude 2 (Anthropic, 52 billion parameters), LLaMA 2 (Meta, 70 billion parameters), Azure OpenAI Service (GPT-4) (Microsoft, integrated with OpenAI's technology), LLaMA 1 (11B) (Meta, 11 billion parameters), PaLM 2 (Google, 100 billion parameters), GPT-NeoX-20B (EleutherAI, 20 billion parameters), Command R+ (Cohere, 20 billion parameters), Stable Diffusion 2.1 (Stability AI, 1 billion parameters), Mistral 7B (60 billion parameters), LLaMA (13B) (Meta, 13 billion parameters), DALL-E 2 (OpenAI, utilizes a combination of GANs and transformers), Gato (DeepMind, 1.2 billion parameters), Bard (Google, with a variable parameter count), GPT-J (EleutherAI, 6 billion parameters), Flan-T5 (Google, up to 11 billion parameters), OPT (Meta, up to 175 billion parameters), BLOOM (BigScience, 176 billion parameters), and Claude 1 (Anthropic, 52 billion parameters) just to name a few. Today's largest generative models exceed several trillions of parameters, hundreds of layers, and thousands of attention heads. These behemoth models are trained on vast datasets, processing terabytes of information in ways that smaller architectures simply cannot. But with great size comes great complexity and managing these large-scale transformers efficiently is becoming one of the central challenges in machine intelligence research.

This chapter focuses on Extremely Large Transformers and tackles this issue head-on. As models balloon in size, incorporating hundreds of attention heads and dozens of experts in mixture-of-experts architectures, understanding their mathematical behavior becomes critical to their optimization and scalability. This chapter offers a revisit into how transformers behave when pushed to their size limits and introduces new game-theoretic insights into managing their performance. One of the key ideas introduced is the concept of the mean-field limit transformer. This approach helps simplify the analysis of transformers with thousands of layers and complex attention mechanisms by focusing on their asymptotic behavior. When models scale to the extremes, individual interactions between data points and layers and attention heads become difficult to track. Instead, the mean-field limit offers a way to understand how these systems behave as a collective entity, similar to how economists study market behavior by looking at aggregate trends rather than individual actions. We also explore how transformers with trillions of data points can be optimized using a simplified version of the Boltzmann-Gibbs architecture. With layers interacting as players in a large, complex game, the mean-field approach offers a strategic lens through which to analyze how decisions made at one level affect outcomes across the entire network. This is especially crucial in models with numerous experts, some of the latest architectures feature up to 128 experts, each contributing to different parts of the computation, further complicating the interaction dynamics. But it is not just about size, performance at scale is key. We explain how transformers with an immense number of attention heads (as many as 96 heads per layer in some models) can be viewed through the game-theory framework, where each attention head plays a role in a larger coordination game. The balance between them must be carefully maintained to ensure that each head contributes optimally without overwhelming the system or creating redundant processes.

One of the most innovative discussions in this chapter focuses on the mixture-of-experts architecture. This approach, now common in some of the largest transformers, divides the computational load across multiple expert networks, with specific parts of the data routed to different experts. The authors analyze how this setup functions as a multi-agent game, with each expert operating as a strategic player tasked with optimizing its own performance while contributing to the overall goal of the system. In some of the latest models, up to 64 experts operate simultaneously, creating a need for coordination that can be improved by viewing the problem through a game-theoretic lens. As models grow, the training process itself becomes a strategic challenge. The chapter explains how mean-field convergence can help streamline training in models with extremely large datasets, making it possible to train even the most massive architectures efficiently by focusing on collective behavior rather than individual training steps. This approach is crucial as datasets continue to grow into the petabyte scale, and transformers are expected to process vast amounts of information in real time. For companies and industries investing in machine intelligence, the insights in this chapter are essential. As generative models reach new heights in size and complexity, understanding their underlying dynamics will be crucial for optimizing performance, reducing costs, and achieving scalability. Whether you are dealing with models featuring 128 experts or transformers with 96 attention heads per layer, the key takeaway is clear: it is not just about adding more layers or parameters; It is about strategically managing their interactions to unlock the full potential of machine intelligence. As the arms race for bigger and more powerful machine intelligence models continues, the frameworks provided in this chapter will help guide researchers and developers toward creating more efficient, effective systems at scale.

3.1 Mean-Field Limit Transformer

In the extremely large sequence length emerges a mean-field limit transformer. Instead of processing lengthy sequences, the mean-field limit transformer processes a single tensor and a collection of mean-field limit distributions that are frozen in the sense that a change in a single data point does not affect the infinite population mean-field limit. The mean-field limit should be computed using a mean-field limit attention block. This leads to an interaction between a single data point and mean-field limit of the normalized data ate each layer. The procedure is mean-field self-attention mechanism.

We examine the interactions at the self-attention mechanism when D goes to infinity.

3.1.1 Extremely Large Data and Asymptotics

As D grows to infinity, the size of the collection of the tensors $(X_1, \ldots, X_D)$ explodes. The normalization layer is done component-wise. We choose a generic component, say o, and its position s_o in the list. This will allow capturing auto-regressive time series and non-anticipative behavior in the analysis.

3.1.1.1 Unmasked Self-Attention

The single-headed unmasked attention mechanism has the following term

$$\sum_{j=1}^{D} \frac{e^{\frac{1}{\sqrt{k}}\langle QX_o, KX_j\rangle}}{\frac{D}{D}\sum_{l=1}^{D} e^{\frac{1}{\sqrt{k}}\langle QX_o, KX_l\rangle}} X_j$$

$$= \frac{1}{D}\sum_{j=1}^{D} \frac{e^{\frac{1}{\sqrt{k}}\langle QX_o, KX_j\rangle}}{\frac{1}{D}\sum_{l=1}^{D} e^{\frac{1}{\sqrt{k}}\langle QX_o, KX_l\rangle}} X_j = \frac{\int e^{\frac{1}{\sqrt{k}}\langle QX_o, KX'\rangle} X' m_D(dX')}{\int e^{\frac{1}{\sqrt{k}}\langle QX_o, KX''\rangle} m_D(dX'')}$$

where $m_D(dX') = \left(\frac{1}{D}\sum_{j=1}^{D}\delta_{X_j}\right)(dX')$ is the empirical measure of the input data set. We discuss below the convergence of m_D in the framework of de Finetti-Hewitt-Savage. It extends the i.i.d case to the asymptotically indistinguishable case. The integrability of these exponential components impose conditions on Q, K, V, W and on the measure m_D. If there is a limiting measure m such that all these integrals are finite, then the multi-headed unmasked self-attention is well-defined and is given by

$$X_o + \frac{1}{\sqrt{H}}\sum_{h=1}^{H} W_h V_h \frac{\int e^{\frac{1}{\sqrt{k}}\langle Q_h X_o, K_h X'\rangle} X' m(dX')}{\int e^{\frac{1}{\sqrt{k}}\langle Q_h X_o, K_h X''\rangle} m(dX'')}.$$

Definition 21. *Given* $m \in \mathcal{P}(\mathcal{H}_0)$, *the mean-field unmasked self-attention is* $mf_{unmasked-att}$:

$$\begin{cases} \mathcal{H}_0 \to \mathcal{H}_0 \\ X \mapsto X + \frac{1}{\sqrt{H}}\sum_{h=1}^{H} W_h V_h \left(\frac{\int_{\mathcal{H}_0} e^{\frac{1}{\sqrt{k}}\langle Q_h X, K_h X'\rangle} X' m(dX')}{\int_{\mathcal{H}_0} e^{\frac{1}{\sqrt{k}}\langle Q_h X, K_h X''\rangle} m(dX'')} \right). \end{cases} \tag{3.1}$$

3.1.1.2 Masked Self-Attention

As D grows to infinity, the size of the sequence of tensors $(X_1, \ldots, X_D) \in \mathcal{H}_0^D$ explodes but now the position of X_o in the list, $o \leq D$ should be encoded. We can capture it through the variable $s_o = \frac{o}{D}$. This describes the set $\{\frac{1}{D}, \frac{2}{D}, \ldots, \frac{D-1}{D}, \frac{D}{D}\}$. As D goes to infinity this becomes the non-empty compact interval $[0,1]$. The normalization layer is done row-wise with the encoding (s_o, X_o). The masked occupancy measure yields

$$\hat{m}_D(ds', dX') = \left[\frac{1}{D}\sum_{j=1}^{D}\delta_{(s_j, X_j)}\right](ds'dX')$$

and is a probability measure over the product space $[0,1] \times \mathcal{H}_0$. We denote its limiting measure (if any) by $\hat{m}(ds', dX')$ and the second marginal as $m(E) = \int_0^1 \int_E \hat{m}(ds', dX')$ will be a probability measure over $\mathcal{H}_0$.

We choose a generic component, say o. $LayerNorm(X_o)$ is well defined for each s_o. The attention mechanism becomes

$$\sum_{j=1}^{o} \frac{e^{\frac{1}{\sqrt{k}}\langle QX_o, KX_j\rangle}}{\frac{D}{D}\sum_{l=1}^{o} e^{\frac{1}{\sqrt{k}}\langle QX_o, KX_l\rangle}} X_j = \frac{1}{D}\sum_{j=1}^{D} \frac{e^{\frac{1}{\sqrt{k}}\langle QX_o, KX_j\rangle}\mathbb{I}_{\frac{j}{D}\leq\frac{o}{D}}}{\frac{1}{D}\sum_{l=1}^{D} e^{\frac{1}{\sqrt{k}}\langle QX_o, KX_l\rangle}\mathbb{I}_{\frac{l}{D}\leq\frac{o}{D}}} X_j$$

$$= \frac{\int_0^1 \int_{\mathcal{H}_0} e^{\frac{1}{\sqrt{k}}\langle QX_o, KX'\rangle}\mathbb{I}_{s\leq s_o} X' \hat{m}_D(dsdX')}{\int_0^1 \int_{\mathcal{H}_0} e^{\frac{1}{\sqrt{k}}\langle QX_o, KX''\rangle}\mathbb{I}_{s'\leq s_o} \hat{m}_D(ds'dX'')}.$$

If there is a limiting measure $\hat{m}$ such that all these integrable are finite then, the one-headed masked self-attention is well-defined and is given by

$$\frac{\int_0^1 \int_{\mathcal{H}_0} e^{\frac{1}{\sqrt{k}}\langle QX_o, KX'\rangle} \mathbb{I}_{s\leq s_o} X' \hat{m}(dsdX')}{\int_0^1 \int_{\mathcal{H}_0} e^{\frac{1}{\sqrt{k}}\langle QX_o, KX''\rangle} \mathbb{I}_{s'\leq s_o} \hat{m}(ds'dX'')}.$$

Thus, the multi-headed masked self-attention is

$$X_o + \frac{1}{\sqrt{H}} \sum_{h=1}^{H} (W_h V_h) \frac{\int_0^1 \int_{\mathcal{H}_0} e^{\frac{1}{\sqrt{k}}\langle Q_h X_o, K_h X'\rangle} \mathbb{I}_{s\leq s_o} X' \hat{m}(dsdX')}{\int_0^1 \int_{\mathcal{H}_0} e^{\frac{1}{\sqrt{k}}\langle Q_h X_o, K_h X''\rangle} \mathbb{I}_{s'\leq s_o} \hat{m}(ds'dX'')}.$$

Definition 22. *Given $\hat{m} \in \mathcal{P}([0,1] \times \mathcal{H}_0)$, the mean-field masked self-attention is $mf_{masked-att}$:*

$$\begin{cases} [0,1] \times \mathcal{H}_0 \to [0,1] \times \mathcal{H}_0 \\ (s, X) \mapsto \left(s, X + \frac{1}{\sqrt{H}} \sum_{h=1}^{H} W_h V_h \frac{\int_0^1 \int_{\mathcal{H}_0} e^{\frac{1}{\sqrt{k}}\langle Q_h X, K_h X'\rangle} \mathbb{I}_{s'\leq s} X' \hat{m}(ds'dX')}{\int_0^1 \int_{\mathcal{H}_0} e^{\frac{1}{\sqrt{k}}\langle Q_h X, K_h X''\rangle} \mathbb{I}_{s''\leq s} \hat{m}(ds''dX'')} \right). \end{cases} \tag{3.2}$$

3.1.1.3 Sigmoid Self-Attention

An alternative is the sigmoid unmasked self-attention given by

$$X_o + \frac{1}{\sqrt{H}} \sum_{h=1}^{H} (W_h V_h) \sum_{j=1}^{D} \frac{e^{\frac{1}{\sqrt{k}}\langle Q_h X_o, K_h X_j\rangle}}{D + e^{\frac{1}{\sqrt{k}}\langle Q_h X_o, K_h X_j\rangle}} X_j$$

$$= X_o + \frac{1}{\sqrt{H}} \sum_{h=1}^{H} (W_h V_h) \frac{1}{D} \sum_{j=1}^{D} \frac{D e^{\frac{1}{\sqrt{k}}\langle Q_h X_o, K_h X_j\rangle}}{D + e^{\frac{1}{\sqrt{k}}\langle Q_h X_o, K_h X_j\rangle}} X_j,$$

which is rewritten

$$X_o + \frac{1}{\sqrt{H}} \sum_{h=1}^{H} (W_h V_h) \frac{1}{D} \sum_{j=1}^{D} \frac{e^{\frac{1}{\sqrt{k}}\langle Q_h X_o, K_h X_j\rangle}}{1 + \frac{1}{D} e^{\frac{1}{\sqrt{k}}\langle Q_h X_o, K_h X_j\rangle}} X_j$$

$$= X_o + \frac{1}{\sqrt{H}} \sum_{h=1}^{H} (W_h V_h) \int \frac{e^{\frac{1}{\sqrt{k}}\langle Q_h X_o, K_h X'\rangle}}{1 + \frac{1}{D} e^{\frac{1}{\sqrt{k}}\langle Q_h X_o, K_h X'\rangle}} X' m_D(dX'),$$

For large D, the term $\frac{1}{D} e^{\frac{1}{\sqrt{k}}\langle Q_h X_o, K_h X'\rangle} \to 0$. Thus, the asymptotic sigmoid unmasked self-attention is

$$X_o + \frac{1}{\sqrt{H}} \sum_{h=1}^{H} W_h V_h \int e^{\frac{1}{\sqrt{k}}\langle Q_h X_o, K_h X'\rangle} X' m(dX').$$

Definition 23. *Given $m \in \mathcal{P}(\mathcal{H}_0)$, the mean-field sigmoid unmasked self-attention is $mfSigmoid_{unmasked-att}$:*

$$\begin{cases} \mathcal{H}_0 \to \mathcal{H}_0 \\ X \mapsto \left(X + \frac{1}{\sqrt{H}} \sum_{h=1}^{H} W_h V_h \int_{\mathcal{H}_0} e^{\frac{1}{\sqrt{k}}\langle Q_h X, K_h X'\rangle} X' m(dX') \right). \end{cases} \tag{3.3}$$

The sigmoid masked self-attention is

$$X_o + \frac{1}{\sqrt{H}} \sum_{h=1}^{H} (W_h V_h) \sum_{j=1}^{o} \frac{e^{\frac{1}{\sqrt{k}} \langle Q_h X_o, K_h X_j \rangle}}{D + e^{\frac{1}{\sqrt{k}} \langle Q_h X_o, K_h X_j \rangle}} X_j.$$

It follows that the asymptotic sigmoid masked self-attention is

$$X_o + \frac{1}{\sqrt{H}} \sum_{h=1}^{H} W_h V_h \int_0^1 \int_{\mathcal{H}_0} e^{\frac{1}{\sqrt{k}} \langle Q_h X_o, K_h X' \rangle} \mathbb{I}_{s \leq s_o} X' \hat{m}(dsdX').$$

Definition 24. *Given $\hat{m} \in \mathcal{P}([0,1] \times \mathcal{H}_0)$, the mean-field sigmoid masked self-attention mechanism is $mfSigmoid_{masked-att}$:*

$$\begin{cases} [0,1] \times \mathcal{H}_0 \to [0,1] \times \mathcal{H}_0 \\ (s, X) \mapsto \left(s, X + \frac{1}{\sqrt{H}} \sum_{h=1}^{H} W_h V_h \int_0^1 \int_{\mathcal{H}_0} e^{\frac{1}{\sqrt{k}} \langle Q_h X, K_h X' \rangle} \mathbb{I}_{s' \leq s} X' \hat{m}(ds'dX')\right). \end{cases} \tag{3.4}$$

3.1.1.4 Projected Boltzmann-Gibbs Self-Attention

Lemma 30. *Let $e \neq 0$ and $C_1 = \{w \in \mathcal{H}^k,\ \langle w, e \rangle = 0\}$.*
The projection into C_1 is given by

$$Proj_{C_1}(X) = X - \frac{\langle X, e \rangle}{\langle e, e \rangle} e.$$

Proof: This assertion is immediate. □

Let $X_o \in \mathcal{H}_0$ be a non-zero vector.

$$\text{proj}_{X_o^\perp}(Y_o) := Y_o - \langle Y_o, \frac{X_o}{\langle X_o, X_o \rangle} \rangle X_o.$$

Then,

$$\begin{aligned} &\|\text{proj}_{X_o^\perp}(Y_o)\|^2 = \|Y_o - \langle Y_o, \tfrac{X_o}{\langle X_o, X_o \rangle} \rangle X_o\|^2 \\ &= \langle Y_o - \tfrac{\langle Y_o, X_o \rangle}{\langle X_o, X_o \rangle} X_o, Y_o - \tfrac{\langle Y_o, X_o \rangle}{\langle X_o, X_o \rangle} X_o \rangle \\ &= \langle Y_o - \tfrac{\langle Y_o, X_o \rangle}{\langle X_o, X_o \rangle} X_o, Y_o \rangle - \langle Y_o - \tfrac{\langle Y_o, X_o \rangle}{\langle X_o, X_o \rangle} X_o, \tfrac{\langle Y_o, X_o \rangle}{\langle X_o, X_o \rangle} X_o \rangle \\ &= \langle Y_o, Y_o \rangle - \tfrac{(\langle Y_o, X_o \rangle)^2}{\langle X_o, X_o \rangle} \leq \langle Y_o, Y_o \rangle \end{aligned} \tag{3.5}$$

This means $\|\text{proj}_{X_o^\perp}(Y_o)\| \leq \|Y_o\|$. From Lemma 30, the operator $\text{proj}_{X_o^\perp}(Y_o)$ is indeed a projection into the orthogonal space to X_o.

Let $\epsilon > 0$. The Boltzmann-Gibbs self-attention works with the parameter $\epsilon > 0$ which allow us to examine its relationship to argmax: $\arg\max_{j \in \{1,2,\ldots,D\}} \langle Q_h X_o, K_h X_j \rangle$ as ϵ goes to zero, i.e., $\frac{1}{\epsilon}$ goes to infinity. Note that $\frac{1}{\sqrt{k}}$ does not go to infinity.

The projected Boltzmann-Gibbs unmasked self-attention

$$X_o + \text{proj}_{X_o^\perp}\left(\frac{1}{\sqrt{H}} \sum_{h=1}^{H} W_h V_h \frac{\sum_{l=1}^{D} e^{\frac{1}{\epsilon} \langle Q_h X_o, K_h X_j \rangle} X_j}{\sum_{l=1}^{D} e^{\frac{1}{\epsilon} \langle Q_h X_o, K_h X_l \rangle}}\right),$$

which is

$$X_o + \text{proj}_{X_o^\perp}\left(\frac{1}{\sqrt{H}}\sum_{h=1}^{H} W_h V_h \frac{\int e^{\frac{1}{\epsilon}\langle Q_h X_o, K_h X'\rangle} X' m_D(dX')}{\int e^{\frac{1}{\epsilon}\langle Q_h X_o, K_h X''\rangle} m_D(dX'')}\right).$$

With H finite and D going to infinity, one has the projected mean-field Boltzmann-Gibbs unmasked self-attention as

$$X_o + \text{proj}_{X_o^\perp}\left(\frac{1}{\sqrt{H}}\sum_{h=1}^{H} W_h V_h \frac{\int e^{\frac{1}{\epsilon}\langle Q_h X_o, K_h X'\rangle} X' m(dX')}{\int e^{\frac{1}{\epsilon}\langle Q_h X_o, K_h X''\rangle} m(dX'')}\right).$$

Definition 25. *Given $m \in \mathcal{P}(\mathcal{H}_0)$, the mean-field Boltzmann-Gibbs unmasked self-attention is $pmfBG_{unmaskedatt}$:*

$$\begin{cases} \mathcal{H}_0 \to \mathcal{H}_0 \\ X \mapsto X + proj_{X^\perp}\left(\frac{1}{\sqrt{H}}\sum_{h=1}^{H} W_h V_h \frac{\int_{\mathcal{H}_0} e^{\frac{1}{\epsilon}\langle Q_h X, K_h X'\rangle} X' m(dX')}{\int_{\mathcal{H}_0} e^{\frac{1}{\epsilon}\langle Q_h X, K_h X''\rangle} m(dX'')}\right). \end{cases} \tag{3.6}$$

Similarly, the projected mean-field Boltzmann-Gibbs masked self-attention is given by

$$X_o + \text{proj}_{X_o^\perp}\left(\frac{1}{\sqrt{H}}\sum_{h=1}^{H} W_h V_h \frac{\sum_{j=1}^{o} e^{\frac{1}{\epsilon}\langle Q_h X_o, K_h X_j\rangle} X_j}{\sum_{l=1}^{o} e^{\frac{1}{\epsilon}\langle Q_h X_o, K_h X_l\rangle}}\right).$$

It follows that projected mean-field Boltzmann-Gibbs masked self-attention is

$$X_o + \text{proj}_{X_o^\perp}\left(\frac{1}{\sqrt{H}}\sum_{h=1}^{H} W_h V_h \frac{\int_0^1\int_{\mathcal{H}_0} e^{\frac{1}{\epsilon}\langle Q_h X, K_h X'\rangle} \mathbb{I}_{s\leq s_o} X' \hat{m}(dsdX')}{\int_0^1\int_{\mathcal{H}_0} e^{\frac{1}{\epsilon}\langle Q_h X_o, K_h X''\rangle} \mathbb{I}_{s'\leq s_o} \hat{m}(ds'dX'')}\right).$$

Definition 26. *Given $\hat{m} \in \mathcal{P}([0,1]\times\mathcal{H}_0)$, the projected mean-field Boltzmann-Gibbs masked self-attention is $pmfBG_{masked-att}$:*

$$\begin{cases} [0,1]\times\mathcal{H}_0 \to [0,1]\times\mathcal{H}_0 \\ (s,X) \mapsto (s, X + proj_{X^\perp}(a)) \\ a = \frac{1}{\sqrt{H}}\sum_{h=1}^{H} W_h V_h \frac{\int_0^1\int_{\mathcal{H}_0} e^{\frac{1}{\epsilon}\langle Q_h X, K_h X'\rangle} \mathbb{I}_{s'\leq s} X' \hat{m}(ds'dX')}{\int_0^1\int_{\mathcal{H}_0} e^{\frac{1}{\epsilon}\langle Q_h X, K_h X''\rangle} \mathbb{I}_{s''\leq s} \hat{m}(ds''dX'')}. \end{cases} \tag{3.7}$$

3.1.2 Mean-Field Limit Boltzmann-Gibbs Transformer with a Simpler Normalization

Here we change the mean-variance normalization with a simpler normalization to the unit ball. It involves less parameters and extends to Hilbert spaces. Note that for Hilbert spaces, using the mean does not necessarily make sense as the infinite sum may be infinite.

$$
\begin{aligned}
&\text{Parameters } : L, d, k, k', q, H \in \mathbb{N},\\
&\quad 1 \le l \le L :\\
&\qquad 1 \le h \le H : Q_{lh}, K_{lh} \in \mathcal{L}(\mathcal{H}^d, \mathcal{H}^k), V_{lh} \in \mathcal{L}(\mathcal{H}^d, \mathcal{H}^{k'}), W_{lh} \in \mathcal{L}(\mathcal{H}^{k'}, \mathcal{H}^d),\\
&\qquad W_{1l} \in \mathcal{L}(\mathcal{H}^d, \mathcal{H}^q), b_{1l} \in \mathcal{H}^q, W_{2l} \in \mathcal{L}(\mathcal{H}^q, \mathcal{H}^d), b_{2l} \in \mathcal{H}^d,\\
&\text{Input } : s_i, y_0 = x_0 \in \mathcal{H}^d, \hat{m}_0 \in \mathcal{P}([0,1] \times \mathcal{H}^d), \hat{m} = (\hat{m}_0, \hat{m}_1, \dots, \hat{m}_{L-1})\\
&1 \le l \le L :\\
&\quad \text{Input at } l : \; s_i, x_l = y_{l-1} \in \mathcal{H}^d, \hat{m},\\
&\qquad u_{li} = \tfrac{x_{li}}{1+\|x_{li}\|},\\
&\qquad \tilde{u}_{li} = \tfrac{1}{\sqrt{H}} \textstyle\sum_{h=1}^H W_{lh} \frac{\int_0^1 \int_{\mathcal{H}^d} e^{\frac{1}{\sqrt{k}}\langle Q_{lh} u_{li}, K_{lh} u'_{lj}\rangle} V_{lh} \mathbb{I}_{s' \le s_i} u'_{lj} \hat{m}_l(ds' du'_{lj})}{\int_0^1 \int_{\mathcal{H}^d} e^{\frac{1}{\sqrt{k}}\langle Q_{lh} u_{li}, K_{lh} u'_{lj'}\rangle} \mathbb{I}_{s'' \le s_i} \hat{m}_l(ds'' du'_{lj'})},\\
&\qquad \hat{u}_{li} = x_{li} + \tilde{u}_{li},\\
&\qquad \hat{y}_{li} = \tfrac{\hat{u}_{li}}{1+\|\hat{u}_{li}\|},\\
&\qquad \tilde{y}_{li} = W_{2l} r_l(W_{1l} \hat{y}_{li} + b_{1l}) + b_{2l},\\
&\qquad y_{li} = \hat{u}_{li} + \tilde{y}_{li},\\
&\quad \text{Output at } l : y_l\\
&\text{Return } y_L \in \mathcal{H}^d
\end{aligned}
\tag{3.8}
$$

3.1.3 Mean-Field Limit Sigmoid Transformer with a Simpler Normalization

Here we replace Boltzmann-Gibbs with sigmoid self-attention.

$$
\begin{aligned}
&\text{Parameters } : L, d, k, k', q, H \in \mathbb{N},\\
&\quad 1 \le l \le L :\\
&\qquad 1 \le h \le H : Q_{lh}, K_{lh} \in \mathcal{L}(\mathcal{H}^d, \mathcal{H}^k), V_{lh} \in \mathcal{L}(\mathcal{H}^d, \mathcal{H}^{k'}), W_{lh} \in \mathcal{L}(\mathcal{H}^{k'}, \mathcal{H}^d),\\
&\qquad W_{1l} \in \mathcal{L}(\mathcal{H}^d, \mathcal{H}^q), b_{1l} \in \mathcal{H}^q, W_{2l} \in \mathcal{L}(\mathcal{H}^q, \mathcal{H}^d), b_{2l} \in \mathcal{H}^d,\\
&\text{Input } : s_i, y_0 = x_0 \in \mathcal{H}^d, \hat{m}_0 \in \mathcal{P}([0,1] \times \mathcal{H}^d), \hat{m} = (\hat{m}_0, \hat{m}_1, \dots, \hat{m}_{L-1})\\
&1 \le l \le L :\\
&\quad \text{Input at } l : \; x_l = y_{l-1} \in \mathcal{H}^d, \hat{m}\\
&\qquad u_{li} = \tfrac{x_{li}}{1+\|x_{li}\|},\\
&\qquad \tilde{u}_{li} = \tfrac{1}{\sqrt{H}} \textstyle\sum_{h=1}^H W_{lh} \int_0^1 \int_{\mathcal{H}^d} e^{\frac{1}{\sqrt{k}}\langle Q_{lh} u_{li}, K_{lh} u'_{lj}\rangle} V_{lh} \mathbb{I}_{s' \le s_i} u'_{lj} \hat{m}_l(ds' du'_{lj})\\
&\qquad \hat{u}_{li} = x_{li} + \tilde{u}_{li},\\
&\qquad \hat{y}_{li} = \tfrac{\hat{u}_{li}}{1+\|\hat{u}_{li}\|},\\
&\qquad \tilde{y}_{li} = W_{2li} r_l(W_{1li} \hat{y}_{li} + b_{1li}) + b_{2li},\\
&\qquad y_{li} = \hat{u}_{li} + \tilde{y}_{li},\\
&\quad \text{Output at } l : s_i, y_l, \hat{m}\\
&\text{Return } s_i, y_L \in \mathcal{H}^d, \hat{m}
\end{aligned}
\tag{3.9}
$$

3.1.4 Mean-Field Limit Transformer

In all these self-attention operations, the number of heads H can be relatively big but not infinite. In practice the data size can be large (e.g., $D = 200000$) but still not infinite. At each layer the number of operations can be large but not infinite. In this context, instead of applying row by row this huge collection of tensors, one may think of taking a generic row and applying a mean-field effect. This leads to a generic decision-maker interacting with the mean-field at each layer l. As the learning rate vanishes, one gets a mean-field difference equation while the mean-field is frozen (because a generic decision-maker does not affect the mean-field in the infinite population regime). In this case, the outcomes of $\mathbb{TGN}_{mf}$ are the steady state of the difference equation and the mean-field solves measure pushforward equation.

The other operators are done only for X_o leading to

$$(Id + \mathcal{O}_{l,ff,o} \circ \mathcal{O}_{l,nn,o}) \circ (Id + \mathcal{O}_{l,att,o} \circ \mathcal{O}_{l,nn,o})(Y_{l-1,o}, m).$$

There is no other row to operate but the measure m is supposed to be learnable to compute the averaging terms of the attention mechanism. Instead of the infinite data $\{(X_o, Y_o), o \in \{1, \dots, \infty\}\}$, one has

$$\left(X_o, Y_o,\ \tilde{m}(dX'dY') = \lim_{D\to\infty} \left[\frac{1}{D-1} \sum_{j\neq o} \delta_{(X_j, Y_j)} \right] (dX'dY') \right)$$

from which we extract the marginal distributions $m = \mathbb{P}_X$ and $\mathbb{P}_Y$.

In the infinite data mean-field limit, the input is (X_o, m) and the same m is transferred at each layer and each generic neural unit where the computation of the attention mechanism is required. The output is (Y_o, m_*).

Definition 27. *The mean-field limit transformer-based neural network $\mathbb{TGN}_{mf}$ specification is given by the family*

$$(X_o, m, L, H, \mathcal{H}_0, (\mathcal{H}_l, \mathcal{O}_{l,ff}, \mathcal{O}_{l,nn}, \mathcal{O}_{l,att}, r_l, Q_{lh}, K_{lh}, V_{lh}, W_{lh}, W_{1l}, W_{2l}, b_{1l}, b_{2l})_{1\leq l\leq L}, \lambda),$$

where $(X_o \in \mathcal{H}_0$ is generic initial tensor drawn according to $m \in \mathcal{P}(\mathcal{H}_0)$.

Note that there is no D in the mean-field limit transformer $\mathbb{TGN}_{mf}$, It is replaced by a distribution m. Input signal space moved from $\mathcal{H}_0^D$ to $(\mathcal{H}_0 \times \mathcal{P}(\mathcal{H}_0))$. However, to run, m is frozen, and we will need to pass it layer by layer the mean-fields $(m^{(1)}, m^{(2)}, \dots, m^{(L-1)})$, where $m^{(l)} = \mathbb{P}_{\mathcal{O}_{l,nn}(Y_*^{(l-1)})}$. The consistency check now requires only a double fixed-point equation:

$$\begin{cases} X_{*,o} = \hat{\mathcal{O}}_{L,o} \circ \hat{\mathcal{O}}_{L-1,o} \circ \dots \circ \hat{\mathcal{O}}_{1,o}(X_{*,o}, m_*), \\ \mathbb{P}_{X_{*,o}} \,|m_* = \mathcal{L}(X_{*,o}\,|m_*) = m_*. \end{cases} \tag{3.10}$$

We rename $\hat{\mathcal{O}}_{l,o}$ as $\hat{\mathcal{F}}_l \circ \hat{\mathcal{F}}_{l-\frac{1}{2}}\,|m_*^{(l-1)}$ where the explicit dependence on m_* appears in $\hat{\mathcal{F}}_{l-\frac{1}{2}}\,|m_*^{(l-1)}$ where the mean-field self-attention mechanism is applied and where $\hat{\mathcal{F}}_l = (Id + \mathcal{O}_{l,ff} \circ \mathcal{O}_{l,nn})$, $\hat{\mathcal{F}}_{l-\frac{1}{2}}\,|m_*^{(l-1)} = (Id + \mathcal{O}_{l,att} \circ \mathcal{O}_{l,nn})$ are now defined on $\mathcal{H}_0 \times \mathcal{P}(\mathcal{H}_0)$.

If these two fixed-point equations are fulfilled, then

$$
\begin{aligned}
&l = 1,\\
&Y^{(\frac{1}{2})} = \hat{\mathcal{F}}_{\frac{1}{2}}(X_*, m_*),\\
&Y^{(1)} = \hat{\mathcal{F}}_1 \circ \hat{\mathcal{F}}_{\frac{1}{2}}(X_*, m_*),\\
&\ldots\\
&l \in \{2, \ldots, L-1\}:\\
&Y^{(l-\frac{1}{2})} = \hat{\mathcal{F}}_{l-\frac{1}{2}\ |m_*^{(l-1)}} \circ \hat{\mathcal{F}}_{l-1} \circ \hat{\mathcal{F}}_{l-\frac{3}{2}\ |m_*^{(l-2)}} \circ \ldots \circ \hat{\mathcal{F}}_1 \circ \hat{\mathcal{F}}_{\frac{1}{2}}(X_*, m_*),\\
&Y^{(l)} = \hat{\mathcal{F}}_l \circ \hat{\mathcal{F}}_{l-\frac{1}{2}\ |m_*^{(l-1)}} \circ \ldots \circ \hat{\mathcal{F}}_1 \circ \hat{\mathcal{F}}_{\frac{1}{2}}(X_*, m_*),\\
&\ldots\\
&Y^{(L-\frac{1}{2})} = \hat{\mathcal{F}}_{L-\frac{1}{2}\ |m_*^{(L-1)}} \circ \hat{\mathcal{F}}_{L-1} \circ \ldots \circ \hat{\mathcal{F}}_1 \circ \hat{\mathcal{F}}_{\frac{1}{2}}(X_*, m_*),\\
&X_* = Y^{(L)} = \hat{\mathcal{F}}_L \circ \hat{\mathcal{F}}_{L-\frac{1}{2}\ |m_*^{(L-1)}} \circ \ldots \circ \hat{\mathcal{F}}_1 \circ \hat{\mathcal{F}}_{\frac{1}{2}}(X_*, m_*),
\end{aligned}
\tag{3.11}
$$

where the dependence on m_* is in the operator $\hat{\mathcal{F}}_{l-\frac{1}{2}} = \hat{\mathcal{F}}_{l-\frac{1}{2}\ |m_*}$ and the resulting output

$$(Y^{(\frac{1}{2})}, Y^{(1)}, Y^{(\frac{3}{2})}, Y^{(2)}, \ldots, Y^{(L-\frac{1}{2})}, Y^{(L)} = X_*, m_*)$$

is the outcome of $\mathbb{TGN}_{mf}$.

> In the infinite population regime, the mean-field limit transformer has an interaction between a generic data point and the infinite population mean-field. The interaction is captured by the mean-field self-attention mechanism. This is exactly what happens in infinite population sequential (or hierarchical) mean-field games where a generic decision-maker reacts to the mean-field.

Proposition 19. *The infinite population* $\mathbb{TGN}_{mf}$ *is a mean-field game and the outcomes of* $\mathbb{TGN}_{mf}$ *are mean-field equilibria.*

Proof: In an infinite population hierarchical mean-field game, the infinite population mean-field object is frozen. Here, the frozen object is $(m^{(0)}, m^{(1)}, m^{(2)}, \ldots, m^{(L-1)})$. Consider a well-designed infinite population $\mathbb{TGN}_{mf}$,, the operator $\hat{\mathcal{O}}_l$ at each layer l can be viewed as the best-response function of decision-maker l to $(Y^{(l-1)}, m^{(l-1)})$. This implies that the output $Y^{(l)}$ is an optimal response to the previous layer's outputs and to the frozen mean-field. The choice of the layer decision-maker does not affect the mean-field as it is frozen. Thus, each layer functions as a decision-maker that optimizes its action based on the action of the preceding decision-maker and the mean-field, as in the best response dynamics to mean-field. This defines an infinite population hierarchical mean-field game.

Given that each layer in the $\mathbb{TGN}_{mf}$ represents a decision-maker optimizing their action based on the infinite population mean-field of actions from previous layers, the sequence of outputs $(Y_*^{(1)}, Y_*^{(2)}, \ldots, Y_*^{(L)})$ and frozen mean-field vector $(m_*^{(1)}, m_*^{(2)}, \ldots, m_*^{(L-1)})$ constitutes a mean-field-type Nash equilibrium if it is the best-response and the resulting action will reproduce m_* as its probability distribution, i.e., $\mathcal{L}(\mathcal{O}_{l,nn}(Y_*^{(l-1)})) = m_*^{(l-1)}$ creating a fixed-point. This hierarchical mean-field equilibrium of $\mathbb{TGN}_{mf}$ is characterized by each generic decision-maker choosing $Y_*^{(l)}$ as a response to $(Y_*^{(l-1)}, m_*^{(l-1)})$, such that the resulting distribution $m_*^{(l)}$ is consistent with the responses of all other generic decision-makers. □

This outcome is also an infinite population mean-field Nash equilibrium of the hierarchical mean-field game. Note that, here, the mean-field is a infinite population mean-field of actions in the sense that m is a distribution of the other decision-makers' actions. Due to the infinite population assumption, X_o which is an action of a generic decision-maker does not influence the population mean-field limit m. We will see this has strong limitations from an interaction perspective where each operation may affect the distribution, which is discussed in the next section.

This type of discrete-time mean-field games with one layer were introduced by Jovanovic 1982 [117] followed by Jovanovic-Rosenthal 1988 [118].

The training problem at this infinite population regime is to move from signal input distribution to an output distribution subject to the difference equation constraint:

$$\begin{array}{l} \mathcal{D} = \{\mathbb{P}\}, \mathbb{P} \in \mathcal{P}(\mathcal{H}_0 \times \mathcal{H}_L) \\ \inf_{\theta\in\Theta} E\Phi(Y_T, \hat{Y}), \\ \text{such that} \\ Y_0 = X_0 = \hat{X} \sim \mathbb{P}_1 = m_0, \hat{Y} \sim \mathbb{P}_2 \\ Y_{t+1} = Y_t + \lambda_t(\hat{\mathcal{F}}_L \circ \hat{\mathcal{F}}_{L-\frac{1}{2}} |_{m_t^{(L-1)}} \circ \ldots \circ \hat{\mathcal{F}}_1 \circ \hat{\mathcal{F}}_{\frac{1}{2}, m_t^{(0)}}(Y_t) - Y_t), \ t > 0 \\ \\ m_{t+1} = ((1-\lambda_t)Id + \lambda_t \hat{\mathcal{F}}_{L-\frac{1}{2}} |_{m_t^{(L-1)}} \circ \ldots \circ \hat{\mathcal{F}}_1 \circ \hat{\mathcal{F}}_{\frac{1}{2}, m_t^{(0)}} |_{\theta=\theta_*})\# m_t \end{array} \tag{3.12}$$

where Φ is a real-valued risk-aware cost function, T is the number of iterations of transformer $\mathbb{TGN}_{mf}$.

The training problem of the mean-field limit transformer as formulated in (3.12) is not a control problem; it is fixed-point problem as the resulting θ_* appears in the distribution m. This is another mean-field game which we refer to as the training mean-field game in discrete time.

A pair (θ, m) solution of (3.12) is also a mean-field Nash equilibrium of another mean-field game where m is frozen and the control action θ is restricted to the class of time-and state-independent controls. Here, Y_t is a state of time t, m_t is the infinite population mean-field of states of the all other decision-makers. Note that generically the mean-field Nash equilibrium here is a function of (Y_t, m_t), which means that the restriction to constant control may lose optimality.

The existence of a unique θ_* is not guaranteed anymore, as the drift function is non-convex and therefore the existence of mean-field equilibria is an open issue. However, one can work with the difference inclusion as in classical dynamic games where the best response is a set.

On the Learnability of the Input Distribution

The distribution m can be obtained as a marginal of $\mathbb{P}(dX_0 dY^{(L)})$ with respect to the first variable. If $\mathbb{P}(dX_0 dY^{(L)})$ is unknown, m can be approximated from m_D which is the empirical measure. In that case the approximated measure m should not be considered as exact. One can use a distributionally robust approach to take into consideration all measures around the learned one.

Beyond Independent and Identically Distributed Data Set

Mean field convergence goes far beyond independent and identically distributed data set input data. It includes indistinguishability, multi-class indistinguishability, pairwise indistinguishability, k-wise indistinguishability, asymptotically k-wise indistinguishability. These situations allow us to cover a broader class of data set in generative machine intelligence where the i.i.d assumption does not hold.

3.2 Small Learning Rate Regime of Mean-Field Limit Transformer

We now examine the dynamics of transformer-based neural networks when λ_t is very small. One can relate it to a pseudo-trajectory of ordinary differential system or stochastic differential system. This is a stylized network case. In practice the data size can be large but still not infinite. At each layer, the number of operations can be large but not infinite. In this context, instead of applying this huge collection of tensors row by row, one may think of taking a generic row and applying a mean-field effect. This leads to a generic decision-maker interacting with the mean-field at each layer l. As the learning rate vanishes, one gets an mean-field ODE while the mean-field is frozen (because a generic decision-maker does not affect the mean-field in the infinite population regime). In this case, the outcomes of $\mathbb{TGN}$ are steady states of some ODEs and mean-field solves the Fokker-Planck equation. The training problem at this infinite population regime is to move from signal input distribution to an output distribution subject to the ODE constraint:

$$\begin{array}{l}
\mathcal{D} = \{\mathbb{P}\}, \mathbb{P} \in \mathcal{P}(\mathcal{H}_0 \times \mathcal{H}_L) \\
\inf_{\theta \in \Theta} E\Phi(Y(T), \hat{Y}), \\
\text{such that} \\
Y(0) = \hat{X} \sim \mathbb{P}_1 = m_0, \hat{Y} \sim \mathbb{P}_2 \\
\dot{Y} = \hat{\mathcal{F}}_L \circ \hat{\mathcal{F}}_{L-\frac{1}{2}} |_{m^{(L-1)}} \circ \ldots \circ \hat{\mathcal{F}}_1 \circ \hat{\mathcal{F}}_{\frac{1}{2}, m^{(0)}}(Y) - Y, \ T > t > 0 \\
\\
\dot{m}(t, Y) + div(m(t, Y)Z) = 0 \\
Z = (\hat{\mathcal{F}}_L \circ \hat{\mathcal{F}}_{L-\frac{1}{2}} |_{m^{(L-1)}(t,.)} \circ \ldots \circ \hat{\mathcal{F}}_1 \circ \hat{\mathcal{F}}_{\frac{1}{2}, m^{(0)}(t,.)}|_{\theta=\theta_*}(Y) - Y)
\end{array} \tag{3.13}$$

where Φ is a real-valued risk-aware cost function.

> The training problem of the mean-field limit transformer as formulated in (3.13) is not a control problem, it is fixed-point problem as the resulting θ_* appears in the distribution m. This is another mean-field game which we refer to as the training mean-field game in continuous time.

A pair (θ, m) solution of (3.13) is also a mean-field Nash equilibrium while restricted to the class of time-and-state-independent controls. Note that generically the mean-field Nash equilibrium here is a function of $(Y(t), m(t))$ which means that the restriction to constant control may lose optimality. We refer to [181, 7, 180, 102, 103, 93, 167, 124, 159, 202, 6, 206, 203, 207, 205, 119, 141, 208, 199, 204, 182, 183, 196, 195, 139, 123, 197, 175, 59, 190] for more details on continuous-time infinite population mean-field games.

3.3 Mean-Field Convergence

The extension of (i) the law of large numbers, (ii) central limit theorem, and (iii) large deviation principle, from independent random variables to sequences of indistinguishable random variables has been widely studied since the appearance in Blum, Chernoff and co-authors [40]. Below we state some well-known results and explain how they can be used in the McKean-Vlasov context with the $L^{\alpha}-$norm.

3.3.1 Indistinguishability

The notion of indistinguishability (or exchangeability or interchangeability) is introduced in order to discuss the existence of a limiting measure and mean-field convergence of the empirical measure of virtual particle states in the framework of de Finetti-Hewitt-Savage [55, 53, 54, 105, 3, 5, 4]. The indistinguishability property is called exchangeability in the probability literature.

Let $\mathcal{X}$ be a separable complete and metrizable topological space (Polish space).

Definition 28 (Indistinguishability). *A collection $(X_1, X_2, \ldots, X_I)$ of $\mathcal{X}-$valued random variables/processes, is indistinguishable (or exchangeable) if the joint law is invariant by permutation over the index set $\{1, \ldots, I\}$, i.e., for any permutation σ over the set $\{1, 2, \ldots, I\}$, one has*

$$\mathcal{L}(X_1, X_2, \ldots, X_I) = \mathcal{L}(X_{\sigma(1)}, \ldots, X_{\sigma(I)}), \tag{3.14}$$

where $\mathcal{L}(X)$ denotes the law of the random variable X. An infinite family of random variables/processes $(X_1, X_2, \ldots)$ is indistinguishable if for every finite I the family $(X_1, X_2, \ldots, X_I)$ is indistinguishable.

This says that the order (position) of the random variable in the family does not change the joint distribution. From (3.14) we also have that, for any measurable operator O,

$$\begin{aligned} &\mathcal{L}\left(X_1, X_2, \ldots, X_I,\ O(\frac{1}{I}\sum_{i=1}^{I}\delta_{X_i})\right) \\ &= \mathcal{L}\left(X_{\sigma(1)}, \ldots, X_{\sigma(I)}, O(\frac{1}{I}\sum_{i=1}^{I}\delta_{X_{\sigma(i)}})\right), \end{aligned} \tag{3.15}$$

where we do not permute the last component.

For indistinguishable random variables/processes, the convergence of the empirical measure $m_I := \frac{1}{I}\sum_{i=1}^{I}\delta_{X_i}$ has been widely studied. This sits at the intersection of group theory and probability theory. The symmetry group properties have been used to derive some properties of the distributions of the processes. The *de Finetti-Hewitt-Savage* theory provides the mean-field convergence of such a measure-valued process.

When studying convergence of measures, an important issue is the choice of the probability metric. In order to measure the gap between two probability measures, we introduce the Wasserstein (Vasershtein) metric (also called Monge-Kantorovich metric) d_α of order $\alpha \geq 1$.

Definition 29 (Wasserstein). *The following is called the Wasserstein metric:*

$$d_\alpha^\alpha(\mu, \nu) = \inf\left\{\int_{(X,Y)} d_0(X, Y)^\alpha \gamma(dX, dY);\ \gamma \in \mathcal{P}(\mathcal{X}^2),\ \mathbb{P}_X = \mu,\ \mathbb{P}_Y = \nu\right\},$$

where $\mathbb{P}_X$ denotes the marginal with respect to the $X-$component, where d_0 is a reference metric on $\mathcal{X}$ (such a metric exists because $\mathcal{X}$ is assumed to be metrizable).

The Kantorovich-Rubinstein 1958 theorem provides a dual representation of d_1 in terms of a Lipschitz-Bounded metric:

$$d(\mu, \nu) := d_1(\mu, \nu) = \sup\left\{\int \phi d(\mu - \nu);\ \|\phi\|_{Lip} \leq 1\right\},$$

where $\|\phi\|_{Lip} = \|\phi\|_\infty + \sup_{X\neq Y} \frac{|\phi(X)-\phi(Y)|}{d_0(X,Y)}$ is the Lipschitz-norm of ϕ. It can be shown that d_α is a metric (a true distance in a topological sense), i.e., it satisfies the axioms of a metric. For Polish spaces $\mathcal{X}$, the Wasserstein distance d_1 is known to metrize the weak topology over $\mathcal{X}$. As stated in [212] the Wasserstein distance has the following properties: for any $1 \leq \alpha < +\infty$,

$$\lim_I d_\alpha(m_I, m) = 0$$

implies, in particular, that

- m_I converges to m in distribution (weak convergence of probability measures), i.e.,

$$E_{m_I}[\phi] := \int \phi dm_I \to E_m[\phi],$$

  as $I \to +\infty$, for any measurable bounded and Lipschitz functions ϕ.

- $\int d_0^\alpha(X, Y) m_I(dY) < +\infty$ for some $X \in \mathcal{X}$.

Thanks to these nice properties, the Wasserstein distance d_α is an appropriate candidate for the convergence of the empirical measure in the weak sense.

Theorem 3 (de Finetti-Hewitt-Savage). *Let $X_1, X_2, \ldots,$ be an indistinguishable sequence of $\mathcal{X}-$valued random variables, where $\mathcal{X}$ is a Polish space. Then, there is a $\mathcal{P}(\mathcal{X})-$valued random measure m such that*

$$m = \lim_{I\to\infty} \frac{1}{I} \sum_{i=1}^{I} \delta_{X_i}, \text{ almost surely,}$$

where $\mathcal{P}(\mathcal{X})$ denotes the space of probability measures on $\mathcal{X}$. Conditioned on m, the random variables $X_1, X_2, \ldots$ are i.i.d. with distribution m, that is, for each measurable bounded function ϕ,

$$E\left(\phi(X_1, X_2, \ldots, X_k) \mid m\right) = \int \phi(Y_1, \ldots, Y_k) m(dY_1) \ldots m(dY_k).$$

In addition, if the moments of X_i are finite then

$$d_1\left(m, \frac{1}{I}\sum_{i=1}^{I} \delta_{X_i}\right) \leq \frac{C_1}{\sqrt{I}} = O\left(\frac{1}{\sqrt{I}}\right),$$

where $C_1 > 0$ and $d = d_1$ denotes the Wasserstein metric of order one.

Note that the convergence in Theorem 3 is in the weak sense since the Monge-Kantorovich distance d_1 metrizes the weak topology. Theorem 3 has been proved by de Finetti (1931, [55]) for infinite binary sequences and has been extended by Hewitt and Savage (1955, [105]) to continuous and compact state spaces. A simple and elegant proof can be found in Aldous (1985, [3]), pp. 18-22, for the general state space. The rate of convergence for the Monge-Kantorovich distance is obtained following the line of the law of large numbers of interacting systems. Theorem 3 was initially used for static (time-independent) maps. Then, several applications in mathematical physics and biology with dynamical models entered the picture. These are dynamically interacting particles, genes, molecules or nodes. Theorem 3 was then extended to the dynamical case in at least two ways: (i) pathwise (up to a certain time step T), and (ii) at each time step t.

3.3.2 State-Action k-wise Indistinguishability

In the context of game theory, one may need not just the dependence of (X_i, m_I) but also the dependence on (X_{i_1}, X_{i_2}, m_I), $(X_{i_1}, X_{i_2}, X_{i_3}, m_I)$, ..., $(X_{i_1}, X_{i_2}, X_{i_3}, \dots, X_{i_k}, m_I)$ in the quantities of interests. These will require pairwise-indistinguishability, 3-wise, ..., k-wise indistinguishability, which correspond the indistinguishability of the family $\{(X_{i_1}, X_{i_2}) \mid (i_1, i_2) \in \mathcal{I}^2, i_1 \neq i_2\}$, $\{(X_{i_1}, X_{i_2}, X_{i_3}) \mid (i_1, i_2, i_3) \in \mathcal{I}^3, i_1 \neq i_2 \neq i_3\}$, $\{(X_{i_1}, X_{i_2}, X_{i_3}, \dots, X_{i_k}) \mid (i_1, i_2, i_3, \dots, i_k) \in \mathcal{I}^k, i_1 \neq i_2 \neq i_3 \dots \neq i_k\}$ respectively. This leads to the introduction of the empirical measures

$$\begin{cases} \frac{1}{I(I-1)} \sum_{(i_1,i_2)\in\mathcal{I}^2, i_1\neq i_2} \delta_{(X_{i_1},X_{i_2})} \\ \frac{1}{I(I-1)(I-2)} \sum_{(i_1,i_2,i_3)\in\mathcal{I}^3, i_1\neq i_2\neq i_3} \delta_{(X_{i_1},X_{i_2},X_{i_3})} \\ \dots \\ \frac{1}{I(I-1)(I-2)\dots(I-k)} \sum_{(i_1,i_2,i_3,\dots,i_k)\in\mathcal{I}^k, i_1\neq i_2\neq i_3\neq i_k} \delta_{(X_{i_1},X_{i_2},X_{i_3},\dots,X_{i_k})} \end{cases} \tag{3.16}$$

These higher order interactions occur when $k \geq 2$.

In game theory, indistinguishability refers to a situation in which decision-makers cannot distinguish between strategic choices or outcomes based on available information. Therefore, the decisions of interacting decision-making should appear in the quantities of interests (state dynamics, payoffs) and hence the process X_i of decision-maker o will now be replaced by the new process $Y_i = (X_i, a_i)$.

This leads to the introduction of the following empirical measures:

$$\begin{cases} \frac{1}{I} \sum_{i\in\mathcal{I}} \delta_{Y_i}, \\ \frac{1}{I(I-1)} \sum_{(i_1,i_2)\in\mathcal{I}^2, i_1\neq i_2} \delta_{(Y_{i_1},Y_{i_2})}, \\ \frac{1}{I(I-1)(I-2)} \sum_{(i_1,i_2,i_3)\in\mathcal{I}^3, i_1\neq i_2\neq i_3} \delta_{(Y_{i_1},Y_{i_2},Y_{i_3})}, \\ \dots \\ \frac{1}{I(I-1)(I-2)\dots(I-k)} \sum_{(i_1,i_2,i_3,\dots,i_k)\in\mathcal{I}^k, i_1\neq i_2\neq i_3\neq i_k}, \delta_{(Y_{i_1},Y_{i_2},Y_{i_3},\dots,Y_{i_k})}, \\ k \geq 3 \end{cases} \tag{3.17}$$

Pairwise indistinguishability in this context means within the family, any two distinct state-actions of the decision-makers are indistinguishable from each other when evaluated in the quantities of interests.

Similarly k-wise indistinguishability in this context means within the family, any k distinct state-action elements of decision-makers are indistinguishable from each other when evaluated in the quantities of interests.

When the state-action is indistinguishable, the mean-field convergence implies

$$\begin{cases} \frac{1}{I} \sum_{i\in\mathcal{I}} \delta_{Y_i} \implies \mu, \\ \frac{1}{I(I-1)} \sum_{(i_1,i_2)\in\mathcal{I}^2, i_1\neq i_2} \delta_{(Y_{i_1},Y_{i_2})} \implies \mu \otimes \mu, \\ \frac{1}{I(I-1)(I-2)} \sum_{(i_1,i_2,i_3)\in\mathcal{I}^3, i_1\neq i_2\neq i_3} \delta_{(Y_{i_1},Y_{i_2},Y_{i_3})} \implies \mu \otimes \mu \otimes \mu, \\ \dots \\ \frac{1}{I(I-1)(I-2)\dots(I-k)} \sum_{(i_1,i_2,i_3,\dots,i_k)\in\mathcal{I}^k, i_1\neq i_2\neq i_3\neq i_k}, \delta_{(Y_{i_1},Y_{i_2},Y_{i_3},\dots,Y_{i_k})} \\ \implies \underbrace{\mu \otimes \mu \otimes \mu \dots \otimes \mu}_{k\ times}, \\ k \geq 3 \end{cases} \tag{3.18}$$

This last property is important when examining population games, evolutionary games, crowding games, and mean-field games with higher-order interactions. In particular, in the finite population regime, when one decision-maker deviates the indistinguishability property may fail.

4

Mean-Field-Type Transformers

As machine intelligence continues to advance, the complexity of the models we employ grows exponentially. This chapter presents one of the most promising frameworks for understanding and optimizing these architectures. The chapter sets itself apart by distinguishing between mean-field-type transformers and the traditional mean-field limit, offering fresh insights into how self-attention mechanisms can evolve to enhance model performance. At its core, the chapter introduces the concept of mean-field-type transformers, where the self-attention mechanisms are influenced not just by an individual's position in the input sequence or the data point itself but also by the broader distribution of data points within the sequence. This nuanced approach allows for a McKean-Vlasov interaction between layers, a sophisticated mathematical framework that captures the interplay between the layers in a transformer model. By incorporating distributional effects, mean-field-type transformers offer a more robust way to understand the interactions occurring within each layer, paving the way for improved learning and decision-making processes. This framework is particularly valuable in variance-aware transformers, where the number of decision-makers (or agents) is finite but exhibits mean-field behavior due to a variance term that is non-linear in the measure. In simpler terms, while each decision-maker operates within its own finite constraints, the distribution behavior mimics that of its own mean-field system, where the interactions among the agents produce a distributional effect on itself and on the other layers. This allows for more accurate representations of uncertainty and variance, leading to enhanced predictive capabilities. The chapter discusses how this mean-field-type approach enables the development of self-attention mechanisms that are not only more flexible but also capable of handling real-world scenarios where data distributions may vary widely. By embracing these distributional dynamics, mean-field-type transformers can adapt to the underlying structure of the data, improving their performance in tasks that require nuanced understanding, such as natural language processing or complex decision-making. It presents a compelling case for the adoption of mean-field-type transformers in the evolving landscape of machine intelligence. By bridging the gap between self-attention mechanisms and mean-field-type game theory, this chapter not only expands our understanding of transformer architectures but also sets the stage for the next generation of models that can navigate the complexities of data with greater finesse and accuracy. For practitioners and researchers alike, embracing these insights is essential for pushing the boundaries of what machine intelligence can achieve. In Mean-Field-Type Transformers, the self-attention mechanism is modeled as a mean-field-type interaction between agents. The chapter explores how this framework can be extended to federated learning environments, where multiple agents (transformers) collaborate across different locations without sharing data. The chapter also describes how diffusion models can be incorporated into transformer architectures, providing explicit solutions to problems in tensor-graph networks. This chapter emphasizes how game theory provides a powerful framework for understanding the strategic interactions within these models. We present the mathematical foundations of generative machine intelligence and link them with mean-field-type game theory. The key interaction mechanism is self-attention, which exhibits aggregative properties similar to those found in mean-field-type game theory. It is not necessary to have an infinite number of neural units to handle mean-field-type terms.

For instance, the variance reduction in error within generative machine intelligence is a mean-field-type problem and does not involve an infinite number of decision-makers. Based on this insight, we construct mean-field-type transformers that operate on data that are not necessarily identically distributed and evolve over several layers using McKean-Vlasov transition kernels. We demonstrate that the outcomes of these mean-field-type transformers correspond exactly to the mean-field-type equilibria of a hierarchical mean-field-type game. Due to the non-convexity of the operators' composition, gradient-based methods alone are insufficient. To distinguish a global minimum from other extrema such as local minima, local maxima, global maxima, and saddle points, higher-order conditions or alternative methods that exploit hidden convexities in the self-attention mechanisms and activation functions are required. Furthermore, the chapter discusses the integration of blockchain technologies into machine intelligence, facilitating an incentive design loop for all contributors and enabling blockchain token economics for each system participant. This feature is especially relevant for ensuring the integrity of factual data, legislative information, medical records, and scientifically published references that should remain immutable by generative machine intelligence.

$$\left\{\begin{array}{l|l|l} & \text{TGN} & \text{MFTT} \\ \hline \text{Input:} & \mathcal{H}_0 \times \mathcal{H}_0^{D-1} & \mathcal{H}_0 \times \mathcal{P}(\mathcal{H}_0) \\ \text{LayerNorm:} & \mathcal{H}_l \times \mathcal{H}_l^{D-1} & \mathcal{H}_l \times \mathcal{P}(\mathcal{H}_l) \\ \text{Attention:} & \mathcal{H}_l \times \mathcal{H}_l^{D-1} & \mathcal{H}_l \times \mathcal{P}(\mathcal{H}_l) \\ \text{Feedforward:} & \mathcal{H}_l \times \mathcal{H}_l^{D-1} & \mathcal{H}_l \times \mathcal{P}(\mathcal{H}_l) \end{array}\right. \tag{4.1}$$

We then introduce mean-field-type transformer architecture, where self-attention mechanisms are treated as mean-field-type interactions. We explore the properties and outcomes of these architectures and develop a comprehensive training methodology for mean-field-type transformers. Equation (4.1) connects the classical transformer with the MFTT (see Figure 4.1)

Special attention is given to the small learning rate regime, where both finite data sequences and continuous-time mean-field limits are presented. This work provides a foundation for integrating blockchain technologies into Tensor-Graph Neural Networks, offering

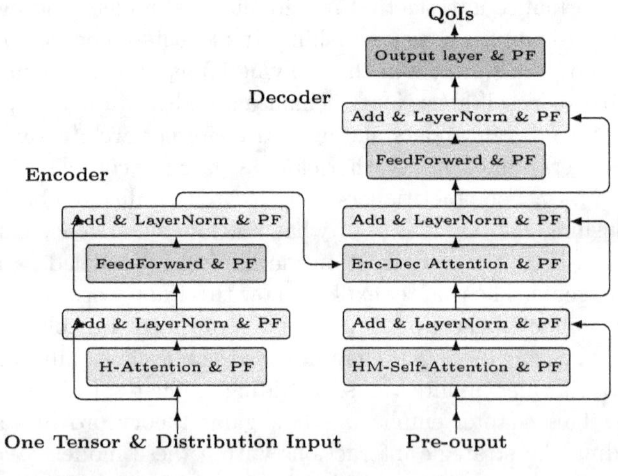

FIGURE 4.1
Mean-field-type Ttransformers. PF is the pushforward operator. We do not have D. Only one tensor and a distribution. Input: (X_0, μ_0) and output is $(Y^{(L)}, \nu^{(L)})$.

a framework for secure, transparent, and scalable machine learning systems. Furthermore, it opens up new avenues for strategic learning in multi-agent systems and the development of decentralized machine intelligence frameworks. The integration of blockchain incentivizes contributors and ensures the immutability of critical data, making it highly relevant for applications in law, medicine, and scientific research. We also present the baseline of diffusion-transformer in Tensor-Graph Networks. The diffusion-transformer architecture is analyzed, particularly through density-based optimal control and time-reversed diffusion processes, which offer alternative mechanisms for enhancing performance and robustness in large-scale learning environments. A mean-field-type diffusion-transformer is introduced, extending the mean-field-type framework to diffusion-based systems.

Federated transformers, focusing on the principles of federated deep training are investigated. This framework allows multiple decentralized nodes to collaboratively train a transformer model without sharing local data, enhancing privacy and security in distributed training settings. The concept of the federated training transformer is presented, outlining its architecture and the challenges associated. We also highlight a framework for multiple machine intelligence agents that interact within a tensor-graph network. We discuss strategic large learning models including both decision-maker-specific architectures and shared neural network architectures. Special attention is given to strategic learning, where multiple agents optimize their decisions within a shared learning environment, modeled as discrete-time mean-field-type games. The dynamics of strategic interactions between machine intelligence agents are analyzed, offering a game-theoretic perspective on multi-agent cooperation and competition.

The chapter is structured as follows. We present mean-field type attention mechanisms for MFTT in Section 4.1. Section 4.2 focuses on already implemented transformers and put them in the framework of MFTT. The outcomes of MFTT are discussed in Section 4.3. Section 4.4 focuses on training of MFTT. Section 4.5 provides limiting behavior under small learning rates. Section 4.6 examines mean-field-type diffusion transformers (transfusions). Mean-field-type federated transformers are presented in Section 4.7. Section 4.8 examines unlearning in MFTT. Federated Unlearning or Untraining in MFTT are presented in Section 4.9. Section 4.10 discusses the suboptimality of a constant queries, keys, values, weights, and biases in MFTT.

4.1 Self-and-Mean-Field Type

Traditional transformers, especially those relying on pairwise attentions such as Boltzmann-Gibbs, Coulomb, Sigmoid, Cucker-Smale, Kuramoto, Krause, consensus-based masked and unmasked self-attention, face scalability issues as the dimensionality or sequence length increases. MFTTs circumvent this by working with distributions rather than pairwise interactions. This means that MFTTs can efficiently handle high-dimensional data without the need to explicitly consider all pairwise interactions, which significantly reduces computational complexity. MFTTs are inherently designed to work with probability distributions, making them particularly well-suited for continuous data or scenarios where the input and output spaces are not discrete. This flexibility allows MFTTs to model more complex data distributions and interactions compared to traditional transformers that are often limited to discrete representations. By incorporating mean-field-type interactions, MFTTs can capture more subtle dependencies in the data. Traditional transformers might struggle to model such dependencies due to their fixed-dimensional representations and distribution-independent attention mechanisms. MFTTs, on the other hand, can adapt their representations based on the distribution of the inputs and outputs, potentially leading to improved performance on tasks that involve risk awareness and structured data. Mean-field-type game theory often provides insights into the distribution-dependent quantities-of-interest of systems with several interacting components. In the context of MFTTs, this means that the model can generalize better from observed data distributions to unseen data, improving robustness and adaptability across a range of tasks and data distributions.

In the previous chapter, the distribution of the data was frozen and fixed across layers. However, during the operation of generative machine intelligence, the distribution of the new data need not be the same across layers. This leads to a mean-field-type game problem and a mean-field-type transformer. In this setting we do not have an infinite number of decision-maker per sub-player. There is no n, D. There is an indeed a initial distribution of data μ_0. Then an initial signal input X_0 is generated according to μ_0. With some abuse of notation, one can consider an $\mathcal{H}_0$-valued random variable X_0 with probability law $\mu_0 = \mu_{X_0} = \mu_{Y_0}$. We equip the Hilbert space with the inner product $E(\langle X_1, X_2\rangle)$, where E denotes the expected value operator. The underlying processes are hence assumed to have a finite variance. The layer mean-field object is computed using a pushforward operator as in standard Mckean-Vlasov kernels.

The earlier implementations of transformers operate on fixed-dimensional representations, and their attention mechanisms were designed to handle the interactions between all pairs of elements within a sequence. Mean-field-type interactions include not only pairwise interaction but also interactions with the distribution of signal inputs and or outputs of the preceding layer. Mean-mield-type transformer extends the traditional transformer architecture by incorporating the concept of mean-field-type game theory, allowing it to work directly with distributions and providing a framework for handling interactions in a more scalable and flexible manner.

4.1.1 Mean-Field-Type Self-Attention Mechanisms

We provide below several examples of mean-field type self-attention mechanisms.

Example 17. *In addition to the mean-field-type interactions provided in Subsection 3.1.1 below are some other mean-field-type attentions*

- *The mean-field-type unmasked (Boltzmann-Gibbs) self-attention is*

$$mft_{unmasked-att}:$$

$$\left\{\begin{array}{l} \mathcal{H}^d \times \mathcal{P}(\mathcal{H}^d) \rightarrow \mathcal{H}^d \\ (x,\mu) \mapsto x + \frac{1}{\sqrt{H}}\sum_{h=1}^{H} W_h V_h \left(\frac{\int_{\mathcal{H}^d} e^{\frac{1}{\sqrt{k}}\langle Q_h x, K_h x'\rangle} x' \mu(dx')}{\int_{\mathcal{H}^d} e^{\frac{1}{\sqrt{k}}\langle Q_h x, K_h x''\rangle} \mu(dx'')}\right) \end{array}\right. . \tag{4.2}$$

- *The mean-field-type masked (Boltzmann-Gibbs) self-attention is*

$$mft_{masked-att}:$$

$$\left\{\begin{array}{l} [0,1] \times \mathcal{H}^d \times \mathcal{P}([0,1]\times\mathcal{H}^d) \rightarrow [0,1]\times\mathcal{H}^d \\ (s,x,\hat{\mu}) \mapsto (s, x + j_2), \\ j_2 = \frac{1}{\sqrt{H}}\sum_{h=1}^{H} W_h V_h \frac{\int_0^1\int_{\mathcal{H}^d} e^{\frac{1}{\sqrt{k}}\langle Q_h x, K_h x'\rangle} \mathbb{I}_{s'\leq s} x' \hat{\mu}(ds'dx')}{\int_0^1\int_{\mathcal{H}^d} e^{\frac{1}{\sqrt{k}}\langle Q_h x, K_h x''\rangle} \mathbb{I}_{s''\leq s}\hat{\mu}(ds''dx'')}. \end{array}\right. \tag{4.3}$$

- *The mean-field-type unmasked sigmoid self-attention is*

$$mftSigmoid_{unmasked-att}:$$

$$\begin{cases} \mathcal{H}^d \times \mathcal{P}(\mathcal{H}^d) \to \mathcal{H}^d \\ (x,\mu) \mapsto (x + j_2), \\ j_2 = \frac{1}{\sqrt{H}} \sum_{h=1}^H W_h V_h \int_{\mathcal{H}^d} e^{\frac{1}{\sqrt{k}}\langle Q_h x, K_h x'\rangle} x' \mu(dx'). \end{cases} \tag{4.4}$$

- *The mean-field-type masked sigmoid self-attention is*

$$mfSigmoid_{masked-att}:$$

$$\begin{cases} [0,1] \times \mathcal{H}^d \times \mathcal{P}([0,1] \times \mathcal{H}^d) \to [0,1] \times \mathcal{H}^d \\ (s, x, \hat{\mu}) \mapsto (s, x + j_2), \\ j_2 = \frac{1}{\sqrt{H}} \sum_{h=1}^H W_h V_h \int_0^1 \int_{\mathcal{H}^d} e^{\frac{1}{\sqrt{k}}\langle Q_h x, K_h x'\rangle} \mathbb{I}_{s' \le s} x' \hat{\mu}(ds'dx') \end{cases} \tag{4.5}$$

- *The projected mean-field unmasked Boltzmann-Gibbs self-attention is*

$$pmftBG_{unmaskedatt}:$$

$$\begin{cases} \mathcal{H}^d \times \mathcal{P}(\mathcal{H}^d) \to \mathcal{H}^d \\ (x,\mu) \mapsto x + proj_{x^\perp}(j_2) \\ j_2 = \left(\frac{1}{\sqrt{H}} \sum_{h=1}^H W_h V_h \frac{\int_{\mathcal{H}^d} e^{\frac{1}{\epsilon}\langle Q_h x, K_h x'\rangle} x' \mu(dx')}{\int_{\mathcal{H}^d} e^{\frac{1}{\epsilon}\langle Q_h x, K_h x''\rangle} \mu(dx'')} \right). \end{cases} \tag{4.6}$$

- *The projected mean-field masked Boltzmann-Gibbs self-attention is*

$$pmftBG_{masked-att}:$$

$$\begin{cases} [0,1] \times \mathcal{H}^d \times \mathcal{P}([0,1] \times \mathcal{H}^d) \to [0,1] \times \mathcal{H}^d \\ (s, x, \hat{\mu}) \mapsto (s, x + proj_{x^\perp}(j_2)), \\ j_2 = \left(\frac{1}{\sqrt{H}} \sum_{h=1}^H W_h V_h \frac{\int_0^1 \int_{\mathcal{H}^d} e^{\frac{1}{\epsilon}\langle Q_h x, K_h x'\rangle} \mathbb{I}_{s' \le s} y \hat{\mu}(ds'dx')}{\int_0^1 \int_{\mathcal{H}^d} e^{\frac{1}{\epsilon}\langle Q_h x, K_h x''\rangle} \mathbb{I}_{s'' \le s} \hat{\mu}(ds''dx'')} \right). \end{cases} \tag{4.7}$$

- *(masked) normalized distance:*

$$\begin{cases} [0,1] \times \mathcal{H}_0 \times \mathcal{P}([0,1] \times \mathcal{H}_0) \to [0,1] \times \mathcal{H}_0 \\ (s, X, \hat{\mu}) \mapsto (s, X + proj_{X^\perp}(j_2)), \\ j_2 = \left(\frac{1}{\sqrt{H}} \sum_{h=1}^H W_h V_h \int_0^1 \int_{\mathcal{H}_0} \phi_h(\|(X - X')\|) X' \mathbb{I}_{s' \le s} \hat{\mu}(ds'dX') \right). \end{cases} \tag{4.8}$$

where ϕ is a nonnegative continuous bounded function, one can choose for example $\frac{\|X\|}{1+\|X\|}$.

- *(masked) Vicsek and Projected Cucker-Smale:*

$$\begin{cases} [0,1] \times \mathcal{H}_0^2 \times \mathcal{P}([0,1] \times \mathcal{H}_0^2) \to [0,1] \times \mathcal{H}_0^2 \\ (s, X_1, X_2, \hat{\mu}) \mapsto (s, X_1 + X_2, X_2 + proj_{X_2^\perp}(\frac{1}{\sqrt{H}} \sum_{h=1}^H W_h V_h \int_0^1 \int_{\mathcal{H}_0} j_2)), \\ j_2 = \phi_h(\|Q_h K_h^\dagger (X_1 - X_1')\|)(X_2' - X_2) \mathbb{I}_{s' \le s} \hat{\mu}(ds'dX_1'dX_2'), \end{cases} \tag{4.9}$$

where ϕ is a smooth nonnegative function vanishing at infinity. Here (X_1, X_2) is interpreted as position-velocity.

- *(masked) Projected Newton:*

$$\begin{cases} [0,1] \times (\mathcal{H}_0)^2 \times \mathcal{P}([0,1] \times \mathcal{H}_0^2) \to [0,1] \times (\mathcal{H}_0)^2 \\ (s, X_1, X_2, \hat{\mu}) \mapsto (s, X_1 + X_2, X_2 - proj_{X_2^\perp} j_2), \\ j_2 = \left(\frac{1}{\sqrt{H}} \sum_{h=1}^H W_h V_h \int_0^1 \int_{\mathcal{H}_0} \nabla V_2(\|Q_h K_h^\dagger (X_1 - X_1')\|) \mathbb{I}_{s' \le s} \hat{\mu}(ds'dX_1') \right), \end{cases} \tag{4.10}$$

where V_2 is a smooth repulsive function. Here (X_1, X_2) is interpreted as position-velocity.

- *(masked) Projected Hegselmann-Krause:*

$$\begin{cases} [0,1] \times \mathcal{H}_0 \times \mathcal{P}([0,1] \times \mathcal{H}_0) \to [0,1] \times \mathcal{H}_0 \\ (s, X, \hat{\mu}) \mapsto (s, X + proj_{X^\perp}(j_2), \\ j_2 = \frac{1}{\sqrt{H}} \sum_{h=1}^{H} W_h V_h \frac{\int_0^1 \int_{\mathcal{H}_0} \phi(\|Q_h K_h^\dagger (X - X')\|) \mathbb{I}_{s' \leq s}(X' - X) \hat{\mu}(ds' dX')}{\int_0^1 \int_{\mathcal{H}_0} \phi(\|Q_h K_h^\dagger (X - X'')\|) \mathbb{I}_{s'' \leq s} \hat{\mu}(ds'' dX'')}. \end{cases} \tag{4.11}$$

- *(masked) Gradient:* $-\nabla V_1(X) - \int_0^1 \int_{\mathcal{H}_0} \nabla V_2(X - X') \mathbb{I}_{s' \leq s} \hat{\mu}(ds' dX')$.
- *(masked) Kuramoto: gradient model with* V_1, V_2 *in one dimension. (with sine function).*
- *(masked) Kernel-based:*

$$\begin{cases} [0,1] \times \mathcal{H}_0 \times \mathcal{P}([0,1] \times \mathcal{H}_0) \to [0,1] \times \mathcal{H}_0 \\ (s, X, \hat{\mu}) \mapsto (s, X + proj_{X^\perp}(j_2), \\ j_2 = \frac{1}{\sqrt{H}} \sum_{h=1}^{H} W_h V_h \int_0^1 \int_{\mathcal{H}_0} ker(h; X, X') \mathbb{I}_{s' \leq s} \hat{\mu}(ds' dX'). \end{cases} \tag{4.12}$$

- *Biot-Savart* $ker_1(X) = \frac{X^\perp}{\langle X, X \rangle}$ *where* $ker(X, X') = ker_1(X - X')$.
- *Platlak-Keller-Segel:* $ker_2(X) = \frac{X}{\langle X, X \rangle}$.
- *Coulomb:* $ker_3(X) = \nabla \phi(X),\ \phi(X) = -\log(\|X\|)\mathbb{I}_{d=2} + \frac{1}{\|X\|^{d-2}} \mathbb{I}_{d \geq 3}$.
- *(masked) Higher-order interaction:* $(k+1)$*-wise interaction:*

$$\begin{cases} [0,1] \times \mathcal{H}_0 \times \mathcal{P}([0,1] \times \mathcal{H}_0^k) \to [0,1] \times \mathcal{H}_0 \\ (s, X, \hat{\mu}^{\otimes k}) \mapsto (s, X + proj_{X^\perp}(\frac{1}{\sqrt{H}} \sum_{h=1}^{H} W_h V_h \int_0^1 \int_{\mathcal{H}_0} j_2)), \\ j_2 = ker(h; X, X_1', \ldots, X_k') \mathbb{I}_{s' \leq s} \hat{\mu}^{\otimes k}(ds' dX_1' \ldots dX_k'). \end{cases} \tag{4.13}$$

- *(masked) Localized mean-field-type interaction:* X *interacts with* X' *only when* X' *is a certain neighborhood of* X *such as* $X' \in \mathcal{B}_\rho(X)$*. This is a counting-based measure and is not a global interaction. It is a localized effect in the sense that it is limited to the neighborhood of the action.*

$$\begin{cases} [0,1] \times \mathcal{H}_0 \times \mathcal{P}([0,1] \times \mathcal{H}_0) \to [0,1] \times \mathcal{H}_0 \\ (s, X, \hat{\mu}) \mapsto (s, X + proj_{X^\perp}\left(\frac{1}{\sqrt{H}} \sum_{h=1}^{H} W_h V_h \int_0^1 \int_{X' \in B_\rho(X)} j_2\right)), \\ j_2 = ker(h; X, X') \mathbb{I}_{s' \leq s} \hat{\mu}(ds' dX'). \end{cases} \tag{4.14}$$

Based on these masked self-attention mechanisms of mean-field type, we build MFTTs.

4.1.2 Properties of Mean-Field-Type Self-Attention

We examine the properties of MFTTs. The topological structure plays an important role in establishing bounds. The space $\mathcal{H}_0 \times \mathcal{P}(\mathcal{H}_0)$ is equipped with the metric

$$((x, \mu), (x', \mu')) \mapsto \|x - x'\| + d_\alpha(\mu, \mu'),$$

where d_α is the α-Wasserstein metric with $\alpha \geq 1$. We study the properties of the mapping

$$(\mathcal{H}_0 \times \mathcal{P}(\mathcal{H}_0)) \ni (x, \mu) \mapsto att(x, \mu) \in \mathcal{H}_0.$$

The first-order (or pairwise) interaction kernels used in self-attention have the following shapes:

$$(\mathcal{H}_0 \times \mathcal{P}(\mathcal{H}_0)) \ni (x, \mu) \mapsto \int \kappa_2(x, x')x'\mu(dx') \in \mathcal{H}_0$$

with $\kappa_2(x, x')$ being real-valued and the mapping

$$(\mathcal{H}_0 \times \mathcal{P}(\mathcal{H}_0)) \ni (x, \mu) \mapsto \frac{\int \kappa_2(x, x')x'\mu(dx')}{\int \kappa_2(x, x'')\mu(dx'')} \in \mathcal{H}_0$$

with $\kappa_2(x, x') > 0, \forall (x, x') \in \mathcal{H}_0^2$.

The k-th order (or $(k+1)$-wise) interaction kernels have the following

$$(\mathcal{H}_0 \times \mathcal{P}(\mathcal{H}_0)) \ni (x, \mu) \mapsto \int \kappa_{k+1}(x, y^{(1)}, y^{(2)}, \ldots, y^{(k)}) \prod_{l=1}^{k} \mu(dy^{(l)}) \in \mathcal{H}_0$$

where $\kappa_{k+1}(x, y^{(1)}, y^{(2)}, \ldots, y^{(k)})$ is $\mathcal{H}_0$-valued.

Regularity of Kernel on the Product of Unit Balls

Due to the quadratic structure, the Boltzmann-Gibbs and Sigmoid self-attention mechanisms of mean-field type are not globally Lipschitz over $\mathcal{H}_0 \times \mathcal{P}(\mathcal{H}_0))$ but they are Lipschitz continuous when restricted to $\mathcal{B}_1 \times \mathcal{P}(\mathcal{B}_1))$. Since the range of the normalization operator is the unit ball and the self-attention operator takes only already normalized inputs, the global existence holds for the measure.

Lemma 31. *Let $\kappa_2 : \mathcal{H}_0^2 \to \mathbb{R}_{++}$ be continuously differentiable on $\mathcal{H}_0^2$. The restricted map*

$$(\mathcal{B}_1 \times \mathcal{P}(\mathcal{B}_1)) \ni (x, \mu) \mapsto \frac{\int \kappa_2(x, x')\mathbb{I}_{\mathcal{B}_1^2}(x, x')x'\mu(dx')}{\int \kappa_2(x, x')\mathbb{I}_{\mathcal{B}_1^2}(x, x'')\mu(dx'')} \in \mathcal{B}_1$$

is Lipschitz.

Corollary 4. *The following statements hold:*

- *The mean-field-type Boltzmann-Gibbs self-attention is continuous.*
- *The mean-field-type Boltzmann-Gibbs self-attention is not globally Lipschitz.*
- *The mean-field-type Boltzmann-Gibbs self-attention is Gâteaux differentiable.*
- *The mean-field-type Boltzmann-Gibbs self-attention is locally Lipschitz.*
- *The mean-field-type Boltzmann-Gibbs self-attention is Lipschitz when restricted on $\mathcal{B}_1 \times \mathcal{P}(\mathcal{B}_1)$ where B_1 is the closed unit ball of $\mathcal{H}_0$.*

Lemma 32. *Let $\kappa_2 : \mathcal{H}_0^2 \to \mathbb{R}_{+}$ be continuously differentiable on $\mathcal{H}_0^2$. The restricted map*

$$(\mathcal{B}_1 \times \mathcal{P}(\mathcal{B}_1)) \ni (x, \mu) \mapsto \int \kappa_2(x, x')x'\mu(dx') \in \mathcal{H}_0$$

is Lipschitz.

Proof: As κ_2 is continuously differentiable in the entire space $\mathcal{H}_0^2$, the restriction on the bounded and closed domain $\mathcal{B}_1^2$ has a bounded differential norm. This means that $\mathcal{B}_1^2 \ni (x,x') \mapsto \kappa_2(x,x')x'\mathbb{I}_{\mathcal{B}_1^2}(x,x')$ is Lipschitz on $\mathcal{B}_1^2$ with Lipschitz constant denoted by L_{κ_2}. Let $(x_1,\mu_1),(x_2,\mu_2)$ be two elements in $(\mathcal{B}_1 \times \mathcal{P}(\mathcal{B}_1))^2$.

$$\begin{array}{l} \| \int \kappa_2(x_1,x_1')x_1'\mu_1(dx_1') - \int \kappa_2(x_2,x_2')x_2'\mu_2(dx_2')\| \\ = \| \int \kappa_2(x_1,x_1')x_1'\mu_1(dx_1') - \int \kappa_2(x_2,x')x'\mu_1(dx') \\ + \int \kappa_2(x_2,x')x'\mu_1(dx') - \int \kappa_2(x_2,x_2')x_2'\mu_2(dx_2')\| \\ \leq \| \int \kappa_2(x_1,x_1')x_1'\mu_1(dx_1') - \int \kappa_2(x_2,x')x'\mu_1(dx')\| \\ + \| \int \kappa_2(x_2,x')x'\mu_1(dx') - \int \kappa_2(x_2,x_2')x_2'\mu_2(dx_2')\| \\ \leq L_{\kappa_2}(\|x_1 - x_2\| + d_1(\mu_1,\mu_2)). \end{array} \tag{4.15}$$

This completes the proof. □

As a corollary we obtain:

Corollary 5. *The following statements hold:*

- *The mean-field-type Sigmoid self-attention is continuous.*
- *The mean-field-type Sigmoid self-attention is not globally Lipschitz.*
- *The mean-field-type Sigmoid self-attention is Gâteaux differentiable.*
- *The mean-field-type Sigmoid self-attention is locally Lipschitz.*
- *The mean-field-type Sigmoid self-attention is Lipschitz when restricted on* $\mathcal{B}_1 \times \mathcal{P}(\mathcal{B}_1)$.

The above applies to the masked case as well: due to the quadratic structure, the Boltzmann-Gibbs and Sigmoid self-attention mechanisms of mean-field type are not globally Lipschitz over $[0,1] \times \mathcal{H}_0 \times \mathcal{P}([0,1] \times \mathcal{H}_0))$ but they are Lipschitz continuous when restricted to $[0,1] \times \mathcal{B}_1 \times \mathcal{P}([0,1] \times \mathcal{B}_1))$ where the space is equipped with the sum of the distances:

$$distance((s,X,\hat{\mu}),(s',X',\hat{\mu}')) = |s-s'| + \sqrt{\langle X-X', X-X'\rangle} + d_1(\hat{\mu},\hat{\mu}').$$

Since the range of the normalization operator is the unit ball and the self-attention operator takes only already normalized inputs, the global existence holds.

4.2 Some Implemented Mean-Field-Type Transformers

Based on these masked self-attention mechanisms of mean-field type, we build MFTTs. The MFTT will require a signal input and a distribution under which the signal was generated (Figure 4.2). Then, the signal will be updated as well as its new distribution in a multi-step operation layer-by-layer. The distribution update may require another block in the MFTT depending on the goal. This can also be a state- and mean-field-type feedback as we will see below.

$$
\begin{aligned}
&\text{Parameters } : L, d, k, k', q, H \in \mathbb{N},\\
&\quad 1 \leq l \leq L :\\
&\quad\quad 1 \leq h \leq H :\\
&\quad\quad Q_{lh}, K_{lh} \in \mathcal{L}(\mathcal{H}^d, \mathcal{H}^k), V_{lh} \in \mathcal{L}(\mathcal{H}^d, \mathcal{H}^{k'}), W_{lh} \in \mathcal{L}(\mathcal{H}^{k'}, \mathcal{H}^d),\\
&\quad\quad W_{1l} \in \mathcal{L}(\mathcal{H}^d, \mathcal{H}^q), b_{1l} \in \mathcal{H}^q, W_{2l} \in \mathcal{L}(\mathcal{H}^q, \mathcal{H}^d), b_{2l} \in \mathcal{H}^d,\\
&\text{Input } : s_i = s, y_0 = x_0 \in \mathcal{H}^d, \hat{\mu}_0 = \hat{\nu}_0 \in \mathcal{P}([0,1] \times \mathcal{H}^d),\\
&1 \leq l \leq L :\\
&\quad \text{Input at } l: \ s, x_l = y_{l-1} \in \mathcal{H}^d, \hat{\nu}_{l-1} \in \mathcal{P}([0,1] \times \mathcal{H}^d)\\
&\\
&\quad\quad u_l = \text{ Normalized } (x_l), \ \hat{\nu}_{l-\frac{3}{4}} = \mathbb{P}_{u_l}\\
&\quad\quad \tilde{u}_l = \frac{1}{\sqrt{H}} \sum_{h=1}^H W_{lh} \int_0^1 \int_{\mathcal{H}^d}\\
&\quad\quad \text{Attention}(s', Q_{lh}u_l, K_{lh}u_l', \hat{\nu}_{l-\frac{3}{4}}) V_{lh} \mathbb{I}_{s' \leq s} u_l' \hat{\nu}_{l-\frac{3}{4}}(ds' du_l')\\
&\quad\quad \hat{u}_l = x_l + \tilde{u}_l, \ \hat{\nu}_{l-\frac{1}{2}} = \mathbb{P}_{\hat{u}_l},\\
&\\
&\quad\quad \hat{y}_l = \text{ Normalized } (\hat{u}_l),\\
&\quad\quad \tilde{y}_l = W_{2l} r_l (W_{1l} \hat{y}_l + b_{1l}) + b_{2l},\\
&\quad\quad y_l = \hat{u}_l + \tilde{y}_l, \ \hat{\nu}_l = \mathbb{P}_{y_l},\\
&\\
&\quad \text{Output at } l : s, y_l, \hat{\nu}_l\\
&\text{Return: } s, y_L \in \mathcal{H}^d, \hat{\nu}_L \in \mathcal{P}([0,1] \times \mathcal{H}^d)
\end{aligned}
\tag{4.16}
$$

4.2.1 Mean-Field-Type Boltzmann-Gibbs Transformer

$$
\begin{aligned}
&\text{Parameters } : L, d, k, k', q, H \in \mathbb{N},\\
&\quad 1 \leq l \leq L :\\
&\quad\quad 1 \leq h \leq H : Q_{lh}, K_{lh} \in \mathcal{L}(\mathcal{H}^d, \mathcal{H}^k), V_{lh} \in \mathcal{L}(\mathcal{H}^d, \mathcal{H}^{k'}), W_{lh} \in \mathcal{L}(\mathcal{H}^{k'}, \mathcal{H}^d),\\
&\quad\quad W_{1l} \in \mathcal{L}(\mathcal{H}^d, \mathcal{H}^q), b_{1l} \in \mathcal{H}^q, W_{2l} \in \mathcal{L}(\mathcal{H}^q, \mathcal{H}^d), b_{2l} \in \mathcal{H}^d,\\
&\text{Input } : s_i = s, y_0 = x_0 \in \mathcal{H}^d, \hat{\mu}_0 = \nu_0 \in \mathcal{P}([0,1] \times \mathcal{H}^d),\\
&1 \leq l \leq L :\\
&\quad \text{Input at } l: \ s, x_l = y_{l-1} \in \mathcal{H}^d, \nu_{l-1} \in \mathcal{P}([0,1] \times \mathcal{H}^d),\\
&\\
&\quad\quad s, u_l = \frac{x_l}{1+\|x_l\|}, \hat{\nu}_{l-\frac{3}{4}}\\
&\quad\quad \tilde{u}_l = \frac{1}{\sqrt{H}} \sum_{h=1}^H W_{lh} \frac{\int_0^1 \int_{\mathcal{H}^d} e^{\frac{1}{\sqrt{k}} \langle Q_{lh}u_l, K_{lh}u_l' \rangle} V_{lh} \mathbb{I}_{s' \leq s} u_l' \hat{\nu}_{l-\frac{3}{4}}(ds' du_l')}{\int_0^1 \int_{\mathcal{H}^d} e^{\frac{1}{\sqrt{k}} \langle Q_{lh}u_l, K_{lh}u_{l'}'' \rangle} \mathbb{I}_{s'' \leq s} \hat{\nu}_{l-\frac{3}{4}}(ds'' du_l'')},\\
&\quad\quad \hat{u}_l = x_{li} + \tilde{u}_l, \ \hat{\nu}_{l-\frac{1}{2}},\\
&\quad\quad \hat{y}_l = \frac{\hat{u}_l}{1+\|\hat{u}_l\|},\\
&\quad\quad \tilde{y}_l = W_{2l} r_l (W_{1l} \hat{y}_l + b_{1l}) + b_{2l},\\
&\quad\quad y_l = \hat{u}_l + \tilde{y}_l, \ \hat{\nu}_l = \mathbb{P}_{y_l},\\
&\\
&\quad \text{Output at } l : s, y_l, \hat{\nu}_l\\
&\text{Return } s, y_L \in \mathcal{H}^d, \hat{\nu}_L \in \mathcal{P}([0,1] \times \mathcal{H}^d)
\end{aligned}
\tag{4.17}
$$

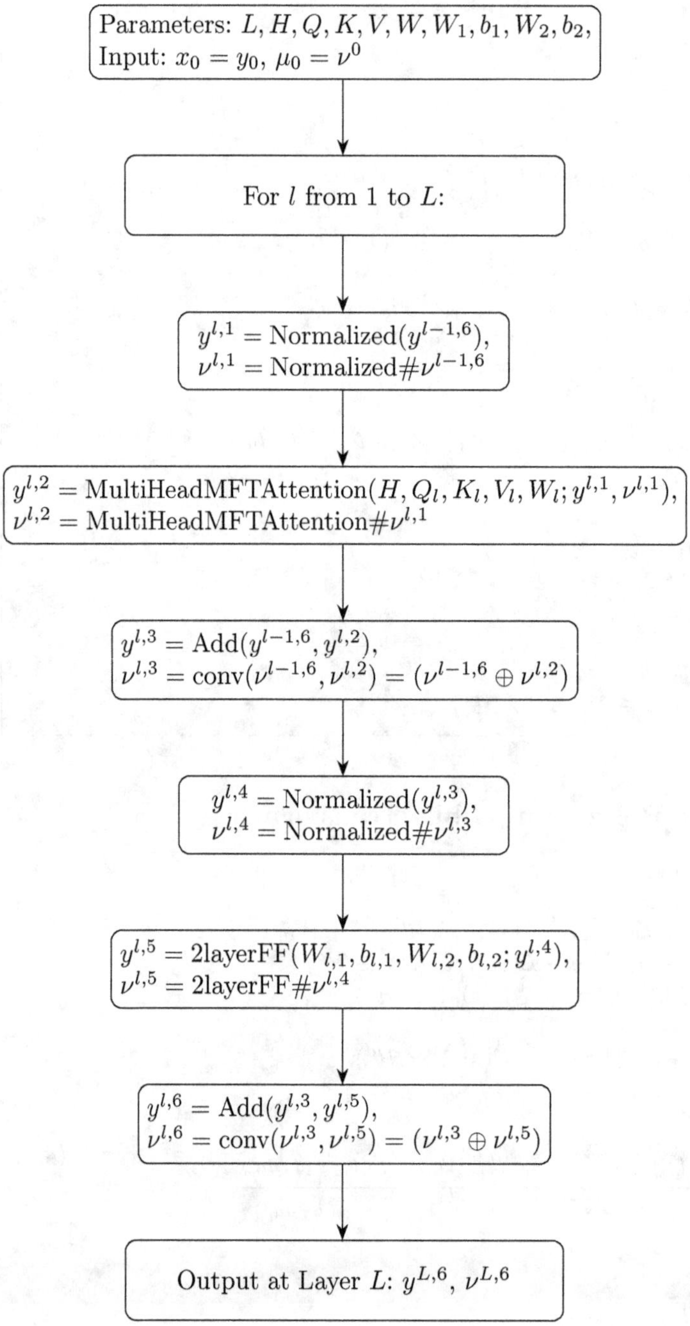
Parameters: $L, H, Q, K, V, W, W_1, b_1, W_2, b_2,$
Input: $x_0 = y_0$, $\mu_0 = \nu^0$
For l from 1 to L:
$y^{l,1} = \text{Normalized}(y^{l-1,6}),$
$\nu^{l,1} = \text{Normalized}\#\nu^{l-1,6}$
$y^{l,2} = \text{MultiHeadMFTAttention}(H, Q_l, K_l, V_l, W_l; y^{l,1}, \nu^{l,1}),$
$\nu^{l,2} = \text{MultiHeadMFTAttention}\#\nu^{l,1}$
$y^{l,3} = \text{Add}(y^{l-1,6}, y^{l,2}),$
$\nu^{l,3} = \text{conv}(\nu^{l-1,6}, \nu^{l,2}) = (\nu^{l-1,6} \oplus \nu^{l,2})$
$y^{l,4} = \text{Normalized}(y^{l,3}),$
$\nu^{l,4} = \text{Normalized}\#\nu^{l,3}$
$y^{l,5} = \text{2layerFF}(W_{l,1}, b_{l,1}, W_{l,2}, b_{l,2}; y^{l,4}),$
$\nu^{l,5} = \text{2layerFF}\#\nu^{l,4}$
$y^{l,6} = \text{Add}(y^{l,3}, y^{l,5}),$
$\nu^{l,6} = \text{conv}(\nu^{l,3}, \nu^{l,5}) = (\nu^{l,3} \oplus \nu^{l,5})$
Output at Layer L: $y^{L,6}$, $\nu^{L,6}$

FIGURE 4.2
Pointwise mean-field-type transformer (P-MFTT) block

4.2.2 Mean-Field-Type Sigmoid Transformer

Here we replace Boltzmann-Gibbs with sigmoid self-attention.

$$\begin{array}{l}
\text{Parameters }: L, d, k, k', q, H \in \mathbb{N}, \\
\quad 1 \le l \le L: \\
\qquad 1 \le h \le H: Q_{lh}, K_{lh} \in \mathcal{L}(\mathcal{H}^d, \mathcal{H}^k), V_{lh} \in \mathcal{L}(\mathcal{H}^d, \mathcal{H}^{k'}), W_{lh} \in \mathcal{L}(\mathcal{H}^{k'}, \mathcal{H}^d), \\
\qquad W_{1l} \in \mathcal{L}(\mathcal{H}^d, \mathcal{H}^q), b_{1l} \in \mathcal{H}^q, W_{2l} \in \mathcal{L}(\mathcal{H}^q, \mathcal{H}^d), b_{2l} \in \mathcal{H}^d, \\
\text{Input }: s_i = s, y_0 = x_0 \in \mathcal{H}^d, \hat{\mu}_0 = \hat{\nu}_0 \in \mathcal{P}([0,1] \times \mathcal{H}^d), \\
1 \le l \le L: \\
\quad \text{Input at } l: \ s, x_l = y_{l-1} \in \mathcal{H}^d, \hat{\nu}_{l-1} \in \mathcal{P}([0,1] \times \mathcal{H}^d)
\end{array}$$

$$\begin{array}{l}
u_l = \frac{x_l}{1+\|x_l\|}, \ \hat{\nu}_{l-\frac{3}{4}} \\
\tilde{u}_l = \frac{1}{\sqrt{H}} \sum_{h=1}^H W_{lh} \int_0^1 \int_{\mathcal{H}^d} e^{\frac{1}{\sqrt{k}} \langle Q_{lh} u_l, K_{lh} u_l' \rangle} V_{lh} \mathbb{I}_{s' \le s} u_l' \hat{\nu}_{l-\frac{3}{4}}(ds' du_l') \\
\hat{u}_l = x_l + \tilde{u}_l, \ \hat{\nu}_{l-\frac{1}{2}},
\end{array} \tag{4.18}$$

$$\begin{array}{l}
\hat{y}_l = \frac{\hat{u}_l}{1+\|\hat{u}_l\|}, \\
\tilde{y}_l = W_{2l} r_l (W_{1l} \hat{y}_l + b_{1l}) + b_{2l}, \\
y_l = \hat{u}_l + \tilde{y}_l, \ \hat{\nu}_l = \mathbb{P}_{y_l},
\end{array}$$

$$\begin{array}{l}
\quad \text{Output at } l: s, y_l, \hat{\nu}_l \\
\text{Return: } s, y_L \in \mathcal{H}^d, \hat{\nu}_L \in \mathcal{P}([0,1] \times \mathcal{H}^d)
\end{array}$$

4.2.3 Mixture of Experts - Mean-Field-Type Boltzmann-Gibbs Transformer

$$\begin{array}{l}
\text{Parameters }: \mathcal{E} = \{1, \ldots, E\}, L, D, d, k, k', q, H \in \mathbb{N}, \\
\quad 1 \le l \le L: \\
\qquad 1 \le h \le H: Q_{lh}, K_{lh} \in \mathcal{L}(\mathcal{H}^d, \mathcal{H}^k), V_{lh} \in \mathcal{L}(\mathcal{H}^d, \mathcal{H}^{k'}), W_{lh} \in \mathcal{L}(\mathcal{H}^{k'}, \mathcal{H}^d), \\
\qquad W_{1l,e} \in \mathcal{L}(\mathcal{H}^d, \mathcal{H}^q), b_{1l,e} \in \mathcal{H}^q, W_{2l,e} \in \mathcal{L}(\mathcal{H}^q, \mathcal{H}^d), b_{2l,e} \in \mathcal{H}^d, e \in \mathcal{E} \\
\qquad W_{3l} \in \mathcal{L}(\mathcal{H}^d, \mathbb{R}^E), b_{3l} \in \mathbb{R}^E, \\
\text{Input }: s_i, y_0 = x_0 \in \mathcal{H}^d, \hat{\mu}_0 = \nu_0 \in \mathcal{P}([0,1] \times \mathcal{H}^d), \\
1 \le l \le L: \\
\quad \text{Input at } l: \ s_i = s, x_l = y_{l-1} \in \mathcal{H}^d, \nu_{l-1} \in \mathcal{P}([0,1] \times \mathcal{H}^d),
\end{array}$$

$$\begin{array}{l}
s, u_l = \frac{x_l}{1+\|x_l\|}, \ \hat{\nu}_{l-\frac{3}{4}} \\
\tilde{u}_l = \frac{1}{\sqrt{H}} \sum_{h=1}^H W_{lh} \frac{\int_0^1 \int_{\mathcal{H}^d} e^{\frac{1}{\sqrt{k}} \langle Q_{lh} u_l, K_{lh} u_l' \rangle} V_{lh} \mathbb{I}_{s' \le s} u_l' \hat{\nu}_{l-\frac{3}{4}}(ds' du_l')}{\int_0^1 \int_{\mathcal{H}^d} e^{\frac{1}{\sqrt{k}} \langle Q_{lh} u_l, K_{lh} u_l'' \rangle} \mathbb{I}_{s'' \le s} \hat{\nu}_{l-\frac{3}{4}}(ds'' du_l'')}, \\
\hat{u}_l = x_l + \tilde{u}_l, \ \hat{\nu}_{l-\frac{1}{2}}, \\
\hat{y}_l = \frac{\hat{u}_l}{1+\|\hat{u}_l\|}, \\
\text{Choose } \mathcal{E}_i, \\
\quad \eta_{le} = \frac{e^{(\tilde{W}_{3l} \hat{y}_l + \tilde{b}_{3l})_e}}{\sum_{e' \in \mathcal{E}_i} e^{(\tilde{W}_{3l} \hat{y}_l + \tilde{b}_{3l})_{e'}}} \mathbb{I}_{e \in \mathcal{E}_i}, \\
\quad e = E+1: \tilde{y}_{l,e} = W_{2l,e} r_l (W_{1l,e} \hat{y}_{l,e} + b_{1l,e}) + b_{2l,e}, \\
\quad 1 \le e \le E_i: \ \tilde{y}_{l,e} = W_{2l,e} r_l (W_{1l,e} \hat{y}_{l,e} + b_{1l,e}) + b_{2l,e}, \\
\quad \tilde{y}_l = \eta_{l,E+1} \tilde{y}_{l,E+1} + \sum_{e \in \mathcal{E}_i} \eta_{le} \tilde{y}_{l,e}, \\
y_l = \hat{u}_l + \tilde{y}_l, \ \hat{\nu}_l = \mathbb{P}_{y_l},
\end{array} \tag{4.19}$$

$$\begin{array}{l}
\quad \text{Output at } l: s, y_l, \hat{\nu}_l, \\
\text{Return } s, y_L \in \mathcal{H}^d, \hat{\nu}_L \in \mathcal{P}([0,1] \times \mathcal{H}^d)
\end{array}$$

4.2.4 Mixture of Experts - Mean-Field-Type Sigmoid Transformer

$$
\begin{array}{l}
\text{Parameters } : \mathcal{E}=\{1,\ldots,E\}, L, D, d, k, k', q, H \in \mathbb{N},\\
\quad 1\leq l \leq L :\\
\qquad 1\leq h\leq H : Q_{lh}, K_{lh} \in \mathcal{L}(\mathcal{H}^d,\mathcal{H}^k), V_{lh}\in \mathcal{L}(\mathcal{H}^d,\mathcal{H}^{k'}), W_{lh}\in\mathcal{L}(\mathcal{H}^{k'},\mathcal{H}^d),\\
\qquad W_{1l,e}\in\mathcal{L}(\mathcal{H}^d,\mathcal{H}^q), b_{1l,e}\in\mathcal{H}^q, W_{2l,e}\in\mathcal{L}(\mathcal{H}^q,\mathcal{H}^d), b_{2l,e}\in\mathcal{H}^d, e\in\mathcal{E}\\
\qquad W_{3l}\in\mathcal{L}(\mathcal{H}^d,\mathbb{R}^E), b_{3l}\in\mathbb{R}^E,\\
\text{Input } : s_i=s, y_0=x_0\in\mathcal{H}^d, \hat{\mu}_0=\nu_0\in\mathcal{P}([0,1]\times\mathcal{H}^d),\\
1\leq l\leq L :\\
\quad \text{Input at } l: \ s, x_l=y_{l-1}\in\mathcal{H}^d, \nu_{l-1}\in\mathcal{P}([0,1]\times\mathcal{H}^d),\\
\\
\quad s_i, u_l=\frac{x_l}{1+\|x_l\|}, \ \hat{\nu}_{l-\frac{3}{4}},\\
\quad \tilde{u}_l=\frac{1}{\sqrt{H}}\sum_{h=1}^H W_{lh}\int_0^1\int_{\mathcal{H}^d} e^{\frac{1}{\sqrt{k}}\langle Q_{lh}u_l, K_{lh}u_l'\rangle} V_{lh}\mathbb{I}_{s'\leq s}u_l'\hat{\nu}_{l-\frac{3}{4}}(ds'du_l')\\
\quad \hat{u}_l=x_l+\tilde{u}_l, \ \hat{\nu}_{l-\frac{1}{2}},\\
\quad \hat{y}_l=\frac{\hat{u}_l}{1+\|\hat{u}_l\|},\\
\quad \text{Choose } \mathcal{E},\\
\qquad \eta_{le}=\frac{e^{(\tilde{W}_{3l}\hat{y}_l+\tilde{b}_{3l})_e}}{\sum_{e'\in\mathcal{E}_i} e^{(\tilde{W}_{3l}\hat{y}_l+\tilde{b}_{3l})_{e'}}}\mathbb{I}_{e\in\mathcal{E}_i},\\
\qquad e=E+1 : \tilde{y}_{l,e}=W_{2l,e}r_l(W_{1l,e}\hat{y}_{l,e}+b_{1l,e})+b_{2l,e},\\
\qquad 1\leq e\leq E_i : \ \tilde{y}_{l,e}=W_{2l,e}r_l(W_{1l,e}\hat{y}_{l,e}+b_{1l,e})+b_{2l,e},\\
\qquad \tilde{y}_l=\eta_{l,E+1}\tilde{y}_{l,E+1}+\sum_{e\in\mathcal{E}_i}\eta_{le}\tilde{y}_{l,e},\\
\quad y_l=\hat{u}_l+\tilde{y}_l=\mathbb{P}_{y_l},\\
\\
\quad \text{Output at } l : s, y_l, \hat{\nu}_l\\
\text{Return: } s, y_L\in\mathcal{H}^d, \hat{\nu}_L\in\mathcal{P}([0,1]\times\mathcal{H}^d)
\end{array}
\tag{4.20}
$$

4.2.5 Mean-Field-Type Transformer with Difference Attention

$$
\begin{array}{l}
\text{Parameters } : L, d, k, k', q, H\in\mathbb{N}, \chi\in\mathbb{R}\\
\quad 1\leq l\leq L :\\
\qquad 1\leq h\leq H : Q_{lh}^{(1)}, K_{lh}^{(1)}, Q_{lh}^{(2)}, K_{lh}^{(2)}\in\mathcal{L}(\mathcal{H}^d,\mathcal{H}^k),\\
\qquad\quad V_{lh}\in\mathcal{L}(\mathcal{H}^d,\mathcal{H}^{k'}), W_{lh}\in\mathcal{L}(\mathcal{H}^{k'},\mathcal{H}^d),\\
\qquad W_{1l}\in\mathcal{L}(\mathcal{H}^d,\mathcal{H}^q), b_{1l}\in\mathcal{H}^q, W_{2l}\in\mathcal{L}(\mathcal{H}^q,\mathcal{H}^d), b_{2l}\in\mathcal{H}^d,\\
\text{Input } : s_i=s, y_0=x_0\in\mathcal{H}^d, \hat{\mu}_0=\hat{\nu}_0\in\mathcal{P}([0,1]\times\mathcal{H}^d),\\
1\leq l\leq L :\\
\quad \text{Input at } l: \ s, x_l=y_{l-1}\in\mathcal{H}^d, \hat{\nu}_{l-1}\in\mathcal{P}([0,1]\times\mathcal{H}^d)\\
\\
\quad u_l=\text{Normalized}(x_l), \ \hat{\nu}_{l-\frac{3}{4}}=\mathbb{P}_{u_l},\\
\quad \tilde{u}_l=\frac{1}{\sqrt{H}}\sum_{h=1}^H W_{lh}\int_0^1\int_{\mathcal{H}^d} V_{lh}\mathbb{I}_{s'\leq s}u_l'\hat{\nu}_{l-\frac{3}{4}}(ds'du_l')\\
\quad (\ \text{Att}(s', Q_{lh}^{(1)}u_l, K_{lh}^{(1)}u_l', \hat{\nu}_{l-\frac{3}{4}})-\chi\ \text{Att}(s', Q_{lh}^{(2)}u_l, K_{lh}^{(2)}u_l', \hat{\nu}_{l-\frac{3}{4}}))\\
\quad \hat{u}_l=x_l+\tilde{u}_l, \ \hat{\nu}_{l-\frac{1}{2}},\\
\\
\quad \hat{y}_l=\text{Normalized}(\hat{u}_l),\\
\quad \tilde{y}_l=W_{2l}r_l(W_{1l}\hat{y}_l+b_{1l})+b_{2l},\\
\quad y_l=\hat{u}_l+\tilde{y}_l, \ \hat{\nu}_l=\mathbb{P}_{y_l},\\
\\
\quad \text{Output at } l : s, y_l, \hat{\nu}_l=\mathbb{P}_{y_l},\\
\text{Return: } s, y_L\in\mathcal{H}^d, \hat{\nu}_L\in\mathcal{P}([0,1]\times\mathcal{H}^d)
\end{array}
\tag{4.21}
$$

4.2.5.1 Mean-Field-Type Boltzmann-Gibbs Difference Transformer

$$\begin{aligned}
&\text{Parameters } : L, d, k, k', q, H \in \mathbb{N}, \chi \in \mathbb{R},\\
&\quad 1 \leq l \leq L:\\
&\quad 1 \leq h \leq H : Q^{(1)}_{lh}, K^{(1)}_{lh}, Q^{(2)}_{lh}, K^{(2)}_{lh} \in \mathcal{L}(\mathcal{H}^d, \mathcal{H}^k), V_{lh} \in \mathcal{L}(\mathcal{H}^d, \mathcal{H}^{k'}),\\
&\qquad W_{lh} \in \mathcal{L}(\mathcal{H}^{k'}, \mathcal{H}^d),\\
&\quad W_{1l} \in \mathcal{L}(\mathcal{H}^d, \mathcal{H}^q), b_{1l} \in \mathcal{H}^q, W_{2l} \in \mathcal{L}(\mathcal{H}^q, \mathcal{H}^d), b_{2l} \in \mathcal{H}^d,\\
&\text{Input } : s_i = s, y_0 = x_0 \in \mathcal{H}^d, \hat{\mu}_0 = \nu_0 \in \mathcal{P}([0,1] \times \mathcal{H}^d),\\
&1 \leq l \leq L:\\
&\quad \text{Input at } l: \ s, x_l = y_{l-1} \in \mathcal{H}^d, \nu_{l-1} \in \mathcal{P}([0,1] \times \mathcal{H}^d),\\
&\quad s, u_l = \frac{x_l}{1+\|x_l\|}, \hat{\nu}_{l-\frac{3}{4}}\\
&\quad \tilde{u}_l = \frac{1}{\sqrt{H}} \sum_{h=1}^{H} W_{lh} \int_0^1 \int_{\mathcal{H}^d} V_{lh} \mathbb{I}_{s' \leq s} u'_l \hat{\nu}_{l-\frac{3}{4}}(ds' du'_l)\\
&\quad \left(\frac{e^{\frac{1}{\sqrt{k}} \langle Q^{(1)}_{lh} u_l, K^{(1)}_{lh} u'_l \rangle}}{\int_0^1 \int_{\mathcal{H}^d} e^{\frac{1}{\sqrt{k}} \langle Q^{(1)}_{lh} u_l, K^{(1)}_{lh} u''_l \rangle} \mathbb{I}_{s'' \leq s} \hat{\nu}_{l-\frac{3}{4}}(ds'' du''_l)} \right.\\
&\quad \left. -\chi \frac{e^{\frac{1}{\sqrt{k}} \langle Q^{(2)}_{lh} u_l, K^{(2)}_{lh} u'_l \rangle}}{\int_0^1 \int_{\mathcal{H}^d} e^{\frac{1}{\sqrt{k}} \langle Q^{(2)}_{lh} u_l, K^{(2)}_{lh} u''_l \rangle} \mathbb{I}_{s'' \leq s} \hat{\nu}_{l-\frac{3}{4}}(ds'' du''_l)} \right),\\
&\quad \hat{u}_l = x_{li} + \tilde{u}_l, \ \hat{\nu}_{l-\frac{1}{2}},\\
&\quad \hat{y}_l = \frac{\hat{u}_l}{1+\|\hat{u}_l\|},\\
&\quad \tilde{y}_l = W_{2l} r_l (W_{1l} \hat{y}_l + b_{1l}) + b_{2l},\\
&\quad y_l = \hat{u}_l + \tilde{y}_l, \ \hat{\nu}_l = \mathbb{P}_{y_l},\\
&\quad \text{Output at } l: s, y_l, \hat{\nu}_l\\
&\text{Return } s, y_L \in \mathcal{H}^d, \hat{\nu}_L \in \mathcal{P}([0,1] \times \mathcal{H}^d)
\end{aligned} \tag{4.22}$$

4.2.5.2 Mean-Field-Type Sigmoid Difference Transformer

Here we replace Boltzmann-Gibbs with sigmoid difference self-attention.

$$\begin{aligned}
&\text{Parameters } : L, d, k, k', q, H \in \mathbb{N}, \chi \in \mathbb{R},\\
&\quad 1 \leq l \leq L:\\
&\quad 1 \leq h \leq H : Q^{(1)}_{lh}, K^{(1)}_{lh}, Q^{(2)}_{lh}, K^{(2)}_{lh} \in \mathcal{L}(\mathcal{H}^d, \mathcal{H}^k),\\
&\qquad V_{lh} \in \mathcal{L}(\mathcal{H}^d, \mathcal{H}^{k'}), W_{lh} \in \mathcal{L}(\mathcal{H}^{k'}, \mathcal{H}^d),\\
&\quad W_{1l} \in \mathcal{L}(\mathcal{H}^d, \mathcal{H}^q), b_{1l} \in \mathcal{H}^q, W_{2l} \in \mathcal{L}(\mathcal{H}^q, \mathcal{H}^d), b_{2l} \in \mathcal{H}^d,\\
&\text{Input } : s_i = s, y_0 = x_0 \in \mathcal{H}^d, \hat{\mu}_0 = \hat{\nu}_0 \in \mathcal{P}([0,1] \times \mathcal{H}^d),\\
&1 \leq l \leq L:\\
&\quad \text{Input at } l: \ s, x_l = y_{l-1} \in \mathcal{H}^d, \hat{\nu}_{l-1} \in \mathcal{P}([0,1] \times \mathcal{H}^d)\\
&\quad u_l = \frac{x_l}{1+\|x_l\|}, \ \hat{\nu}_{l-\frac{3}{4}}\\
&\quad O_{lh} = \left(e^{\frac{1}{\sqrt{k}} \langle Q^{(1)}_{lh} u_l, K^{(1)}_{lh} u'_l \rangle} - \chi e^{\frac{1}{\sqrt{k}} \langle Q^{(2)}_{lh} u_l, K^{(2)}_{lh} u'_l \rangle} \right)\\
&\quad \tilde{u}_l = \frac{1}{\sqrt{H}} \sum_{h=1}^{H} W_{lh} \int_0^1 \int_{\mathcal{H}^d} O_{lh} V_{lh} \mathbb{I}_{s' \leq s} u'_l \hat{\nu}_{l-\frac{3}{4}}(ds' du'_l)\\
&\quad \hat{u}_l = x_l + \tilde{u}_l, \ \hat{\nu}_{l-\frac{1}{2}},
\end{aligned}$$

$$\hat{y}_l = \frac{\hat{u}_l}{1+\|\hat{u}_l\|},$$
$$\tilde{y}_l = W_{2l} r_l (W_{1l}\hat{y}_l + b_{1l}) + b_{2l},$$
$$y_l = \hat{u}_l + \tilde{y}_l, \ \hat{\nu}_l = \mathbb{P}_{y_l}, \tag{4.23}$$

$$\text{Output at } l : s, y_l, \hat{\nu}_l,$$
$$\text{Return: } s, y_L \in \mathcal{H}^d, \hat{\nu}_L \in \mathcal{P}([0,1] \times \mathcal{H}^d)$$

4.2.6 Mean-Field-Type Transformer with HoloNorm

Let $(\mathcal{X}, \langle \cdot, \cdot \rangle)$ and $(\mathcal{Y}, \langle \cdot, \cdot \rangle)$ be separable Hilbert spaces. Denote by $\mathcal{P}_2(\mathcal{X})$ the set of Borel probability measures on $\mathcal{X}$ with finite second moments. Let $\mu(0, x, y) \in \mathcal{P}_2(\mathcal{X} \times \mathcal{Y})$ denote the input-output joint distribution, and define its marginal in x as $\nu(0, x) \in \mathcal{P}_2(\mathcal{X})$. Initialize:

$$\nu^{(0,6)} := \nu(0, \cdot).$$

Changes at Layer l

We define the evolution of the MFTT layer-by-layer, for $l = 1$ to L.

HoloNorm pushforward. Define the HoloNorm function defined in $\mathcal{X} \to B_{\mathcal{X}}(0, 1)$

$$\text{hn}(x) := \frac{x}{1 + \|x\|},$$

then apply

$$\nu^{(l,1)} := \text{hn} \, \# \nu^{(l-1,6)}.$$

The **Pushforward:**

$$\text{hn}_{\#} \, \nu \in \mathcal{P}_2(\mathcal{X}), \quad \text{for } \nu \in \mathcal{P}_2(\mathcal{X})$$

$T : \mathcal{X} \to \mathcal{Y}$ be a measurable function. The *pushforward measure* $T_{\#}\mu \in \mathcal{P}_2(\mathcal{Y})$ is defined by

$$(T_{\#}\mu)(B) := \mu\left(T^{-1}(B)\right) \quad \text{for all Borel sets } B \subseteq \mathcal{Y}.$$

Equivalently, for any bounded measurable test function $\varphi : \mathcal{Y} \to \mathbb{R}$, we have:

$$\int_{\mathcal{Y}} \varphi(y) \, (T_{\#}\mu)(dy) = \int_{\mathcal{X}} \varphi(T(x)) \, \mu(dx).$$

Multi-head attention pushforward. For each head $h \in \{1, \ldots, H\}$, let

$$Q_h^{(\ell)}, K_h^{(\ell)} : \mathcal{X} \to \mathcal{K}, \quad V_h^{(\ell)} : \mathcal{X} \to \mathcal{V}, \quad W_h : \mathcal{V} \to \mathcal{X}$$

be bounded linear maps, with $\mathcal{K}$ and $\mathcal{V}$ Hilbert spaces.

Define the exponential kernel:

$$\alpha_h^{(l)}(x, y) := \exp\left(\langle Q_h^{(l)} x, K_h^{(l)} y \rangle_{\mathcal{K}} \right).$$

Then the mean-field-type attention map defined on $\mathcal{X} \times \mathcal{P}_2(\mathcal{X})$ by

$$\text{mha}^{(l)}(x, \nu) := \frac{1}{\sqrt{H}} \sum_{h=1}^{H} W_h^{(l)} \left(\int_{\mathcal{X}} \alpha_h^{(l)}(x, y) V_h^{(l)} y \, \nu(dy) \right).$$

Pushforward the measure:

$$\nu^{(l,2)} := \text{mha}^{(l)}(\cdot, \nu^{(l,1)}) \# \nu^{(l,1)}.$$

Convolution. Define a symmetric convolution between measures:

$$\nu^{(l,3)} := \mathrm{conv}(\nu^{(l-1,6)}, \nu^{(l,2)}).$$

HoloNorm again.

$$\nu^{(l,4)} := \mathrm{hn}\,\#\nu^{(l,3)}.$$

2-layered Feedforward pushforward. Let $W_1^{(l)}, W_2^{(l)} : \mathcal{X} \to \mathcal{X}$ be bounded linear maps and $b_1^{(l)}, b_2^{(l)} \in \mathcal{X}$ be biases. Define the two-layered map:

$$2\mathrm{ff}^{(l)}(x) := W_2^{(l)} \cdot \mathrm{hn}(W_1^{(l)}x + b_1^{(l)}) + b_2^{(l)},$$

then pushforward:

$$\nu^{(l,5)} := 2\mathrm{ff}^{(l)}\,\#\nu^{(l,4)}.$$

Final convolution.

$$\nu^{(l,6)} := \mathrm{conv}(\nu^{(l,3)}, \nu^{(l,5)}).$$

Final Output

Return the final measure: $\nu^{(L,6)}$

This last measure represents the outcome of the transformer applied in distributional space and can be expressed as $\nu^{(L,6)} = (o^{(L)}_{\theta(L)} \circ \ldots \circ o^{(1)}_{\theta(1)})\#\nu(0,.)$ with the ℓ-th operator $o^{(\ell)}_{\theta(\ell)} = (Id + 2ff \circ hn) \circ (Id + mha \circ hn)$. Algorithm 1 displays the steps.

Algorithm 1 Mean-Field-Type Transformer (MFTT)

Require: Initial measure $\nu^{(0)}$ on Hilbert space $\mathcal{X}$ from marginal of input-output distribution $\mu(0, x, y)$.

Initialize $\nu^{(0,6)} \leftarrow \nu^{(0)}$
for $\ell = 1$ to L **do**
 $\nu^{(\ell,1)} \leftarrow hn_{\#}\nu^{(\ell-1,6)}$ {Apply HoloNorm pushforward}
 $\nu^{(\ell,2)} \leftarrow mha^{(\ell)}(\cdot, \nu^{(\ell,1)})_{\#}\nu^{(\ell,1)}$ {Pushforward through multi-head attention}
 $\nu^{(\ell,3)} \leftarrow \mathrm{conv}(\nu^{(\ell-1,6)}, \nu^{(\ell,2)})$ {Convolution merge}
 $\nu^{(\ell,4)} \leftarrow hn_{\#}\nu^{(\ell,3)}$ {HoloNorm projection}
 $\nu^{(\ell,5)} \leftarrow 2ff^{(\ell)}_{\#}\nu^{(\ell,4)}$ {2-layer feedforward pushforward}
 $\nu^{(\ell,6)} \leftarrow \mathrm{conv}(\nu^{(\ell,3)}, \nu^{(\ell,5)})$ {Final residual update}
end for
return Final measure $\nu^{(L,6)}$

Each function in the HoloNorm MFTT architecture is designed to be simple, and compatible with probability measures. The *HoloNorm* function $\mathrm{hn}(x) = \frac{x}{1+\|x\|}$ is smooth, norm-bounded, and contractive, mapping elements of $\mathcal{X}$ into a open unit ball (bounded subset). It ensures numerical stability and is computationally inexpensive, involving only a norm computation and a scalar division. The *multi-head attention* mechanism $\mathrm{mha}(x, \nu)$ computes weighted averages over the measure ν, using exponential attention scores based on inner products $\langle Q_h x, K_h y\rangle$. This mechanism is linear in both the integration and matrix operations, and attention heads are naturally parallelizable (given ν), enabling efficient computation. The *two-layer feedforward network* $2\mathrm{ff}(x) = W_2 \cdot \mathrm{hn}(W_1 x + b_1) + b_2$ is a composition of affine maps and a HoloNorm nonlinearity, yielding low computational complexity while maintaining smoothness and stability.

4.2.7 Mean-Field-Type Boltzmann-Gibbs Transformer with Mixture-of-Heads

$$
\begin{aligned}
&\text{Parameters } : L, d, k, k', q, H \in \mathbb{N},\\
&\quad 1 \leq l \leq L:\\
&\quad\quad 1 \leq h \leq H : Q_{lh}, K_{lh} \in \mathcal{L}(\mathcal{H}^d, \mathcal{H}^k), V_{lh} \in \mathcal{L}(\mathcal{H}^d, \mathcal{H}^{k'}), W_{lh} \in \mathcal{L}(\mathcal{H}^{k'}, \mathcal{H}^d),\\
&\quad\quad W_{1l} \in \mathcal{L}(\mathcal{H}^d, \mathcal{H}^q), b_{1l} \in \mathcal{H}^q, W_{2l} \in \mathcal{L}(\mathcal{H}^q, \mathcal{H}^d), b_{2l} \in \mathcal{H}^d,\\
&\text{Input } : s_i = s, y_0 = x_0 \in \mathcal{H}^d, \hat{\mu}_0 = \nu_0 \in \mathcal{P}([0,1] \times \mathcal{H}^d),\\
&1 \leq l \leq L:\\
&\quad \text{Input at } l: \ s, x_l = y_{l-1} \in \mathcal{H}^d, \nu_{l-1} \in \mathcal{P}([0,1] \times \mathcal{H}^d),\\
&\\
&\quad\quad s, u_l = \frac{x_l}{1+\|x_l\|}, \hat{\nu}_{l-\frac{3}{4}}\\
&\quad\quad \text{SelectH} \subseteq \{1, \ldots, H\}, p_l \in \mathbb{R}^H\\
&\\
&\quad\quad O_{lh} = \frac{\int_0^1 \int_{\mathcal{H}^d} e^{\frac{1}{\sqrt{k}}\langle Q_{lh} u_l, K_{lh} u_l' \rangle} V_{lh} \mathbb{I}_{s' \leq s} u_l' \hat{\nu}_{l-\frac{3}{4}}(ds' du_l')}{\int_0^1 \int_{\mathcal{H}^d} e^{\frac{1}{\sqrt{k}}\langle Q_{lh} u_l, K_{lh} u_l'' \rangle} \mathbb{I}_{s'' \leq s} \hat{\nu}_{l-\frac{3}{4}}(ds'' du_l'')},\\
&\quad\quad \tilde{u}_l = \frac{1}{\sqrt{H}} \sum_{h=1}^H p_{lh} \mathbb{I}_{h \in SelectH} W_{lh} O_{lh},\\
&\quad\quad \hat{u}_l = x_{li} + \tilde{u}_l, \ \hat{\nu}_{l-\frac{1}{2}},\\
&\quad\quad \hat{y}_l = \frac{\hat{u}_l}{1+\|\hat{u}_l\|},\\
&\quad\quad \tilde{y}_l = W_{2l} r_l (W_{1l} \hat{y}_l + b_{1l}) + b_{2l},\\
&\quad\quad y_l = \hat{u}_l + \tilde{y}_l, \ \hat{\nu}_l = \mathbb{P}_{y_l},\\
&\\
&\quad \text{Output at } l: s, y_l, \hat{\nu}_l\\
&\text{Return } s, y_L \in \mathcal{H}^d, \hat{\nu}_L \in \mathcal{P}([0,1] \times \mathcal{H}^d)
\end{aligned}
\tag{4.24}
$$

4.2.8 Mean-Field-Type Sigmoid Transformer with Mixture-of-Heads

Here we replace Boltzmann-Gibbs with sigmoid of mixture-of-heads self-attention.

$$
\begin{aligned}
&\text{Parameters } : L, d, k, k', q, H \in \mathbb{N},\\
&\quad 1 \leq l \leq L:\\
&\quad\quad 1 \leq h \leq H : Q_{lh}, K_{lh} \in \mathcal{L}(\mathcal{H}^d, \mathcal{H}^k), V_{lh} \in \mathcal{L}(\mathcal{H}^d, \mathcal{H}^{k'}), W_{lh} \in \mathcal{L}(\mathcal{H}^{k'}, \mathcal{H}^d),\\
&\quad\quad W_{1l} \in \mathcal{L}(\mathcal{H}^d, \mathcal{H}^q), b_{1l} \in \mathcal{H}^q, W_{2l} \in \mathcal{L}(\mathcal{H}^q, \mathcal{H}^d), b_{2l} \in \mathcal{H}^d,\\
&\text{Input } : s_i = s, y_0 = x_0 \in \mathcal{H}^d, \hat{\mu}_0 = \hat{\nu}_0 \in \mathcal{P}([0,1] \times \mathcal{H}^d),\\
&1 \leq l \leq L:\\
&\quad \text{Input at } l: \ s, x_l = y_{l-1} \in \mathcal{H}^d, \hat{\nu}_{l-1} \in \mathcal{P}([0,1] \times \mathcal{H}^d)\\
&\\
&\quad\quad u_l = \frac{x_l}{1+\|x_l\|}, \ \hat{\nu}_{l-\frac{3}{4}}\\
&\quad\quad \text{SelectH} \subseteq \{1, \ldots, H\}, p_l \in \mathbb{R}^H\\
&\quad\quad \tilde{u}_l = \frac{1}{\sqrt{H}} \sum_{h=1}^H p_{lh} \mathbb{I}_{h \in SelectH} W_{lh} j_2\\
&\quad\quad\quad j_2 = \int_0^1 \int_{\mathcal{H}^d} e^{\frac{1}{\sqrt{k}}\langle Q_{lh} u_l, K_{lh} u_l' \rangle} V_{lh} \mathbb{I}_{s' \leq s} u_l' \hat{\nu}_{l-\frac{3}{4}}(ds' du_l')\\
&\quad\quad \hat{u}_l = x_l + \tilde{u}_l, \ \hat{\nu}_{l-\frac{1}{2}},\\
&\\
&\quad\quad \hat{y}_l = \frac{\hat{u}_l}{1+\|\hat{u}_l\|},\\
&\quad\quad \tilde{y}_l = W_{2l} r_l (W_{1l} \hat{y}_l + b_{1l}) + b_{2l},\\
&\quad\quad y_l = \hat{u}_l + \tilde{y}_l, \ \hat{\nu}_l = \mathbb{P}_{y_l},\\
&\\
&\quad \text{Output at } l: s, y_l, \hat{\nu}_l\\
&\text{Return: } s, y_L \in \mathcal{H}^d, \hat{\nu}_L \in \mathcal{P}([0,1] \times \mathcal{H}^d)
\end{aligned}
\tag{4.25}
$$

4.2.9 Mean-Field-Type Sigmoid Transformer with Mixture-of-Heads and Mixture-of-Experts

What we save in the mean-field-type transformer from the previous architecture is D presented in Subsection 2.8.3 and all the operations repeated D times. We instead have to compute the probability measure using a pushforward operator.

$$
\begin{array}{l}
\text{Parameters } : L, d, k, k', q, H \in \mathbb{N}, \\
\quad 1 \le l \le L : \\
\quad\quad 1 \le h \le H : Q_{lh}, K_{lh} \in \mathcal{L}(\mathcal{H}^d, \mathcal{H}^k), V_{lh} \in \mathcal{L}(\mathcal{H}^d, \mathcal{H}^{k'}), W_{lh} \in \mathcal{L}(\mathcal{H}^{k'}, \mathcal{H}^d), \\
\quad\quad W_{1l} \in \mathcal{L}(\mathcal{H}^d, \mathcal{H}^q), b_{1l} \in \mathcal{H}^q, W_{2l} \in \mathcal{L}(\mathcal{H}^q, \mathcal{H}^d), b_{2l} \in \mathcal{H}^d, p, \lambda \\
\text{Input } : s_i = s, y_0 = x_0 = y_{0,6} \in \mathcal{H}^d, \hat{\mu}_0 = \hat{\nu}_0 = \hat{\nu}_{0,6} \in \mathcal{P}([0,1] \times \mathcal{H}^d), \\
t \in \{0, \dots, T-1\} : \\
\quad 1 \le l \le L : \\
\quad\quad \text{Input at } l : \ s, y_{l-1,6} \in \mathcal{H}^d, \hat{\nu}_{l-1,6} = \mathbb{P}_{y_{l-1,6}} \in \mathcal{P}([0,1] \times \mathcal{H}^d) \\
\\
\quad\quad\quad y_{l,1} = \frac{y_{l-1,6}}{1+\|y_{l-1,6}\|} =: \text{Normalized}(y_{l-1,6}), \ \hat{\nu}_{l,1} = \text{Normalized}\#\hat{\nu}_{l-1,6} \\
\quad\quad\quad \text{SelectH} \subseteq \{1, \dots, H\}, p_l \in \mathbb{R}_+^H \\
\quad\quad\quad y_{l,2} = \frac{1}{\sqrt{H}} \sum_{h=1}^H p_{lh} \mathbb{I}_{h \in SelectH} W_{lh} \int_0^1 \int_{\mathcal{H}^d} j_2 \\
\quad\quad\quad\quad j_2 = e^{\frac{1}{\sqrt{k}} \langle Q_{lh} y_{l,1}, K_{lh} y_l' \rangle} V_{lh} \mathbb{I}_{s' \le s} y_l' \hat{\nu}_{l,1}(ds' dy_l') \\
\quad\quad\quad \hat{\nu}_{l,2} = \ \text{MultiHead-MoH-MFTAttention } \#\hat{\nu}_{l,1} \\
\quad\quad\quad y_{l,3} = Add(y_{l-1,6}, y_{l,2}) = y_{l-1,6} + y_{l,2}, \ \hat{\nu}_{l,2} = conv(\hat{\nu}_{l-1,6}, \hat{\nu}_{l,2}), \\
\\
\quad\quad\quad y_{l,4} = \frac{y_{l,3}}{1+\|y_{l,3}\|} = \text{Normalized}(y_{l,3}), \ \hat{\nu}_{l,4} = \text{Normalized}\#\hat{\nu}_{l,3}, \\
\quad\quad\quad \text{Choose } \mathcal{E}, \\
\quad\quad\quad\quad \eta_{le} = \frac{e^{(\tilde{W}_{3l} y_{l,4} + \tilde{b}_{3l})_e}}{\sum_{e' \in \mathcal{E}_i} e^{(\tilde{W}_{3l} y_{l,4} + \tilde{b}_{3l})_{e'}}} \mathbb{I}_{e \in \mathcal{E}_i}, \\
\quad\quad\quad\quad e = E+1 : \tilde{y}_{l,e} = W_{2l,e} r_l(W_{1l,e} y_{l,4} + b_{1l,e}) + b_{2l,e}, \\
\quad\quad\quad\quad 1 \le e \le E_i : \ \tilde{y}_{l,e} = W_{2l,e} r_l(W_{1l,e} y_{l,4} + b_{1l,e}) + b_{2l,e}, \\
\quad\quad\quad\quad y_{l,5} = \eta_{l,E+1} \tilde{y}_{l,E+1} + \sum_{e \in \mathcal{E}_i} \eta_{le} \tilde{y}_{l,e}, \\
\quad\quad\quad\quad \hat{\nu}_{l,5} = \ \text{2LayerFF-MoE}\#\hat{\nu}_{l,4}, \\
\quad\quad\quad y_{l,6} = \text{Add}(y_{l,5}, y_{l,3}) = y_{l,5} + y_{l,3}, \ \hat{\nu}_{l,6} = \hat{\nu}_{l,3} \oplus \hat{\nu}_{l,5} = conv(\hat{\nu}_{l,3}, \hat{\nu}_{l,5}) \\
\\
\quad\quad \text{Output at } l : s, y_{l,6}, \hat{\nu}_{l,6} \\
\quad \text{Return: } s, y_{L,6} \in \mathcal{H}^d, \hat{\nu}_{L,6} \in \mathcal{P}([0,1] \times \mathcal{H}^d) \\
\quad s, x_{t+1} = [(1-\lambda_t) x_t + \lambda_t y_{L,6}], \ \mu_{t+1} = [(1-\lambda_t) Id + \lambda_t y_{L,6}]\#\mu_t, \\
\quad \text{Return: } s, x_T, \mu_T
\end{array}
\tag{4.26}
$$

4.3 Outcomes of Mean-Field-Type Transformer

The MFTT operates by iteratively processing data through a sequence of layers, where each layer includes an operator that combines the input with the distribution of inputs (or preceding outputs). Specifically, the operator at each layer, denoted $\hat{O}_l$, maps an input and a measure (or distribution) over that input to an output and a new measure. Formally, at layer l, the operator $\hat{\mathcal{O}}_l$ is defined as: $\hat{\mathcal{O}}_l : \mathcal{H}_{l-1} \times \mathcal{P}(\mathcal{H}_{l-1}) \to \mathcal{H}_l$, where $\mathcal{H}_{l-1}$ represents the space of pre-outputs at the previous layer which serves as pre-inputs at layer l, and $\mathcal{P}(\mathcal{H}_{l-1})$ represents the space of probability measures over $\mathcal{H}_{l-1}$.

Given an initial pre-input $X^{(0)}$ with distribution $\mu^{(0)}$, one gets the input $(X^{(0)}, \mu^{(0)})$. The MFTT processes data through the following sequence of operations:

- **Layer 1:** $Y^{(1)} = \hat{\mathcal{O}}_1(X^{(0)}, \mu^{(0)})$, and $\nu^{(1)} = \hat{\mathcal{O}}_1(. \mid \mu^{(0)})\#\mu^{(0)}$. Here, $Y^{(1)}$ is the pre-output of the first layer, and $\nu^{(1)}$ is the measure induced by applying $\hat{\mathcal{O}}_1$ to $\mu^{(0)}$.
- **Subsequent Layers:** At layer l: $Y^{(l)} = \hat{\mathcal{O}}_l(Y^{(l-1)}, \nu^{(l-1)})$, and $\nu^{(l)} = \hat{\mathcal{O}}_l(. \mid \nu^{(l-1)})\#\nu^{(l-1)}$. This process continues until layer L.
- **Final Output:** The output of the MFTT is $(Y^{(L)}, \nu^{(L)})$, which represents the result after processing through all L layers.

Note that here the collection $(\nu^{(1)}, \dots, \nu^{(L-1)})$ is not frozen; it will be computed layer by layer together with the output of the preceding layer.

Definition 30. *The Mean-Field-Type Transformer (MFTT) is formally defined by the tuple* $(X^{(0)}, \mu^{(0)}, L, H, d, (\mathcal{H}_{l-1} \times \mathcal{P}(\mathcal{H}_{l-1}), \hat{\mathcal{O}}_l, \theta_l)_{l\in\{1,\dots,L\}}, \mathcal{H}_L)$.

Let $Y^{(l-\frac{3}{4})} = u_l = \mathcal{O}_{l,nn}(Y^{(l-1)})$, and $\mu_{Y^{(l-\frac{3}{4})}} = \mathbb{P}_{Y^{(l-\frac{3}{4})}} = \mathcal{L}(Y^{(l-\frac{3}{4})})$ be the probability law of $Y^{(l-\frac{3}{4})}$.

$$
\begin{aligned}
& X^{(0)} \sim \mu^{(0)}, \\
& l = 1, \\
& Y^{(\frac{1}{4})} = \mathcal{O}_{1,nn}(X^{(0)}), \mu_{Y^{(\frac{1}{4})}}, \\
& Y^{(\frac{1}{2})} = \hat{\mathcal{F}}_{\frac{1}{2},\mu_{Y^{\frac{1}{4}}}}(X^{(0)}), \\
& Y^{(1)} = \hat{\mathcal{F}}_1(Y^{(\frac{1}{2})}), \quad \text{compute: } \mu_{Y^{(1)}} \\
& \dots \\
& \\
& l \in \{2, \dots, L-1\} : \\
& Y^{(l-\frac{3}{4})} = \mathcal{O}_{l,nn}(Y^{(l-1)}), \mu_{Y^{(l-\frac{3}{4})}}, \\
& \\
& Y^{(l-\frac{1}{2})} = \hat{\mathcal{F}}_{l-\frac{1}{2},\mu_{Y^{(l-\frac{3}{4})}}}(Y^{(l-1)}), \\
& Y^{(l)} = \hat{\mathcal{F}}_l(Y^{(l-\frac{1}{2})}), \quad \text{compute: } \mu_{Y^{(l)}} \\
& \\
& \dots \\
& \\
& Y^{(L-\frac{3}{4})} = \mathcal{O}_{L,nn}(Y^{(L-1)}), \mu_{Y^{(L-\frac{3}{4})}}, \\
& Y^{(L-\frac{1}{2})} = \hat{\mathcal{F}}_{L-\frac{1}{2},\mu_{Y^{(l-\frac{3}{4})}}}(Y^{(l-1)}), \\
& X_* = Y^{(L)} = \hat{\mathcal{F}}_L(Y^{(L-\frac{1}{2})}), \quad \text{compute: } \mu_{Y^{(L)}}
\end{aligned}
\tag{4.27}
$$

The mean-field-type fixed-point system yields

$$
\begin{aligned}
&l = 1: \\
&Y^{(\frac{1}{4})} = \mathcal{O}_{1,nn}(X_*), \ \ \mu_{Y^{(\frac{1}{4})}}, \\
&Y^{(\frac{1}{2})} = \hat{\mathcal{F}}_{\frac{1}{2},\mu_{Y^{(\frac{1}{4})}}}(X_*), \\
&Y^{(1)} = \hat{\mathcal{F}}_1 \circ \hat{\mathcal{F}}_{\frac{1}{2},\mu_{Y^{(\frac{1}{4})}}}(X_*), \ \ \text{compute: } \mu_{Y^{(1)}} \\
&\ldots \\
&l \in \{2, \ldots, L-1\}: \\
&Y^{(l-\frac{3}{4})} = \mathcal{O}_{l,nn}(Y^{(l-1)}), \mu_{Y^{(l-\frac{3}{4})}}, \\
&Y^{(l-\frac{1}{2})} = \hat{\mathcal{F}}_{l-\frac{1}{2}} |_{\mu_{Y^{(l-\frac{3}{4})}}} \circ \hat{\mathcal{F}}_{l-1} \circ \hat{\mathcal{F}}_{l-\frac{3}{2}} |_{\mu_{Y^{(l-\frac{7}{4})}}} \circ \ldots \circ \hat{\mathcal{F}}_1 \circ \hat{\mathcal{F}}_{\frac{1}{2},\mu_{Y^{(\frac{1}{4})}}}(X_*), \\
&Y^{(l)} = \hat{\mathcal{F}}_l \circ \hat{\mathcal{F}}_{l-\frac{1}{2}} |_{\mu_{Y^{(l-\frac{3}{4})}}} \circ \ldots \circ \hat{\mathcal{F}}_1 \circ \hat{\mathcal{F}}_{\frac{1}{2},\mu_{Y^{(\frac{1}{4})}}}(X_*), \\
&\text{compute: } \mu_{Y^{(l)}} \\
&\ldots \\
&Y^{(L-\frac{3}{4})} = \mathcal{O}_{L,nn}(Y^{(L-1)}), \mu_{Y^{(L-\frac{3}{4})}}, \\
&Y^{(L-\frac{1}{2})} = \hat{\mathcal{F}}_{L-\frac{1}{2}} |_{\mu_{Y^{(L-\frac{3}{4})}}} \circ \hat{\mathcal{F}}_{L-1} \circ \ldots \circ \hat{\mathcal{F}}_1 \circ \hat{\mathcal{F}}_{\frac{1}{2},\mu_{Y^{(\frac{1}{4})}}}(X_*), \\
&X_* = Y^{(L)} = \hat{\mathcal{F}}_L \circ \hat{\mathcal{F}}_{L-\frac{1}{2}} |_{\mu_{Y^{(L-\frac{3}{4})}}} \circ \ldots \circ \hat{\mathcal{F}}_1 \circ \hat{\mathcal{F}}_{\frac{1}{2},\mu_{Y^{(\frac{1}{4})}}}(X_*), \\
&\text{compute: } \mu_{Y^{(L)}}
\end{aligned}
\tag{4.28}
$$

Iterating the MFTT, the resulting output (if any) provides

$$(X_*, \mu_{X_*}, Y^{(\frac{1}{2})}, Y^{(1)}, \mu_{Y^{(1)}}, \ldots, Y^{(L-\frac{1}{2})}, Y^{(L)} = X_*, \mu_{Y^{(L)}} = \mu_{X_*})$$

when $\mathcal{H}_L = \mathcal{H}_0$

We will show how this outcome is connected to mean-field-type Nash equilibria. Note that, here, the mean-field is the distribution of own-action of the decision-maker called individual mean-field.

The distributional mean-field-type transformer block is displayed in Figure 4.3 and distribution-wise mean-field-type transformer (D-MFTT) in Figure 4.4.

The following result summarizes the evolution of the distribution layer by layer in the MFTT.

Lemma 33. *Distribution-wise, the mean-field-type transformer-based deep neural network*

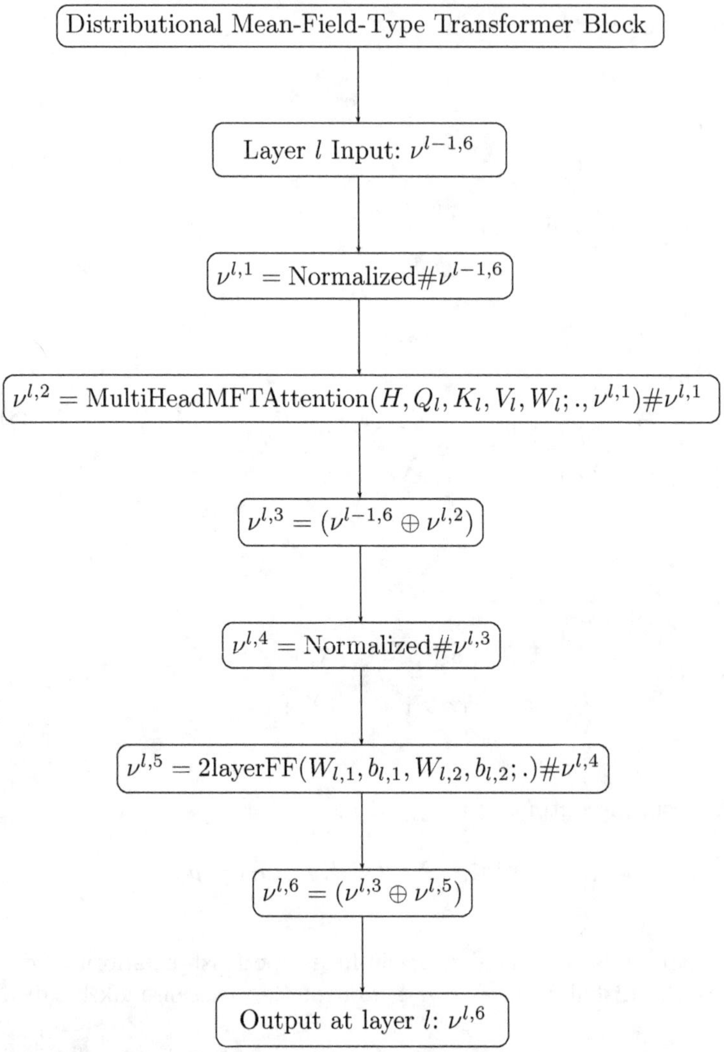

FIGURE 4.3
Distributional mean-field-type transformer block

is given by

$$
\begin{array}{l}
\mu_0 \in \mathcal{P}(\mathcal{H}_0)\\
l = 1,\\
\nu^{(\frac{3}{4})} = \mathcal{O}_{1,nn}\#\mu_0,\\
\nu^{(\frac{1}{2})} = \hat{\mathcal{F}}_{\frac{1}{2},\nu^{(\frac{3}{4})}}\#\mu_0,\\
\nu^{(1)} = \hat{\mathcal{F}}_1\#\nu^{(\frac{1}{2})},\\
\dots\\
l \in \{2, \dots, L-1\}:\\
\nu^{(l-\frac{3}{4})} = \mathcal{O}_{l,nn}\#\nu^{(l-1)},\\
\nu^{(l-\frac{1}{2})} = \hat{\mathcal{F}}_{l-\frac{1}{2},\nu^{(l-\frac{3}{4})}}\#\nu^{(l-1)},\\
\nu^{(l)} = \hat{\mathcal{F}}_l\#\nu^{(l-\frac{1}{2})},\\
\dots\\
\nu^{(L-\frac{3}{4})} = \mathcal{O}_{L,nn}\#\nu^{(L-1)},\\
\nu^{(L-\frac{1}{2})} = \hat{\mathcal{F}}_{L-\frac{1}{2},\nu^{(L-\frac{3}{4})}}\#\nu^{(L-1)},\\
\nu^{(L)} = \hat{\mathcal{F}}_L\#\nu^{(L-\frac{1}{2})},
\end{array}
\qquad (4.29)
$$

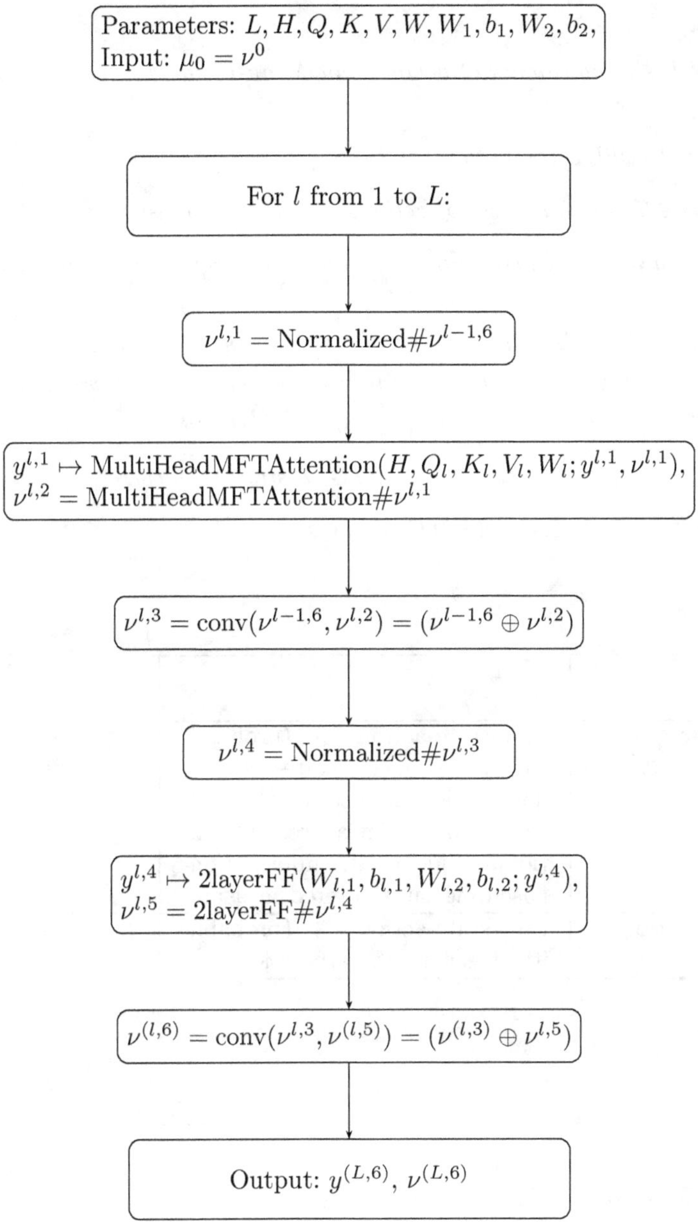
Parameters: $L, H, Q, K, V, W, W_1, b_1, W_2, b_2$,
Input: $\mu_0 = \nu^0$
For l from 1 to L:
$\nu^{l,1} = \text{Normalized}\#\nu^{l-1,6}$
$y^{l,1} \mapsto \text{MultiHeadMFTAttention}(H, Q_l, K_l, V_l, W_l; y^{l,1}, \nu^{l,1})$,
$\nu^{l,2} = \text{MultiHeadMFTAttention}\#\nu^{l,1}$
$\nu^{l,3} = \text{conv}(\nu^{l-1,6}, \nu^{l,2}) = (\nu^{l-1,6} \oplus \nu^{l,2})$
$\nu^{l,4} = \text{Normalized}\#\nu^{l,3}$
$y^{l,4} \mapsto \text{2layerFF}(W_{l,1}, b_{l,1}, W_{l,2}, b_{l,2}; y^{l,4})$,
$\nu^{l,5} = \text{2layerFF}\#\nu^{l,4}$
$\nu^{(l,6)} = \text{conv}(\nu^{l,3}, \nu^{(l,5)}) = (\nu^{(l,3)} \oplus \nu^{l,5})$
Output: $y^{(L,6)}$, $\nu^{(L,6)}$

FIGURE 4.4
Distribution-wise mean-field-type transformer (D-MFTT).

The algorithm iterates as follows:

$$\mu_{t+1} = ((1-\lambda_t)Id + \lambda_t \hat{\mathcal{F}}_L \circ \hat{\mathcal{F}}_{L-\frac{1}{2}\ |\nu_t^{(L-\frac{3}{4})}} \circ \ldots \circ \hat{\mathcal{F}}_1 \circ \hat{\mathcal{F}}_{\frac{1}{2},\nu_t^{(\frac{1}{4})}})\#\mu_t, \tag{4.30}$$

starting from $\mu^{(0)} \in \mathcal{P}(\mathcal{H}_0)$.

4.3.1 One-Shot MFTG

Definition 31. *[58, 57] The basic ingredients of a One-Shot Mean-Field-Type Game are given by*

- *The set of decision-makers* $\mathcal{I}$*, with cardinality* $|\mathcal{I}| \geq 2$.
- *For every decision-maker* $i \in \mathcal{I}$*, there is a set of actions* $\mathcal{A}_i$ *which is non-empty.*
- *Every decision-maker i has a preference structure that can be represented by an instant performance functional*

  $g_i : \prod_{j\in\mathcal{I}} \mathcal{A}_j \times \mathcal{P}\left(\prod_{j\in\mathcal{I}} \mathcal{A}_j\right) \to \mathbb{R}$, $(a, D_a) \mapsto g_i(a, D_a)$, *with* $a = (a_1, \ldots, a_I)$, *and* $D_a = \mathbb{P}_a$ *the probability distribution of a. The expected performance functional of decision-maker i is given by* $E[g_i(a, D_a)] := \int_{b\in\prod_{j\in\mathcal{I}} \mathcal{A}_j} g_i(b, D_a)D_a(db)$.

The collection $\mathcal{G} = (\mathcal{I}, (\mathcal{A}_i, E[g_i])_{i\in\mathcal{I}})$*, is called an MFTG in strategic form.*

Tables 4.1, 4.2, 4.3, 4.4 and 4.5, 4.6 present MFTG features.

TABLE 4.1
What is an MFTG?

Quantities-of-Interest	classical	MFTG
Instant Payoffs	time-state-action	time-state-action + (probability) measure of state-action : density-dependence, local, localized, entire measure. Conditional, quenched, information-based measure, filtered measure
Instant Coefficients	time-state-action	time-state-action + (probability) measure of state-action

TABLE 4.2
MFTG Features

Quantities-of-Interest	classical	MFTG
Expected Instant Payoffs	linear in the measure of state-action	possibly non-linear in the measure of state-action : point-wise density-dependence, local, localized, entire measure, conditional, quenched, information-based measure, filtered measure
Expected Instant Coefficients	linear in the measure of state-action	possibly non-linear in the measure of state-action

TABLE 4.3
Infinite population MFTG Features

Quantities-of-Interest	Classical	MFTG
Expected Instant Payoffs	linear in the individual mean-field of state-action	possibly non-linear in the individual measure of state-action. The state can be common, individual or mixture.
Expected Instant Coefficients	linear in the individual mean-field of state-action	possibly non-linear in the individual measure of state-action

TABLE 4.4
Instant Quantities-of-Interest (QoI) in Mixture of Atomic and Nonatomic MFTG Features [194]

classical	Mixture of Atomic and Nonatomic MFTG
For atomic decision-makers: time, common state, atomic state-action profiles, non-atomic multi-population state-action. For nonatomic decision-makers: time, common state, atomic state-action profiles, non-atomic multi-population state-action, generic individual state-action	For atomic decision-makers: time, common state, atomic state-action profiles, non-atomic multi-population state-action, measure of common state, measure of atomic state-action profiles For nonatomic decision-makers: time, common state, atomic state-action profiles, non-atomic multi-population state-action, generic individual state-action, measure of common state, measure of atomic state-action profiles, measure of the generic individual state-action dependence: density-based, mass/occupancy, local, localized, entire measure. Conditional, quenched, information-based measure, filtered measure
Liouville	McKean-Vlasov, Nemytskii-type
	can be noncooperative, cooperative, coalitional, coopetitive or mixture of these

4.3.2 A Mean-Field-Type Transformer is a Mean-Field-Type Game

We demonstrate that any well-trained Mean-Field-Type Transformer can be represented as a Mean-Field-Type Game with multiple hierarchies.

- At Initial layer, the input $(X^{(0)}, \mu^{(0)})$ is seen as parameter of the game.
- Each hidden layer l represents a decision-maker.
- The action of the decision-maker l is given by the output $Y^{(l)}$.
- The distribution $\nu^{(l)}$ of $Y^{(l)}$ is analogous to the distribution of actions in an MFTG.

TABLE 4.5
Difference between infinite vs finite population MFTG

Properties	Infinite population MFTG	MFTG
Model Focus	Strategic interactions influenced by population state-mf, population action-mf, individual state-mf, individual action-mf	Broader range, including aggregate effects of state-mf, action-mf of all decision-makers
Equilibrium Concept	mean-field-type equilibria in individual state, individual-mf, population state-mf, population action-mf behavioral strategies	mean-field-type equilibria in type, state, state-mf, action profile, action-mf, behavioral strategies
Population Size	Typically infinite	Flexible, accommodates finite and large populations
Population Structure	one or multiple infinite populations	Flexible, each decision-maker has its distinguished features, accommodates individuals, teams, adversaries finite and large populations
Equilibrium principles	Relies on sPIDEs, backward-forward SIDE	Direct method, DPP, SMP, Wiener Chaos Expansion

In the MFTT, the output $Y^{(l)}$ at each layer l can be interpreted as the action of decision-maker l in an MFTG. This output depends on the previous output $Y^{(l-1)}$ and the distribution $\nu^{(l-1)}$ of the outputs from the preceding layer. Similarly, in an MFTG, each decision-maker's action $Y^{(l)}$ is influenced by the distribution of actions from previous stages, which corresponds to $\nu^{(l-1)}$ in the MFTT. Each decision-maker in the MFTG reacts to the state-actions and distribution of state-actions of others, including their own-state distribution. In the MFTT, the output $Y^{(l)}$ at each layer l is determined by $\nu^{(l-1)}$, the distribution of the previous layer's outputs. Thus, the MFTT's approach to processing distributions directly aligns with the MFTG's framework, where actions are dependent on the distribution of all actions in the game. Here the MFTG is given by $(\mathcal{L}, (\mathcal{H}_l \times \mathcal{P}(\mathcal{H}_l), g_l)_{l\in\mathcal{L}})$ where $\text{argmin}_z Eg_l(z) = \hat{\mathcal{O}}_l$ with the hierarchical structure of decision l at the hierarchical layer l and the information structure that decision-maker l observes the decision and the distribution of the decision-maker $l-1$.

While restricting to $(\mathcal{B}_1 \times \mathcal{P}(\mathcal{B}_1)^L)$ the MFTG admits a mean-field-type equilibrium, which is again a corollary of the Brouwer-Schauder fixed-point theorem.

Subgame Perfect Mean-Field-Type Nash equilibrium

In a well-designed MFTT, the operator $\hat{\mathcal{O}}_l$ at each layer l can be viewed as the best-response function of decision-maker l to the distribution $\nu^{(l-1)}$. This implies that the output $Y^{(l)}$ is an optimal response to the distribution of the previous layer's outputs. Thus, each layer functions as a decision-maker that optimizes its action based on the action and distribution of action from the previous layer, as in the best-response dynamics in an MFTG. Given that each layer in the MFTT represents a decision-maker optimizing its action based on the

TABLE 4.6
Difference between infinite vs finite population MFTG (continued)

Properties	**Infinite population MFTG**	**MFTG**
Applications	Economics, Finance, traffic management, biology, etc.	Risk Engineering, Blockchain, Energy, Crowd, Transportation, Food, Water, Risk AI, Social Networks, Marriage, Psychology, Economics, Finance, network routing, social dynamics, etc.
Congestion Effects	coming from the crowd	coming from individual decisions
Asymptotic Indistinguishability per class	Considered	Flexibly treated, allowing for diverse modeling with atomic, non-atomic or mixture of both
Discrete time Noises	fairly general	fairly general
Continuous time Noises	Brownian motion, Poisson process, regime switching, Fractional Brownian motion, Gauss-Volterra, Rosenblatt	Brownian motion, Poisson process, regime switching, Fractional Brownian motion, Gauss-Volterra, Rosenblatt, mixture of these
Model Complexity	Involves solving infinite dimensional sPIDE in the space of measures	Involves PIDEs and backward-forward SIDEs

distribution of actions from previous layers, the sequence of outputs $(Y^{(1)}, Y^{(2)}, \ldots, Y^{(L)})$ and distributions $(\nu^{(1)}, \nu^{(2)}, \ldots, \nu^{(L)})$ constitutes a mean-field-type Nash equilibrium. This equilibrium is characterized by each decision-maker choosing $Y^{(l)}$ as a response to $\nu^{(l-1)}$, such that the resulting distribution $\nu^{(l)}$ is consistent with the responses of all other decision-makers. Therefore, the MFTT framework models a hierarchical best-response process that results in a mean-field-type Nash equilibrium.

Mean-Field-Type Stackelberg Solution

The hierarchical structure of MFTT (Table 4.7) and of decision-makers across layers can be viewed as a special case of a Stackelberg solution in mean-field-type games. In a multi-layer Stackelberg game, a leader makes a strategic decision first, and then followers react optimally to this decision. Here there is one leader per layer. Leader 1, who is decision-maker 1, acts first. Then follows leader 2 based on the choice of leader 1. Then follows leader 3 based on the choices of leaders 1 and 2, and so on. This sequential decision-making process inherently influences the equilibrium since the leaders' choices affect the followers' responses, and the equilibrium thus reflects this hierarchical structure. In an MFTT, each layer represents a decision-maker in a hierarchical sequence where the output of one layer (or decision-maker) serves as the input for the next. The crucial insight here is that while the current layer's pre-output $Y^{(l)}$ is the best response to the action and distribution $(Y^{(l-1)}, \nu^{(l-1)})$ of the previous layer's outputs, the decision-making process at each layer does not depend on future

TABLE 4.7
Hierarchy in MFTGs.

MFTG	individual mean-field	infinite population mean-field	Mixture of atomic and nonatomic decision-makers
1-layer	simultaneous MFTG	simultaneous population MFTG	
2-layer	Stackelberg MFTG, Inverse Stackelberg, double Stackelberg	Stackelberg population MFTG	
L-layer hierarchy	L-hierarchy MFTG	population L-hierarchy MFTG	

layers' choices but rather on the distribution induced by the preceding layer. This structure means that each decision-maker's reaction does not depend on all subsequent decisions and distributions $(Y^{(l')}, \nu^{(l')}, l' \geq l+1)$. The preceding leaders anticipate the future reactions. That anticipation coincides with the correct reaction due to the structure of the MFTT. This can be seen from the fact $\hat{\mathcal{O}}_l(Y, \nu) = \hat{\mathcal{O}}_l(Y^{(l-1)}, \nu^{(l-1)})$. Thus, the resulting outcome of the MFTT, where each layer's output constitutes a mean-field-type Nash equilibrium, also qualifies as a Stackelberg equilibrium of the hierarchical mean-field-type game.

Limitations of MFTT

In order to get a true computational advantage of MFTT over the infinite data size $\mathbb{TGN}$ discussed earlier, the computation of the mean-field-type interaction term $\hat{\mathcal{O}}_l(Y^{(l-1)}, \nu^{(l-1)})$ should be done in a very efficient way. However, such efficiency needs to be examined in detail depending on the form of interaction (pairwise, three-wise, k-wise, local, localized, global, etc). The efficient computation of quenched measure or McKean-Vlasov measure is still an open issue. One may need another sub-layer dedicated to computation of the interaction term in the same way as the (finite) self-attention was introduced.

4.3.3 How to Compute the Mean-Field-Type Terms

Global mean-field-type terms

The mean-field-type self-attention mechanisms involve a kernel and a measure μ. It has the following form $\int \mathcal{K}(X, X')\mu(dX')$.

Localized mean-field-type terms

Consider the localized term: $\int \mathcal{K}(X, X')\mathbb{I}_{X' \in B_\rho(X)}\mu(dX')$.

Point-density-based mean-field-type terms

Through the Kolmogorov forward equation or pushforward operator or kernel-based density estimation (KDE), the point density can be approximated. Using this computational approach could be more efficient than the calculation of an initial aggregative term in the attention mechanism. The answer is unclear as it depends on the numerical scheme used and the bandwidth parameter used in the KDE.

Let $(\Omega, \mathcal{F}, \mu)$ be a probability space and $f : \Omega \to \mathcal{Y}$ be a measurable function. The **pushforward measure** $f_{\#}\mu$ on $\mathcal{Y}$ is defined by

$$(f_{\#}\mu)(A) = \mu\left(f^{-1}(A)\right) \quad \text{for all measurable } A \subseteq \mathcal{Y}.$$

This describes the distribution of the random variable $f(X)$ when $X \sim \mu$.

As an example, let $X \sim \mathcal{N}(0,1)$ follow a standard normal distribution, and $f(x) = x^2$. The pushforward measure $f_{\#}\mu$ corresponds to

$$f_{\#}\mu \sim \chi_1^2,$$

where χ_1^2 denotes the chi-squared distribution with 1 degree of freedom.

In the mean-field-type transformer, the pushforward measure appears at each step. For example, at the normalization operator, the pushforward becomes:

- Input signal: $y_{l-1} \in \mathcal{H}^d, \hat{\nu}_{l-1} \in \mathcal{P}([0,1] \times \mathcal{H}^d)$
- Normalization: $y_{l,1} = \frac{y_{l-1}}{1+\|y_{l-1}\|}$
- Measure update: $\hat{\nu}_{l,1} = \text{Normalized}\#\hat{\nu}_{l-1}$

Here, Normalized# denotes the pushforward of the measure $\hat{\nu}_{l-1}$ under the normalization function $y_{l-1} \mapsto \frac{y_{l-1}}{1+\|y_{l-1}\|}$.

4.4 Training of Mean-Field-Type Transformers

> We formulate the training of Mean-Field-Type Transformers as a fully cooperative mean-field-type game.

Given the data $\mathcal{D} = \{\mathbb{P}\}, \mathbb{P} \in \mathcal{P}(\mathcal{H}_0 \times \mathcal{H}_L)$, with marginals $\mathbb{P}_1$ and $\mathbb{P}_2$, the training problem is to find the Queries, Keys, Values, weights, biases such that

$$\begin{aligned}
&\inf_{\theta\in\Theta} E\Phi(Y_T, \hat{Y}),\\
&\text{such that}\\
&Y_0 = \hat{X} \sim \mathbb{P}_1, \hat{Y} \sim \mathbb{P}_2\\
&t \in \{1,\ldots,T-1\}:\\
&Y_{t+1} = Y_t + \lambda_t(\hat{\mathcal{F}}_L \circ \hat{\mathcal{F}}_{L-\frac{1}{2}\ |\nu_t^{(L-\frac{3}{4})}} \circ \ldots \circ \hat{\mathcal{F}}_2 \circ \hat{\mathcal{F}}_{\frac{3}{2},\nu_t^{(\frac{5}{4})}} \circ \hat{\mathcal{F}}_1 \circ \hat{\mathcal{F}}_{\frac{1}{2},\mu_t^{(\frac{1}{4})}}(Y_t) - Y_t),\\
&\\
&\mu_0 = \mathbb{P}_1,\\
&\mu_{t+1} = ((1-\lambda_t)Id + \lambda_t\hat{\mathcal{F}}_L \circ \hat{\mathcal{F}}_{L-\frac{1}{2}\ |\nu_t^{(L-\frac{3}{4})}} \circ \ldots \circ \hat{\mathcal{F}}_2 \circ \hat{\mathcal{F}}_{\frac{3}{2},\nu_t^{(\frac{5}{4})}} \circ \hat{\mathcal{F}}_1 \circ \hat{\mathcal{F}}_{\frac{1}{2},\nu_t^{(\frac{1}{4})}})\#\mu_t,\\
&l \in \{1,\ldots,L-1\}:\\
&\nu_t^{(l-\frac{3}{4})} = \mathcal{O}_{l,nn}\#\nu_t^{(l-1)},\\
&\nu_t^{(l-\frac{1}{2})} = \hat{\mathcal{F}}_{l-\frac{1}{2},\nu_t^{(l-\frac{3}{4})}}\#\nu_t^{(l-1)},\\
&\nu_t^{(l)} = \hat{\mathcal{F}}_l\#\nu_t^{(l-\frac{1}{2})},\\
&\nu_t^{(0)} := \mu_t
\end{aligned} \tag{4.31}$$

The training problem can be seen as another hierarchical mean-field-type game where the state is Y and the control action θ. Here μ_t is an individual mean-field object, i.e., the distribution of the individual state. Thus, μ_t depends on θ. Here each decision-maker at each layer has a significant effect on the mean-field.

If Θ is restricted to a non-empty and compact set, the training problem admits at least one solution as we deal with Lipschitz continuous drift. The cost is of a terminal type and is given by $E\Phi(Y_T, \hat{Y})$. This is a degenerated cost function from the perspective of θ. One can add a strong regularization term to the cost or use a singular mean-field-type game framework to identify best-response sequences.

We also work directly on the distributions.

$$
\begin{array}{l}
\inf_{\theta\in\Theta} \hat{\Phi}(\mu_T, \mathbb{P}_2),\\
\text{such that}\\
\mu_0 = \mathbb{P}_1,\\
t \in \{1, \ldots, T-1\}:\\
\mu_{t+1} = ((1-\lambda_t)Id + \lambda_t \hat{\mathcal{F}}_L \circ \hat{\mathcal{F}}_{L-\frac{1}{2}\,|\nu_t^{(L-\frac{3}{4})}} \circ \ldots \circ \hat{\mathcal{F}}_2 \circ \hat{\mathcal{F}}_{\frac{3}{2},\nu_t^{(\frac{5}{4})}} \circ \hat{\mathcal{F}}_1 \circ \hat{\mathcal{F}}_{\frac{1}{2},\nu_t^{(\frac{1}{4})}})\#\mu_t,\\
l \in \{1, \ldots, L-1\}:\\
\nu_t^{(l-\frac{3}{4})} = \mathcal{O}_{l,nn}\#\nu_t^{(l-1)},\\
\nu_t^{(l-\frac{1}{2})} = \hat{\mathcal{F}}_{l-\frac{1}{2},\nu_t^{(l-\frac{3}{4})}}\#\nu_t^{(l-1)},\\
\nu_t^{(l)} = \hat{\mathcal{F}}_l\#\nu_t^{(l-\frac{1}{2})},\\
\nu_t^{(0)} := \mu_t.
\end{array}
\tag{4.32}
$$

4.5 Small Learning Rate Regime of MFTT

We now examine the dynamics of MFTT when λ_t is very small. One can relate this to a pseudo-trajectory of ordinary differential system or stochastic differential system. This leads to a continuous-time Mean-Field-Type Transformer. The training problem in the mean-field-type case is

$$
\begin{array}{l}
\mathcal{D} = \{\mathbb{P}\}, \mathbb{P} \in \mathcal{P}(\mathcal{H}_0 \times \mathcal{H}_L)\\
\inf_{\theta\in\Theta} E\Phi(Y(T), \hat{Y}),\\
\text{such that}\\
Y(0) = \hat{X} \sim \mathbb{P}_1, \hat{Y} \sim \mathbb{P}_2\\
\dot{Y} = (\hat{\mathcal{F}}_L \circ \hat{\mathcal{F}}_{L-\frac{1}{2}\,|\nu_t^{(L-\frac{3}{4})}} \circ \ldots \circ \hat{\mathcal{F}}_2 \circ \hat{\mathcal{F}}_{\frac{3}{2},\nu_t^{(\frac{5}{4})}} \circ \hat{\mathcal{F}}_1 \circ \hat{\mathcal{F}}_{\frac{1}{2},\nu_t^{(\frac{1}{4})}})(Y) - Y,\\
\\
\mu(0,.) = \mathbb{P}_1,\\
\dot{\mu} + div(\mu(\hat{\mathcal{F}}_L \circ \hat{\mathcal{F}}_{L-\frac{1}{2}\,|\nu^{(L-\frac{3}{4})}} \circ \ldots \circ \hat{\mathcal{F}}_2 \circ \hat{\mathcal{F}}_{\frac{3}{2},\nu^{(\frac{5}{4})}} \circ \hat{\mathcal{F}}_1 \circ \hat{\mathcal{F}}_{\frac{1}{2},\nu^{(\frac{1}{4})}}(Y) - Y)) = 0,,\\
l \in \{1, \ldots, L-1\}:\\
\nu^{(l-\frac{3}{4})}(t,.) = \mathcal{O}_{l,nn}\#\nu^{(l-1)}(t,.),\\
\nu^{(l-\frac{1}{2})}(t,.) = \hat{\mathcal{F}}_{l-\frac{1}{2},\nu^{(l-\frac{3}{4})}(t,.)}\#\nu^{(l-1)}(t,.),\\
\nu^{(l)}(t,.) = \hat{\mathcal{F}}_l\#\nu^{(l-\frac{1}{2})}(t,.),\\
\nu^0(t,.) := \mu(t,.)
\end{array}
\tag{4.33}
$$

The main difference between systems (3.13) and (4.33) is that in the infinite population regime in (3.13) is the population mean-field m is frozen and all the computations at all layers are done while m is frozen. In the mean-field-type case (4.33) the mean-field is an own-state mean-field which is coming from the own-Fokker-Planck equation. In the mean-field-type case, there is a finite number of decision-makers at each layer.

Note that the minimization is not for all the joint distribution $\mathbb{P}$ as in the Wasserstein formulation. Here the minimization is on the time- and state-independent control action θ. In this MFTG each decision-maker at each layer contribute to the mean-field μ in a non-negligible way. It is clear that a θ that is time- and state-independent may not be globally optimal. It may not be a global best response. Typically, for adjoint functional p, if the non-integrated Hamiltonian term

$$\theta \mapsto \langle p, (\hat{\mathcal{F}}_L \circ \hat{\mathcal{F}}_{L-\frac{1}{2}\ |\nu_t^{(L-\frac{3}{4})}} \circ \ldots \circ \hat{\mathcal{F}}_2 \circ \hat{\mathcal{F}}_{\frac{3}{2},\nu_t^{(\frac{5}{4})}} \circ \hat{\mathcal{F}}_1 \circ \hat{\mathcal{F}}_{\frac{1}{2},\nu_t^{(\frac{1}{4})}})(Y)\rangle$$

admits a minimizer, it may depend on $Y, \mu_t, \nu_t^{(1)}, \ldots, \nu_t^{(L-1)}$ which is called state- and mean-field-type control strategy.

In some casex, we may prefer to work directly with the distribution:

$$\begin{array}{l}
\mathcal{D} = \{\mathbb{P}\}, \mathbb{P} \in \mathcal{P}(\mathcal{H}_0 \times \mathcal{H}_L) \\
\inf_{\theta\in\Theta} \hat{\Phi}(\mu(T), \mathbb{P}_2), \\
\text{such that} \\
\mu(0,.) = \mathbb{P}_1, \\
\dot{\mu} + div(\mu(\hat{\mathcal{F}}_L \circ \hat{\mathcal{F}}_{L-\frac{1}{2}\ |\nu^{(L-\frac{3}{4})}} \circ \ldots \circ \hat{\mathcal{F}}_1 \circ \hat{\mathcal{F}}_{\frac{1}{2},\nu^{(\frac{1}{4})}}(Y) - Y)) = 0, \\
l \in \{1, \ldots, L-1\} : \\
\nu^{(l-\frac{3}{4})}(t,.) = \mathcal{O}_{l,nn} \# \nu^{(l-1)}(t,.) \\
\nu^{(l-\frac{1}{2})}(t,.) = \hat{\mathcal{F}}_{l-\frac{1}{2},\nu^{(l-\frac{3}{4})}(t,.)} \# \nu^{(l-1)}(t,.), \\
\nu^{(l)}(t,.) = \hat{\mathcal{F}}_l \# \nu^{(l-\frac{1}{2})}(t,.), \\
\nu^{(0)}(t,.) := \mu(t,.).
\end{array} \tag{4.34}$$

4.6 Mean-Field-Type Diffusion-Transformer

This section introduces some basic notions of diffusion models together with transformers. This combination is called transfusion. Diffusion transformers are a generative modeling approach that combines diffusion models (Table 4.8) with transformer architectures, leveraging the strengths of both for handling structured data like images or text. To describe this process step by step, we begin with the forward diffusion process, which gradually corrupts the data by adding noise over time. This forward process is governed by a stochastic differential equation. Let $x(0)$ denote the initial, clean data at time $t = 0$, and over time, noise is progressively added, transforming the data into a noisy version by the time $t = T$. The forward SDE is expressed as:

TABLE 4.8

Noises examined in MFTGs.

Noises	Gaussian	Non-Gaussian
Markovian noise	Brownian motion	Poisson Jump, Regime Switching
non-Markovian noise	Fractional Brownian motion, Gauss-Volterra	Rosenblatt noise

$$dx = b(x,t)\,dt + \sqrt{2\epsilon}\,dB, \quad x(0) \sim \mu(0,.),$$

where x is the data, $b(x,t)$ is the drift term that represents the deterministic part of the process, $\sqrt{2\epsilon}$ is the diffusion coefficient that controls the noise level, and B is the standard Brownian motion. As time t increases from 0 to T, the data evolves from its clean state $x(0)$ to a highly noisy state $x(T)$, which resembles random Gaussian noise.

The goal of the generative model is to reverse this process, recovering the original data from noise. This is achieved by solving the reverse-time SDE, which effectively denoises the noisy data and retrieves the clean data distribution. The reverse-time SDE is formulated as

$$dx = [b(x,t) - 2\epsilon\nabla_x \log\mu(t,x)]\,dt + \sqrt{2\epsilon}dB, \quad x(T) \sim \mu(T,.),$$

where $\nabla_x \log\mu(t,x)$ is the gradient of the log-probability of the data at time t. This score function guides the process by indicating how to reverse the noise and restore the data. The starting point of the reverse process is the noisy data at the final time T. This noisy data, denoted by $x(T)$, is typically a sample drawn from a standard Gaussian distribution. The reverse process begins from this noisy state $x(T)$ and aims to recover the clean data $x(0)$ by progressively denoising it over time, moving backward from $t = T$ to $t = 0$. Since the true score function $\nabla_x \log\mu(t,x)$ is unknown, we approximate it by training a neural network. In diffusion transformers, the transformer architecture is used in place of the traditional convolutional network. Transformers use self-attention mechanisms. During the reverse diffusion process, the transformer predicts either the score function $\nabla_x \log\mu(t,x)$ or the denoised data $x(t)$ at each reverse step.

4.6.1 Density-based System

Proposition 20 (Density-based system). *Let $z_t \in \mathbb{R}^d$ be a flow defined by the push-forward map*

$$z_t = T(t,z_0), \qquad \partial_t T(t,x) = f\big(t, T(t,x)\big),$$

and let $\rho(t,x)$ be its density satisfying the continuity equation

$$\partial_t\rho(t,x) + \nabla_x\cdot\big(\rho(t,x)\,f(t,x)\big) = 0\,.$$

Define along the trajectory z_t

$$\ell_t = \log\rho\big(t,z_t\big), \quad s_t = \nabla_z \log\rho\big(t,z_t\big),$$

Then the following system holds:

$$\partial_t z_t = f\big(t,z_t\big), \tag{4.35a}$$
$$\partial_t \ell_t = -\nabla_z\cdot f\big(t,z_t\big), \tag{4.35b}$$
$$\partial_t s_t = -\nabla_z f\big(t,z_t\big)^{\top} s_t \;-\; \nabla_z\big(\nabla_z\cdot f\big(t,z_t\big)\big), \tag{4.35c}$$

Moreover, if $\tilde{\ell}_t = \rho(t,z_t)$, then

$$\partial_t\tilde{\ell}_t = -\left(\nabla_z\cdot f(t,z_t)\right)\tilde{\ell}_t.$$

Proof. **State dynamic** (4.35a). By definition of the push-forward map,

$$\partial_t z_t = \partial_t T\big(t,z_0\big) = f\big(t, T(t,z_0)\big) = f\big(t,z_t\big).$$

Density along trajectory and ℓ_t dynamic (4.35b)**.** Set $\tilde{\ell}_t = \rho(t, z_t)$. Then by the chain rule and the continuity equation,

$$\frac{d}{dt}\rho(t, z_t) = \partial_t \rho(t, z_t) + \nabla_z \rho(t, z_t)^\top \partial_t z_t = -\nabla_z \cdot (\rho f)(t, z_t) + \nabla_z \rho(t, z_t)^\top f(t, z_t)$$

$$\frac{d}{dt}\rho(t, z_t) = -(\nabla_z \cdot f(t, z_t))\, \rho(t, z_t).$$

Hence $\partial_t \tilde{\ell}_t = -(\nabla_z \cdot f(t, z_t))\, \tilde{\ell}_t$ and since $\ell_t = \log \tilde{\ell}_t$,

$$\partial_t \ell_t = \frac{1}{\tilde{\ell}_t}\, \partial_t \tilde{\ell}_t = -\,\nabla_z \cdot f(t, z_t).$$

First-order score s_t dynamic (4.35c)**.** By definition $s_t = \nabla_z \ell_t$. Differentiating in time and using the chain rule,

$$\partial_t s_t = \nabla_z(\partial_t \ell_t) = \nabla_z[-\nabla_z \cdot f(t, z_t)] - \nabla_z f(t, z_t)^\top \nabla_z \ell_t,$$

which is exactly (4.35c). □

4.6.2 Explicit Solutions to Diffusion Systems

Given a measurable subset S at time t, we would like to learn the mean-field term

$$m(t, S) = \mathbb{P}_{x(t)}(S) = \mathbb{P}(x(t) \in S)$$

and compute specific quantities-of-interests. We start with the stochastic optimal control problem with $\epsilon > 0$:

$$(P1) = \begin{cases} \inf_u E[h(T, x(T)) + \int_0^T \frac{1}{2}\|u(t)\|^2 dt], \\ dx = u dt + \sqrt{2\epsilon} dB, \\ x(0) = x_0 \sim m_0 \end{cases}$$

$$(P2) = \begin{cases} v_t + \inf_u \langle u, v_x \rangle + \frac{1}{2}\|u\|^2 + \epsilon \text{ trace}(v_{xx}) = 0, \\ v(T, x) = h(T, x), \\ m_t + div(um) - \epsilon \text{ trace}(m_{xx}) = 0, \\ m(0) = m_0 \end{cases} \quad (4.36)$$

$$(P3) = \begin{cases} v_t - \frac{1}{2}\|v_x\|^2 + \epsilon \text{ trace}(v_{xx}) = 0, \\ v(T, x) = h(T, x), \\ m_t - div_x(v_x m) - \epsilon \text{trace}(m_{xx}) = 0, \\ m(0, dx) = m_0(dx) \end{cases}$$

We use Hopf-Cole transformation.

Lemma 34. *Let*

$$(HC) = \begin{cases} v_t + b\|v_x\|^2 + a \;\; trace(v_{xx}) = 0, \\ v(T, x) = h(T, x). \end{cases} \quad (4.37)$$

Then $v = \frac{a}{b}\log(w)$ *with*

$$(HC2) = \begin{cases} w_t + a \;\; trace(w_{xx}) = 0, \\ w(T, x) = e^{\frac{b}{a} h(T, x)}. \end{cases} \quad (4.38)$$

Proof. Use the transformation $w(t,x) = \phi(v(t,x))$, then $w_t = \phi'(v)(-b\|v_x\|^2 - a\ \text{trace}(v_{xx}))$, $w_x = \phi'(v)v_x$, and $w_{xx} = \phi''(v)v_x(v_x)' + \phi'(v)v_{xx}$

This implies that $w_t = -\phi'(v)b\|v_x\|^2 - a\ \text{trace}(w_{xx} - \phi''(v)v_x(v_x)')$, which reduces to $w_t + a\ \text{trace}(w_{xx}) = 0$ if we choose $-\phi'(v)b + a\phi''(v) = 0$. Hence, the transformation map $\phi(z) = e^{\frac{b}{a}z}$ will reduce the original system to the following system

$$\begin{cases} w_t + a\ \text{trace}(w_{xx}) = 0, \\ w(T,x) = e^{\frac{b}{a}h(T,x)}, \end{cases} \tag{4.39}$$

which is explicitly solved by Laplace transform, Fourier transform or by separation of variables $a(x)b(t)$.

□

Using the above transformation with $a = \epsilon, b = -\frac{1}{2}$, we obtain $v(t,x) = -2\epsilon \log(w(t,x))$. Thus, $v_x(t,x) = -2\epsilon\frac{w_x(t,x)}{w(t,x)}$. The distribution of x at time t solves

$$(P4) = \begin{cases} v(t,x) = -2\epsilon \log(w(t,x)), \\ w_t + \epsilon\ \text{trace}(w_{xx}) = 0, \\ w(T,x) = e^{-\frac{1}{2\epsilon}h(T,x)}, \\ m_t + 2\epsilon\ \text{div}_x(\frac{w_x}{w}m) - \epsilon\ \text{trace}(m_{xx}) = 0, \\ m(0,dx) = m_0(dx). \end{cases} \tag{4.40}$$

Note that $\text{trace}(m_{xx}) = \text{div}_x(\frac{m_x}{m}.m) = \text{div}_x((\log m)_x.m)$. This means that the distribution solves

$$\begin{cases} m_t - div_x\left((v_x + \epsilon(\log m)_x)m\right) = 0, \\ m(0,dx) = m_0(dx). \end{cases} \tag{4.41}$$

As T gets larger, and $\frac{1}{T}h(T,x) \to 0$, we obtain

$$(P5) = \begin{cases} \lambda + \|U\|^2 = 2\epsilon\ \text{trace}(U_x), \\ m_x = -Um \\ \int m(dx) = 1. \end{cases} \tag{4.42}$$

In one dimension, this provides $m(x) = \frac{e^{-U(x)}}{\int e^{-U(y)}dy}$, i.e., the Boltzmann-Gibbs distribution, or the softargmin of U. Note that in order to obtain m we do not need v but v_x. The transformer architecture can be used to estimate a θ that approximate $(\log m)_x$.

4.6.2.1 Diffusion Transformers

We examine diffusion transformers under the assumption of Brownian motion. It covers various types of data, including time-series, image, audio, and video data. Diffusion transformers under Brownian motion face significant limitations due to their inability to capture long-range dependencies and higher-order statistical properties inherent in many real-world datasets. Brownian motion which is characterized by short-range dependence and a lack of memory, leads to models that often fail to account for the non-Gaussian behaviors observed in practice.

Consider the Ornstein-Uhlenbeck process $x(t) = x_0 + \int_0^t \theta(m - x(t'))dt' + \int_0^t \sigma dB(t')$ with constant x_0, θ, m, σ. Then

$$x(t) = e^{-\theta t}x_0 + (1 - e^{-\theta t})m + \frac{\sigma}{\sqrt{2\theta}}B_{1-e^{-2\theta t}}.$$

It means that, starting from any constant x_0 we end up at time T with $\mathcal{N}(e^{-\theta T}x_0 + (1 - e^{-\theta T})m, \frac{\sigma^2}{2\theta}(1 - e^{-2\theta T})$. By properly choosing σ, θ, m we can match with any Gaussian $\mathcal{N}(m_T, \sigma_T^2)$ at time T. To do so, we match $e^{-\theta T}x_0 + (1 - e^{-\theta T})m = m_T$ i.e., $m =$

$\frac{(m_T - e^{-\theta T} x_0)}{(1-e^{-\theta T})}$. Similarly we solve $\frac{\sigma^2}{2\theta}(1 - e^{-2\theta T}) = \sigma_T^2$ to freely match with $\sigma = \sqrt{\frac{2\theta\sigma_T^2}{(1-e^{-2\theta T})}}$. It follows that for any x_0 and $\theta > 0$ the stochastic process driven by Brownian motion

$$x(t) = x_0 + \int_0^t \theta(\frac{(m_T - e^{-\theta T} x_0)}{(1 - e^{-\theta T})} - x(t'))dt' + \int_0^t \sqrt{\frac{2\theta\sigma_T^2}{(1 - e^{-2\theta T})}} dB(t')$$

will have the distribution of the Gaussian $\mathcal{N}(m_T, \sigma_T^2)$ at time T. The parameter θ represents the speed rate at which x_0 is being transferred to m_T. From this perspective, one may want to simply consider the process $y(t) = (1 - \frac{t}{T})x_0 + \frac{t}{T}\mathcal{N}(m_T, \sigma_T^2)$, which also transfer the initial signal x_0 at time 0 to the process $\mathcal{N}(m_T, \sigma_T^2)$ at time T.

One of the key differences between these two solutions is their path. The first one has a speed $\theta(m - x(t))$ while the second one has a constant speed $\frac{(m_T - x_0)}{T}$ and of course the variances are also different which can make a difference when it comes to risk-aware decision-making between the curves. Optimal transport theory deals with such transport map and path cost criteria.

When x_0 is an input signal of a transformer, which could be a **Time-Series**, **Text**, **Image**, **Audio**, **Video**, etc., we can still adjust the above methodology to be vectors or matrices. This leads to

- **Time-Series** Diffusion Transformers under Brownian motion
- **Image** Diffusion Transformers under Brownian motion
- **Audio** Diffusion Transformers under Brownian motion
- **Video** Diffusion Transformers under Brownian motion

The final output of the diffusion with distribution $\mathcal{N}(m_T, \sigma_T^2)$ can be seen as a pure mask. We now reverse the process to find the original signal. Given the terminal process $x_T = x(T)$ and using the martingale property, we aim to identify the reverse process $\hat{x}(t) = x(T-t)$. It starts at time $t = 0$ with a mask $\mathcal{N}(m_T, \sigma_T^2)$ and ends up with a high-quality signal output x_0. With that in mind, one can identify high-quality text, time-series, image, audio, video as final output and then train the neural network to arrive at such top quality outcomes at time T. We observe that the process $\hat{x}(t) = (1 - \frac{t}{T})\mathcal{N}(m_T, \sigma_T^2) + \frac{t}{T}x_0$ starting a mask $\mathcal{N}(m_T, \sigma_T^2)$ at time 0 and discover progressively the signal x_0 at time T.

The time-reversed OU process is

$$x(t) = x_T - \int_t^T \theta(m - x(t'))dt' - \int_t^T \sigma dB(t').$$

The modified drift of $\hat{x}(t)$ is $-\theta(m - \hat{x}(t)) + \sigma^2 \frac{(\hat{x}(t) - \mu(T-t))}{v^2(T-t)}$. The mean value is $\mu(T-t) = e^{-\theta(T-t)}x_0 + (1 - e^{-\theta(T-t)})m$ and the variance is $v^2(T-t) = \frac{\sigma^2}{2\theta}(1 - e^{-2\theta(T-t)})$. The forward drift starting from x_T is given by $-\theta(m - \hat{x}(t)) + 2\theta\frac{(\hat{x}(t) - e^{-\theta(T-t)}x_0 - (1-e^{-\theta(T-t)})m)}{(1-e^{-2\theta(T-t)})}$. The time-reverse process starting with a mask $mask \sim \mathcal{N}(m_T, \sigma_T^2)$ and solves

$$\hat{x}(t) = mask + \int_0^t -\theta(m - \hat{x}(t')) + 2\theta\frac{(\hat{x}(t') - e^{-\theta(T-t')}x_0 - (1 - e^{-\theta(T-t')})m)}{(1 - e^{-2\theta(T-t')})}dt'$$

$$+ \int_t^T \sigma dB(t').$$

The Gaussian distribution will be approaching the Dirac measure concentrated at x_0 as t goes to T. Here, **ignoring longer-range memory may lead to flawed generative machine intelligence models as the output is a pure Gaussian independently of the input.**

4.6.2.2 Sub/Super Diffusion Transformers

Our method builds upon an explicit solution of a fractional Ornstein–Uhlenbeck Process with time-varying drift and mean.

Proposition 21. *Let $x(t)$ satisfy the stochastic differential equation*

$$dx(t) = \theta(t)\big(\bar{m}(t) - x(t)\big)\,dt + \sigma(t)\,dB^H(t), \quad x(0) = x_0,$$

where $B^H(t)$ is a fractional Brownian motion with Hurst parameter $H \in (\frac{1}{2}, 1)$, $\theta(t), \bar{m}(t), \sigma(t) \in C^1([0,T])$ are real-valued functions, and $\sigma(t)$ is Hölder continuous with exponent $\delta > 1 - H$.

Then the unique solution $x(t)$ to this SDE exists and is given by

$$x(t) = e^{-\Phi(t)}x_0 + e^{-\Phi(t)}\int_0^t e^{\Phi(t')}\theta(t')\bar{m}(t')\,dt' + e^{-\Phi(t)}\int_0^t e^{\Phi(t')}\sigma(t')\,dB^H(t'),$$

where $\Phi(t) := \int_0^t \theta(t')\,dt'$, and the stochastic integral is understood in the Young sense [218].

Proof: We begin by rewriting the SDE in linear form:

$$dx(t) + \theta(t)x(t)\,dt = \theta(t)\bar{m}(t)\,dt + \sigma(t)\,dB^H(t).$$

Define the integrating factor

$$\mu(t) := e^{\int_0^t \theta(t')\,dt'} = e^{\Phi(t)}.$$

Multiplying both sides of the equation by $\mu(t)$, we obtain

$$\mu(t)dx(t) + \mu(t)\theta(t)x(t)\,dt = \mu(t)\theta(t)\bar{m}(t)\,dt + \mu(t)\sigma(t)\,dB^H(t),$$

which can be recognized as the total differential

$$d[\mu(t)x(t)] = \mu(t)\theta(t)\bar{m}(t)\,dt + \mu(t)\sigma(t)\,dB^H(t).$$

Integrating both sides from 0 to t, we get

$$\mu(t)x(t) = x_0 + \int_0^t \mu(t')\theta(t')\bar{m}(t')\,dt' + \int_0^t \mu(t')\sigma(t')\,dB^H(t').$$

Solving for $x(t)$ yields

$$x(t) = e^{-\Phi(t)}x_0 + e^{-\Phi(t)}\int_0^t e^{\Phi(t')}\theta(t')\bar{m}(t')\,dt' + e^{-\Phi(t)}\int_0^t e^{\Phi(t')}\sigma(t')\,dB^H(t').$$

We now justify the existence of the Young integral. Since $B^H(t)$ has Hölder continuous paths of order $\gamma < H$, and $\sigma(t)$ is Hölder continuous of order $\delta > 1 - H$, the composition $e^{\Phi(t')}\sigma(t')$ is also Hölder continuous of order $> 1 - H$, ensuring that the integral

$$\int_0^t e^{\Phi(t')}\sigma(t')\,dB^H(t')$$

is well-defined in the Young sense.

Finally, the linearity of the SDE and regularity assumptions imply the uniqueness of the solution.

□

Proposition 22. *Let $x(t)$ be the solution to the fractional stochastic differential equation*

$$dx(t) = \theta(t)(\bar{m}(t) - x(t))\,dt + \sigma(t)\,dB^H(t), \quad x(0) = x_0,$$

where $\theta(t), \bar{m}(t), \sigma(t)$ are continuous deterministic functions on $[0, T]$, and $B^H(t)$ is a fractional Brownian motion with Hurst parameter $H > \frac{1}{2}$. Then $x(t)$ is a Gaussian process with mean

$$m(t) = e^{-\Phi(t)}x_0 + e^{-\Phi(t)} \int_0^t e^{\Phi(t')}\theta(t')\bar{m}(t')\,dt',$$

and variance

$$v^2(t) = e^{-2\Phi(t)}H(2H-1)\int_0^t \int_0^t e^{\Phi(t')}\sigma(t')e^{\Phi(s')}\sigma(s')|t'-s'|^{2H-2}\,dt'\,ds',$$

where $\Phi(t) = \int_0^t \theta(s)\,ds$.

Proof: The solution $x(t)$ admits the explicit form

$$x(t) = e^{-\Phi(t)}x_0 + e^{-\Phi(t)}\int_0^t e^{\Phi(t')}\theta(t')\bar{m}(t')\,dt' + e^{-\Phi(t)}\int_0^t e^{\Phi(t')}\sigma(t')\,dB^H(t').$$

The first two terms are deterministic, so the randomness comes solely from the last term, which is a Young integral with respect to B^H. Since the integrand is deterministic and B^H is Gaussian, the integral is Gaussian as well, and hence $x(t)$ is Gaussian.

The mean of $x(t)$ is thus

$$\mathbb{E}[x(t)] = e^{-\Phi(t)}x_0 + e^{-\Phi(t)}\int_0^t e^{\Phi(t')}\theta(t')\bar{m}(t')\,dt' =: m(t),$$

and its variance is

$$\mathrm{Var}(x(t)) = \mathbb{E}\left[\left(e^{-\Phi(t)}\int_0^t e^{\Phi(t')}\sigma(t')\,dB^H(t')\right)^2\right].$$

Since for $H > \frac{1}{2}$, the Young integral satisfies:

$$\mathbb{E}\left[\left(\int_0^t f(t')\,dB^H(t')\right)^2\right] = H(2H-1)\int_0^t\int_0^t f(t')f(s')|t'-s'|^{2H-2}\,dt'\,ds',$$

we substitute $f(t') = e^{\Phi(t')}\sigma(t')$ and factor out $e^{-2\Phi(t)}$, yielding

$$v^2(t) = e^{-2\Phi(t)}H(2H-1)\int_0^t\int_0^t e^{\Phi(t')}\sigma(t')e^{\Phi(s')}\sigma(s')|t'-s'|^{2H-2}\,dt'\,ds'.$$

□

Forward process that ends up with a mask

To ensure accurate terminal constraints critical for conditional generation, we design a tailored process that begins at x_0 and precisely reaches a terminal distribution $\mathcal{N}(m_T^*, (\sigma_T^*)^2)$ at time $t = T$. This process incorporates an adaptive drift:

Proposition 23. *Let $T > 0$ be fixed and let $m_T^* \in \mathbb{R}$, $\sigma_T^* > 0$. Then for any deterministic, continuous, and strictly positive function $\theta(t)$ on $[0, T]$, and for any initial condition $x(0) = x_0$, there exist deterministic, continuous functions $\bar{m}(t)$ and $\sigma(t)$, such that the solution $x(t)$ to the fractional SDE*

$$dx(t) = \theta(t)(\bar{m}(t) - x(t))\, dt + \sigma(t)\, dB^H(t), \quad x(0) = x_0,$$

satisfies

$$x(T) \sim \mathcal{N}(m_T^*, (\sigma_T^*)^2).$$

Proof: Let us define:

$$\Phi(t) := \int_0^t \theta(s)\, ds.$$

We know from the explicit solution of the SDE that

$$x(T) = e^{-\Phi(T)}x_0 + e^{-\Phi(T)} \int_0^T e^{\Phi(t')}\theta(t')\bar{m}(t')\, dt' + e^{-\Phi(T)} \int_0^T e^{\Phi(t')}\sigma(t')\, dB^H(t').$$

Mean constraint:
To match the mean m_T^*, define $\bar{m}(t)$ such that

$$m_T^* = e^{-\Phi(T)}x_0 + e^{-\Phi(T)} \int_0^T e^{\Phi(t')}\theta(t')\bar{m}(t')\, dt'.$$

Rewriting, we require

$$\int_0^T e^{\Phi(t')}\theta(t')\bar{m}(t')\, dt' = e^{\Phi(T)}(m_T^* - e^{-\Phi(T)}x_0).$$

This is a linear integral equation for $\bar{m}(t')$, which admits a solution under broad conditions. In particular, since $\theta(t')$ and $e^{\Phi(t')}$ are strictly positive, a continuous $\bar{m}(t')$ exists.

Solving for $\bar{m}$ to match $\mathbb{E}[x(T)] = m_T^*$ gives

$$\bar{m} = \frac{m_T^* - x_0 e^{-\Phi(T)}}{1 - e^{-\Phi(T)}}.$$

Variance constraint:
Let $f(t') := e^{\Phi(t')}\sigma(t')$. To achieve $\text{Var}(x(T)) = (\sigma_T^*)^2$, we require

$$(\sigma_T^*)^2 = e^{-2\Phi(T)}H(2H-1) \int_0^T \int_0^T f(t')f(s')|t' - s'|^{2H-2}\, dt'\, ds'.$$

Let us define:

$$K := H(2H-1) \int_0^T \int_0^T f(t')f(s')|t' - s'|^{2H-2}\, dt'\, ds'.$$

Then

$$K = e^{2\Phi(T)}(\sigma_T^*)^2.$$

This is a homogeneous quadratic form in $f(t')$. Since the kernel $|t' - s'|^{2H-2}$ is symmetric and positive definite for $H > 1/2$, a solution for $f(t')$ (and hence $\sigma(t') = f(t')e^{-\Phi(t')}$) exists, for instance, by choosing:

$$f(t') := C, \quad \text{with } C = \left(\frac{e^{2\Phi(T)}(\sigma_T^*)^2}{H(2H-1)\int_0^T \int_0^T |t' - s'|^{2H-2} dt' ds'} \right)^{1/2}.$$

Thus, a constant $f(t')$ yields an explicit solution:

$$\sigma(t') = Ce^{-\Phi(t')}.$$

The double integral is symmetric and evaluates as:

$$\int_0^T \int_0^T |t' - s'|^{2H-2} dt' ds' = 2 \int_0^T \int_0^{t'} (t' - s')^{2H-2} ds' dt' = \frac{2T^{2H}}{(2H-1)(2H)}.$$

$$C = \sigma_T^* T^{-H} e^{\Phi(T)},$$

Hence, both the mean and variance constraints can be satisfied simultaneously by choosing appropriate $\bar{m}(t)$ and $\sigma(t)$. □

Corollary 6. *This result shows you can steer a linear SDE in fractional noise exactly to any Gaussian target. It is especially useful in diffusion-based generative modeling, where sampling from a terminal Gaussian is desired. We define the dynamics of $x(t) \in \mathbb{R}$ by the fractional stochastic differential equation:*

$$dx(t) = \theta(t) \left(\frac{m_T^* - x_0 e^{-\Phi(T)}}{1 - e^{-\Phi(T)}} - x(t) \right) dt + \sigma_T^* T^{-H} e^{\Phi(T) - \Phi(t)} dB^H(t), x(0) = x_0,$$

where $\Phi(t) := \int_0^t \theta(t')\, dt'$, and $B^H(t)$ is a fractional Brownian motion with Hurst parameter $H > \frac{1}{2}$, $\theta(t) > 0$ is deterministic and continuous, and $x_0, m_T^*, \sigma_T^* \in \mathbb{R}$ are given constants. Under this dynamics, the process $x(t)$ satisfies $x(T) \sim \mathcal{N}(m_T^*, (\sigma_T^*)^2)$.

This construction ensures that the diffusion process not only adheres to the target endpoint distribution but also embeds the long-memory structure characteristic of fractional noise. By exploiting the non-local properties of fBm, **fractional diffusion, super diffusion, and subdiffusion** provide a principled way to enrich generative models across modalities spanning time-series, vision, audio, and video with a richer stochastic representation of uncertainty and correlation.

Explicit score function

Despite the recent surge in popularity of score-based generative modeling, a common misconception persists: that one must always train a neural network to approximate the score function of the forward process. This belief overlooks fundamental properties of Gaussian processes.

Proposition 24. *Let $x(t)$ be a stochastic process such that for every time t,*

$$x(t) \sim \mathcal{N}(m(t), v^2(t)),$$

where $m(t)$ and $v^2(t) > 0$ are deterministic, differentiable functions of time. Then the score function (i.e., the gradient of the log-density with respect to x) at time t is given analytically by

$$\nabla_x \log p(x(t)) = -\frac{x(t) - m(t)}{v^2(t)}.$$

Thus, no neural network is required to estimate the score function when the marginal distribution is Gaussian with known mean and variance.

Proof: By definition, the probability density function (pdf) of a Gaussian random variable $x(t) \sim \mathcal{N}(m(t), v^2(t))$ is:

$$p(x(t)) = \frac{1}{\sqrt{2\pi v^2(t)}} e^{-\frac{(x(t)-m(t))^2}{2v^2(t)}}.$$

The log-density is then

$$\log p(x(t)) = -\frac{1}{2}\log(2\pi v^2(t)) - \frac{(x(t)-m(t))^2}{2v^2(t)}.$$

Differentiating with respect to $x(t)$:

$$\nabla_x \log p(x(t)) = \frac{\partial}{\partial x(t)}\left(-\frac{(x(t)-m(t))^2}{2v^2(t)}\right)$$

$$= -\frac{1}{2v^2(t)} \cdot 2(x(t)-m(t)) = -\frac{x(t)-m(t)}{v^2(t)}.$$

Hence, the score function is

$$\nabla_x \log p(x(t)) = -\frac{x(t)-m(t)}{v^2(t)}.$$

□

This says that the score function of a Gaussian is explicit, *linear* function in y. This form is derived directly from the analytical expression of the Gaussian density and requires **no** learning or approximation when the mean and variance are known. Therefore, in purely Gaussian settings, the score is **exactly known by construction**, and using deep neural networks to estimate it introduces unnecessary complexity, computational burden, and potential approximation error. In any Gaussian process, the form of the score is always this linear expression. What may be unknown in practice is the mean, which might be time-dependent. The variance, which could also vary with time or depend on latent dynamics. The true modeling difficulty lies not in estimating the score structure, but in estimating (or tracking) these parameters over time especially in non-stationary or parameter-varying settings. The misconception in many score-based diffusion models is in treating the form of the score as unknown when it is the parameter evolution that is uncertain or unobserved.

This insight extends naturally to more general Gaussian processes, including those governed by *fractional Brownian motion (fBm)* and *Gauss-Volterra processes.* In both subdiffusion ($H < 0.5$) and superdiffusion ($H > 0.5$) regimes, the marginal distributions at each time point remain Gaussian due to the underlying linear dynamics and Gaussian noise. Consequently, their corresponding score functions also admit closed-form expressions, preserving the linearity in the state variable.

We introduce a mean-field-type fractional diffusion generative model. The mean-field-type terms are mainly involved in the backward process where the denoising pass requires the mean and the variance to match the original process. We do not need the entire mean-field state distribution in this particular case; we only need the first two moments.

Fractional Forward Dynamics

$$dx(t) = \theta(t)(\bar{m}(t) - x(t))\,dt + \sigma(t)\,dB^H(t), \quad x(0) = x_0,$$

Distribution of $x(t)$

The solution $x(t)$ admits the explicit form

$$x(t) = e^{-\Phi(t)}x_0 + e^{-\Phi(t)} \int_0^t e^{\Phi(t')}\theta(t')\bar{m}(t')\,dt' + e^{-\Phi(t)} \int_0^t e^{\Phi(t')}\sigma(t')\,dB^H(t').$$

where $\Phi(t) := \int_0^t \theta(t')\,dt'$. In this deterministic coefficient setting and deterministic initial signal x_0, the random variable $x(t)$ has the same distribution as the Gaussian $\mathcal{N}(m(t), v^2(t))$. where

$$\begin{array}{l} m(t) = e^{-\Phi(t)}x_0 + e^{-\Phi(t)} \int_0^t e^{\Phi(t')}\theta(t')\bar{m}(t')\,dt', \\ v^2(t) = e^{-2\Phi(t)}H(2H-1)\int_0^t \int_0^t e^{\Phi(t')}\sigma(t')e^{\Phi(s')}\sigma(s')|t'-s'|^{2H-2}\,dt'\,ds', \end{array} \tag{4.43}$$

Endpoint Conditioning

$$dx(t) = \theta(t)\left(\frac{m_T^* - x_0 e^{-\Phi(T)}}{1-e^{-\Phi(T)}} - x(t)\right)dt + \sigma_T^* T^{-H} e^{\Phi(T)-\Phi(t)} dB^H(t), x(0) = x_0,$$

which satisfies $x(T) \sim \mathcal{N}(m_T^*, (\sigma_T^*)^2)$.

Reverse-Time fractional SDE:

Given the forward process is Gaussian, the reverse-time dynamics are governed by

$$\begin{array}{l} dx(t) = \left[\theta\left(\frac{m_T^* - e^{-\Phi(T)}x_0}{1-e^{-\Phi(T)}} - x(t)\right) + \frac{x(t)-m(t)}{t}2H\right]dt \\ +(\sigma_T^*)T^{-H}e^{\Phi(T)-\Phi(t)}\,d\bar{B}^H(t), \\ \\ m(t) = e^{-\Phi(t)}x_0 + (1-e^{-\Phi(t)})(\frac{m_T^* - e^{-\Phi(T)}x_0}{1-e^{-\Phi(T)}}), \\ v^2(t) = e^{2\Phi(T)-2\Phi(t)}(\frac{t}{T})^{2H}(\sigma_T^*)^2, \end{array} \tag{4.44}$$

where $\bar{B}^H(t)$ is a time-reversed fractional Brownian motion, and $\tilde{\sigma}(t)$ adjusts for non-Markovian memory in the reverse direction.

Score Function

Because this process remains Gaussian, the exact score function is known analytically:

$$\nabla_y \log p_{x(T-t)}(y) = -\frac{y - m(T-t)}{v^2(T-t)}. \tag{4.45}$$

4.6.2.3 Super-Diffusion Transformers under Rosenblatt Process

We introduce super-diffusion transformers driven by the Rosenblatt process. This approach incorporates longer-term context-awareness and non-Gaussianity which significantly enhances the performance of generative machine intelligence models. The types of data covered include time-series, image, audio, and video. The Rosenblatt process is particularly suited for capturing the non-Gaussian dependencies observed in real-world data. $x(t) = x_0 + \int_0^t \theta(m - x(t'))dt' + \int_0^t \sigma dR^H(t')$. Then

$$x(t) = e^{-\theta t}x_0 + (1-e^{-\theta t})m + \sigma \int_0^t e^{-\theta(t-t')}dR^H(t').$$

As H approaches one, this last process have the same distribution as

$$x(t) = e^{-\theta t}x_0 + (1-e^{-\theta t})m + \frac{\sigma}{\theta\sqrt{2}}(1-e^{-\theta t})(Z^2-1)$$

where (Z^2-1) is the centred chi-square random variable, and $Z \sim \mathcal{N}(0,1)$. The variance of the process is

$$var\left(\int_0^t e^{\theta t'}dR^H(t')\right) = H(2H-1)\int_0^t\int_0^t e^{\theta t'}e^{\theta t''}|t'-t''|^{2H-2}dt'dt'',$$

which highlights the strong long-run dependence and non-Gaussianity.

- **Time-Series** Rosenblatt Super-Diffusion Transformers
- **Image** Rosenblatt Super-Diffusion Transformers
- **Audio** Rosenblatt Super-Diffusion Transformers
- **Video** Rosenblatt Super-Diffusion Transformers

Incorporating longer-term context-awareness and non-Gaussianity (if any) leads to improved generative machine intelligence models.

4.6.3 Diffusion-Transformer

We start with an open issue with a very local (point-based) density-based estimation.

4.6.3.1 Density-based Optimization

In the context of infinite population games, we have used non-atomic terms in the quantity of interest in the point mean-field $m(t, x(t))$ i.e., the density at the point $x(t)$. However, here there is no infinite number of decision-makers in the Tensor-Graph Neural network. We aim to compute $m(t, x(t))$ for the aim to examine the output data density $m(t, y)$ at y at time t. To date we do not have a stochastic maximum principle for functionals such as:

$$g(t, x(t), m(t, x(t)), u(t, x(t), m(t, x(t))))$$

where $x(t)$ is the input data at time t, u is the decision variable at time t and g is the performance metric (could be a score function or error function). When the function m is twice continuous differentiable, one can apply Itô's formula to $m(t, x(t))$. In the context of mean-field-type games, one can consider $\frac{m(t,B_\epsilon(x))}{vol(B_\epsilon)}$ which is localized around the entire ball as an approximation of the density at x. One can also consider a generic smooth regularization kernel with a convolution with the probability density function. The standard methods such as calculus of variations, maximum principle, infinite dimensional dynamic programming apply for the regularized forms. The question then is to know the limit as ϵ goes to zero. As we know, approximating first and then optimizing may differ from optimizing first and then approximating.

4.6.3.2 Time-Reversed Diffusion Process

We consider the input signal diffusion process $(x(t),\ 0 \le t \le T)$ with $x(0) \sim m_0$.

$$\begin{cases} dx = bdt + \sqrt{2\epsilon}dB, \\ x(0) = X_0 \sim m_0, \\ m(t, dx) = \mathcal{L}(x(t)), \\ \\ m_t + div_x\left((b - \epsilon(\log m)_x)m\right) = 0, \\ m(0, dx) = m_0(dx) \end{cases} \tag{4.46}$$

Now consider the time-reversed process $(y(t) = x(T-t),\ 0 \leq t \leq T)$ which starts with $y(0) = x(T) \sim m(T-0, dx)$. The probability law of y is $\tilde{m}(t, dy) = \mathcal{L}(y(t)) = \mathcal{L}(x(T-t)) = m(T-t, dy)$. We write the dynamics of $\tilde{m}$ from the Fokker-Planck-Kolmogorov equation in m. The time-derivative of $\tilde{m}$ is

$$\begin{aligned}\tilde{m}_t(t,y) &= -m_t(T-t,y) = div_y\left((b(T-t,y) - \epsilon(\log m(T-t,y))_y)m(T-t,y)\right)\\ &= div_y\left((b(T-t,y) - \epsilon(\log \tilde{m})_y)\tilde{m}\right).\end{aligned}$$

Putting it all together we arrive at

$$\begin{cases} dx = bdt + \sqrt{2\epsilon}dB, \\ x(0) = x_0 \sim m_0, \\ m(t,dx) = \mathcal{L}(x(t)), \\ \\ m_t + div_x\left((b - \epsilon(\log m)_x)m\right) = 0, \\ m(0,dx) = m_0(dx), \\ \\ \tilde{m}_t(t,y) + div_y\left((-b(T-t,y) + \epsilon(\log \tilde{m}(t,y))_y)\tilde{m}(t,y)\right) = 0, \\ \tilde{m}(0,dy) = \nu_1(dy) = m(T-0,dy), \end{cases} \tag{4.47}$$

Knowing that $-\sigma dB$ and σdB have the same distribution, the process $(y(t),\ 0 \leq t \leq T)$ is a diffusion process with drift $\tilde{b}(t,y) = -b(T-t,y) + \epsilon(\log \tilde{m}(t,y))_y$. This shows a first main result:

$$\tilde{b}(T-t,y) + b(t,y) = \epsilon(\log m(t,y))_y. \tag{4.48}$$

The associated reversed ODE is

$$\begin{cases} dz = (b(t,z) - \epsilon(\log m(t,z))_z)dt, \\ z(T) \sim \nu_1. \end{cases} \tag{4.49}$$

The drift coefficient, however, need to carefully designed to match with shift in the drift: $b(t,y) - \epsilon(\log m(t,y))_y = (b(t,y) - 2\epsilon(\log m(t,y))_y) + \epsilon(\log m(t,y))_y$. Hence, the time-reversed process is given by

$$\begin{cases} dy = (b(t,y) - 2\epsilon(\log m(t,y))_y)dt + \sqrt{2\epsilon}dB, \\ y(T) \sim \nu_1. \end{cases} \tag{4.50}$$

The idea is to use this latter equation to sample m_0. To do so, one needs to compute $(\log m(t,y))_y$. Thus, one needs to design a deep neural network to estimate it or to solve the Fokker-Planck-Kolmogorov in an efficient way.

4.6.4 Mean-Field-Type Diffusion-Transformer

When the drift is of mean-field-type as in MFTT:

$$b(y,\mu,\nu^{(1)},\ldots,\nu^{(L-1)}) = (\hat{\mathcal{F}}_L \circ \hat{\mathcal{F}}_{L-\frac{1}{2}\ |\nu^{(L-\frac{5}{4})}(t,.)} \circ \ldots \circ \hat{\mathcal{F}}_1 \circ \hat{\mathcal{F}}_{\frac{1}{2},\nu^{(\frac{1}{4})}(t,.)})(y) - y,$$

the ODE with initial stochastic (marginal) data μ_0 is given by

$$\begin{aligned} dy &= b(y,\mu(t,.),\nu^{(1)}(t,.),\ldots,\nu^{(L-1)}(t,.))dt, \\ y_0 &\sim \mu_0(.). \end{aligned} \tag{4.51}$$

The backward ODE with terminal stochastic (marginal) data ν_1 is given by

$$\begin{aligned} dz &= (b(z,\mu(t,.),\nu^{(1)}(t,.),\ldots,\nu^{(L-1)}(t,.)) - \epsilon(\log \mu(t,z))_z)dt, \\ z(T) &\sim \mu(T,.). \end{aligned} \tag{4.52}$$

where

$$
\begin{array}{l}
l \in \{1,\ldots,L-1\}: \\
\nu^{(l-\frac{3}{4})}(t,.) = \mathcal{O}_{l,nn}\#\nu^{(l-1)}(t,.), \\
\nu^{(l-\frac{1}{2})}(t,.) = \hat{\mathcal{F}}_{l-\frac{1}{2},\nu^{(l-\frac{3}{4})}(t,.)}\#\nu^{(l-1)}(t,.), \\
\nu^{(l)}(t,.) = \hat{\mathcal{F}}_{l}\#\nu^{(l-\frac{1}{2})}(t,.), \\
\nu^{(0)}(t,.) := \mu(t,.).
\end{array}
\tag{4.53}
$$

The mean-field-type diffusion-transformer is therefore given by

$$
\begin{array}{l}
dz = (b(z,\mu(t,.),\nu^{(1)}(t,.),\ldots,\nu^{(L-1)}(t,.)) - 2\epsilon(\log\mu(t,z))_z)dt + \sqrt{2\epsilon}dB, \\
z(T) \sim \mu(T,.).
\end{array}
\tag{4.54}
$$

In practice, the forward process is used to add noise at each step and backward process is used as a denoising process. Using this process one can get a top-quality image at the end starting from white noise. However, one needs to be careful with model collapse if the denoising process does not match as indicated in (4.54).

The mean-field-type diffusion-transformer training problem belongs to the ones detailed in the remarks below.

Remark 14. *Let*

$$
\left\{
\begin{array}{l}
T > 0, U \textit{ nonempty convex set}, \\
(\Omega, \mathbb{P}, \{\mathcal{F}_t\}_{0\le t\le T}), \\
\mathcal{A}dm = \{\theta : [0,T] \to \Theta,\ \theta(t)\ \mathcal{F}_t - \textit{measurable}\}, \\
c: [0,T] \times \mathcal{X} \times \Theta \times \mathbb{R}_+ \times \mathcal{P}(\mathcal{X}) \to \mathbb{R}_+ \\
(t,x,\theta,pd,\mu) \mapsto c(t,x,\theta,pd,\mu), \\
\\
h : \mathcal{X} \times \mathbb{R}_+ \times \mathcal{P}(\mathcal{X}) \to \mathbb{R}_+ \\
\\
\textit{state probability density: pd: } \mu(t,dy) = pd(t,y)dy, \\
\textit{state probability distribution: } \mu(t,A) = \mathbb{P}_{x^{x_0,\theta}(t)}(A),\ \ A \subset \mathcal{X}
\end{array}
\right.
\tag{4.55}
$$

Consider

$$
\left\{
\begin{array}{l}
\inf_{\theta\in\mathcal{A}dm} \mathbb{E}\{h(x^{x_0,\theta}(T), pd(x^{x_0,\theta}(T)), \mu(T)) \\
\quad + \int_0^T c(t, x^{x_0,\theta}(t), \theta(t), pd(t, x^{x_0,\theta}(t)), \mu(t))dt\}, \\
\\
\mu(t,dy) = pd(t,y)dy, \\
\mu(t,A) = \mathbb{P}_{x^{x_0,\theta}(t)}(A), \\
\mu(0) = \mu_0 \in \mathcal{P}(\mathcal{X}), \\
x^{x_0,\theta}(t) =: x(t) \textit{ solves:} \\
x(t) = x_0 + \int_0^t b(t', x(t'), \theta(t'), pd(t', x(t')), \mu(t'))dt' \\
\quad + \int_0^t \sigma(t', x(t'), \theta(t'), pd(t', x(t')), \mu(t'))dB(t'), \\
\theta(t) \in \Theta,\ \mathbb{P} - a.s,
\end{array}
\right.
\tag{4.56}
$$

The standard formulation is for the entire distribution μ. Here, there is in addition a (point) probability density distribution at the corresponding state.

Example: $c = \langle\theta(t), \theta(t)\rangle - \epsilon\log(pd(x^{x_0,\theta}(t)))$ used in generative machine intelligence.

Remark 15. *The above can be extended to density of control and probability of control as well.*

$$
\left\{
\begin{array}{l}
c: [0,T] \times \mathcal{X} \times \Theta \times \mathbb{R}_+ \times \mathcal{P}(\mathcal{X}) \times \mathbb{R}_+ \times \mathcal{P}(\Theta) \to \mathbb{R}_+ \\
(t,x,\theta,pd,\mu,\tilde{pd},\tilde{\mu}) \mapsto c(t,x,\theta,pd,\mu,\tilde{pd},\tilde{\mu}), \\
\\
\textit{control probability density: } \tilde{\mu}(t,d\theta) = \tilde{pd}(t,\theta)d\theta, \\
\textit{control probability distribution: } \tilde{\mu}(t,\Theta') = \mathbb{P}_{\theta(t)}(\Theta'), \Theta' \subset \Theta,
\end{array}
\right.
\tag{4.57}
$$

$$\begin{cases} \inf_{\theta\in\mathcal{A}dm} \mathbb{E}\{h(x^{x_0,\theta}(T), pd(x^{x_0,\theta}(T)), \mu(T)) \\ + \int_0^T c(t, x^{x_0,\theta}(t), u(t), pd(t, x^{x_0,\theta}(t)), \mu(t), \tilde{pd}(t,\theta(t)), \tilde{\mu}(t))dt\}, \\ \\ \mu(t,dy) = pd(t,y)dy, \\ \mu(t,A) = \mathbb{P}_{x^{x_0,\theta}(t)}(A), \\ \mu(0) = \mu_0 \in \mathcal{P}(\mathcal{X}), \\ \tilde{\mu}(t,d\theta) = \tilde{pd}(t,\theta)d\theta, \\ \tilde{\mu}(t,\Theta') = \mathbb{P}_{\theta(t)}(\Theta'), \\ x^{x_0,\theta}(t) =: x(t) \ \textit{solves:} \\ x(t) = x_0 + \int_0^t b(t', x(t'), \theta(t'), pd(t', x(t')), \mu(t'), \tilde{pd}(t', \theta(t')), \tilde{\mu}(t'))dt' \\ \qquad + \int_0^t \sigma(t', x(t'), \theta(t'), pd(t', x(t')), \mu(t'), \tilde{pd}(t', \theta(t')), \tilde{\mu}(t'))dB(t'), \\ u(t) \in U, \ \mathbb{P} - a.s, \end{cases} \tag{4.58}$$

4.6.5 Training Mean-Field-Type Diffusion Tensor-Graph Transformer

Let the process $Z_t = \lambda_t Y_{LT} + (1-\lambda_t)(B_{t+1} - B_t)$ with $\lambda_0 = 0$ and $\lambda_{LT} = 1$. One can choose, for example, a uniform time scaling: $\lambda_t = \frac{t}{LT}$. The process Z starts with white noise, iterates LT times and ends up with the process Y_{LT}. Following this simple procedure, one can start with a noise and get a very nice and top-quality picture or movie at time LT. Let us choose Y_{LT} to have the distribution $\mathbb{P}_2$, i.e., the final output distribution.

Scaling to continuous time, the process Z starts with standard Brownian motion $Z(0) = B(0) = 0$ and evolves to end up at Y_{LT} at time $t = LT$. Let $Z(t) = \hat{\lambda}_1(t)Y_{LT} + \hat{\lambda}_2(t)B(t)$, with $\hat{\lambda}_1(LT) = 1$ and $\hat{\lambda}_2(LT) = 0$, and $\hat{\lambda}_1(0) = 0$ and $\hat{\lambda}_2(0) = 1$. Z solves a backward SDE and its score function can be explicitly computed from the dynamics without using any extra neural network. $Z(t) = Y_{LT} - \int_t^{LT} pdt + \int_t^{LT} qdB$ where the pair process (p, q) is to be determined. The process p is easily obtained from the time derivative conditional expectation. This identification of p comes from the matching with the drift of Z in the backward SDE, thus identifying the score function defined in (4.52).

4.7 Mean-Field-Type Federated Transformers

We are now interested in collaborative interaction between several server owners and local clients who are data owners.

4.7.1 Federated Training in Tensor-Graph Transformer

Federated training is a specialized distributed machine learning paradigm where multiple devices or nodes such as smartphones, IoT devices, or edge servers, collaboratively train a shared model without exchanging their local data. In the context of transformer-based generative machine intelligence, where large transformer architectures are utilized for tasks like speech generation, video generation, text generation, image synthesis, or sequence modeling for backcasting, nowcasting and forecasting, federated training enables decentralized model training. This allows devices to keep their data private while still contributing to a shared global model's improvement.

Key components of federated training, when applied to transformer-based models, include:

In federated training, local devices (clients) do not share raw data $\mathcal{D}^{s,c} := (\hat{X}_o^{s,c}, \hat{Y}_o^{s,c})_{1\leq o \leq D^{s,c}}$. Instead, they perform computations on their local datasets $\mathcal{D}^{s,c}$ and share only the resulting updates $\theta^{s,c}$ to the model parameters such as queries, keys, values, weights, biases, and other parameters of the transformer model with a central server.

The central server s aggregates these updates from multiple clients and combines them to update the global model. In transformer-based generative machine intelligence, this setup allows devices to train transformer-based generative models locally, preserving privacy while simultaneously improving the global model's performance and generalization.

The central server s initializes and distributes a global transformer model to each client c. This model may include components like multi-head self-attention, feedforward layers, and layer normalization modules.

Each client trains the global model locally on its own data, adjusting only the model's parameters. Trained local models are not shared; instead, only the parameter updates are sent to the server, minimizing the risk of exposing sensitive local data.

After each local training iteration, clients send model parameter updates to the central server. These updates often include gradients or updated weights/biases for the transformer model's layers (including parameters like queries, keys, values, and attention weights).

The server aggregates the updates, typically through Federated Averaging, where it computes a weighted average of the clients' model updates based on their respective data sizes. This process ensures that no raw data is shared, only model parameters. Efficient parameter exchange minimizes data leakage risks and reduces communication overhead.

A significant advantage of federated training is that raw data $\mathcal{D}^{s,c}$ never leaves the local device. Each client performs local computations on their data and only sends the model updates to the server. These updates include various components of the transformer model, such as queries, keys, values, weights, biases, normalization parameters, errors and gradients.

After local training, clients send their model updates to the central server. The central server aggregates these updates and then distributes the updated global model back to the clients. The clients use this updated global model as a starting point for the next round of local training.

The Federated Transformer Training has several regional clusters of servers. The set of regional cluster servers is denoted by $\mathcal{S}$ which is a finite (and non-empty) set. Each regional cluster server s interacts with several eligible clients $\mathcal{C}$. Server s has a neural network model

$$\mathcal{M}^s := (L^s, D^s, (\mathcal{H}_0^s)^{D^s}, (V_s', E_s'), (\mathcal{H}_l^s, R_l^s, \mathcal{O}_{l,ff}^s, \mathcal{O}_{l,nn}^s, \mathcal{O}_{l,att}^s)_{1\leq l\leq L^s}, \lambda),$$

and a query-key-value-weight-bias parameter

$$\theta^s = ((Q_{l,h}^s, K_{l,h}^s, V_{l,h}^s, W_{l,h}^s)_{1\leq h\leq H^s}, W_{l,2}^s, b_{l,2}^s, W_{l,1}^s, b_{l,1}^s)_{1\leq l\leq L^s}.$$

The federated transformer training is as follows. Let $X^{(s,0)} \in (\mathcal{H}_0^s)^D$ be the input of the tensor-graph transformer, $\{\lambda_t^s\}_{t\geq 0}$ be a non-negative sequence and $\hat{\mathcal{O}}_{l,t}^s$ given by

$$\begin{aligned} \hat{\mathcal{O}}_{l,t}^s :\ & \mathcal{H}_{l-1}^s \to \mathcal{H}_l^s \\ & X^s \mapsto (Id + \mathcal{O}_{l,ff}^s \circ \mathcal{O}_{l,nn}^s) \circ (Id + \mathcal{O}_{l,att}^s \circ \mathcal{O}_{l,nn}^s)(X^s) \end{aligned}$$

and iterate layer by layer and by timestep the following maps:

$$\left| \begin{array}{l} Parameters \\ Input : X_0^s = X^{(s,0)} \\ \text{for } t \in \{0,1,2,\ldots\} : \\ \quad Y_t^{(s,1)} = \hat{\mathcal{O}}_{1,t}^s(X_t^s) \\ \quad \text{for } l \in \{2,\ldots,L^s\} :\ Y_t^{(s,l)} = \hat{\mathcal{O}}_{l,t}^s(Y_t^{(s,l-1)}) \\ \quad X_{t+1}^s = X_t^s + \lambda_t^s(Y_t^{(s,L^s)} - X_t^s), \end{array} \right. \tag{4.59}$$

which means that $X^s_{t+1} = X^s_t + \lambda^s_t(\hat{\mathcal{O}}^s_{L^s,t} \circ \ldots \circ \hat{\mathcal{O}}^s_{1,t}(X^s_t) - X^s_t)$ starting from X^s_0.

The server s cannot train such a network on real data. A regional server s cannot run the algorithm because it is missing the raw data:

$$\mathcal{D} = \cup_{c\in\mathcal{C}^s}\{(X^{s,c}_o, Y^{s,c}_o),\ o \in \{1,\ldots,D^s\}\}.$$

There is a set of eligible clients $\mathcal{C}^s_t \subset \mathcal{C}$ to server s to collaboratively train the network. After selecting and securing the authentication procedure, each selected client $c \in \mathcal{C}^s_t$ has τ time slots to complete a local training on its own device and fully controlled own-data. Client c receives the network architecture model from server s $\mathcal{M}^s_t$ and c trains by finding the queries, keys, values, weights and biases θ^s_t as follows. Let $X^{(s,c)}_0 \in (\mathcal{H}^s_0)^{D^s}$ be the input of the transformer-based tensor-graph network, $\{\lambda^{s,c}_t\}_{t\geq 0}$ be a non-negative sequence and $\hat{\mathcal{O}}^{s,c}_l$ given by

$$\begin{aligned}\hat{\mathcal{O}}^{s,c}_l :\ & \mathcal{H}^s_{l-1} \to \mathcal{H}^s_l \\ & X^{s,c} \mapsto (Id + \mathcal{O}^{s,c}_{l,ff} \circ \mathcal{O}^{s,c}_{l,nn}) \circ (Id + \mathcal{O}^{s,c}_{l,att} \circ \mathcal{O}^{s,c}_{l,nn})(X^{s,c})\end{aligned}$$

and iterate layer by layer and by timestep the following maps:

$$\left|\begin{array}{l} Parameters \\ Input : X^{s,c}_0 \\ \text{for } t \in \{0,1,2,\ldots\} : \\ \quad Y^{(s,c,1)}_t = \hat{\mathcal{O}}^{s,c}_{1,t}(X^{s,c}_t) \\ \quad \text{for } l \in \{2,\ldots,L^s\} :\ Y^{(s,c,l)}_t = \hat{\mathcal{O}}^{s,c}_{l,t}(Y^{(s,c,l-1)}_t) \\ \quad X^{s,c}_{t+1} = X^{s,c}_t + \lambda^{s,c}_t(Y^{(s,c,L^s)}_t - X^{s,c}_t), \end{array}\right. \tag{4.60}$$

which means that $X^{s,c}_{t+1} = X^{s,c}_t + \lambda^{s,c}_t(\hat{\mathcal{O}}^{s,c}_{L^s,t} \circ \ldots \circ \hat{\mathcal{O}}^{s,c}_{1,t}(X^{s,c}_t) - X^{s,c}_t)$ starting from $X^{s,c}_0$.

From a training perspective, client c updates $\theta^{s,c}_t$ to $\theta^{s,c}_{t+1}$ starting from $\theta^{s,c}_0$. This implies that c trains its local model by updating the queries, keys, values, weights and biases

$$\theta^{s,c}_t = ((Q^{s,c}_{l,h,t}, K^{s,c}_{l,h,t}, V^{s,c}_{l,h,t}, W^{s,c}_{l,h,t})_{1\leq h\leq H^s}, W^{s,c}_{l,2,t}, b^{s,c}_{l,2,t}, W^{s,c}_{l,1,t}, b^{s,c}_{l,1,t})_{1\leq l\leq L^s}$$

to $\theta^{s,c}_{t+1}$.

The training of the model on c consists of solving the following variational inequality. Given the local input-output (real) data $(X^{s,c}_t, Y^{s,c}_t)$ the client finds a vector θ^* such that

$$\sum_{l=1}^{L^{s,c}} \omega^{s,c}_l \langle (\hat{\mathcal{O}}^{s,c}_l[\theta^*] - y^{s,c}_l), A^{s,c}_l(\theta - \theta^*)\rangle \geq 0,\ \forall\theta,$$

where $\omega^{s,c}_l > 0$, $\sum_{l=1}^{L} \omega^{s,c}_l = 1$. The update is $\theta_{t+1} = \theta_t + \lambda_t \sum_{l=1}^{L^s} \omega^{s,c}_l (A^{s,c}_l)^*(\hat{\mathcal{O}}^{s,c}_l[\theta^*] - Y^{s,c}_l)$ with compatible Hilbert spaces.

At the end of the time slot $t = k$, client c sends to the server s the trained queries, keys, values, weights, biases $\theta^{s,c}_t$ through a secure protocol. The server s does not have access to the data $(X^{s,c}_t, Y^{s,c}_t)_t$ of client c. It is the server s which aggregates the model to $\mathcal{M}^s_{t+1}$ from the collected partial models $\{\theta^{s,c}_k\}_{k\leq t, c\in\mathcal{C}^s_t}$ from eligible clients in $\cup_t\mathcal{C}^s_t$.

There are different ways of aggregating the models such as

- averaging the parameters:

$$\theta^s_{t+1} := \frac{1}{|\mathcal{C}^s_t|}\sum_{c\in\mathcal{C}^s_t} \theta^{s,c}_t.$$

For example, $b^s_{l,t+1} := \frac{1}{|\mathcal{C}^s_t|}\sum_{c\in\mathcal{C}^s_t} b^{s,c}_{l,t}$, $W^s_{l,t+1} := \frac{1}{|\mathcal{C}^s_t|}\sum_{c\in\mathcal{C}^s_t} W^{s,c}_{l,t}$, for every layer of the network of the server.

- averaging the activation operator $R^s_{l,t+1} := \frac{1}{|C^s_t|}\sum_{c\in C^s_t} R^{s,c}_{l,t}$ or a weighted average $\sum_{c\in C^s_t} \omega^{s,c} R^{s,c}_{l,t}$ with $\omega^{s,c} \geq 0$, $\sum_c \omega^{s,c} = 1$, which is a γ-averaged operator for some suitable $\gamma \in (0, 1]$.

By applying the arguments above, we have the following:

Proposition 25. *The outcomes of the federated neural networks $\mathcal{M}^s$ are exactly the Nash equilibria of a non-zero sum game between regional servers and clients.*

Proposition 26. *The asymptotic federated training of queries, keys, values, weights and biases are exactly the solutions of the variational inequality involving regional servers and clients.*

4.7.2 Mean-Field-Type Federated Transformers

Mean-field-type federated transformer training is an approach to distributed large training models, where regional servers collaborate with multiple clients to train a shared model without accessing the true joint distribution $\mathbb{P}$ of input-output pairs (X, Y). Instead, the training operates on inputs (X, μ), where μ represents the first marginal of the joint distribution $\mathbb{P}$.

Key Components of Mean-Field-Type Federated Transformer Training include:

Client-Server Collaboration: In this framework, the regional server coordinates with multiple clients, each operating on local data, to iteratively train the transformer model. The server initializes the global model and sends it to the clients, who compute model updates on their local datasets. Importantly, the server does not have access to the true joint distribution (X, Y), relying on the clients to apply mean-field-type updates based on their localized data.

Normalization and Pushforward Operator: Clients locally normalize their input X and apply a pushforward operator to μ, the distribution of state of X. This transforms the pair (X, μ) into (X', μ'), which is then processed through a mean-field-type self-attention mechanism. This transformation and attention mechanism allow the model to capture interactions between the input and the distribution, enhancing its generalization ability across varied client datasets.

Mean-Field-Type Self-Attention: Unlike conventional attention mechanisms, which focus on pairwise token interactions, the mean-field-type self-attention processes the pair (X', μ'). This mechanism emphasizes the relationship between the input and the global state distribution μ', ensuring the model accounts for both local and McKean-Vlasov dynamics. After this attention step, the updated pair is passed through a normalized and pushforword followed by a two-layer feedforward network, where further transformations are applied to the input X'', and its distribution μ'' remains part of the modeling process.

Iterative Layer Updates: The model is trained iteratively over a specified number of layers, with the server setting the architecture depth. At each layer, the same sequence of steps is repeated: normalization of X, application of the pushforward operator to μ, mean-field-type attention, and feedforward transformations.

Reward-based Participation: To encourage client participation, a reward system is employed. Clients receive incentives based on the success of the global model training. This reward-based system ensures that clients remain motivated to contribute high-quality local updates, which ultimately benefit the federated model. The server uses these contributions to continuously update and improve the global model.

Preservation of Data Privacy: Similar to traditional federated training, mean-field-type federated transformer training ensures that raw data never leaves the local devices.

Clients share only the updated model parameters θ. This minimizes the risk of data exposure and allows for privacy-preserving model updates.

A regional server s cannot run the following algorithm because it is missing the raw data:

$$\mathcal{D} = \cup_{c \in \mathcal{C}^s} \{(X^{s,c}, Y^{s,c}, \mathbb{P}^{s,c})\}\}$$

where $\mathbb{P}^{s,c}$ is the joint probability distribution of $(X^{s,c}, Y^{s,c})$.

$$\begin{array}{l}
\mu_0^s = \nu_0^{(s,0)} =\in \mathcal{P}(\mathcal{H}_0^s)\\
t \in \{0, \ldots, T-1\}:\\
\quad l = 1,\\
\quad \nu_t^{(s,\frac{1}{4})} = \mathcal{O}_{l,nn} \# \mu_t^s,\\
\quad Y_t^{(s,\frac{1}{2})} = \hat{\mathcal{F}}_{\frac{1}{2}, \nu_t^{(s,\frac{1}{4})}}(X_t),\\
\quad \nu_t^{(s,\frac{1}{2})} = \hat{\mathcal{F}}_{\frac{1}{2}, \nu^{(s,\frac{1}{4})_t}} \# \mu_t^s,\\
\quad Y_t^{(s,1)} = \hat{\mathcal{F}}_1(Y_t^{(s,\frac{1}{2})}), \qquad \nu_t^{(s,1)} = \hat{\mathcal{F}}_1 \# \nu_t^{(s,\frac{1}{2})},\\
\quad \ldots\\
\\
\quad l \in \{2, \ldots, L^s - 1\}:\\
\qquad \nu_t^{(s,l-\frac{3}{4})} = \mathcal{O}_{l,nn} \# \nu_t^{(s,l-1)},\\
\qquad Y_t^{(s,l-\frac{1}{2})} = \hat{\mathcal{F}}_{l-\frac{1}{2}, \nu_t^{(s,l-\frac{3}{4})}}(Y_t^{(s,l-1)}),\\
\qquad \nu_t^{(s,l-\frac{1}{2})} = \hat{\mathcal{F}}_{l-\frac{1}{2}, \nu_t^{(s,l-\frac{3}{4})}} \# \nu_t^{(s,l-1)},\\
\qquad Y_t^{(s,l)} = \hat{\mathcal{F}}_l(Y_t^{(s,l-\frac{1}{2})}), \qquad \nu_t^{(s,l)} = \hat{\mathcal{F}}_l \# \nu_t^{(s,l-\frac{1}{2})},\\
\\
\quad \ldots\\
\\
\quad \nu_t^{(s,L-\frac{3}{4})} = \mathcal{O}_{L,nn} \# \nu_t^{(s,L-1)},\\
\quad Y_t^{(s,L^s-\frac{1}{2})} = \hat{\mathcal{F}}_{L^s-\frac{1}{2}, \nu_t^{(s,L^s-\frac{3}{4})}}(Y_t^{(s,L^s-1)}),\\
\quad \nu_t^{(s,L^s-\frac{1}{2})} = \hat{\mathcal{F}}_{L^s-\frac{1}{2}, \nu_t^{(s,L^s-\frac{3}{4})}} \# \nu_t^{(s,L^s-1)},\\
\quad Y_t^{(s,L^s)} = \hat{\mathcal{F}}_L(Y_t^{(s,L^s-\frac{1}{2})}), \qquad \nu_t^{(s,L^s)} = \hat{\mathcal{F}}_{L^s} \# \nu_t^{(s,L^s-\frac{1}{2})},\\
\\
\quad \mu_{t+1}^s = ((1-\lambda_t)Id + \lambda_t \hat{\mathcal{F}}_{L^s} \circ \hat{\mathcal{F}}_{L^s-\frac{1}{2} \; |\nu_t^{(s,L^s-\frac{3}{4})}} \circ \ldots \circ \hat{\mathcal{F}}_1 \circ \hat{\mathcal{F}}_{\frac{1}{2}, \mu_t^{(s,\frac{1}{4})}}) \# \mu_t^s.
\end{array} \tag{4.61}$$

Instead, server s does the following:

$$\begin{array}{l}
\quad \text{Server } s \text{ randomly picks and initializes the parameters } : \theta_0^s\\
\quad \text{Server } s \text{ collects the building blocs of the model } : \mathcal{M}^s\\
c \in\in \mathcal{C}^s: \quad \text{Server } s \text{ sends the model } : \mathcal{M}^s \text{ to client } c\\
t \in \{T, 2T \ldots, kT\}:\\
\quad c \in \mathcal{C}_t^s:\\
\qquad\qquad \text{Server } s \text{ receives the updated parameters from } c : \theta_t^{(s,c)}\\
\quad \text{Server } s \text{ collects only the blockchain-approved parameters } \mathcal{C}_{t,approved}^s\\
\quad \text{Server } s \text{ aggregates all the validated parameters:}\\
\qquad \theta_t^{(s)} = \sum_{c \in \mathcal{C}_{t,approved}^s} \theta_t^{(s,c)}\\
\quad c \in\in \mathcal{C}^s: \quad \text{Server } s \text{ sends the aggregated parameters to } c : \theta_t^{(s)}
\end{array} \tag{4.62}$$

Let

$$\theta_t^{(s,c)} = ((Q_{lh,t}^{s,c}, K_{lh,t}^{s,c}, V_{lh,t}^{s,c}, W_{lh,t}^{s,c})_{1\leq h\leq H}, W_{l,2,t}^{s,c}, b_{l,2,t}^{s,c}, W_{l,1,t}^{s,c}, b_{l,1,t}^{s,c})_{1\leq l\leq L}.$$

Each client c of the regional server s runs the following:

$$\begin{array}{l}
\text{Client } c \text{ receives the transformer model from server } s: \mathcal{M}^s \\
\mu_0^{s,c} = \nu_0^{(s,c,0)} \in \mathcal{P}(\mathcal{H}_0^s) \\
t \in \{0,\ldots,T-1\}: \\
\quad l = 1, \\
\quad \nu_t^{(s,c,\frac{1}{4})} = \mathcal{O}_{l,nn}\#\mu_t^{s,c}, \\
\quad Y_t^{(s,c,\frac{1}{2})} = \hat{\mathcal{F}}_{\frac{1}{2},\nu_t^{(s,c,\frac{1}{4})}}(X_t), \qquad \nu_t^{(s,c,\frac{1}{2})} = \hat{\mathcal{F}}_{\frac{1}{2},\nu^{(s,c,\frac{1}{4})}t}\#\mu_t^s, \\
\quad Y_t^{(s,c,1)} = \hat{\mathcal{F}}_1(Y_t^{(s,c,\frac{1}{2})}), \qquad \nu_t^{(s,c,1)} = \hat{\mathcal{F}}_1\#\nu_t^{(s,c,\frac{1}{2})}, \\
\quad \ldots \\
\quad l \in \{2,\ldots,L^s-1\}: \\
\qquad \nu_t^{(s,c,l-\frac{3}{4})} = \mathcal{O}_{l,nn}\#\nu_t^{(s,c,l-1)}, \\
\qquad Y_t^{(s,c,l-\frac{1}{2})} = \hat{\mathcal{F}}_{l-\frac{1}{2},\nu_t^{(s,c,l-\frac{3}{4})}}(Y_t^{(s,l-1)}), \\
\qquad \nu_t^{(s,c,l-\frac{1}{2})} = \hat{\mathcal{F}}_{l-\frac{1}{2},\nu_t^{(s,c,l-\frac{3}{4})}}\#\nu_t^{(s,c,l-1)}, \\
\qquad Y_t^{(s,c,l)} = \hat{\mathcal{F}}_l(Y_t^{(s,c,l-\frac{1}{2})}), \qquad \nu_t^{(s,c,l)} = \hat{\mathcal{F}}_l\#\nu_t^{(s,c,l-\frac{1}{2})}, \\
\quad \ldots \\
\quad \nu_t^{(s,c,L-\frac{3}{4})} = \mathcal{O}_{L,nn}\#\nu_t^{(s,c,L-1)}, \\
\quad Y_t^{(s,c,L^s-\frac{1}{2})} = \hat{\mathcal{F}}_{L^s-\frac{1}{2},\nu_t^{(s,L^s-\frac{3}{4})}}(Y_t^{(s,c,L^s-1)}), \\
\quad \nu_t^{(s,c,L^s-\frac{1}{2})} = \hat{\mathcal{F}}_{L^s-\frac{1}{2},\nu_t^{(s,c,L^s-\frac{3}{4})}}\#\nu_t^{(s,c,L^s-1)}, \\
\quad Y_t^{(s,c,L^s)} = \hat{\mathcal{F}}_L(Y_t^{(s,c,L^s-\frac{1}{2})}), \qquad \nu_t^{(s,c,L^s)} = \hat{\mathcal{F}}_{L^s}\#\nu_t^{(s,c,L^s-\frac{1}{2})}, \\
\quad \mu_{t+1}^{s,c} = \\
\quad ((1-\lambda_t)Id + \lambda_t\hat{\mathcal{F}}_{L^s}\circ\hat{\mathcal{F}}_{L^s-\frac{1}{2}\,|\nu_t^{(s,c,L^s-\frac{3}{4})}}\circ\ldots\circ\hat{\mathcal{F}}_1\circ\hat{\mathcal{F}}_{\frac{1}{2},\nu_t^{(s,c,\frac{1}{4})}})\#\mu_t^{s,c}. \\
\quad \text{Client } c \text{ updates parameters from server } s: \theta_{t+1}^{(s,c)} \\
\text{Client } c \text{ sends the updated parameters from server } s: \theta_T^{(s,c)}
\end{array} \tag{4.63}$$

As displayed in (4.62), this federated transformer training is not necessarily incentive compatible. There may be some free-riders: those who do not train the model but who want to benefit from the work of the other contributors. In order to reduce such behaviors, an incentive design will be included in the blockchain rewarding system. The blockchain verifies and validates only the high-quality parameters. Then contributors of high-quality parameters are rewarded with blockchain tokens and those who send low-quality parameters or non-eligible parameters are penalized: their blockchain tokens are reduced and they will be excluded from the training process if their total tokens in the blockchain are below is a certain threshold. The benefits of a server s is measured by the quality of the machine intelligence deployed at the end of the training process. These benefits are then obtained from the services it offers to subscribers to that technology. This will in turn reward each contributor to the technology through their token valuation.

All the regional servers will work together for number of regulation updates, international laws and machine intelligence ethics, making a global interactive system.

Putting all together we arrive at:

Proposition 27. *The outcomes (if any) of the mean-field-type federated transformer training are exactly the Nash equilibria of an hierarchical mean-field-type game between regional servers and clients.*

Proposition 28. *The mean-field-type federated transformer training of queries, keys, values, weights and biases are exactly the solutions of the variational inequality involving the aggregates of regional servers and clients.*

4.8 Unlearning in Mean-Field-Type Transformers

As the capabilities of generative machine intelligence soar, the need for self-regulating machine intelligence systems has become critical, particularly in navigating an era where data privacy, authenticity, and regulatory compliance are paramount. Foundational unlearning offers an elegant solution for addressing this challenge across the entire machine intelligence lifecycle from initial design and data collection to deployment. By integrating unlearning mechanisms, these systems can autonomously discard outdated or irrelevant data, enhancing data integrity by removing inputs that no longer reflect current knowledge or ethical standards. Consider a design phase where unlearning models can actively ignore obsolete specifications or insights, allowing for architecture that adapts to the latest regulatory requirements and scientific advancements. In terms of data and information integrity, foundational unlearning allows systems to retain only verified, accurate data, minimizing the risk of bias and misinformation, a particularly valuable asset in industries where accuracy is non-negotiable, like healthcare or finance. During pre-training and training, unlearning-based mechanisms make it possible for generative models to shed outdated patterns or incorrect associations, ensuring that the model evolves alongside advancements in legislation or scientific understanding without requiring a complete overhaul. This capacity for selective memory loss could even empower machine integrity by refining the way models self-correct and adapt based on continuous feedback, allowing for the preservation of core capabilities while jettisoning unreliable or misleading data. When ready for deployment, a model equipped with foundational unlearning can better meet regulatory standards, as it is able to retain only the latest compliance guidelines and discard legacy information that no longer aligns with current laws. The result is a generative machine intelligence that is not only more flexible and responsive to change but also poised to keep pace with ever-evolving requirements, scientific breakthroughs, and multi-scale ethical considerations, laying the groundwork for machine learning systems that are as trustworthy as they are intelligent.

Foundational unlearning models offer a glimpse into a future where machine intelligence systems can selectively forget as much as they learn. These emerging architectures are set to transform everything from conventional transformers to systems like diffusion-transformers, liquid transformers, fluid transformers, and even mean-field-type transformers or McKean-Vlasov models, each of which operates on the bleeding edge of machine intelligence research. The concept is deceptively simple: rather than relying solely on vast amounts of new data to update systems, foundational unlearning models can discard outdated or unwanted information selectively, allowing these advanced architectures to become leaner, more efficient, and better suited for real-world applications that demand privacy and compliance. Imagine a healthcare model that can "forget" a patient's outdated medical data once it becomes irrelevant or a financial forecasting system that swiftly unlearns past biases as it adapts to shifting market conditions. While large-scale unlearning remains a powerful, albeit computationally demanding tool, foundational unlearning is a lightweight cousin, enabling these models to adapt on-the-fly with minimal computational overhead. The implications stretch far beyond simple optimization; they offer an entirely new way of thinking about machine intelligence training, where models are no longer a monolithic blend of past knowledge but instead have the flexibility to shed portions of memory in response to new contexts. The allure of foundational unlearning models lies in our potential to drive both adaptability and compliance, making them an attractive bet for industries that require machine intelligence systems to be as nimble and privacy-conscious as they are intelligent.

Several methods have been used in the literature to unlearn from learning models. Their extensions to large learning models remains a challenging task.

- SISA (Sharded, Isolated, Sliced, and Aggregated) Training in Transformer Models

  How it applies: For large transformers, SISA training can be adapted by dividing the training data into independent shards and training each shard on different parts of the model or different subsets of model layers. When unlearning is required, only the relevant shards of data and model components (e.g., attention heads or layers) are retrained.

  Challenges in transformers: Transformers are highly interconnected, and isolating specific parts of the model might be more complex than in traditional neural networks, which could limit the effectiveness of this method.

- Knowledge Editing Methods in Transformers (MEMIT and Others)

  How it applies: In transformer models like GPT, Knowledge Editing Networks (KENs) and Model Editing via Minimal Invasive Transformations (MEMIT) can target specific layers or attention heads to edit or remove learned knowledge without affecting the broader language generation capabilities. Transformers process and store knowledge in distributed representations across multiple layers, making it possible to locate and edit specific knowledge. Model Editing via Minimal Invasive Transformation is an approach in machine unlearning that addresses the need to quickly and effectively "unlearn" specific data points from machine learning models. Traditional methods often require retraining the model from scratch to remove data influence, but MEMIT streamlines this by directly adjusting model parameters. This technique involves minimal edits to the model, adjusting parameters only where necessary to eliminate the influence of targeted data, often by modifying specific layers or neurons rather than re-training the entire model. The MEMIT approach is particularly useful for privacy-focused scenarios where rapid compliance with data deletion requests is essential, such as in compliance with privacy laws. Additionally, MEMIT's efficient editing process means it requires less computational cost and time compared to retraining, making it highly practical for large-scale models. MEMIT is a part of the broader field of machine unlearning, which explores ways to ensure models "forget" data without the need for complete retraining, often using techniques like elastic weight consolidation, which preserves important parameters, or sharded training, which compartmentalizes the model to reduce the impact of deletions. These approaches are essential for improving privacy and maintaining performance without disrupting the integrity of the entire model. MEMIT specifically allows mass editing in transformers by identifying key neurons in the middle layers that encode factual knowledge. These methods are efficient because they do not require full retraining but rather focus on editing parameter space locally. It has been used in LLMs, especially when needing to remove or edit many facts or pieces of knowledge without retraining from scratch. It can modify localized sections of the model while preserving the general linguistic and contextual knowledge.

- Fine-Pruning in Transformer Layers

  How it applies: In transformers, fine-pruning can be applied to prune specific neurons or attention heads that are responsible for encoding undesired knowledge or biases. Each layer in a transformer can focus on different representations of language, meaning specific layers (or even individual attention heads) can be targeted for unlearning.

  Challenges in transformers: Transformers rely on attention mechanisms that distribute the learning across layers, making it harder to isolate exact representations of specific knowledge. Pruning needs to be done carefully to avoid affecting the model's ability to generate coherent text.

- Elastic Weight Consolidation (EWC) in Transformers: EWC can be adapted to transformers by identifying critical weights (e.g., in self-attention layers) and preventing them from changing during unlearning. When removing certain knowledge, the algorithm prioritizes modifying non-essential weights so the model retains overall performance, while targeted unlearning takes place. EWC can be effective in ensuring that important language patterns are preserved during the unlearning process. It helps manage the trade-off between unlearning specific information and maintaining general linguistic abilities in large transformers.

- Retraining with Differential Privacy in Transformers: For transformers, retraining with differential privacy ensures that sensitive data points can be removed from the model without requiring full retraining. This method can be useful for privacy-sensitive applications where specific user data embedded in the model needs to be forgotten. Large transformer models, when trained with differential privacy, might experience a loss in performance due to noise injection required by privacy constraints. However, selectively removing knowledge while retaining the rest of the model can still be accomplished with minimal retraining.

- Projection-Based Unlearning in Transformer Architectures: In transformers, projection-based unlearning can be used to project model parameters (e.g., attention weights) onto a subspace that excludes the undesired knowledge. By identifying which attention heads or layers encode the knowledge that needs to be forgotten, projection-based techniques can remove those representations while preserving the rest of the model. Since transformers use distributed representations, projection-based unlearning allows for fine control over which aspects of the learned knowledge are forgotten without affecting the rest of the model's capabilities. Properly identifying the layers and parameters responsible for specific knowledge is a challenge due to the highly distributed nature of transformers.

- SCRUB (Selectively Forgetting in Neural Networks) in Transformers: SCRUB can be adapted to transformers by identifying the specific parts of the model (layers, attention heads, or neurons) that are most influenced by the data to be unlearned. This can be done using influence functions to trace how particular training points affect the model's predictions. Once identified, only these sections are retrained or updated. SCRUB allows efficient unlearning by focusing on the sections of the model most affected by the data, thus avoiding full retraining. This is particularly useful for transformers where data influence is often spread across many layers, allowing for a targeted and computationally efficient unlearning process. Identifying specific training data influence in transformer models can be more complex than in simpler architectures due to the deep and wide attention mechanisms that distribute learning across the network.

Unlearning via Orthogonal Projection

To further enhance the effectiveness of orthogonal projection-based unlearning in large transformer-based models, a double-check projection strategy can be introduced. In this approach, the unwanted knowledge is removed both at the input layer and at the output layer, ensuring that any trace of the undesired information that might reappear during the internal model processing is also eliminated by the time the output is generated.

Double-Check Projection Strategy

We apply two orthogonal projections:

- At the Input Layer: The input embedding is projected onto the subspace orthogonal to the feature vector, which represents the knowledge to be unlearned.
- At the Output Layer: After the data has passed through all the layers of the transformer, the output is also projected onto the subspace orthogonal to ensure any unwanted knowledge that might have reappeared during the intermediate processing is removed.

By projecting at both the input and output layers, we provide an additional safeguard against the re-encoding of unwanted knowledge, ensuring a more thorough and reliable unlearning process.

Let us break this down step by step with the updated formulation:

- Input Embedding Projection: As before, let this be the input embedding for a given data point, and represents the feature vector corresponding to the knowledge to be unlearned.

  The input embedding is projected onto the subspace orthogonal to

  $x_{\perp z} = x - \frac{\langle x, v_z \rangle}{\langle v_z, v_z \rangle} v_z.$

- Passing Through the Transformer: The modified input is then processed by the transformer model, which applies its series of attention layers and hidden layers.

- Output Projection: After the data has passed through all the layers and produced an output, we apply the same projection onto the subspace orthogonal to at the output layer:

  $o_{\perp z} = o - \frac{\langle o, v_z \rangle}{\langle v_z, v_z \rangle} v_z.$

Why the Double-Check Projection is Effective

- Catching Re-encoded Knowledge: Transformer models are known to distribute knowledge across layers, and despite projecting out the unwanted knowledge at the input layer, the model might re-encode that knowledge in later layers. By performing an additional projection at the output layer, we eliminate any remaining influence of the knowledge that might have been regenerated within the model during its processing.

- Global Impact on Representations: By applying projections at both the start and the end of the model, we ensure that the removal of the unwanted information is not only thorough but also affects all layers uniformly. The input projection prevents the knowledge from being introduced, while the output projection guarantees it doesn't persist by the time the output is generated.

- Reduced Risk of Overfitting: Unlike approaches like MEMIT, which modify internal model parameters and can lead to overfitting or unintended changes to model behavior, this double-check projection strategy minimizes the risk of such side effects. It only modifies the input and output representations while leaving the core model's weights and architecture intact.

Advantages Over Single Projection and MEMIT

- More Thorough Unlearning: By projecting both at the input and output, this method ensures that any re-emergence of the unwanted knowledge within the model is eliminated. MEMIT and other internal-editing methods may miss re-encoded information because they rely on modifying specific parts of the model, whereas double-check projection ensures global removal.

- Computational Simplicity: This approach remains computationally efficient, focusing only on the input and output layers. There's no need for expensive retraining or fine-tuning of the internal parameters of the model, as required by methods like MEMIT.

- Scalability: Since it works at the input and output levels, this method can be easily scaled to large models with minimal overhead. There is no need for identifying and modifying specific parameters within the internal layers, making it easier to apply in practice for models like GPT.

Challenges and mitigation strategies include Identifying the feature vector. As with the input-only projection, accurately identifying the feature vector representing the unwanted knowledge is crucial. This can be addressed using interpretability techniques like Layer-wise Relevance Propagation or Feature Attribution to track down how specific facts are encoded. There is also a risk of over-projecting. Projecting at both the input and output could potentially lead to excessive removal of information, especially if the projection removes more than just the unwanted knowledge. To mitigate this, careful regularization should be applied. We can use a regularization term to prevent excessive projection:

$$\mathcal{L}_{proj}(\theta) = \mathcal{L}(\theta) + \lambda \left(\|x_{\perp z}\|^2 + \|o_{\perp z}\|^2 \right).$$

Even with projections at both ends, intermediate layers could still have re-encoded traces of the unwanted knowledge that do not appear directly in the input or output representations.

To address this, a hybrid method combining this projection approach with minimal internal adjustments (e.g., fine-pruning) could be used.

Potential Improvements Over MEMIT

By applying orthogonal projections at both the input and output layers, this method could surpass MEMIT in several key ways:

- Holistic Unlearning: Instead of focusing on specific layers, this approach ensures that knowledge is removed from the entire process, from input to output.
- Less Risk of Degrading Other Knowledge: MEMIT modifies internal parameters, which could inadvertently disrupt other useful knowledge. Double projection ensures only the unwanted knowledge is removed without disturbing the internal structure of the model.
- Flexibility: This method can be easily extended to other types of knowledge and scales well to large models. You can project out multiple facts by iteratively applying the projection for different feature vectors.

Applying orthogonal projections at both the input and output layers offers a comprehensive and efficient way to unlearn specific knowledge in transformer-based models. This double-check projection ensures that unwanted knowledge is removed before it enters the model at the input stage, and any re-encoded or residual knowledge is eliminated at the output stage. This strategy provides a robust safeguard against the reintroduction of unwanted information and can be computationally cheaper and more scalable than methods like MEMIT, while minimizing the risk of side effects on the model's overall performance. To identify the feature vector representing a specific piece of knowledge from the knowledge itself, you can use a combination of model interpretability techniques and mathematical tools.

Below is a detailed explanation of the process, followed by equations to help with the identification. The piece of knowledge you want to unlearn could be a fact or a specific relationship between concepts encoded in the model. For example, in a language model, might be a fact like "Paris is the capital of France." Determine activation or embedding of that fact. The next step is to identify how the model encodes this knowledge. For transformers, this is often captured in the embeddings or the hidden state activations of specific layers in response to inputs.

Suppose we input a sentence or prompt that invokes the knowledge (e.g., "What is the capital of France?"). The model generates an internal embedding in one or more layers. Let $h_l(z) \in \mathbb{R}^d$. Use an attribution method such as Integrated gradients, layer-wise relevance propagation (LRP), or saliency maps to determine which dimensions of are most responsible for encoding the knowledge. One common approach is integrated gradients, which calculates the gradient of the model's output with respect to the input or hidden states, highlighting which parts of the hidden representation contribute most to the model's output. The integrated gradient for a neuron in a layer is: $IG_i(z) = (h_l(z) - h_l(z_0)) + \int_{\alpha=0}^{1} \frac{\partial \text{output}}{\partial h_l(z_\alpha)} d\alpha$. The result will be a vector where each component corresponds to the importance of each dimension in the output related to the knowledge. Based on the attribution results, construct the feature vector. The simplest approach is to use the hidden state itself, but with a focus on the dimensions most relevant to the knowledge. You can define as the part of that is responsible for encoding. If is the hidden state, and is the importance of each dimension, then you can define as a weighted version of where irrelevant dimensions (those with low) are downweighted: $v_z = \sum_{i=1}^{d} IG_i(z) \cdot h_i$. It is useful to normalize the vector to ensure consistent projection and scaling: $v'_z = \frac{v_z}{\|v_z\|}$. Once it is identified, you can use it in the orthogonal projection to remove the knowledge from both input and output representations. For a vector (input or output), the projection onto the subspace orthogonal is $x_{\perp z} = x - \frac{\langle x, v_z \rangle}{\langle v_z, v_z \rangle} v_z$. For this unlearning method, we use the following: Identify the knowledge to be unlearned. First, filter of the query over a blockchain of facts. If there is green light we use the double-check method. Pass a prompt related to the model to generate a hidden state. Use an attribution method like integrated gradients to calculate the importance of each hidden dimension. Construct the feature vector by weighting the hidden state dimensions by their importance scores. This method effectively identifies the feature vector corresponding to a specific piece of knowledge, which can then be used for orthogonal projection in both input and output layers to unlearn the knowledge from the model.

4.9 Mean-Field-Type Federated Unlearning or Untraining

We now provide an algorithm for Federated Unlearning or Untraining using orthogonal projection of features.

$$
\begin{array}{l}
\quad \text{Server } s \text{ randomly picks and initializes the parameters } : \ \theta_0^s \\
\quad \text{Server } s \text{ collects the building blocs of the model } : \ \mathcal{M}^s \\
c \in\in \mathcal{C}^s : \ \text{Server } s \text{ sends the model } : \ \mathcal{M}^s \text{ to client } c \\
t \in \{T, 2T \ldots, kT\} : \\
\quad\quad c \in \mathcal{C}_t^s : \\
\quad\quad\quad\quad \text{Server } s \text{ receives the updated parameters from } c : \theta_t^{(s,c)} \\
\quad\quad \text{Server } s \text{ collects only the blockchain-approved parameters } \mathcal{C}_{t,approved}^s \\
\quad\quad \text{Server } s \text{ aggregates all the validated parameters: } \theta_t^{(s)} = \sum_{c \in \mathcal{C}_{t,approved}^s} \theta_t^{(s,c)} \\
\quad\quad c \in\in \mathcal{C}^s : \ \text{Server } s \text{ sends the aggregated parameters to } c : \theta_t^{(s)}
\end{array}
\tag{4.64}
$$

Let

$$\theta_t^{(s,c)} = ((Q_{lh,t}^{s,c}, K_{lh,t}^{s,c}, V_{lh,t}^{s,c}, W_{lh,t}^{s,c})_{1 \le h \le H}, W_{l,2,t}^{s,c}, b_{l,2,t}^{s,c}, W_{l,1,t}^{s,c}, b_{l,1,t}^{s,c})_{1 \le l \le L}.$$

Let $\phi_{unlearn,z} : x \mapsto x - \langle x, \frac{v_z}{\langle v_z, v_z \rangle} \rangle v_z$

Each client c of the regional server s runs:

$$
\begin{aligned}
&\text{Client } c \text{ receives the transformer model from server } s: \ \mathcal{M}^s\\
&\mu_0^{s,c} = \nu_0^{(s,c,0)} \in \mathcal{P}(\mathcal{H}_0^s)\\
&Z_0 = \phi_{unlearn,z}(X_0), \nu^{(s,c,0)}_{unlearn,z,0} = \phi_{unlearn,z}\#\nu^{(s,c,0)},\\
&t \in \{0,\ldots,T-1\}:\\
&\quad l = 1,\\
&\quad \nu_t^{(s,c,\frac{1}{4})} = \mathcal{O}_{l,nn}\#\nu^{s,c}_{unlearn,z,t},\\
&\quad Y_t^{(s,c,\frac{1}{2})} = \hat{\mathcal{F}}_{\frac{1}{2},\nu_t^{(s,c,\frac{1}{4})}}(Z_t), \qquad \nu_t^{(s,c,\frac{1}{2})} = \hat{\mathcal{F}}_{\frac{1}{2},\nu^{(s,c,\frac{1}{4})_t}}\#\nu_t^{s,c},\\
&\quad Y_t^{(s,c,1)} = \hat{\mathcal{F}}_1(Y_t^{(s,c,\frac{1}{2})}), \qquad \nu_t^{(s,c,1)} = \hat{\mathcal{F}}_1\#\nu_t^{(s,c,\frac{1}{2})},\\
&\quad Z_t^{(s,c,1)} = \phi_{unlearn,z}(Y_t^{(s,c,1)}), \qquad \nu^{(s,c,1)}_{unlearn,z,t} = \phi_{unlearn,z}\#\nu_t^{(s,c,1)},\\
&\quad \ldots\\
&\quad l \in \{2,\ldots,L^s-1\}:\\
&\qquad \nu_t^{(s,c,l-\frac{3}{4})} = \mathcal{O}_{l,nn}\#\nu^{(s,c,l-1)}_{unlearn,z,t},\\
&\qquad Y_t^{(s,c,l-\frac{1}{2})} = \hat{\mathcal{F}}_{l-\frac{1}{2},\nu_t^{(s,c,l-\frac{3}{4})}}(Z_t^{(s,l-1)}),\\
&\qquad \nu_t^{(s,c,l-\frac{1}{2})} = \hat{\mathcal{F}}_{l-\frac{1}{2},\nu_t^{(s,c,l-\frac{3}{4})}}\#\nu^{(s,c,l-1)}_{unlearn,z,t},\\
&\qquad Y_t^{(s,c,l)} = \hat{\mathcal{F}}_l(Y_t^{(s,c,l-\frac{1}{2})}), \qquad \nu_t^{(s,c,l)} = \hat{\mathcal{F}}_l\#\nu_t^{(s,c,l-\frac{1}{2})},\\
&\qquad Z_t^{(s,c,l)} = \phi_{unlearn,z}(Y_t^{(s,c,l)}), \qquad \nu^{(s,c,l)}_{unlearn,z,t} = \phi_{unlearn,z}\#\nu_t^{(s,c,l)},\\
&\quad \ldots\\
&\quad \nu_t^{(s,c,L^s-\frac{3}{4})} = \mathcal{O}_{L^s,nn}\#\nu^{(s,c,L^s-1)}_{unlearn,z,t},\\
&\quad Y_t^{(s,c,L^s-\frac{1}{2})} = \hat{\mathcal{F}}_{L^s-\frac{1}{2},\nu_t^{(s,L^s-\frac{3}{4})}}(Z_t^{(s,c,L^s-1)}),\\
&\quad \nu_t^{(s,c,L^s-\frac{1}{2})} = \hat{\mathcal{F}}_{L^s-\frac{1}{2},\nu_t^{(s,c,L^s-\frac{3}{4})}}\#\nu^{(s,c,L^s-1)}_{unlearn,z,t},\\
&\quad Y_t^{(s,c,L^s)} = \hat{\mathcal{F}}_L(Y_t^{(s,c,L^s-\frac{1}{2})}), \qquad \nu_t^{(s,c,L^s)} = \hat{\mathcal{F}}_{L^s}\#\nu_t^{(s,c,L^s-\frac{1}{2})},\\
&\quad Z_t^{(s,c,L^s)} = \phi_{unlearn,z}(Y_t^{(s,c,L^s)}), \qquad \nu^{(s,c,L^s)}_{unlearn,z,t} = \phi_{unlearn,z}\#\nu_t^{(s,c,L^s)},\\
&\quad \mu_{t+1}^{s,c} = ((1-\lambda_t)Id + \lambda_t\phi \circ \hat{\mathcal{F}}_{L^s} \circ \hat{\mathcal{F}}_{L^s-\frac{1}{2}}|_{\nu_t^{(s,c,L^s-\frac{3}{4})}} \circ\\
&\quad \ldots \circ \phi \circ \hat{\mathcal{F}}_2 \circ \hat{\mathcal{F}}_{\frac{3}{2},\nu_t^{(s,c,\frac{5}{4})}} \circ \phi \circ \hat{\mathcal{F}}_1 \circ \hat{\mathcal{F}}_{\frac{1}{2},\nu_t^{(s,c,\frac{1}{4})}} \circ \phi)\#\mu_t^{s,c}.\\
&\quad \text{Client } c \text{ updates parameters from server } s: \ \theta_{t+1}^{(s,c)}\\
&\text{Client } c \text{ sends the updated parameters from server } s: \ \theta_T^{(s,c)}
\end{aligned} \tag{4.65}
$$

Note that the server cannot run the federated untraining procedure because it does not have access to the data.

4.10 On the Suboptimality of Constant Parameters in Mean-Field-Type Transformers

In the context of mean-field-type transformers, particularly those that incorporate mean-field-type attention mechanisms, the role of dynamic parameters, specifically, query, key, value, weight, and bias, is crucial for achieving optimal performance. The inherent complexity in these architectures arises from the fact that each data point interacts not only with other data points but also with the distribution of states across the entire dataset. This leads to an increased need for state-and-mean-field-type dependent parameters, which stands in contrast to the suboptimality of constant or static parameterizations.

In such systems, the mean-field-type attention mechanism aggregates information about the distribution of states, meaning each point's interaction is influenced not just by individual signals but by the overall statistical structure of the data. This behavior can be formalized using optimal control frameworks such as the Mean-Field-Type Bellman system or the Master Adjoint Systems. These systems highlight how parameters should be informed both by the current state and the mean-field-type (the distribution of other states) to ensure optimal decision-making. Constant parameters, which do not account for this dynamic feedback, fail to adapt to both individual variations in the input signals and the broader statistical properties of the dataset.

Mean-Field-Type Bellman System and Optimal Feedback Control

The Mean-Field-Type Bellman system, derived from optimal control theory, emphasizes the necessity for feedback-based control laws that consider not only the current state of the system but also the statistical distribution (mean-field-type) of other agents or data points. This principle carries over to mean-field-type transformers, where query, key, value, weight, and bias parameters should be functions of both the state (input signal) and the mean-field-type distribution. In doing so, these parameters adapt dynamically to variations in both the local and global structure of the data.

In contrast, constant parameters, those that are independent of both the input signal (state) and the overall data distribution (mean-field-type), ignore these variations. This leads to suboptimal performance, as static parameterization assumes uniformity across both individual data points and the overall dataset. Constant parameters cannot effectively capture the evolving relationship between a given data point and the shifting mean-field-type, resulting in inefficiencies in the learning process and suboptimal model outputs.

Master Adjoint Systems and Mean-Field-Type Feedback

The Master Adjoint Systems, a tool often used in stochastic mean-field-type game theory, further justify the use of state-and-mean-field-type dependent parameters. In these systems, optimal control policies depend on both the state of an individual system and the overall mean-field-type distribution of states. Applying this to mean-field-type transformers, the optimal parameterization of the query, key, value, and other components should likewise be feedback-driven, informed by both the state and the interaction of that state with the larger data distribution.

By accounting for both local state and global mean-field-type feedback, the network is better equipped to generalize over large, complex datasets, where the relationship between individual data points and the overall structure is crucial. Constant parameters, which remain fixed regardless of changes in data, fail to capture this interaction, leading to suboptimal learning outcomes, especially in high-dimensional, large-scale datasets.

Argument Against State-Independent and Mean-Field-Type Independent Parameters

Although one might argue that simplifying parameterization through constant, state-independent, and mean-field-type independent parameters could reduce computational costs and improve scalability, this comes at the expense of performance. Such parameters do not have the capacity to adjust to the underlying data distribution, leading to poorer generalization, especially when dealing with non-uniform datasets where data points exhibit complex interactions with the mean-field-type.

In high-dimensional neural networks, such as transformers dealing with complex data, adaptive, dynamic parameterization is crucial. By leveraging state-and-mean-field-type feedback, neural networks are more equipped to handle diverse data structures and can better approximate the optimal solution in both learning and inference phases. The suboptimality of constant parameters is thus not just a theoretical result but is evident in the practical limitations they impose on large-scale machine learning tasks.

In mean-field-type transformers, where each data point interacts not just with other points but with the distribution of states, constant or state-independent parameters lead to suboptimal outcomes. From the perspectives of both the Mean-Field-Type Bellman system and the Master Adjoint Systems, optimal parameters must be functions of both the input signal (state) and the distribution of the dataset (mean-field-type). By failing to account for this dynamic feedback, constant parameters hinder a model's ability to adapt to complex, real-world data, limiting the efficiency and effectiveness of the network.

We have considered a tensor-graph neural network combined with mean-field-type attention mechanisms leading to mean-field-type transformers. The mean-field-type transformers that do not need infinite size knowledge of the distribution is available. Then, computability capabilities are used to pre-compute the interaction kernels at each layer and this is done for multiple attention-heads. The framework is connected to mean-field-type game theory where multiple agents interact while facing both individual terms and population mean-field terms. We also highlight connection to mean-field-type diffusion-transformers (transfusions) which is extremely useful for high-quality image and vision in real-world scenarios. The mathematics behind the architectures are key towards understanding the errors and enabling reasoning capabilities in the future.

4.11 Post-Training Without Re-Training

The next result (Theorem 4) establishes a foundational principle: affine corrections in Hilbert spaces yield non-increasing MSEs when calibrated using validation data from the same distribution. This result is highly relevant for functional data analysis, kernel-based time series models, and high-dimensional forecasting tasks, where predictions and ground truth are elements of infinite-dimensional function spaces. It also provides a low-cost, theoretically sound post-training calibration step that is model-agnostic: applicable to any predictive method producing outputs in a Hilbert space. It is computationally efficient: avoids retraining and leverages closed-form expressions. It is statistically grounded: optimal under the squared loss. Such affine corrections can serve as a first layer of a broader optimization framework, possibly followed by dynamic or nonlinear corrections, such as those driven by reinforcement learning or human-in-the-loop strategies.

Theorem 4 (Affine Correction in Hilbert Spaces Reduces MSE). *Let $\mathcal{H}$ be a real Hilbert space with inner product $\langle \cdot, \cdot \rangle$ and induced norm $\| \cdot \|$. Let $Y, \hat{Y} \in \mathcal{H}$ be square-integrable* *random elements representing the ground truth and model predictions, respectively. Define the affine-corrected prediction as*

$$Y^{corrected} := a^*\hat{Y} + b^*,$$

where $a^* \in \mathbb{R}$ *and* $b^* \in \mathcal{H}$ *are chosen to minimize the expected squared error:*

$$\mathbb{E}[\|Y - (a\hat{Y} + b)\|^2].$$

Then, the optimal coefficients are

$$a^* = \frac{\mathrm{Cov}(Y, \hat{Y})}{\mathrm{Var}(\hat{Y})}, \quad b^* = \mathbb{E}[Y] - a^*\mathbb{E}[\hat{Y}],$$

where

$$\mathrm{Cov}(Y, \hat{Y}) := \mathbb{E}\left[\langle Y - \mathbb{E}[Y], \hat{Y} - \mathbb{E}[\hat{Y}]\rangle\right], \quad \mathrm{Var}(\hat{Y}) := \mathbb{E}\left[\|\hat{Y} - \mathbb{E}[\hat{Y}]\|^2\right].$$

Furthermore, this correction strictly reduces or preserves the mean squared error (MSE), i.e.,

$$\mathrm{MSE}_{before} - \mathrm{MSE}_{after} = \left(\sqrt{\mathrm{Var}(\hat{Y})} - \frac{\mathrm{Cov}(Y, \hat{Y})}{\sqrt{\mathrm{Var}(\hat{Y})}}\right)^2 \geq 0.$$

Proof. Define the uncorrected mean squared error as

$$R_{\text{before}} := \mathbb{E}[\|Y - \hat{Y}\|^2],$$

and define the post-correction MSE as

$$R_{\text{after}} := \inf_{a \in \mathbb{R},\ b \in \mathcal{H}} \mathbb{E}[\|Y - a\hat{Y} - b\|^2].$$

Let us define the objective function:

$$L(a, b) := \mathbb{E}[\|Y - a\hat{Y} - b\|^2].$$

Optimize over b. For fixed a, we take the Fréchet derivative with respect to b and set it to zero:

$$\frac{\partial L}{\partial b} = -2\mathbb{E}[Y - a\hat{Y} - b] = 0 \quad \Rightarrow \quad b = \mathbb{E}[Y] - a\mathbb{E}[\hat{Y}].$$

Substitute optimal b into $L(a, b)$. Let $\tilde{Y} := Y - \mathbb{E}[Y]$, $\tilde{\hat{Y}} := \hat{Y} - \mathbb{E}[\hat{Y}]$. Then,

$$L(a) := \mathbb{E}[\|Y - a\hat{Y} - b^*(a)\|^2] = \mathbb{E}[\|\tilde{Y} - a\tilde{\hat{Y}}\|^2].$$

Expanding the square:

$$L(a) = \mathbb{E}[\|\tilde{Y}\|^2] - 2a\mathbb{E}[\langle\tilde{Y}, \tilde{\hat{Y}}\rangle] + a^2\mathbb{E}[\|\tilde{\hat{Y}}\|^2] = \mathrm{Var}(Y) - 2a\,\mathrm{Cov}(Y, \hat{Y}) + a^2\,\mathrm{Var}(\hat{Y}).$$

This is a convex quadratic in a, minimized at

$$a^* = \frac{\mathrm{Cov}(Y, \hat{Y})}{\mathrm{Var}(\hat{Y})}.$$

Compute MSE improvement. We compute:

$$\Delta := R_{\text{before}} - R_{\text{after}}$$
$$= \mathrm{Var}(Y) + \mathrm{Var}(\hat{Y}) - 2\mathrm{Cov}(Y, \hat{Y}) - \left(\mathrm{Var}(Y) - \frac{[\mathrm{Cov}(Y, \hat{Y})]^2}{\mathrm{Var}(\hat{Y})} \right),$$

$$\Delta = \mathrm{Var}(\hat{Y}) - 2\mathrm{Cov}(Y, \hat{Y}) + \frac{[\mathrm{Cov}(Y, \hat{Y})]^2}{\mathrm{Var}(\hat{Y})} = \left(\sqrt{\mathrm{Var}(\hat{Y})} - \frac{\mathrm{Cov}(Y, \hat{Y})}{\sqrt{\mathrm{Var}(\hat{Y})}} \right)^2 \geq 0.$$

□

This result shows that even in infinite-dimensional settings, such as functional data forecasting or kernel-based models, a simple affine correction calibrated on a validation set strictly reduces (or at worst preserves) the mean squared error. The result is fully model-agnostic and requires no retraining.

We prove that for any pair of square-integrable random elements in a Hilbert space, an affine transformation of the forecast strictly reduces or preserves the *expectile risk* of the squared error. We define expectile risk as the minimum asymmetric squared loss and show the corrected forecast achieves strictly better or equal risk. Let $(\Omega, \mathcal{F}, \mathbb{P})$ be a probability space, and let $\mathcal{H}$ be a real Hilbert space with inner product $\langle \cdot, \cdot \rangle$ and norm $\| \cdot \|$. Let $Y, \hat{Y} \in L^2(\Omega; \mathcal{H})$ be square-integrable random elements.

Definition 32 (Asymmetric Quadratic Loss). *Let $\tau \in (0, 1)$. Define the asymmetric loss $\rho_\tau : \mathbb{R} \to \mathbb{R}_+$ by*

$$\rho_\tau(u) = \begin{cases} \tau u^2 & \text{if } u \geq 0, \\ (1 - \tau)u^2 & \text{if } u < 0. \end{cases}$$

Definition 33 (Expectile Value-at-Risk (EVaR)). *Let $\xi \in L^2(\Omega)$ be a real-valued random variable. The* expectile value-at-risk *at level $\tau \in (0, 1)$ is the minimizer*

$$\mathrm{EVaR}_\tau(\xi) := \arg\min_{x \in \mathbb{R}} \mathbb{E}[\rho_\tau(\xi - x)].$$

The associated expectile risk value *is*

$$\mathrm{ER}_\tau(\xi) := \min_{x \in \mathbb{R}} \mathbb{E}[\rho_\tau(\xi - x)].$$

Let the squared errors be

$$\xi_0 := \|Y - \hat{Y}\|^2, \quad \xi_a := \|Y - a\hat{Y} - b_a\|^2, \quad \text{where } b_a := \mathbb{E}[Y] - a\mathbb{E}[\hat{Y}].$$

Theorem 5 (Affine Correction Reduces Expectile Risk). *Let $Y, \hat{Y} \in L^2(\Omega; \mathcal{H})$ and fix $\tau \in (0, 1)$. Define the centered variables*

$$\tilde{Y} := Y - \mathbb{E}[Y], \quad \tilde{\hat{Y}} := \hat{Y} - \mathbb{E}[\hat{Y}].$$

Then the optimal affine transformation minimizing the expectile risk of the squared error is

$$Y^{corr} := a^*\hat{Y} + b^*, \quad \text{where } b^* := \mathbb{E}[Y] - a^*\mathbb{E}[\hat{Y}],$$

and

$$a^* := \arg\min_{a \in \mathbb{R}} \mathrm{ER}_\tau(\|\tilde{Y} - a\tilde{\hat{Y}}\|^2).$$

Moreover, the expectile risk improvement is explicitly given by

$$\mathrm{ER}_\tau(\xi_0) - \mathrm{ER}_\tau(\xi_{a^*}) = \min_x \mathbb{E}[\rho_\tau(\|\tilde{Y} - \tilde{\hat{Y}}\|^2 - x)] - \min_x \mathbb{E}[\rho_\tau(\|\tilde{Y} - a^*\tilde{\hat{Y}}\|^2 - x)] \geq 0,$$

with equality if and only if $a^* = 1$ *and* $\tilde{Y} = \tilde{\hat{Y}}$ *almost surely.*

Proof. Fix $a \in \mathbb{R}$. The optimal shift minimizing the squared error is

$$b_a := \arg\min_b \mathbb{E}[\|Y - a\hat{Y} - b\|^2] = \mathbb{E}[Y] - a\mathbb{E}[\hat{Y}].$$

Substitute this into the squared error

$$\xi_a := \|Y - a\hat{Y} - b_a\|^2 = \|\tilde{Y} - a\tilde{\hat{Y}}\|^2.$$

Define the expectile risk functional

$$R_\tau(a) := \min_{x\in\mathbb{R}} \mathbb{E}[\rho_\tau(\|\tilde{Y} - a\tilde{\hat{Y}}\|^2 - x)].$$

This is convex in a because the mapping $a \mapsto \|\tilde{Y} - a\tilde{\hat{Y}}\|^2$ is quadratic, and ρ_τ is convex.

Let $a^* := \arg\min_{a\in\mathbb{R}} R_\tau(a)$. Then,

$$\mathrm{ER}_\tau(\xi_0) = R_\tau(1), \quad \mathrm{ER}_\tau(\xi_{a^*}) = R_\tau(a^*),$$

and therefore

$$\mathrm{ER}_\tau(\xi_0) - \mathrm{ER}_\tau(\xi_{a^*}) = R_\tau(1) - R_\tau(a^*) \geq 0.$$

Equality holds if and only if $a^* = 1$ and $\|\tilde{Y} - \tilde{\hat{Y}}\|^2 = \|\tilde{Y} - a^*\tilde{\hat{Y}}\|^2$ almost surely, i.e., when $\tilde{Y} = \tilde{\hat{Y}}$. □

The theorem shows that even when minimizing asymmetric tail-sensitive risk, affine post-processing of a forecast yields better performance. This result is model-agnostic and holds in infinite-dimensional Hilbert spaces.

5

Mean-Field-Type Learning

We introduce Mean-Field-Type Learning (MFTL), a novel framework that models collective learning dynamics through fractional Brownian motion (fBM) and mean-field-type interactions. Unlike traditional Markovian processes, fBM introduces memory effects and subdiffusive dynamics, allowing for controlled exploration without oscillatory divergence. By embedding this mechanism within a mean-field-type setting, MFTL captures global structural information while mitigating the convergence issues seen in existing approaches. We demonstrate that MFTL achieves robust convergence across multiple optimization scenarios where adaptive moment estimation may fail. MFTL uses the concept of collective behavior observed in natural systems, such as the coordinated movement of fish schools and bird flocks. By modeling the learning process as a mean-field-type interaction, we enable a collective exploration of the strategic interaction. The mean-field-type interaction facilitates escape from suboptimal strategies and convergence to optimal strategies. MFTL is grounded in the Boltzmann-Gibbs distribution and it ensures that the probability distribution of actions concentrates on the optimal strategies over time.

5.1 Failure of Adaptive Moment Estimation

We discuss the failure of the adaptive moment estimation (ADAM) algorithm in training neural networks. The ADAM algorithm is given by

$$\begin{array}{l} g_{t,i} = \frac{\partial f(\theta_{t-1})}{\partial \theta_{t-1,i}} \quad \text{for } i \in \{1,2,\ldots,d\} \\ m_{t,i} = \beta_1 m_{t-1,i} + (1-\beta_1) g_{t,i} \quad \text{for } i \in \{1,2,\ldots,d\} \\ v_{t,i} = \beta_2 v_{t-1,i} + (1-\beta_2) g_{t,i}^2 \quad \text{for } i \in \{1,2,\ldots,d\} \\ \hat{m}_{t,i} = \frac{m_{t,i}}{1-\beta_1^t} \quad \text{for } i \in \{1,2,\ldots,d\} \\ \hat{v}_{t,i} = \frac{v_{t,i}}{1-\beta_2^t} \quad \text{for } i \in \{1,2,\ldots,d\} \\ \theta_{t,i} = \theta_{t-1,i} - \alpha \frac{\hat{m}_{t,i}}{\sqrt{\hat{v}_{t,i}+\epsilon}} \quad \text{for } i \in \{1,2,\ldots,d\}, \end{array} \tag{5.1}$$

where $g_{t,i}$ represents the i-th component of the gradient vector $\mathbf{g}_t$ at timestep t, $m_{t,i}$ represents the i-th component of the first moment vector $\mathbf{m}_t$ at timestep t, $v_{t,i}$ represents the i-th component of the second moment vector $\mathbf{v}_t$ at timestep t, $x_{t,i}$ represents the i-th component of the parameter vector $\mathbf{x}_t$ at timestep t, and d represents the dimensionality of the parameter vector $\mathbf{x}$.

Another extension of ADAM to Hilbert spaces is given by

$$\begin{aligned} g_t &= \nabla c(\theta_{t-1}) \\ m_t &= \beta_1 m_{t-1} + (1-\beta_1) g_t \\ v_t &= \beta_2 v_{t-1} + (1-\beta_2) \|g_t\|^2 \\ \hat{m}_t &= \tfrac{m_t}{1-\beta_1^t} \\ \hat{v}_t &= \tfrac{v_t}{1-\beta_2^t} \\ \theta_t &= \theta_{t-1} - \alpha \tfrac{\hat{m}_t}{\sqrt{\hat{v}_t}+\epsilon}, \end{aligned} \tag{5.2}$$

where v_t is the averaged squared norm of the gradients. It differs from the entry-wise normalization.

The goal is to design $(T, m_0, v_0, \theta_0, \beta_1, \beta_2, \alpha, \epsilon)$ such that ADAM finds $\theta^* \in \arg\min_\theta c(\theta)$. Most proofs in the literature focus on the convergence to the critical points, i.e., $\{\theta \mid \nabla c(\theta) = 0\}$ which may be different from the set of global minimizers. When the set of critical points $\{\theta \mid \nabla c(\theta) = 0\}$ is different from $\{\theta^* \mid c(\theta^*) = \min_\theta c(\theta)\}$, the question of selection becomes crucial. Therefore showing the convergence of the algorithm to the set $\{\theta \mid \nabla c(\theta) = 0\}$ is not enough for the training problem. It may not solve the training problem.

$$c_0(x) = \begin{cases} 1.0 + (-0.0500)\cdot(x-1) & \text{for} \quad 1 \le x < 2 \\ 0.95 + (-0.0300)\cdot(x-2) & \text{for} \quad 2 \le x < 3 \\ 0.92 + (0.0000)\cdot(x-3) & \text{for} \quad 3 \le x < 4 \\ 0.92 + (0.0600)\cdot(x-4) & \text{for} \quad 4 \le x < 5 \\ 0.98 + (0.0100)\cdot(x-5) & \text{for} \quad 5 \le x < 6 \\ 0.99 + (-0.4400)\cdot(x-6) & \text{for} \quad 6 \le x < 7 \\ 0.55 + (-0.0300)\cdot(x-7) & \text{for} \quad 7 \le x < 8 \\ 0.52 + (0.0000)\cdot(x-8) & \text{for} \quad 8 \le x < 9 \\ 0.52 + (0.1600)\cdot(x-9) & \text{for} \quad 9 \le x < 10 \\ 0.68 + (0.0100)\cdot(x-10) & \text{for} \quad 10 \le x < 11 \\ 0.69 + (-0.4400)\cdot(x-11) & \text{for} \quad 11 \le x < 12 \\ 0.25 + (-0.0300)\cdot(x-12) & \text{for} \quad 12 \le x < 13 \\ 0.22 + (0.0000)\cdot(x-13) & \text{for} \quad 13 \le x < 14 \\ 0.22 + (0.1600)\cdot(x-14) & \text{for} \quad 14 \le x < 15 \\ 0.38 + (0.0100)\cdot(x-15) & \text{for} \quad 15 \le x < 16 \\ 0.39 + (0.3600)\cdot(x-16) & \text{for} \quad 16 \le x < 17 \\ 0.75 + (0.0100)\cdot(x-17) & \text{for} \quad 17 \le x < 18 \\ 0.76 + (-0.0400)\cdot(x-18) & \text{for} \quad 18 \le x < 19 \\ 0.72 + (0.0000)\cdot(x-19) & \text{for} \quad 19 \le x < 20 \\ 0.72 + (0.1600)\cdot(x-20) & \text{for} \quad 20 \le x < 21 \\ 0.88 + (0.0100)\cdot(x-21) & \text{for} \quad 21 \le x < 22 \\ 0.89 + (-0.3000)\cdot(x-22) & \text{for} \quad 22 \le x < 23 \\ 0.59 + (-0.3000)\cdot(x-23) & \text{for} \quad 23 \le x < 24 \\ 0.29 + (-0.0700)\cdot(x-24) & \text{for} \quad 24 \le x < 25 \\ 0.22 + (0.0000)\cdot(x-25) & \text{for} \quad 25 \le x < 26 \\ 0.22 + (0.3700)\cdot(x-26) & \text{for} \quad 26 \le x < 27 \\ 0.59 + (0.0000)\cdot(x-27) & \text{for} \quad 27 \le x < 28 \end{cases} \tag{5.3}$$

In the basic function provided by (5.3, Figure 5.1), the gradient algorithm and the ADAM algorithm fail to find the global minimum when started from the initial $\theta_0 = 5$. Thus, ADAM does not work for this elementary function. It goes to a local minimum and gets trapped there.

One may think that the limitations of ADAM in finding global minima are because the elementary function we have chosen is non-convex function and has several local extrema.

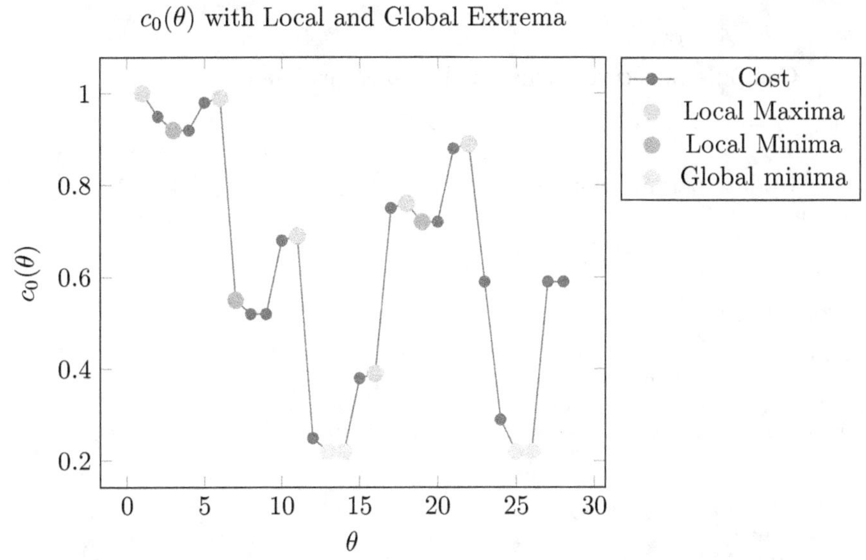

FIGURE 5.1
$c_0(\theta)$ with local maxima, local minima, global minima, global maxima.

We show below that ADAM may not work even for a strong convex if the parameters $(\alpha, \beta_1, \beta_2, \epsilon, \theta_0, v_0, m_0, T)$ are not carefully selected. To illustrate it we choose the strongly convex function $\frac{1}{2}\theta^2$ in one dimension. The adaptive moment estimation is set to the following:

Initialize parameters:

$$\theta_0 = 2.2205\ 10^{-16}$$
$$m_0 = 0 \quad \text{(first moment estimate)}$$
$$v_0 = 0 \quad \text{(second moment estimate)}$$
$$\beta_1 = 0.5 \quad \text{(decay rate for first moment)}$$
$$\beta_2 = 0.6 \quad \text{(decay rate for second moment)}$$
$$\epsilon = 0.01 \quad \text{(small constant to prevent division by zero)}$$
$$\alpha = 0.8 \quad \text{(learning rate)}$$
$$T = 50000\text{(number of iterations)} :$$

For $t = 1, 2, \ldots, T$ (number of iterations):

Compute gradient of cost function:

$$g_t = \frac{\partial c(\theta)}{\partial \theta} = \theta_{t-1}$$

Update biased first moment estimate:

$$m_t = \beta_1 m_{t-1} + (1 - \beta_1) g_t$$

Update biased second moment estimate:

$$v_t = \beta_2 v_{t-1} + (1 - \beta_2) g_t^2$$

Compute bias-corrected first moment estimate:

$$\hat{m}_t = \frac{m_t}{1 - \beta_1^t}$$

Compute bias-corrected second moment estimate:

$$\hat{v}_t = \frac{v_t}{1 - \beta_2^t}$$

Update parameters:

$$\theta_t = \theta_{t-1} - \alpha \frac{\hat{m}_t}{\sqrt{\hat{v}_t} + \epsilon}$$

Return optimized θ

```
import numpy as np
import matplotlib.pyplot as plt
from mpl_toolkits.mplot3d import Axes3D
from matplotlib.colors import LinearSegmentedColormap
import time

# Define the cost function c(\theta) = 0.5 \theta^2
def cost_function(theta):
    return 0.5 * theta**2

# Define the gradient of the cost function: \nabla c(\theta) = \theta
def gradient(theta):
    return theta

# Initialize parameters
theta_0 = 2.2205e-16  # Initial theta
m_0 = 0               # Initial first moment estimate
v_0 = 0               # Initial second moment estimate
beta_1 = 0.5          # Decay rate for first moment
beta_2 = 0.6          # Decay rate for second moment
epsilon = 0.01        # Small constant to prevent division by zero
alpha = 0.8           # Learning rate
T = 100000             # Number of iterations

print(f"Starting Adam optimization with T={T} iterations...")
start_time = time.time()

# Arrays to store values over time
theta_t = np.zeros(T+1)
m_t = np.zeros(T+1)
v_t = np.zeros(T+1)
cost_t = np.zeros(T+1)

# Initialize
theta_t[0] = theta_0
m_t[0] = m_0
v_t[0] = v_0
cost_t[0] = cost_function(theta_0)

# Run Adam optimizer
for t in range(1, T+1):
    # Compute gradient
    g_t = gradient(theta_t[t-1])

    # Update biased first moment estimate
    m_t[t] = beta_1 * m_t[t-1] + (1 - beta_1) * g_t

    # Update biased second moment estimate
    v_t[t] = beta_2 * v_t[t-1] + (1 - beta_2) * (g_t**2)

    # Compute bias-corrected first moment estimate
```

```
    m_hat_t = m_t[t] / (1 - beta_1**t)

    # Compute bias-corrected second moment estimate
    v_hat_t = v_t[t] / (1 - beta_2**t)

    # Update parameters
    theta_t[t] = theta_t[t-1] - alpha * m_hat_t / (np.sqrt(v_hat_t) + epsilon)

    # Store cost
    cost_t[t] = cost_function(theta_t[t])

end_time = time.time()
print(f"Optimization completed in {end_time - start_time:.2f} seconds")

# Print final values
print("Final values after T=100,000 iterations:")
print(f"Final theta: {theta_t[-1]:10e}")
print(f"Final m_t: {m_t[-1]:10e}")
print(f"Final v_t: {v_t[-1]:10e}")
print(f"Initial cost: {cost_t[0]:10e}")
print(f"Final cost: {cost_t[-1]:10e}")

# Create a custom colormap for better visualization
colors = [(0, 'darkblue'), (0.25, 'blue'), (0.5, 'green'), (0.75, 'orange'), (1, 'red')]
custom_cmap = LinearSegmentedColormap.from_list('custom_cmap', colors)

# Create a 3D plot with improved aesthetics
fig = plt.figure(figsize=(14, 12))
ax = fig.add_subplot(111, projection='3d')

# We'll sample points to make the plot more manageable
# For a deep analysis, we'll use logarithmic sampling to capture early dynamics
sample_indices = np.unique(np.logspace(0, np.log10(T), 1000, dtype=int))
sample_indices = np.append(sample_indices, T)  # Make sure to include the final point

# Plot the trajectory (m_t, v_t, theta_t)
scatter = ax.scatter(
    m_t[sample_indices],
    v_t[sample_indices],
    theta_t[sample_indices],
    c=sample_indices,
    cmap=custom_cmap,
    s=10,
    alpha=0.8
)

# Add a colorbar to indicate iteration number
cbar = plt.colorbar(scatter)
cbar.set_label('Iteration (log scale)')

# Set labels and title
ax.set_xlabel('m_t (First Moment)', fontsize=12)
ax.set_ylabel('v_t (Second Moment)', fontsize=12)
ax.set_zlabel('\theta_t (Parameter)', fontsize=12)
ax.set_title('3D Trajectory of Adam Optimizer (T=100,000)', fontsize=14)

# Adjust view angle for better visualization
ax.view_init(elev=30, azim=45)

# Add grid for better depth perception
```

```
ax.grid(True)

plt.tight_layout()
plt.show()

# Create additional plots for analysis
fig, axs = plt.subplots(2, 2, figsize=(14, 10))

# Plot theta over iterations
axs[0, 0].plot(np.arange(T+1), theta_t)
axs[0, 0].set_title('\theta vs Iteration')
axs[0, 0].set_xlabel('Iteration')
axs[0, 0].set_ylabel('\theta')
axs[0, 0].grid(True)

# Plot cost over iterations
axs[0, 1].plot(np.arange(T+1), cost_t)
axs[0, 1].set_title('Cost vs Iteration')
axs[0, 1].set_xlabel('Iteration')
axs[0, 1].set_ylabel('Cost')
axs[0, 1].grid(True)

# Plot m_t over iterations
axs[1, 0].plot(np.arange(T+1), m_t)
axs[1, 0].set_title('m_t vs Iteration')
axs[1, 0].set_xlabel('Iteration')
axs[1, 0].set_ylabel('m_t')
axs[1, 0].grid(True)

# Plot v_t over iterations
axs[1, 1].plot(np.arange(T+1), v_t)
axs[1, 1].set_title('v_t vs Iteration')
axs[1, 1].set_xlabel('Iteration')
axs[1, 1].set_ylabel('v_t')
axs[1, 1].grid(True)

plt.tight_layout()
plt.show()

# Create log-scale plots to better visualize convergence
fig, axs = plt.subplots(2, 2, figsize=(14,10))

# Plot theta over iterations (log x-scale)
axs[0, 0].semilogx(np.arange(1, T+1), theta_t[1:])
axs[0, 0].set_title('\theta vs Iteration (Log Scale)')
axs[0, 0].set_xlabel('Iteration (log scale)')
axs[0, 0].set_ylabel('\theta')
axs[0, 0].grid(True)

# Plot cost over iterations (log-log scale)
axs[0, 1].loglog(np.arange(1, T+1), cost_t[1:])
axs[0, 1].set_title('Cost vs Iteration (Log-Log Scale)')
axs[0, 1].set_xlabel('Iteration (log scale)')
axs[0, 1].set_ylabel('Cost (log scale)')
axs[0, 1].grid(True)

# Plot m_t over iterations (log x-scale)
axs[1, 0].semilogx(np.arange(1, T+1), m_t[1:])
axs[1, 0].set_title('m_t vs Iteration (Log Scale)')
axs[1, 0].set_xlabel('Iteration (log scale)')
axs[1, 0].set_ylabel('m_t')
axs[1, 0].grid(True)

# Plot v_t over iterations (log x-scale)
```

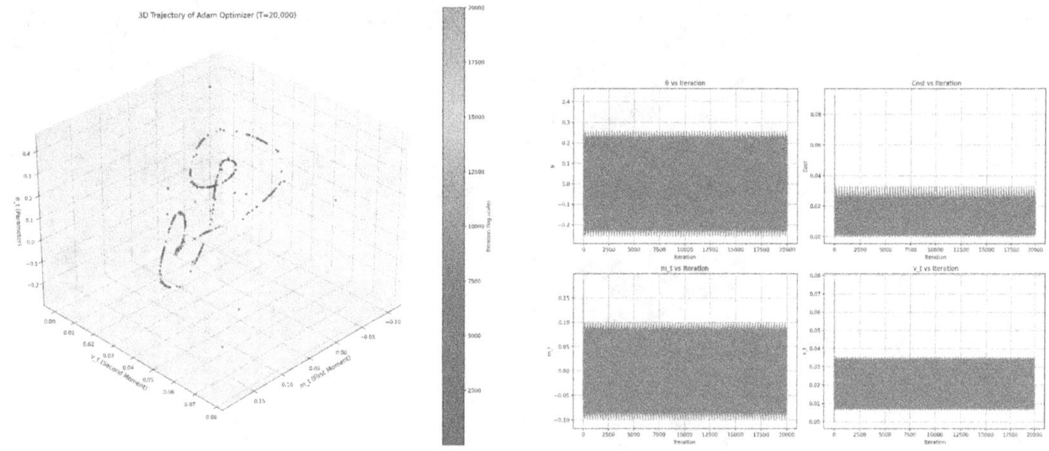

FIGURE 5.2
ADAM fails to find the unique solution even for the basic strongly convex function $c(\theta) = 0.5\theta^2$.

```
axs[1, 1].semilogx(np.arange(1, T+1), v_t[1:])
axs[1, 1].set_title('v_t vs Iteration (Log Scale)')
axs[1, 1].set_xlabel('Iteration (log scale)')
axs[1, 1].set_ylabel('v_t')
axs[1, 1].grid(True)

plt.tight_layout()
plt.show()
```

Non-convergence of ADAM even in a basic convex case

The Python code implements the Adam optimizer for the cost function $c(\theta) = 0.5\theta^2$ and $T = 100000$ iterations. Surprisingly, we observe in Figures 5.3 and 5.2 that the final values of θ, m_t, and v_t are not better than the initial ones. Even worse, the trajectory goes to a limit cycle, running again and again, even for this basic function $\frac{1}{2}\theta^2$. Using a three-dimensional plot of (m_t, v_t, θ_t) over time $t \in \{0, 1, 2, \ldots, T\}$ we observe the emergence of a cycling behavior. This is mainly because the above chosen parameters are not optimized. We illustrate the trajectory of the Adam optimizer in 3D in log-scale representation for visualization. ADAM is not able to find the solution for this range of parameters that we have chosen. Note, however, that the standard gradient quickly converges to the solution with the chosen learning rate 0.8 as $0.2^t x_0$ vanishes very quickly. Thus, gradient converges in this case but not ADAM. This illustrates that one needs to be careful with ADAM even in the convex case. For the same function, MFTL finds the solution very quickly and there is no limit cycle as illustrated in Figure 5.4.

5.2 Mean-Field-Type Learning is Exactly What You Need

The quest for optimal strategies in distributed learning environments is a fundamental challenge in machine intelligence. Traditional deterministic gradient descent methods often get

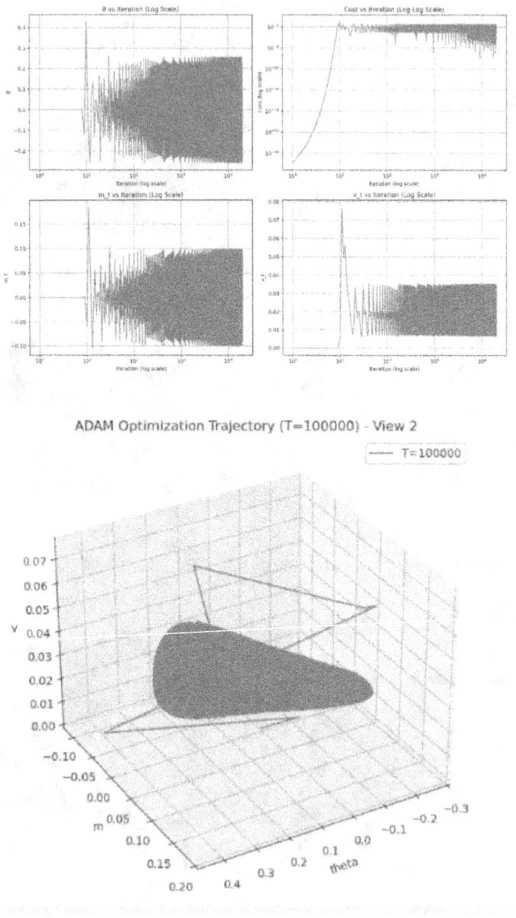

FIGURE 5.3
ADAM fails to find the unique solution even for the basic strongly convex function $c(\theta) = 0.5\theta^2$ (continued)

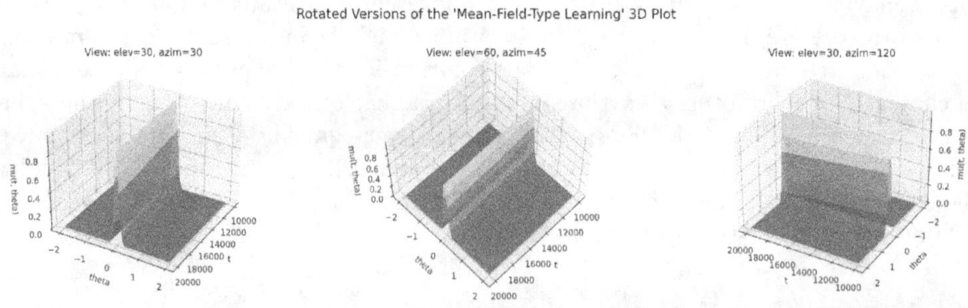

FIGURE 5.4
MFTL very quickly finds the unique solution in the situation where ADAM fails drastically.

trapped in suboptimal strategies, while stochastic gradient descent with fixed noise variance can be hindered by fluctuations and variance-awareness, preventing it from achieving optimal strategic learning. This chapter presents mean-field-type learning as a ground-breaking solution that integrates the strengths of both deterministic and stochastic approaches within a unified framework.

Mean-field-type learning (MFTL) is a framework in which an agent, or a system of agents, optimizes by learning the probability distribution of its optimal states (positions, velocities, decisions, revenues, etc.) over time. In the single-agent case, this involves updating both the state and its evolving probability distribution (the mean-field-type term), capturing not just the optimal trajectory but the broader statistical structure of optimality. In the multi-agent setting, each agent simultaneously learns the probability distribution of the system's optimal states as well as its own mean-field-type terms, while accounting for the mean-field-type contributions of other agents. This ensures that interactions and collective dynamics are explicitly incorporated into the collective learning process, distinguishing it from traditional optimization approaches that neglect distributional dependencies. Mean-field-type learning has emerged as a particularly effective paradigm, demonstrating robustness and convergence where single-agent path-based learning via gradient descent fails. Mean-field-type learning succeeds in scenarios where ADAM, one of the most widely used optimization algorithms for pre-training AI foundation models, fails, as illustrated through a fundamental example. Moreover, it outperforms single-agent stochastic gradient descent with fixed noise, highlighting its broader applicability.

In the context of multi-agent or ensemble-based learning, most existing analyses focus on distributions or probabilities, which is linked to the mean-field representation of agent positions or states. Ensemble learning provides an approximation of mean-field-type learning and it includes methodologies such as particle swarm learning and swarm gradient descent. One of the key advantages of mean-field-type learning lies in its ability to efficiently eliminate local extrema through adaptive learning rate adjustments, a process that can be embedded within the initial design parameters. Crucially, rather than relying on conventional Brownian motion noise or Langevin-type of fixed noise, we demonstrate that optimization is improved through sub-diffusive noise modeled as fractional Brownian motion with a Hurst parameter $H < \frac{1}{2}$. This approach achieves convergence in cases where Langevin-type fixed noise fails. Another important property of mean-field-type learning is that the final equilibrium distribution exhibits a Boltzmann-Gibbs structure, corresponding to a softargmin formulation in error minimization. This result suggests that softargmin is not only sufficient for pre-training AI foundation models but also extends naturally to optimization problems in distributed power networks, electric vehicle coordination, and autonomous systems.

Mean-field-type learning uses the concept of collective behavior observed in natural systems, such as the coordinated movement of fish schools and bird flocks. By modeling the learning process as a mean-field-type interaction, we enable a collective exploration of the strategic interaction, facilitating escape from suboptimal strategies and convergence to optimal strategies. The approach here is grounded in the Boltzmann-Gibbs distribution, ensuring that the probability distribution of strategies concentrates on the optimal strategies over time [107, 174, 9].

Intuitively, MFTL succeeds where ADAM fails mainly because MFTL is built on the distribution, i.e., the cost is non-convex in the parameters but the new mean-field-type cost is linear in the measure (mean-field) and therefore convex in the measure (Figure 5.5). As the two cost functions have the same infimum value, it means that it is better to work directly with the mean-field and find a constructive representation of the mean-field distribution for implementation purposes. If the final representation is one Dirac measure, the purification method, which consists of returning back to the parameters' space, is quite elementary. If not, the interpretation of the mean-field remains an open discussion. The fact that the mixed

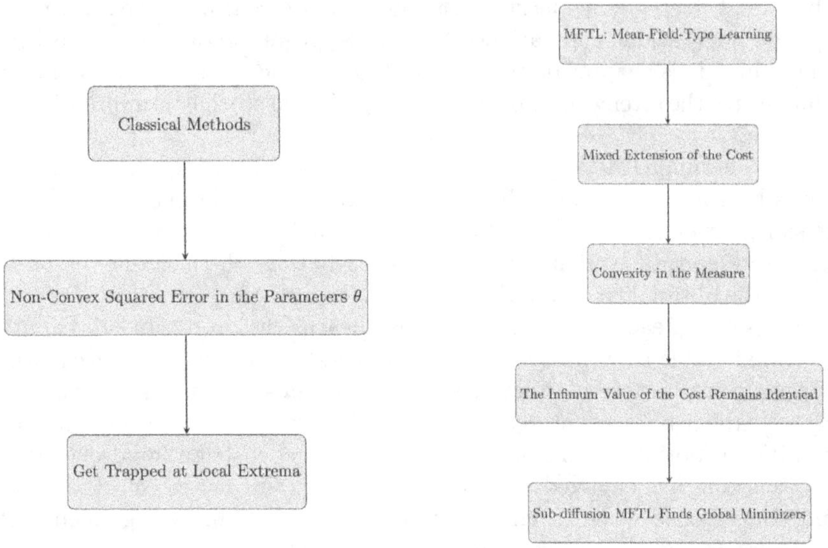

FIGURE 5.5
The reason why MFTL outperforms standard gradient descent methods.

extension works in terms of value is widely known in mixed strategy game theory, relaxed control theory and weak formulation problem. Similarly, machine learning and multi-agent machine learning can take advantage of this method and not be trapped in local extrema. Adding a penalty function to such measure-linear functional such as Bregman divergence, f-divergence or entropy function enables specific selection features of mean-field-type learning. Figures 5.6 and 5.7 illustrate the method in detail. The one-shot convergence to the set of global optima is illustrated in Figure 5.9. Transient phases are illustrated in Figure 5.8.

5.2.1 Boltzmann-Gibbs is Concentrated Only at the Global Minimizers

Let $\epsilon(t) > 0 \ \forall t \geq 0$ and consider the replicator equation

$$\dot{\mu}_i(t) = -\frac{1}{\epsilon(t)}\mu_i(t)\left(c_i - \sum_{j=1}^{I}\mu_j(t)c_j\right), \quad i \in \{1,\dots,I\}, t > 0, \tag{5.4}$$

starting from the interior of the simplex:

$$\mu_i(0) = \mu_{i0} > 0, \quad \sum_{j=1}^{I}\mu_{j0} = 1. \tag{5.5}$$

It can be checked that

$$\mu_i(t) = \frac{\mu_{i0}e^{\left(-\int_0^t \frac{c_i}{\epsilon(t')}\,dt'\right)}}{\sum_{j=1}^{I}\mu_{j0}e^{\left(-\int_0^t \frac{c_j}{\epsilon(t'')}\,dt''\right)}} \tag{5.6}$$

solves the system. Moreover, as $\int_0^t \frac{1}{\epsilon(t')}dt'$ goes to infinity as t goes to infinity, the distribu-

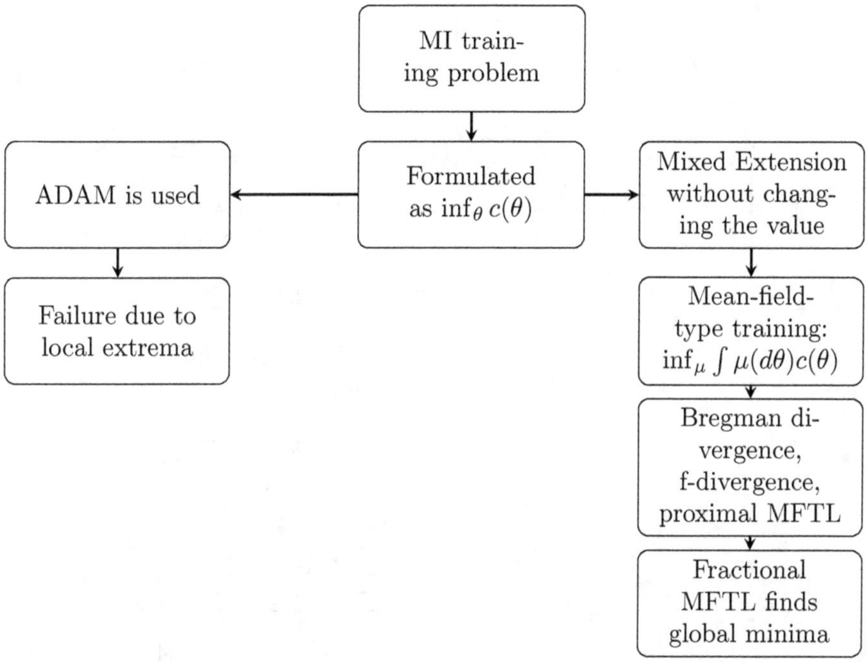

FIGURE 5.6
The reason why MFTL outperforms ADAM methods is due to a careful mean-field-type exploration and exploitation.

tion $\mu(t)$ is concentrated at the set $\arg\min_{i\in\{1,\dots,I\}} c_i$ i.e., $support(\mu_i^*) \subseteq \arg\min_{i\in\{1,\dots,I\}} c_i$. Note that $\min_{i\in\{1,\dots,I\}} c_i = \min_{\{x\in\Delta_{I-1}\}} \sum_{j=1}^{I} x_j c_j$ where $\Delta_{I-1} = \{x \in \mathbb{R}^I \mid x_j \geq 0, \sum_{j=1}^{I} x_j = 1\}$ is the $(I-1)$-dimensional simplex of $\mathbb{R}^I$. This proves that the softargmin distribution (also called Boltzmann-Gibbs distribution) $\sim e^{\left(-(c_i - \min_k c_k)\int_0^t \frac{1}{\epsilon(t')}dt'\right)}$ converges to the set of global minimizers of $y \to \langle y, c\rangle$. Let $\mathcal{I}^* = \{i, c_i = \min_k c_k\} = \{i_1, i_2, \dots, i_k\}$. Then,

$$\lim_{t\to\infty} \mu_i(t) = \begin{cases} \frac{\mu_{i0}}{\sum_{j\in\mathcal{I}^*}\mu_{j0}} & \text{if } i \in \mathcal{I}^*, \\ 0 & \text{otherwise.} \end{cases} \tag{5.7}$$

In the long run, the behavior of $\mu_i(t)$ is dominated by the exponential terms in the numerator and denominator. When $\int_0^t \frac{1}{\epsilon(t')}dt'$ goes to infinity, the exponential growth rates of these terms become crucial. The term $e^{\left(-\int_0^t \frac{c_i}{\epsilon(t')}dt'\right)}$ decays exponentially with rate c_i. If c_i is not the smallest among all c_j, then there exists some c_j such that $c_j < c_i$. Consequently, $e^{\left(-\int_0^t \frac{c_j}{\epsilon(t')}dt'\right)}$ will decay slower than $e^{\left(-\int_0^t \frac{c_i}{\epsilon(t')}dt'\right)}$. The denominator $\sum_{j=1}^{I} \mu_{j0} e^{\left(-\int_0^t \frac{c_j}{\epsilon(t'')}dt''\right)}$ will be dominated by the terms with the smallest c_j. If c_i is not the smallest, its contribution to the sum will become negligible compared to the terms with the minimum c_j.

Since $\mu_i(t)$ is normalized by the sum in the denominator, if c_i is not the smallest, the fraction $\frac{\mu_{i0} e^{\left(-\int_0^t \frac{c_i}{\epsilon(t')}dt'\right)}}{\sum_{j=1}^{I} \mu_{j0} e^{\left(-\int_0^t \frac{c_j}{\epsilon(t'')}dt''\right)}}$ will approach zero. This means that the probability assigned to i will be zero in the long run if c_i is not the minimum. This idea can be used to escape

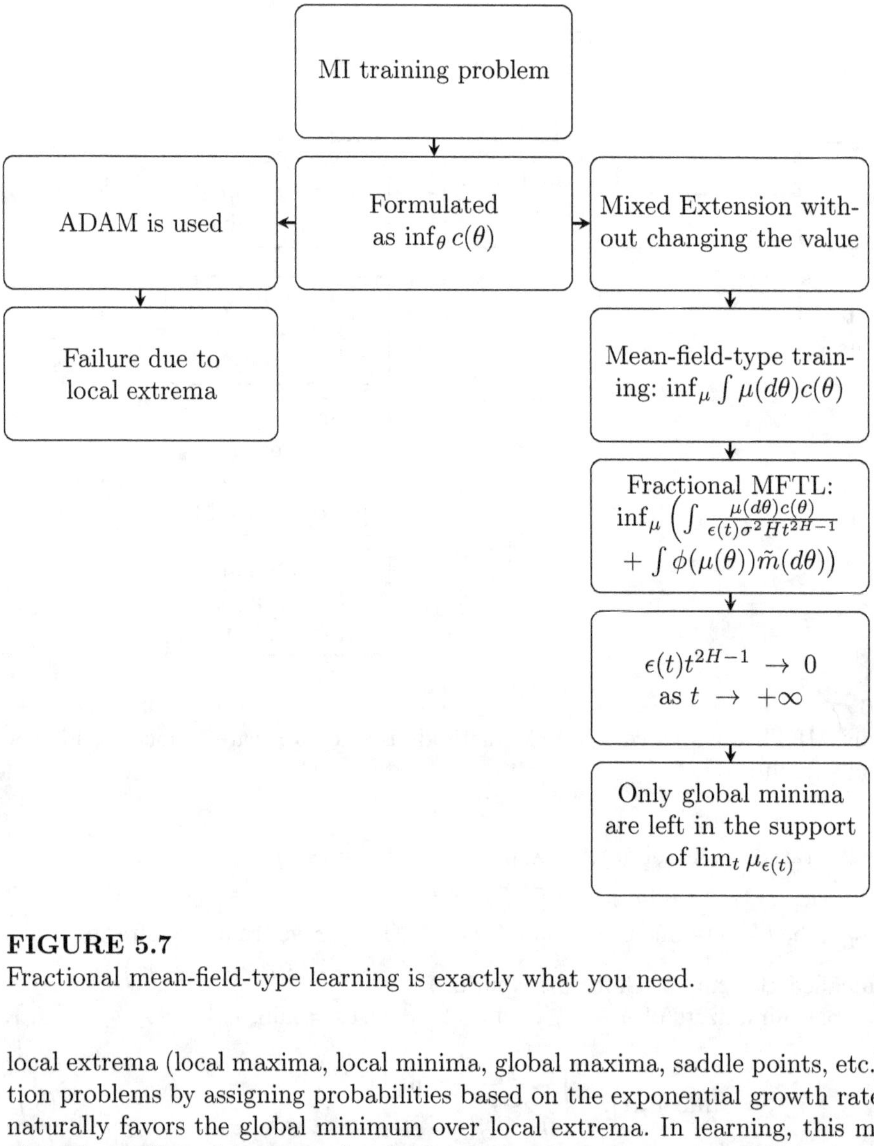

FIGURE 5.7
Fractional mean-field-type learning is exactly what you need.

local extrema (local maxima, local minima, global maxima, saddle points, etc.) in optimization problems by assigning probabilities based on the exponential growth rates, the system naturally favors the global minimum over local extrema. In learning, this means that the algorithm will tend to escape local minima, local maximum, or saddle points and converge towards the global optimum. The gap between the Boltzmann-Gibbs distribution cost and the global minimum cost is

$$\sum_{i=1}^{I} \mu_i(t)c_i - \sum_{i=1}^{I} \mu_i^* c_* \to 0 \text{ as } \epsilon(t) \to 0.$$

Figure 5.10 illustrates a typical example where the classical gradient may fail but the Boltzmann-Gibbs with high rationality is closer to the set of global minimizer of the non-convex function (displayed in Figure 5.10).

We now use the above idea of Boltzmann-Gibbs distribution for continuous variables.

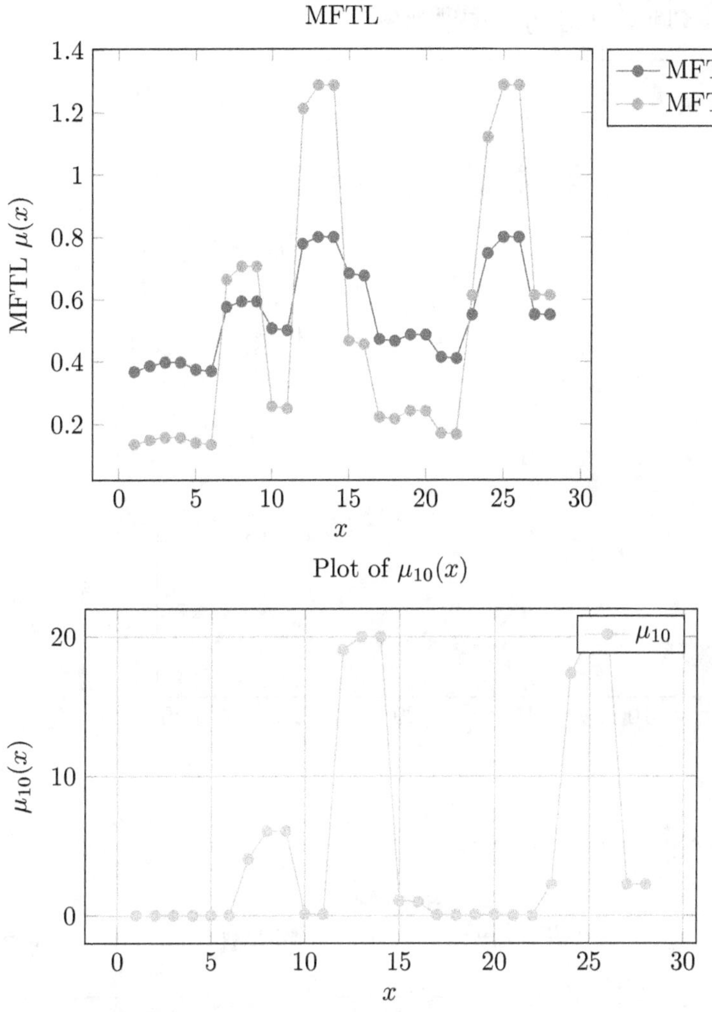

FIGURE 5.8
Transient phases of the set of global optima.

Proposition 29. *Let $c : \mathbb{R}^d \to \mathbb{R}$ be a Borel function such that $c_* = c_{\inf} := essinf(c) = \inf\{y : \lambda_d\{c \leq y\} > 0\} > -\infty$, and $e^{-c} \in L^1(\mathbb{R}^d)$ Let*

$$\mu(t, d\theta) = C^{-1}(t)e^{\left(-\frac{c(\theta)}{\epsilon(t)}\right)}, \quad C(t) = \int_{\mathbb{R}^d} e^{\left(-\frac{c(y)}{\epsilon(t)}\right)} dy \tag{5.8}$$

be the Boltzmann-Gibbs distribution with learning rate $\epsilon(t)$.

Then, for all $\eta > 0$, $\mu(t, \{c \geq c_{\inf} + \eta\}) \to 0$ as $\epsilon(t) \to 0$.

Proof:

Without loss of generality, assume $c_{\inf} = 0$ by replacing c with $c - c_{\inf}$. Let $\eta > 0$. It follows that $c \geq 0$ almost everywhere with respect to the Lebesgue measure λ_d, and $\lambda_d\{c \leq \eta\} > 0$ for every $\eta > 0$.

Since $e^{-c} \in L^1(\mathbb{R}^d)$, we have $\lambda_d\{c \leq \eta/3\} \leq e^{\eta/3} \int_{\mathbb{R}^d} e^{-c} d\lambda_d < \infty$. By dominated convergence, $C^{-1}(t) \downarrow \lambda_d\{c = 0\} < \infty$.

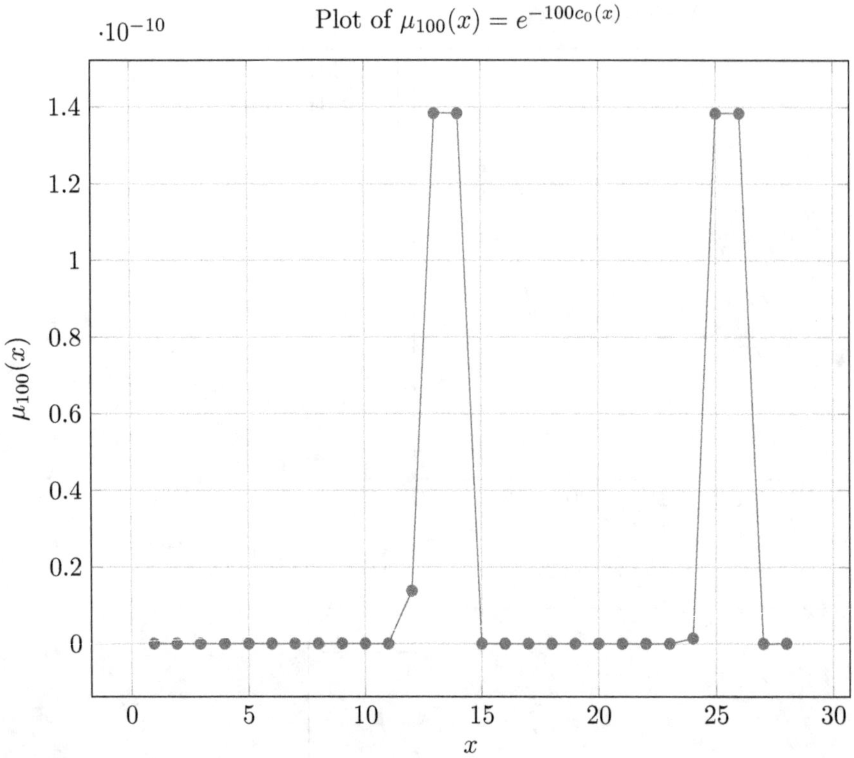

FIGURE 5.9
One-shot convergence to the set of global optima.

We have $C^{-1}(t) \le \left(\int_{\{c\le\eta/3\}} e^{-c(\theta)/\epsilon(t)}d\theta\right)^{-1} \le \left(e^{-\eta/(3\epsilon(t))}\lambda_d\{c\le\eta/3\}\right)^{-1}$. Then, $\mu(t,\{c\ge\eta\}) = C^{-1}(t)\int_{c\ge\eta} e^{-c(\theta)/\eta}d\theta \le \frac{e^{\eta/(3\epsilon(t))}\int_{c\ge\eta} e^{-c(\theta)/\epsilon(t)}d\theta}{\lambda_d\{c\le\eta/3\}} \le \frac{e^{-\eta/(3\epsilon(t))}C^{-1}_{3\epsilon(t)}}{\lambda_d\{c\le\eta/3\}} \to 0$ as $\epsilon(t)\to 0$. This holds because if $c(\theta)\ge\eta$, then $e^{-c(\theta)/\epsilon(t)} \le e^{-2\eta/(3\epsilon(t))}e^{-c(\theta)/(3\epsilon(t))}$, and $C^{-1}_{3\epsilon(t)} \le C^{-1}_1$ if $\epsilon(t)\le 1/3$.

Therefore, for all $\eta>0$, $\mu(t,\{c\ge c_{\text{inf}}+\eta\})\to 0$ as $\epsilon(t)\to 0$. □

In order to identify the limiting distribution of the Boltzmann-Gibbs measure, let us recall a result by [9].

Proposition 30 ([9]). *Let $c:\mathbb{R}^d\to[0,\infty)$ be a measurable function satisfying the following conditions:*

1. $e^{-c}\in L^1(\mathbb{R}^d)$.
2. *For all* $\delta>0$, $\inf\{c(\theta): \|\theta-\theta_i^*\|>\delta, 1\le i\le m\}>0$.
3. *There exist* $(\alpha_{ij})_{1\le i\le m, 1\le j\le d}$ *such that for all* i, j, $\alpha_{ij}\ge 0$ *and for all* i: $\frac{1}{\epsilon}c(\theta_i^*+(\epsilon^{\alpha_{i1}}h_1,\ldots,\epsilon^{\alpha_{id}}h_d))\to g_i(h_1,\ldots,h_d)\in[0,\infty)$ *as* $\epsilon\to 0$.
4. *For all* $i\in\{1,\ldots,m\}$,

$$\int_{\mathbb{R}^d}\sup_{0<\epsilon<1} e^{-\frac{c(\theta_i^*+(\epsilon^{\alpha_{i1}}h_1,\ldots,\epsilon^{\alpha_{id}}h_d))}{\epsilon}}dh_1\ldots dh_d<\infty$$

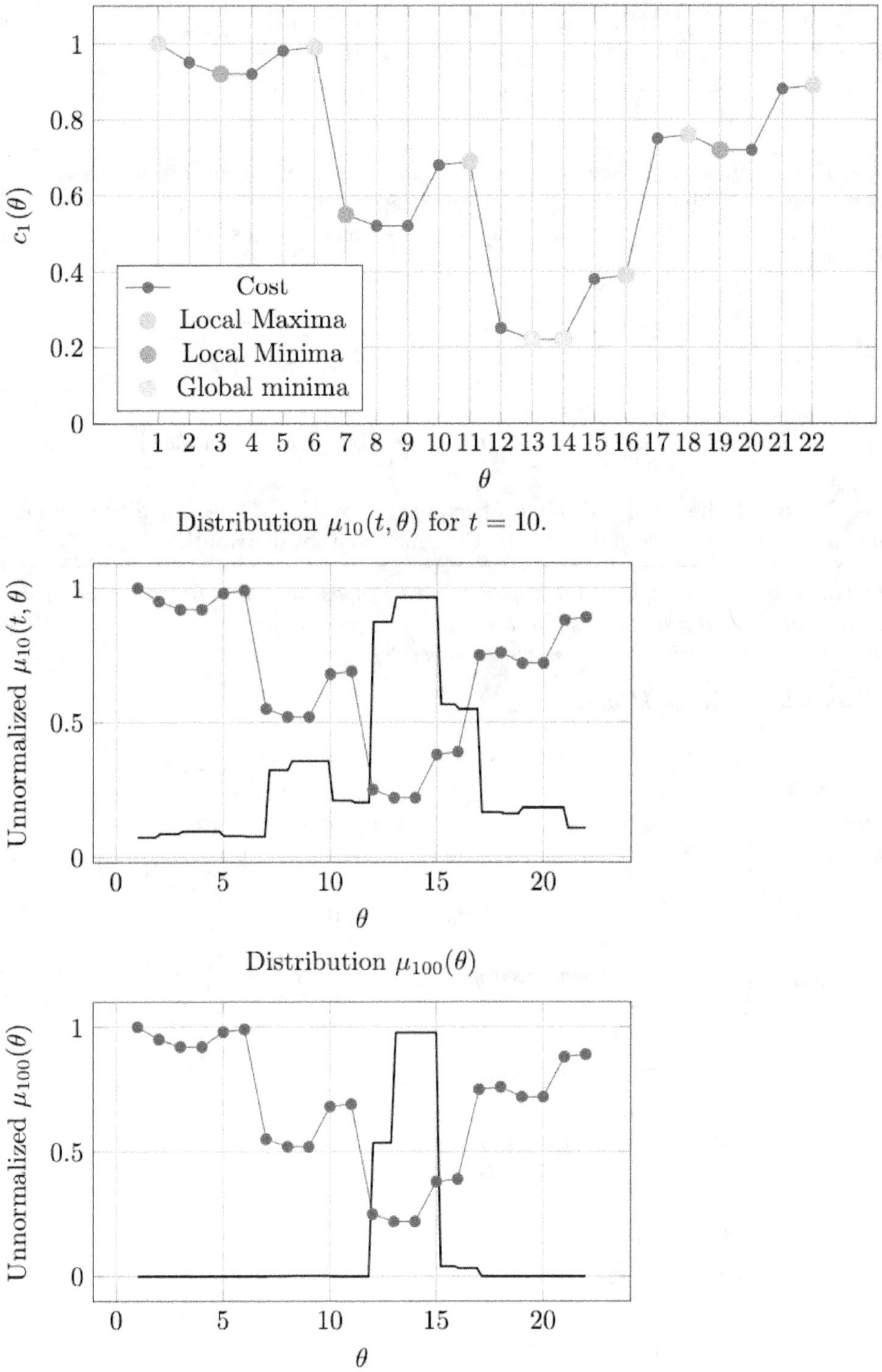

FIGURE 5.10
MFTL very quickly finds the set of global minimizers of the non-convex function.

Let $\alpha := \min_{1\leq i\leq m}\sum_{j=1}^d \alpha_{ij}$ *and let* $J := \{i \in \{1,\ldots,m\} : \sum_{j=1}^d \alpha_{ij} = \alpha\}$. *For* $0<\epsilon<1$, *let* $\theta(t)$ *be a random vector with distribution* $\mu(t,.)$. *Then:*
$\theta(t) \xrightarrow{\mathcal{L}} \mu^* = \frac{1}{\sum_{j\in J}\int_{\mathbb{R}^d} e^{-g_j(\theta)}d\theta}\sum_{i\in J}\int_{\mathbb{R}^d} e^{-g_i(\theta)}d\theta.\delta_{\theta_i^*}$ *as* $\epsilon \to 0$.

Let $(\Omega, \mathcal{F}, \tilde{m})$ is a measure space, a measurable function f and ν an absolutely continuous measure, i.e., there exists $\rho(.) > 0$ such that $\nu(dx) = \rho(x)\tilde{m}(dx)$. Let l be a function such that e^{-l} is $\tilde{m}$-integrable. The Legendre-Fenchel transform of the relative entropy yields:

$$\sup_\nu\left\{-\int l(x)\nu(dx) - \int \log(\rho(x))\nu(dx)\right\} = \log\left(\int_\Omega e^{-l(\omega)}\tilde{m}(d\omega)\right).$$

$$\sup_\nu\{-\int l_i(x)\nu(dx) - \epsilon_i\int \log(\rho(x))\nu(dx)\} = \epsilon_i\log\left(\int_\Omega e^{-\frac{l_i(\omega)}{\epsilon_i}}\tilde{m}(d\omega)\right), \quad (5.9)$$

which is the log of the expected value of exponentiated loss. The ρ achieving the infimum value is of the softargmin i.e., the Boltzmann-Gibbs distribution.

Example 18. *A typical example of the application of Proposition 30 is in training a mean-field-type transformer with sigmoid self-attention. An agent uses a mean-field-type transformer to make decision using the algorithm is given by:*

$$\begin{aligned}
&\textit{Parameters} : L, d, k, k', q, H \in \mathbb{N},\\
&\quad 1 \leq l \leq L:\\
&\quad 1 \leq h \leq H : Q_{lh}, K_{lh} \in \mathcal{L}(\mathbb{R}^d, \mathbb{R}^k), V_{lh} \in \mathcal{L}(\mathbb{R}^d, \mathbb{R}^{k'}), W_{lh} \in \mathcal{L}(\mathbb{R}^{k'}, \mathbb{R}^d),\\
&\quad W_{1l} \in \mathcal{L}(\mathbb{R}^d, \mathbb{R}^q), b_{1l} \in \mathbb{R}^q, W_{2l} \in \mathcal{L}(\mathbb{R}^q, \mathbb{R}^d), b_{2l} \in \mathbb{R}^d, p, \lambda\\
&\textit{Input} : s_i = s, y_0 = x_0 = y_{0,6} \in \mathbb{R}^d, \hat{\mu}_0 = \hat{\nu}_0 = \hat{\nu}_{0,6} \in \mathcal{P}([0,1]\times\mathbb{R}^d),\\
&t \in \{0,\ldots,T-1\}:\\
&1 \leq l \leq L:\\
&\quad \textit{Input at } l: \ s, y_{l-1,6} \in \mathbb{R}^d, \hat{\nu}_{l-1,6} = \mathbb{P}_{y_{l-1,6}} \in \mathcal{P}([0,1]\times\mathbb{R}^d)\\
&\\
&\quad y_{l,1} = \frac{y_{l-1,6}}{1+\|y_{l-1,6}\|} =: \textit{Normalized}(y_{l-1,6}),\\
&\quad \hat{\nu}_{l,1} = \textit{Normalized}\#\hat{\nu}_{l-1,6}\\
&\quad \textit{SelectH} \subseteq \{1,\ldots,H\}, p_l \in \mathbb{R}_+^H\\
&\quad O_{lh} := \int_0^1\int_{\mathbb{R}^d} e^{\frac{1}{\sqrt{k}}\langle Q_{lh}y_{l,1}, K_{lh}y_l'\rangle} V_{lh}\mathbb{I}_{s'\leq s}y_l'\hat{\nu}_{l,1}(ds'dy_l')\\
&\quad y_{l,2} = \frac{1}{\sqrt{H}}\sum_{h=1}^H p_{lh}\mathbb{I}_{h\in \textit{SelectH}}W_{lh}O_{lh}\\
&\quad \hat{\nu}_{l,2} = \ \textit{MultiHead-MoH-MFTAttention}\ \#\hat{\nu}_{l,1}\\
&\quad y_{l,3} = y_{l-1,6} + y_{l,2}, \ \hat{\nu}_{l,2} = \textit{conv}(\hat{\nu}_{l-1,6}, \hat{\nu}_{l,2}),\\
&\\
&\quad y_{l,4} = \frac{y_{l,3}}{1+\|y_{l,3}\|} = \textit{Normalized}(y_{l,3}),\\
&\quad \hat{\nu}_{l,4} = \textit{Normalized}\#\hat{\nu}_{l,3},\\
&\quad \textit{Choose } \mathcal{E},\\
&\quad \eta_{le} = \frac{e^{(\tilde{W}_{3l}y_{l,4}+\tilde{b}_{3l})_e}}{\sum_{e'\in\mathcal{E}_i} e^{(\tilde{W}_{3l}y_{l,4}+\tilde{b}_{3l})_{e'}}}\mathbb{I}_{e\in\mathcal{E}_i},\\
&\quad e = E+1 : \tilde{y}_{l,e} = W_{2l,e}r_l(W_{1l,e}y_{l,4} + b_{1l,e}) + b_{2l,e},\\
&\quad 1 \leq e \leq E_i : \ \tilde{y}_{l,e} = W_{2l,e}r_l(W_{1l,e}y_{l,4} + b_{1l,e}) + b_{2l,e},\\
&\quad y_{l,5} = \eta_{l,E+1}\tilde{y}_{l,E+1} + \sum_{e\in\mathcal{E}_i}\eta_{le}\tilde{y}_{l,e},\\
&\quad \hat{\nu}_{l,5} = \ \textit{2LayerFF-MoE}\#\hat{\nu}_{l,4},\\
&\quad y_{l,6} = y_{l,5} + y_{l,3}, \ \hat{\nu}_{l,6} = \textit{conv}(\hat{\nu}_{l,3}, \hat{\nu}_{l,5})\\
&\quad \textit{Output at } l: s, y_{l,6}, \hat{\nu}_{l,6}
\end{aligned}$$

$$\begin{aligned}&\textit{Return: } s, y_{L,6} \in \mathbb{R}^d, \hat{\nu}_{L,6} \in \mathcal{P}([0,1]\times\mathbb{R}^d)\\ &s, x_{t+1} = [(1-\lambda_t)x_t + \lambda_t y_{L,6}],\\ &\mu_{t+1} = [(1-\lambda_t)Id + \lambda_t y_{L,6}]\#\mu_t,\\ &\textit{Return: } s, x_T, \mu_T,\end{aligned} \tag{5.10}$$

where $\theta = (Q_{lh}, K_{lh}V_{lh}, W_{lh}, W_{1l}, b_{1l}, W_{2l}, b_{2l}, p, \lambda, \tilde{W}_{3l}, \tilde{b}_{3l})_{l,h}$ *is determined given input-output data. During the training of the transformer, the cost* $c = \|x_T - y_T\|_2^2 + W_2(\mu_T, \mathbb{P}_{y_T})$, *which is clearly non-convex in* θ, *needs to be minimized. For the non-convex cost function* $\theta \mapsto c(\theta, x, y)$ *obtained from the transformer-based deep neural network, if the cost satisfies the above assumptions with finitely many global minima, the resulting Boltzmann-Gibbs measure* $\mu(t, d\theta)$ *will have support concentrated only at its global minimizers. All the local minimizers and local/global maximizers and saddle points (if any) will not appear in the limiting support.*

- *The cost function is not convex in the parameter* θ. *The non-convexity is due to the composition of operators as illustrated in Figures 5.11 and 5.12.*
- *The training problem is to find* θ *in* $\{\theta \mid error = 0\}$. *While minimizing the squared error may be a good idea, using a gradient descent to solve it may not be. Indeed, the gradient is error.*∇*error. Thus, for any problem in which* $\{\theta \mid error\nabla error = 0\}$ *differs from* $\{\theta \mid error = 0\}$ *we will have extra zeros to be filtered out and it may require a second-order checking to eliminate them. The second thing is that the critical point of the gradient may include local extrema (including local minima, local maxima, global maxima, saddle points, etc.) which are not needed for the training. Therefore, it seems not a good idea to use the standard gradient descent method (or ADAM) to solve the training problem.*
- *Generically, the sub-diffusion fractional mean-field-type learning provides the set of global minima in one single step in all the computable cases of cost functions with finite number of isolated extrema.*

Example 19. *We will use this property to build a mean-field-type learning algorithm for* $\mu(t, d\theta)$ *for both convex and non-convex functions. In Figures 5.10, 5.13, 5.14, 5.15, 5.16, and 5.17 we plot the behavior of the unnormalized distribution* $\mu(t, d\theta)$ *for different values of time with a properly designed Hurst index making a subdiffusion. We observe that the distribution concentrates very quickly to the set of global minima.*

5.2.2 Building a Mean-Field-Type Learning Algorithm

The mean-field-type learning does not have to be stochastic depending on the tools in hand. One can work, for example, with. the Fokker-Planck-Kolmogorov equation in the simple cases to derive the evolution $\mu(t, d\theta)$ over the horizon $[0, T]$ and estimates asymptotics as T gets larger. However, the Boltzmann-Gibbs distribution can be extremely slow if $\epsilon(t)$ is not properly designed.

Mean-field-type learning with additive Brownian motion

Start with an initial measure density $\mu(0, d\theta) = \mu_0(\theta)d\theta$ and the partial differential equation $\partial_t\mu(t,\theta) - \frac{div_\theta(\mu(t,\theta)c_\theta(\theta))}{\epsilon(t)} - \frac{\sigma^2}{2}trace(\mu_{\theta\theta}(t,\theta)) = 0$.

Using $trace(\mu_{\theta\theta}(t,\theta)) = div_\theta(\mu(t,\theta)(\log\mu(t,\theta))_\theta)$, the latter equation can also be rewritten as $\mu_t(t,\theta) - div_\theta[\mu(t,\theta)(\frac{c_\theta(\theta)}{\epsilon(t)} + \frac{\sigma^2}{2}(\log\mu(t,\theta)))_\theta] = 0$.

We now explain why this Fokker-Planck-Kolmogorov equation is related to the global minimization of c in the entire domain $\mathbb{R}^d$ even for a non-convex function c.

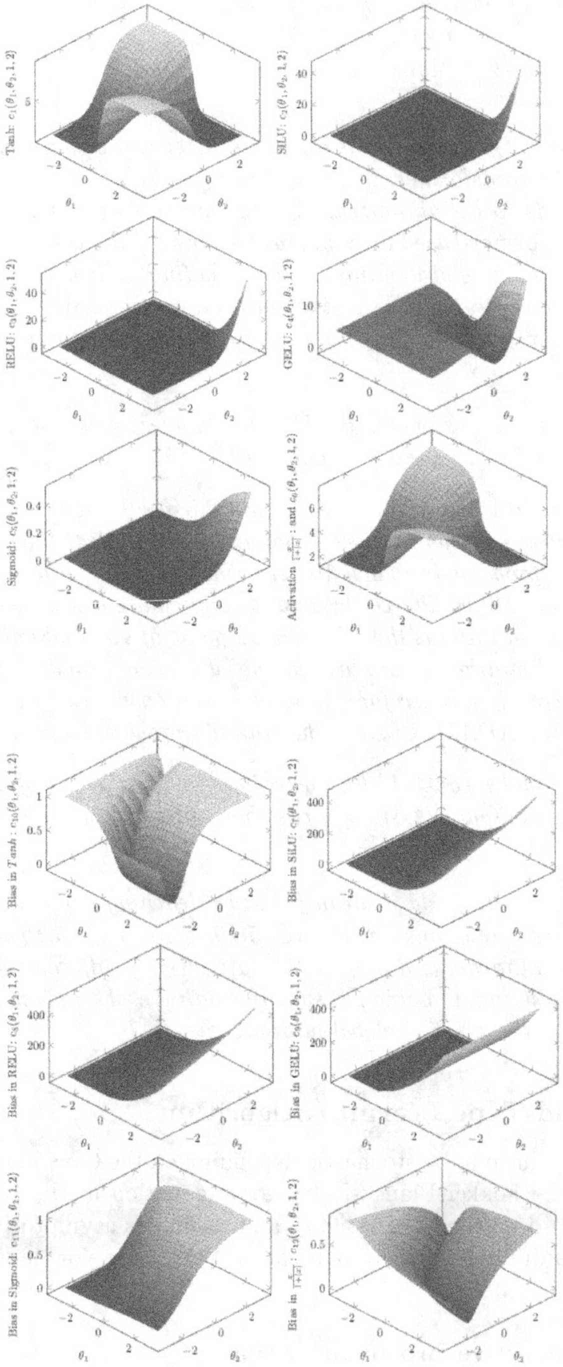

FIGURE 5.11
Cost obtained from Tanh, RELU, SiLU, GELU, Sigmoid, and $\frac{x}{1+\|x\|}$ activations.

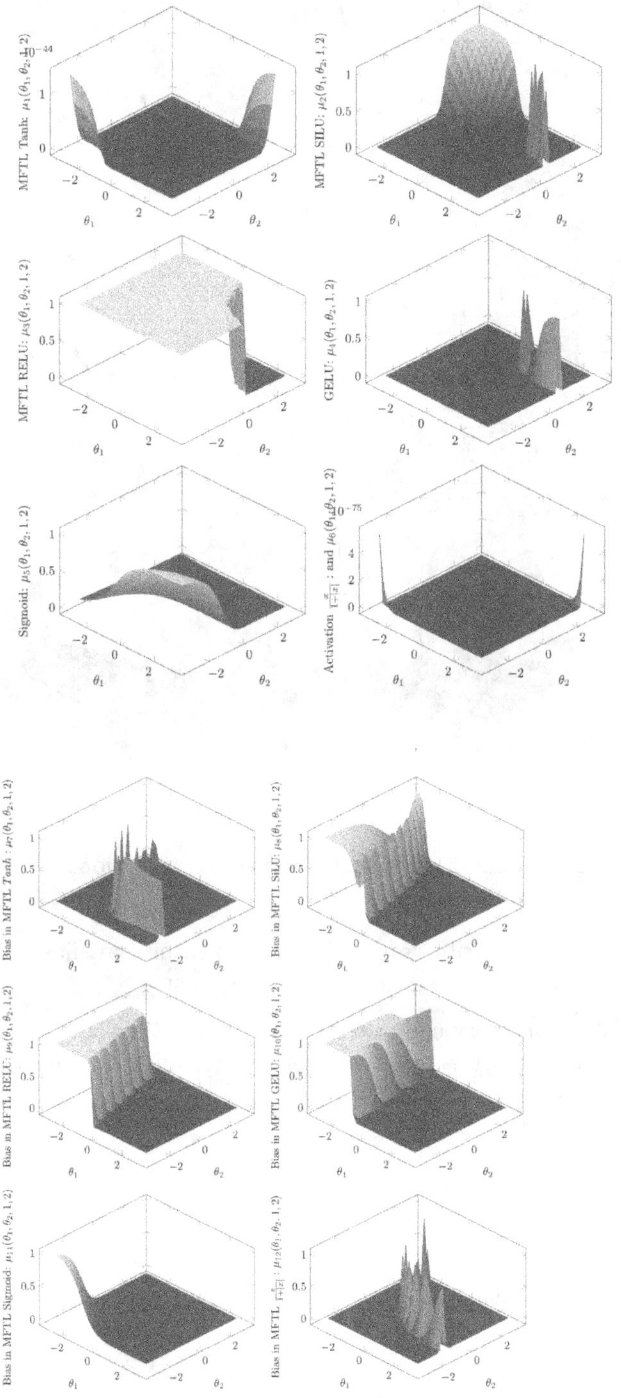

FIGURE 5.12
MFTL finds the set of global minimizers in a single step with cost functions $(f(\theta_2 f(\theta_1 x)) - y)^2$ with Tanh, RELU, SiLU, GELU, Sigmoid, and $\frac{x}{1+\|x\|}$ activations.

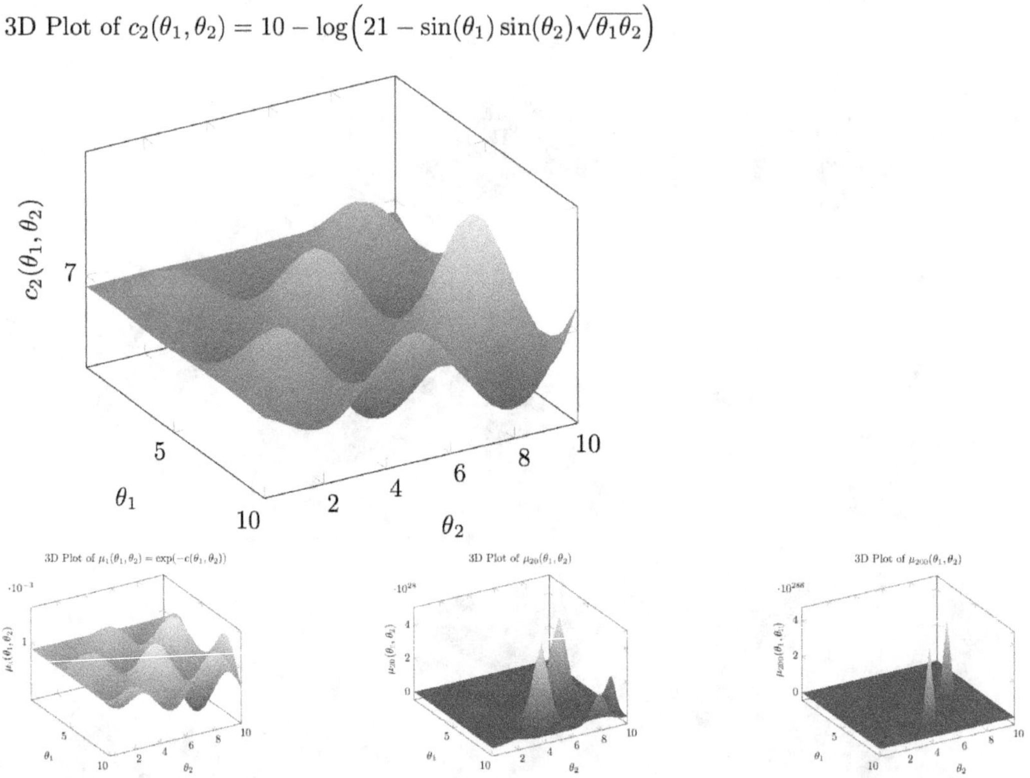

FIGURE 5.13
MFTL very quickly finds the set of global minimizers of the non-convex function.

For $\epsilon(t) = \epsilon$ we connect the stationary distribution to the set of measure satisfies in the weak sense $div_\theta[\mu_*(d\theta)(\frac{c_\theta(\theta)}{\epsilon(t)} + \frac{\sigma^2}{2}(\log \mu_*(\theta)))_\theta] = 0$, which means that if μ is smooth, the function $\frac{c(\theta)}{\epsilon} + \frac{\sigma^2}{2} \log \mu_*(\theta)$ is constant on every connected component of the set $\{\mu_* > 0\}$. There is a unique stationary distribution, it is positive everywhere, and it is given by $\mu_\epsilon(d\theta) = C^{-1} e^{-\frac{2c(\theta)}{\epsilon\sigma^2}} d\theta$, $C = \int_\theta e^{-\frac{2c(\theta)}{\epsilon\sigma^2}} d\theta$. μ_ϵ is a global minimizer of $\mu \mapsto \int \mu(d\theta)[\frac{c(\theta)}{\epsilon} + \log \mu]$. One has

$$\epsilon \left(\int \mu_\epsilon(d\theta)[\frac{c(\theta)}{\epsilon} + \log \mu_\epsilon] \right) - \inf_\theta c(\theta) \to 0$$

as ϵ goes to infinity.

Proposition 31. *The measure μ_ϵ is the global minimizer of $\mu \mapsto c_\epsilon(\mu) := \int \mu(d\theta)[\frac{c(\theta)}{\epsilon} + \frac{\sigma^2}{2} \log \mu(\theta)]$ even when c is non-convex.*

Proof:

$$\begin{aligned}
& c_\epsilon(\mu) - c_\epsilon(\mu_\epsilon) \\
& = \int \mu(dy)[\tfrac{c(\theta)}{\epsilon} + \tfrac{\sigma^2}{2} \log \mu(\theta)] \\
& - \int \mu_\epsilon(d\theta)[\tfrac{c(\theta)}{\epsilon} + \tfrac{\sigma^2}{2} \log \mu_\epsilon(\theta)] \\
& = \int (\mu(d\theta) - \mu_\epsilon(d\theta)) \tfrac{c(\theta)}{\epsilon} \\
& + \tfrac{\sigma^2}{2} \int [\mu(d\theta) \log \mu(\theta) - \mu_\epsilon(d\theta) \log \mu_\epsilon(\theta)]
\end{aligned} \tag{5.11}$$

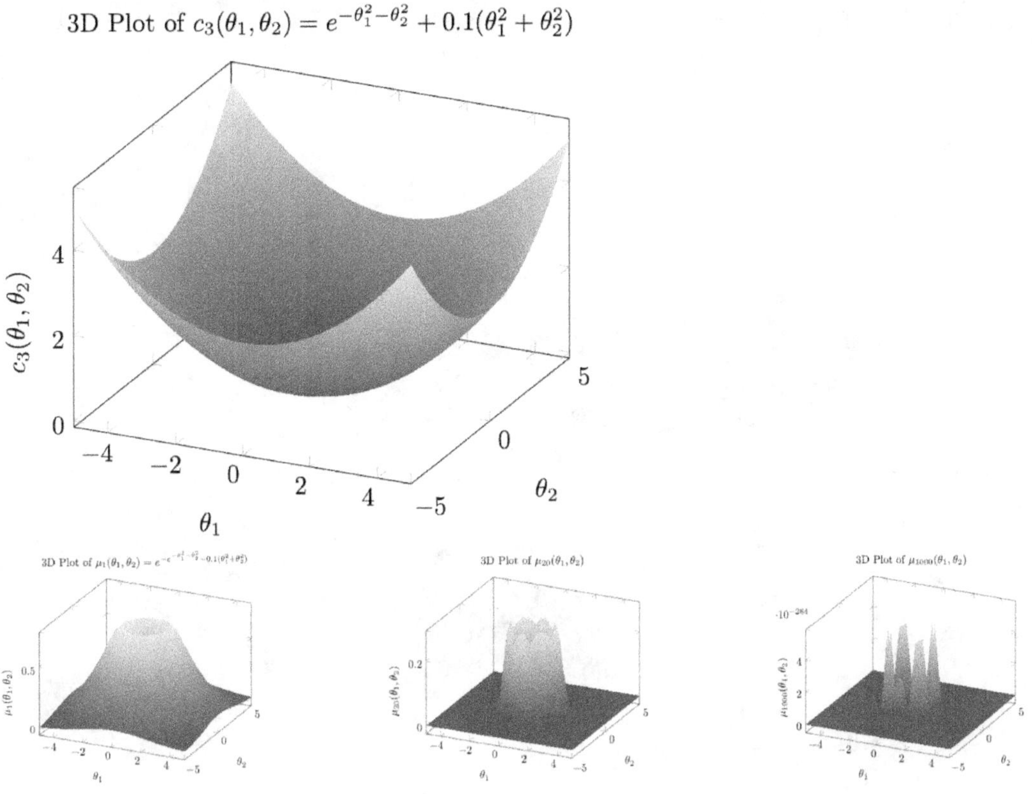

FIGURE 5.14
MFTL very quickly finds the set of global minimizers of the non-convex function.

As $\frac{c(\theta)}{\epsilon} + \frac{\sigma^2}{2} \log \mu_\epsilon(\theta)$ is constant, say equal to λ, it follows that $\frac{c(\theta)}{\epsilon} = \lambda - \frac{\sigma^2}{2} \log \mu_\epsilon(\theta)$.

$$\begin{aligned} c_\epsilon(\mu) - c_\epsilon(\mu_\epsilon) &= \int (\mu(d\theta) - \mu_\epsilon(d\theta)) \left(\lambda - \frac{\sigma^2}{2} \log \mu_\epsilon(\theta)\right) \\ &+ \frac{\sigma^2}{2} \int [\mu(d\theta) \log \mu(\theta) - \mu_\epsilon(d\theta) \log \mu_\epsilon(\theta)]. \end{aligned} \tag{5.12}$$

This leads to

$$\begin{aligned} c_\epsilon(\mu) - c_\epsilon(\mu_\epsilon) &= -\frac{\sigma^2}{2} \int (\mu(d\theta) - \mu_\epsilon(d\theta)) \log \mu_\epsilon(\theta) + \\ &\frac{\sigma^2}{2} \int [\mu(d\theta) \log \mu(\theta) - \mu_\epsilon(d\theta) \log \mu_\epsilon(\theta)], \end{aligned} \tag{5.13}$$

which is rewritten as

$$\begin{aligned} &c_\epsilon(\mu) - c_\epsilon(\mu_\epsilon) \\ &= \frac{\sigma^2}{2} \int [\mu(d\theta) \log \mu(\theta) - \mu_\epsilon(d\theta) \log \mu_\epsilon(\theta) \\ &-(\mu(d\theta) - \mu_\epsilon(d\theta)) \log \mu_\epsilon(\theta)], \end{aligned} \tag{5.14}$$

which is positive whenever $\mu \neq \mu_\epsilon$ by strict convexity of function $\theta \log \theta$ on $\mathbb{R}_+$.
□

The same method is used in [107] to prove convergence when $\epsilon(t)$ depends on time but with carefully designed $\epsilon(t)$. Unfortunately, the dependence in time is crucial in order to get a small error. We want $\epsilon(t)$ to be small as time increases so that the error gap is reduced.

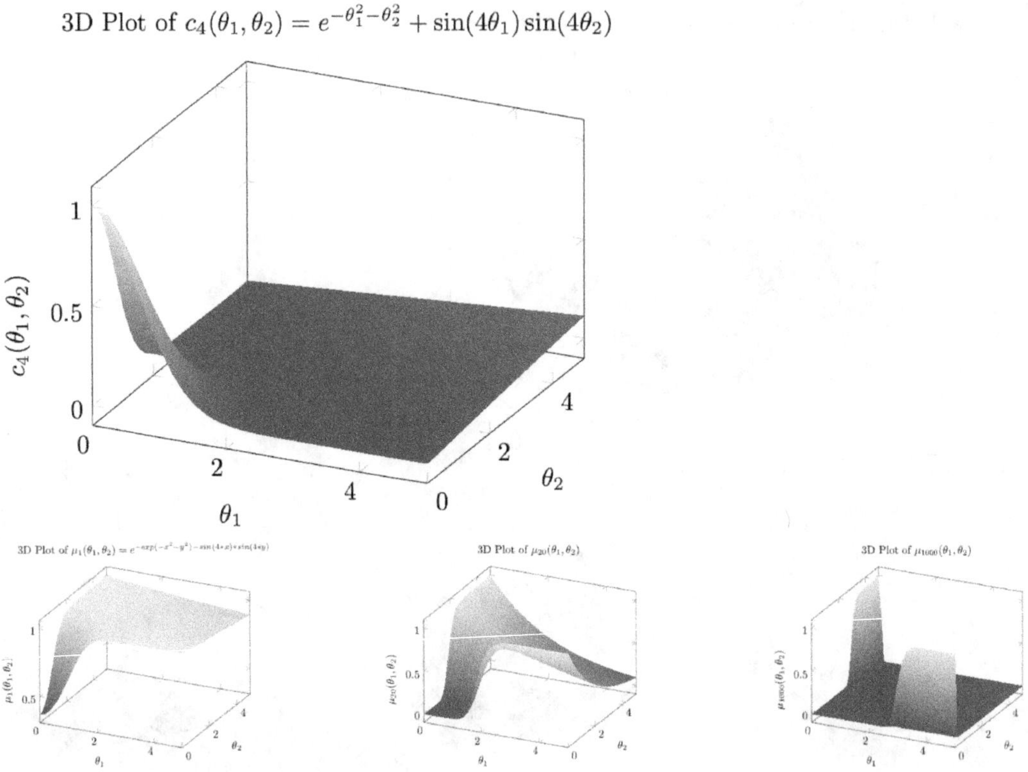

FIGURE 5.15
MFTL very quickly finds the set of global minimizers of the non-convex function.

Typical examples are $\epsilon(t) = \frac{1}{(1+t)^{\frac{1}{3}}}$ or $\epsilon(t) = \frac{1}{\sqrt{\log(e+t)}}$, $t > 0$. The presence of the mean-field-type interaction forces the trajectories to escape faster from non-global minimizers and stay longer in the vicinity of global ones. We use such mean-field-type interaction to design fast convergence to global minimizers. It is also connected to strategic swarm learning with properly designed swarm communication (see, for example, One-Swarm-per-Queen algorithm in [174] for non-convex game problems.) In [43] it is shown that the following mean-field-type learning with multiplicative Brownian motion

$$\begin{array}{l} \mu_t(t,\theta) \\ -div_\theta[\mu(t,\theta)(\frac{\sigma\sigma' c_\theta(\theta)}{\epsilon(t)} - (\sum_{j=1}^d \partial_j(\sigma(\theta)\sigma'(\theta))_{ij})_{i\in\{1,\ldots,d\}})] \\ -\frac{1}{2}trace(\sigma(\theta)\sigma'(\theta)\mu(t,\theta))_{\theta\theta} = 0, \end{array} \tag{5.15}$$

converges to global minimizers of c even with c non-convex but coercive and have elliptic properties outside a certain compact set and c has finitely global minimizers $(\theta_i^*)_{i\in\mathcal{I}}$, $|\mathcal{I}| < +\infty$, with $c_{\theta\theta}(\theta_j^*)$ being positive matrix at each of the these global minimizers. The algorithms with multiplicative noise prove to be faster than the algorithm with constant additive noise. The measure $\mu(t, d\theta) \xrightarrow{\mathcal{L}} \mu^* = \frac{1}{\sum_{j\in\mathcal{I}} \frac{1}{\sqrt{det(c_{\theta\theta}(\theta_j^*))}}} \sum_{i\in\mathcal{I}} \frac{1}{\sqrt{det(c_{\theta\theta}(\theta_i^*))}} \delta_{\theta_i^*}$ as $\epsilon(t) \to 0$.

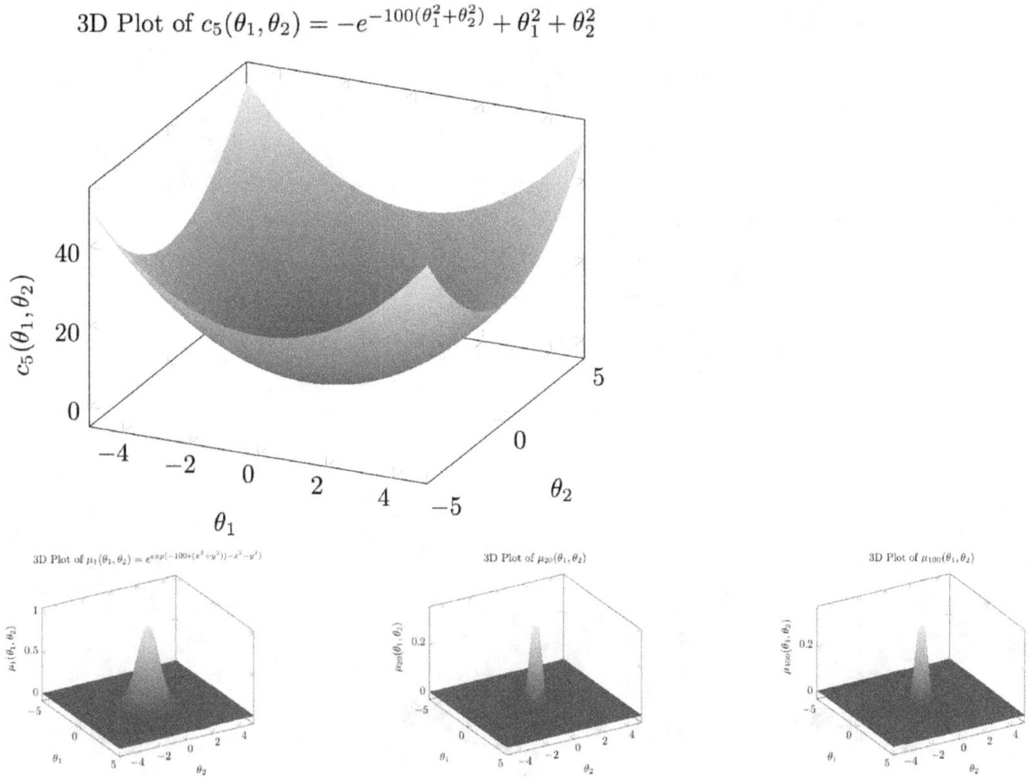

FIGURE 5.16
MFTL very quickly finds the set of global minimizers of the non-convex function.

Mean-field-type learning with fractional Brownian motion

We now include sub-diffusion and super-diffusion ideas in the fractional Fokker-Planck-Kolmogorov equation which reduces to

$$\mu_t(t, \theta) - [\mu(t, \theta)(\tfrac{c_\theta(\theta)}{\epsilon(t)})]_\theta - H\sigma^2 t^{2H-1} trace \mu_{\theta\theta}(t, \theta) = 0.$$

> We consider a time-modulated mean-field-type learning under fractional Brownian motion as follows:
>
> $$\begin{aligned} &\mu_t(t, \theta) - [\mu(t, \theta)(\tfrac{c_\theta(\theta)}{H\epsilon(t)t^{2H-1}\sigma^2} + (\log \mu(t, \theta)_\theta))]_\theta = 0, \\ &\mu(0, d\theta) = \mu_0(\theta)d\theta. \end{aligned} \qquad (5.16)$$

By using the above analysis, if the design parameter $\epsilon(t) = \frac{\epsilon_0}{Ht^{2H-1}\sigma^2\sqrt{\log(e+t)}}$, $t > 0, \epsilon_0 > 0$ fulfills the required convergence properties. Moreover, the learning rate is much smaller in the super-diffusion case $H > \frac{1}{2}$ but higher in the sub-diffusion case $H < \frac{1}{2}$. It means, for example, that for a sub-diffusion with Hurst parameter $H = \frac{1}{4}$, we do not need $\epsilon(t)$ to be vanishing as t goes to infinity. This relaxes the design range and the convergence range.

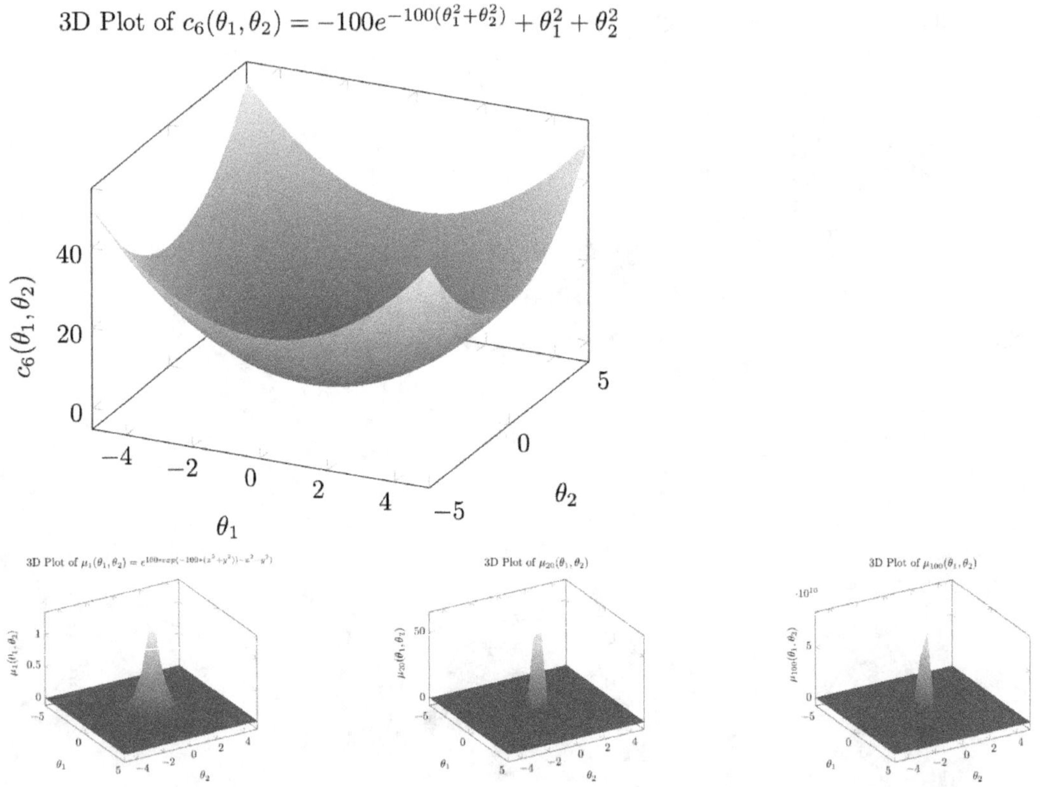

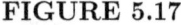

FIGURE 5.17
MFTL very quickly finds the set of global minimizers of the non-convex function.

Mean-field-type learning with penalty

We consider a generalized fractional mean-field-type learning algorithm governed by the following time-inhomogeneous partial differential equation:

$$\begin{aligned} &\mu_t(t,\theta) - div_\theta \left[\mu(t,\theta)\left(\frac{c_\theta(\theta)}{H\epsilon(t)t^{2H-1}\sigma^2} + \phi'_\theta(\mu(t,\theta))\right)\right] = 0, \\ &\mu(0,d\theta) = \mu_0(\theta)d\theta, \end{aligned} \tag{5.17}$$

where $\mu(t,\theta)$ represents the probability density function at time t and position θ. The function $\phi : [0,+\infty) \to \mathbb{R}$ is a convex, twice-differentiable function with $\phi''(\theta) > 0$ and $\phi'(0) = -\infty$, ensuring strong regularization and preventing densities from becoming zero. The parameter $\epsilon(t) > 0$ is a time-varying regularization parameter, and $\mu(0,d\theta) = \mu_0(\theta)d\theta$ is the initial distribution, assumed to be absolutely continuous with respect to the Lebesgue measure and within the domain of the regularization functional. The parameters H and σ are constants related to the fractional Brownian motion.

This equation can be interpreted as a Wasserstein gradient flow, which provides a powerful framework for understanding the evolution of probability measures. To demonstrate this, we introduce the Gateaux derivative, which, when followed by a gradient, yields the variational derivative $\partial_{\theta\mu} = \nabla_\theta \frac{\delta}{\delta\mu}$. We apply this concept in the distributional sense.

Consider the regularized functional $c_{\epsilon(t)}(\mu)$:

$$c_{\epsilon(t)}(\mu) = \int \frac{c(\theta)}{H\epsilon(t)t^{2H-1}\sigma^2}\mu(d\theta) + \int \phi(\mu(\theta))d\theta. \tag{5.18}$$

The first term, $\int \frac{c(y)}{H\epsilon(t)t^{2H-1}\sigma^2}\mu(dy)$, represents a linear energy term (linear in the measure μ) associated with the cost function $c(y)$, scaled by the time-dependent regularization. Minimizing this term with respect to μ yields a constant c_{inf}. The Wasserstein gradient of this term is given by

$$-\text{div}_\theta \left[\mu(t,\theta)\left(\frac{c_\theta(\theta)}{H\epsilon(t)t^{2H-1}\sigma^2}\right)\right], \tag{5.19}$$

which is obtained by identifying the derivative in the sense of distributions.

The second term, $\int \phi(\mu(\theta))d\theta$, represents a regularization term that penalizes deviations from uniform distributions. Its Wasserstein gradient is

$$-\text{div}_\theta \left[\mu(\theta)\phi'_\theta(\mu(\theta))\right]. \tag{5.20}$$

While traditional gradient descent in $\mathbb{R}^d$ may fail to reach global minimizers, especially for non-convex functions, Wasserstein gradient flows in the space of probability measures can often achieve them. This is because the Wasserstein metric accounts for the geometry of probability distributions, allowing for more robust convergence properties. Therefore, fractional mean-field-type learning, when formulated as a Wasserstein gradient flow, offers a promising approach to reach global optima in both convex and non-convex settings. The regularization term ϕ and the time varying regularization parameter $\epsilon(t)$ are crucial to guarantee convergence and a well-defined solution.

Figure 5.13 represents a 3D plot of the non-convex function $c(\theta_1,\theta_2) = 10 - \log\Big(21 - \sin(\theta_1)\sin(\theta_2)\sqrt{\theta_1\theta_2}\Big)$ over the domain $[0,10]^2$ and the associated measures over time $t \in \{1,20,200\}$. As we can see two δ measures emerge as time gets larger. Figure 5.14 represents 3D plot of $c_3(\theta_1,\theta_2) = e^{-\theta_1^2-\theta_2^2} + 0.1(\theta_1^2+\theta_2^2)$ and the associated measures over time $t \in \{1,20,1000\}$. Figure 5.15 represents 3D plot of $c_4(\theta_1,\theta_2) = e^{-\theta_1^2-\theta_2^2} + \sin(4\theta_1)\sin(4\theta_2)$ and the associated measures over time $t \in \{1,20,1000\}$. Figure 5.16 is a 3D plot of $c_5(\theta_1,\theta_2) = -e^{-100(\theta_1^2+\theta_2^2)} + \theta_1^2+\theta_2^2$ and the associated measures over time $t \in \{1,20,100\}$. Figure 5.17 is a 3D plot of $c_6(\theta_1,\theta_2) = -100e^{-100(\theta_1^2+\theta_2^2)} + \theta_1^2+\theta_2^2$ and the associated measures over time $t \in \{1,20,100\}$. Figure 5.18 is a plot of $c_7(\theta_1,\theta_2) = \sin(deg(\pi(\theta_1+3)/4))^2 + ((\theta_1-1)/4)^2(1+10\sin(deg(\pi*(\theta_1+3)/4)+1)) + ((\theta_2-1)/4)^2*(1+\sin(deg(2*\pi*(\theta_2+3)/4)))$. Global minimizer $(1,1)$ and global minimum value $= 0$. Figure 5.19 is a plot of $c_8(\theta_1,\theta_2) = \sin(deg(3\pi\theta_1))^2 + ((\theta_1-1)^2)*(1+\sin(deg(3\pi\theta_2))) + ((\theta_2-1)^2)(1+\sin(deg(2*\pi*\theta_2)))$. Figure 5.20 is a plot of $c_9(\theta_1,\theta_2) = -(1+\cos(12\sqrt{\theta_1^2+\theta_2^2}))/(0.5(\theta_1^2+\theta_2^2)+2)$. Global minimizer $(0,0)$ and global minimum value $= -1$. Figure 5.21 is a plot of $c_{10}(\theta_1,\theta_2) = 100\sqrt{|\theta_2 - 0.01\theta_1^2|} + 0.01|\theta_1+10|$. Global minimizer $(-10,1)$ and global minimum value $= 0$. Figure 5.22 is a plot of $c_{11}(\theta_1,\theta_2) = 20 + 2\theta_1^2 + 2\theta_2^2 - \cos(deg(2\pi\theta_1)) - \cos(deg(2\pi\theta_2))$. Global minimizer $(0,0)$ and global minimum value $= 0$. Figure 5.23 is a plot of $c_{12}(\theta_1,\theta_2) = 0.5 + \frac{\sin(\theta_1^2-\theta_2^2)^2-0.5}{(1+0.0001(\theta_1^2+\theta_2^2))^2}$. Global minimizer $(0,0)$ and global minimum value $= 0$.

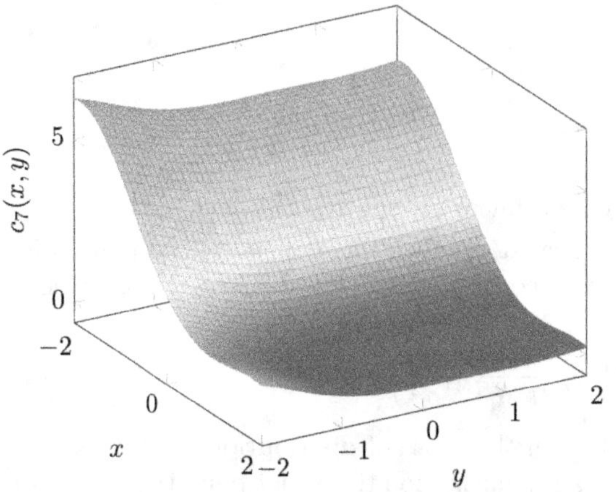

(a) Plot of $c_7(x, y) = \sin(deg(\pi * (x+3)/4))^2 + ((x-1)/4)^2 * (1+10*\sin(deg(\pi * (x+3)/4)+1)) + ((y-1)/4)^2 * (1+\sin(deg(2*\pi*(y+3)/4)))$. Global minimizer $(1, 1)$ and global minimum value $= 0$.

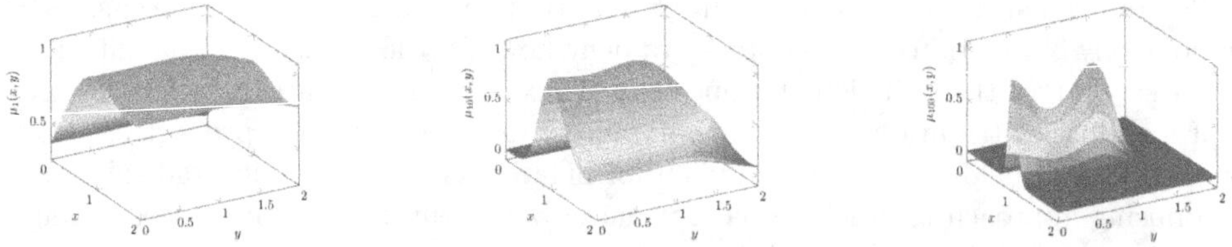

(b) Plot of $\mu_1(x, y)$ (c) Plot of $\mu_{10}(x, y)$ (d) Plot of $\mu_{100}(x, y)$

FIGURE 5.18
MFTL very quickly finds the set of global minimizers of the non-convex function.

5.2.3 Sample Tests

Table 5.1 provides sample test functions in one dimension. These functions have several local extrema and we observe the failure of ADAM and many of its variants. Table 5.2 focuses cost functions in two dimensions. Tables 5.3, 5.4, 5.5, 5.6, 5.7 provide a class of high-dimensional function with 1 billion local extrema, specially designed to feature the difficulties of non-communicating gradient explorers.

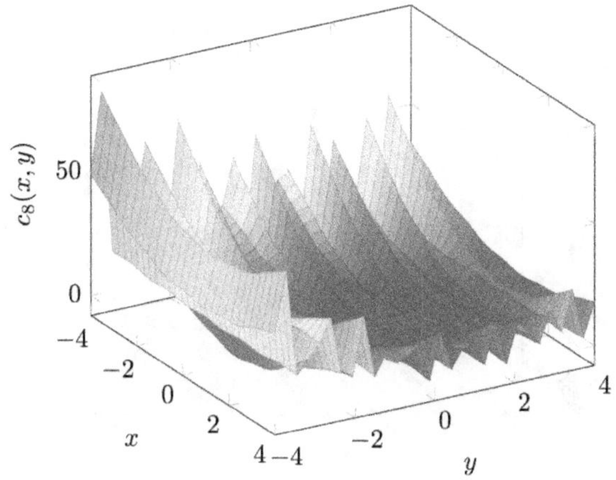

(a) Plot of $c_8(x, y) = \sin(deg(3 * \pi * x))^2 + ((x-1)^2) * (1 + \sin(deg(3 * \pi * y))) + ((y-1)^2) * (1 + \sin(deg(2 * \pi * y)))$.

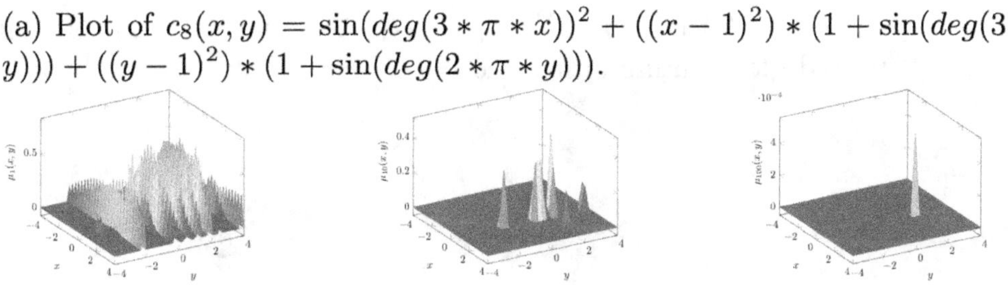

(b) Plot of $\mu_1(x, y)$ (c) Plot of $\mu_{10}(x, y)$ (d) Plot of $\mu_{100}(x, y)$

FIGURE 5.19
MFTL finds very quickly the set of global minimizers of the non-convex function.

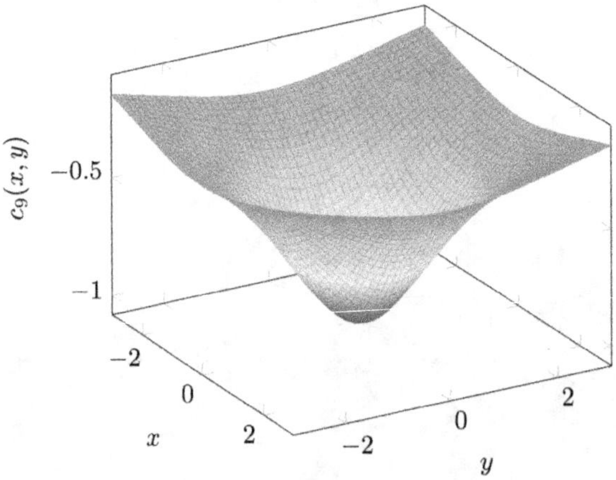

(a) Plot of $c_9(x, y) = -(1 + \cos(12 * \sqrt{x^2 + y^2}))/(0.5 * (x^2 + y^2) + 2)$. Global minimizer $(0, 0)$ and global minimum value $= -1$.

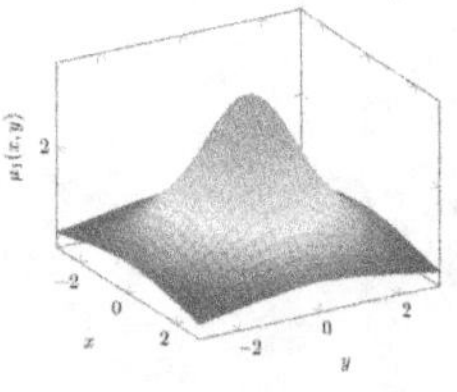

(b) Plot of $\mu_1(x, y)$

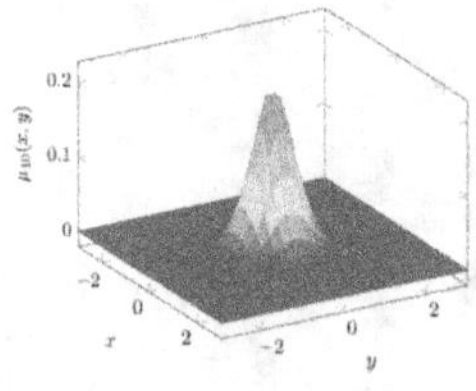

(c) Plot of $\mu_{10}(x, y)$

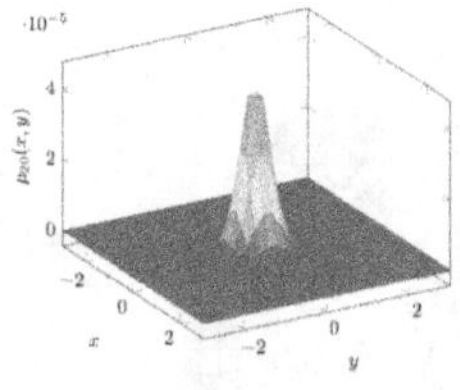

(d) Plot of $\mu_{20}(x, y)$

FIGURE 5.20
MFTL very quickly finds the set of global minimizers of the non-convex function.

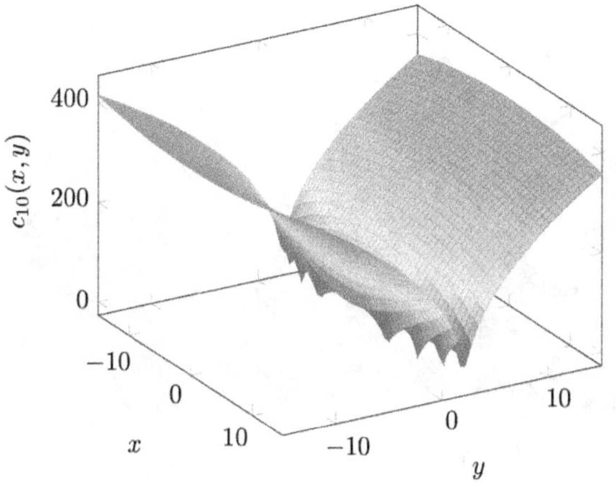

(a) Plot of $c_{10}(x, y) = 100 * \sqrt{|y - 0.01 * x^2|} + 0.01 * |x + 10|$. Global minimizer $(-10, 1)$ and global minimum value $= 0$.

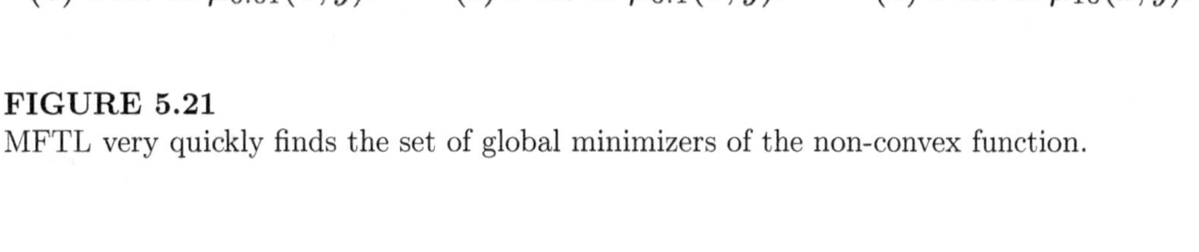

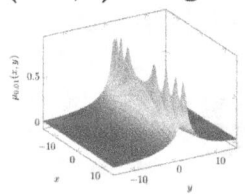

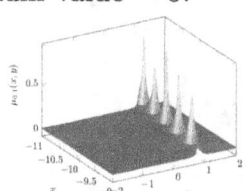

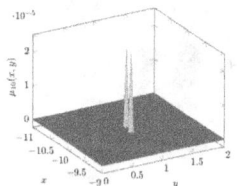

(b) Plot of $\mu_{0.01}(x, y)$. (c) Plot of $\mu_{0.1}(x, y)$. (d) Plot of $\mu_{10}(x, y)$.

FIGURE 5.21
MFTL very quickly finds the set of global minimizers of the non-convex function.

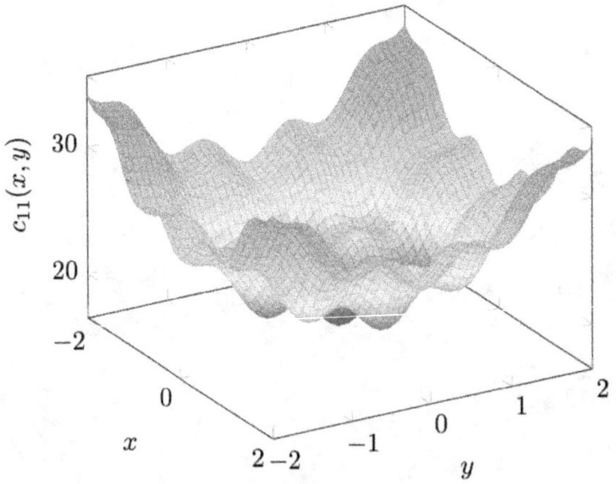

(a) Plot of $c_{11}(x, y) = 20 + 2 * x^2 + 2 * y^2 - \cos(deg(2 * \pi * x)) - \cos(deg(2 * \pi * y))$. Global minimizer $(0, 0)$ and global minimum value $= 0$.

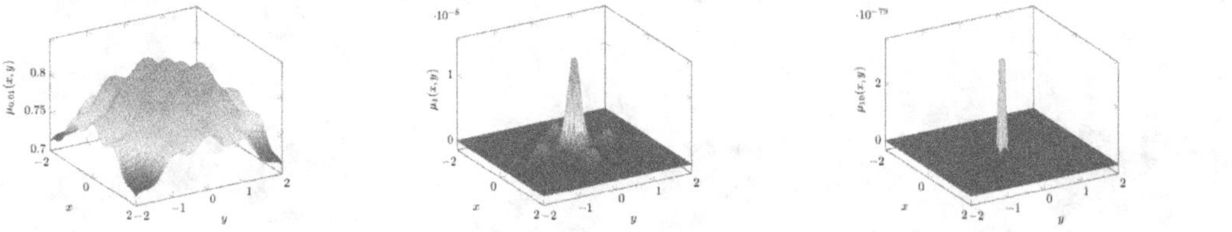

(b) Plot of $\mu_{0.01}(x, y)$. (c) Plot of $\mu_1(x, y)$. (d) Plot of $\mu_{10}(x, y)$.

FIGURE 5.22
MFTL finds very quickly the set of global minimizers of the non-convex function.

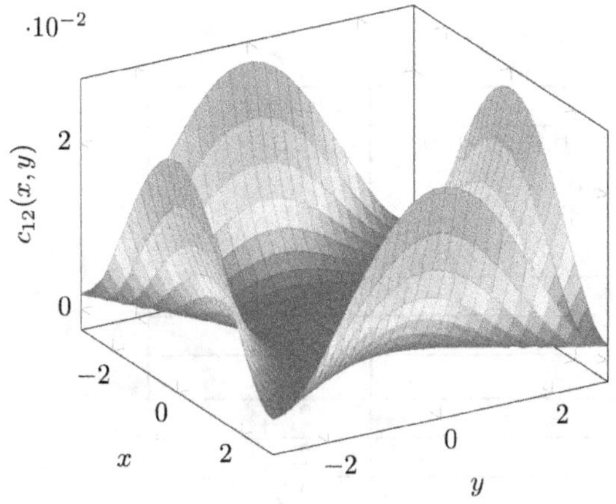

(a) Plot of $c_{12}(x, y) = 0.5 + \frac{\sin(x^2 - y^2)^2 - 0.5}{(1+0.0001*(x^2+y^2))^2}$. Global minimizer $(0, 0)$ and global minimum value $= 0$.

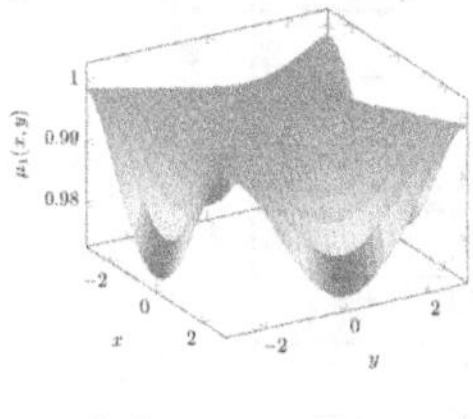

(b) Plot of $\mu_1(x, y)$.

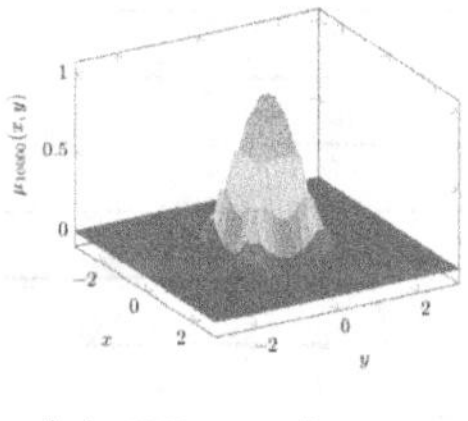

(c) Plot of $\mu_{10000}(x, y)$.

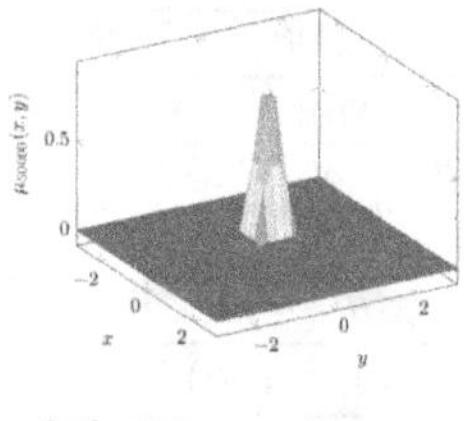

(d) Plot of $\mu_{50000}(x, y)$.

FIGURE 5.23
MFTL very quickly finds the set of global minimizers of the non-convex function.

TABLE 5.1
1D Cost Functions

Function $c(\theta)$	**Domain**
$c_1(\theta)=$ interpolation $(1,1)(2,0.95)(3,0.92)(4,0.92)(5,0.98)$ $(6,0.99)(7,0.55)(8,0.52)(9,0.52)(10,0.68)(11,0.69)$ $(12,0.25)(13,0.22)(14,0.22)(15,0.38)(16,0.39)$ $(17,0.75)(18,0.76)(19,0.72)(20,0.72)$ $(21,0.88)(22,0.89)$	$[0,25]$
$c_2 = \sin(3\pi\theta) + 0.3\theta$	$[-2,2]$
$c_3 = \theta^2 + 10(1 - \cos(\pi\theta))$	$[-3,3]$
$c_4 = -e^{-\theta^2} + 0.1\theta^2$	$[-5,5]$
$c_5 = \theta^6 - 3\theta^4 + 2\theta$	$[-2,2]$
$c_6 = \cos(2\pi\theta) + 0.2\theta^2$	$[-2,2]$
$c_7 = \theta^3 - 3\theta$	$[-2.5,2.5]$
$c_8 = \|\theta\| - 2\cos(\theta)$	$[-4,4]$
$c_9 = \theta^2 - 4\cos(\pi\theta)$	$[-3,3]$
$c_{10} = \sin(\theta) + \sin(\sqrt{2}\theta)$	$[-10,10]$
$c_{11} = \theta^4 - 10\theta^2 + 0.5\theta$	$[-4,4]$
$c_{12} = e^{-\theta}\sin(4\theta)$	$[0,5]$
$c_{13} = \theta^5 - 5\theta^3 + 5\theta$	$[-2.5,2.5]$
$c_{14} = \tanh(\theta) - 0.1\theta^2$	$[-5,5]$
$c_{15} = \log(\theta^2+1) - 0.3\theta$	$[-5,5]$
$c_{16} = \sqrt{\|\theta\|}\cos(5\theta)$	$[-3,3]$
$c_{17} = 0.1\theta^3 + 0.5\theta^2 - 4\theta$	$[-10,10]$
$c_{18} = \theta^2\sin(4\theta)$	$[-3,3]$
$c_{19} = \frac{\theta^4}{4} - \frac{3\theta^2}{2} + \theta$	$[-3,3]$
$c_{20} = \mathrm{sign}(\theta)\sqrt{\|\theta\|} - 0.2\theta$	$[-5,5]$
$c_{21} = \theta^4 - 4\theta^3 + 2\theta^2 + 4\theta$	$[-1,4]$

TABLE 5.2
2D Test Cost Functions

Function $c(\theta_1, \theta_2)$	**Domain**
$c_{22}(\theta_1,\theta_2) = 10 - \log\left(21 - \sin(\theta_1)\sin(\theta_2)\sqrt{\theta_1\theta_2}\right)$	$[0,10]^2$
$c_{23}(\theta) = exp(\sin(2\theta^2)) + 0.1(\theta - \pi/2)^2$	
$c_{24}(\theta_1,\theta_2) = -(1+\cos(12\sqrt{(\theta_1^2+\theta_2^2)}))/(2+0.5(\theta_1^2+\theta_2^2))$,	
$c_{25}(\theta_1,\theta_2) = (1-\theta_1)^2 + 100(\theta_2-\theta_1^2)^2$	
$c_{26} = \cos(2\pi\theta_1) + \cos(2\pi\theta_2) + 0.2(\theta_1^2+\theta_2^2)$	$[-2,2]\times[-2,2]$
$c_{27} = \theta_1^3 + \theta_2^3 - 3\theta_1 - 3\theta_2$	$[-2.5,2.5] \times [-2.5,2.5]$
$c_{28} = \|\theta_1\| + \|\theta_2\| - 2(\cos(\theta_1)+\cos(\theta_2))$	$[-4,4]\times[-4,4]$
$c_{29} = \theta_1^2 + \theta_2^2 - 4(\cos(\pi\theta_1)+\cos(\pi\theta_2))$	$[-3,3]\times[-3,3]$
$c_{30} = \sin(\theta_1) + \sin(\sqrt{2}\theta_1) + \sin(\theta_2) + \sin(\sqrt{2}\theta_2)$	$[-10,10]\times[-10,10]$
$c_{31} = \theta_1^4 + \theta_2^4 - 10\theta_1^2 - 10\theta_2^2 + 0.5\theta_1 + 0.5\theta_2$	$[-4,4]\times[-4,4]$
$c_{32} = e^{-\theta_1^2-\theta_2^2}\sin(4\theta_1)\sin(4\theta_2)$	$[0,5]\times[0,5]$
$c_{33} = \theta_1^5 + \theta_2^5 - 5\theta_1^3 - 5\theta_2^3 + 5\theta_1 + 5\theta_2$	$[-2.5,2.5] \times [-2.5,2.5]$
$c_{34} = \tanh(\theta_1) + \tanh(\theta_2) - 0.1(\theta_1^2+\theta_2^2)$	$[-5,5]\times[-5,5]$
$c_{35} = \log(\theta_1^2+\theta_2^2+1) - 0.3(\theta_1+\theta_2)$	$[-5,5]\times[-5,5]$
$c_{36} = \sqrt{\|\theta_1\|}\cos(5\theta_1) + \sqrt{\|\theta_2\|}\cos(5\theta_2)$	$[-3,3]\times[-3,3]$
$c_{37} = 0.1\theta_1^3 + 0.5\theta_1^2 - 4\theta_1 + 0.1\theta_2^3 + 0.5\theta_2^2 - 4\theta_2$	$[-10,10]\times[-10,10]$
$c_{38} = \theta_1^2\sin(4\theta_1) + \theta_2^2\sin(4\theta_2)$	$[-3,3]\times[-3,3]$
$c_{39} = \frac{\theta_1^4+\theta_2^4}{4} - \frac{3\theta_1^2+3\theta_2^2}{2} + \theta_1 + \theta_2$	$[-3,3]\times[-3,3]$
$c_{40} = \theta_1 sign(\theta_1)\sqrt{\|\theta_1\|} + \theta_1 sign(\theta_2)\sqrt{\|\theta_2\|} - 0.2(\theta_1+\theta_2)$	$[-5,5]\times[-5,5]$
$c_{41} = \theta_1^4 + \theta_2^4 - 4\theta_1^3 - 4\theta_2^3 + 2\theta_1^2 + 2\theta_2^2$	$[-2,4]\times[-2,4]$
$c_{42} = \sin(3\pi\theta_1) + \sin(3\pi\theta_2) + 0.3(\theta_1+\theta_2)$	$[-2,2]\times[-2,2]$
$c_{43} = \theta_1^2 + \theta_2^2 + 10(2 - \cos(\pi\theta_1) - \cos(\pi\theta_2))$	$[-3,3]\times[-3,3]$
$c_{44} = -e^{-\theta_1^2-\theta_2^2} + 0.1(\theta_1^2+\theta_2^2)$	$[-5,5]\times[-5,5]$
$c_{45} = \theta_1^6 + \theta_2^6 - 3\theta_1^4 - 3\theta_2^4 + 2\theta_1 + 2\theta_2$	$[-2,2]\times[-2,2]$

TABLE 5.3
2D test cost functions.

$c_{46}(\theta_1,\theta_2) = 10 - \log\left(21 - \sin(\theta_1)\sin(\theta_2)\sqrt{\theta_1\theta_2}\right)$	$[0,10]^2$
$c_{47}(\theta_1,\theta_2) = e^{-\theta_1^2-\theta_2^2} + 0.1(\theta_1^2+\theta_2^2)$	
$c_{48}(\theta_1,\theta_2) = e^{-\theta_1^2-\theta_2^2} + \sin(4\theta_1)\sin(4\theta_2)$	
$c_{49}(\theta_1,\theta_2) = -e^{-100(\theta_1^2+\theta_2^2)} + \theta_1^2 + \theta_2^2$	
$c_{50}(\theta_1,\theta_2) = -100e^{-100(\theta_1^2+\theta_2^2)} + \theta_1^2 + \theta_2^2$	
$c_{51}(x,y) = \sin(deg(\pi(x+3)/4))^2$ $+((x-1)/4)^2(1+10\sin(deg(\pi(x+3)/4)+1))$ $+((y-1)/4)^2(1+\sin(deg(2\pi(y+3)/4)))$	
$c_{52}(x,y) = \sin(deg(3\pi x))^2$ $+((x-1)^2)(1+\sin(deg(3\pi y)))$ $+((y-1)^2)(1+\sin(deg(2\pi y)))$	
$c_{53}(x,y) = -(1+\cos(12\sqrt{x^2+y^2}))/(0.5(x^2+y^2)+2)$	
$c_{54}(x,y) = 100\sqrt{\lvert y-0.01x^2\rvert} + 0.01\lvert x+10\rvert$	
$c_{55}(x,y) = 20 + 2x^2 + 2y^2 - \cos(deg(2\pi x)) - \cos(deg(2\pi y))$	
$c_{56}(x,y) = 0.5 + \frac{\sin(x^2-y^2)^2-0.5}{(1+0.0001(x^2+y^2))^2}$	

TABLE 5.4
High-dimensional functions with with 1 billion local extrema.

Function $c(\theta_1,\ldots,\theta_I)$	**Domain**
$c_{57} = \frac{1}{I}\sum_{i=1}^{I}\{(x_i-B)^2 - 10\cos(2\pi(x_i-A)) + 10\} + B$	
$c_{58} = -20e^{\frac{0.2}{\sqrt{I}}\sqrt{\sum_{i=1}^{I}\{(x_i-A)^2}} - e^{\frac{1}{I}\sum_{i=1}^{d}\cos(2\pi(x_i-A))} + 20 +$ $e + B$	
$c_{59}(\theta_1,\ldots,\theta_I) = \sum_{i=1}^{I-1}((1-\theta_i)^2 + 100(\theta_{i+1}-\theta_i^2)^2)$	
$c_{60}(\theta_1,\ldots,\theta_I) = \frac{1}{2}\sum_{i=1}^{I}(\theta_i^4 - 16\theta_i^2 + 5\theta_i)$	
$c_{61}(\theta) = \sum_{i=1}^{2.10^{24}}\left[\underbrace{\sin\left(10^6\pi\theta_i\right)}_{\text{High-frequency term}} + \underbrace{0.1\ \theta_i^2}_{\text{Convex bias}}\right],$	$\theta \in \mathbb{R}^{2.10^{24}}$ $\theta_i \in [-1,1]$ $\forall i \in$ $\{1,\ldots,2.10^{24}\}$

TABLE 5.5
High-dimensional functions with several local extrema

Function	Domain
$c_{62}(\theta) = \sum_{i=1}^{100} \sin(1000 \sinh(\theta_i)) + \cos(1000 \cosh(\theta_{i+1}))$	$[-1, 1]^{100}$
$c_{63}(\theta) = \sum_{i=1}^{50} \sin(1000 \tanh(\theta_i)) + \cos(1000 \tanh(\theta_{i+1}))$	$[-5, 5]^{50}$
$c_{64}(\theta) = \sum_{i=1}^{20} \sin(1000\text{erf}(\theta_i)) + \cos(1000\text{erf}(\theta_{i+1}))$	$[-5, 5]^{20}$
$c_{65}(\theta) = \sum_{i=1}^{30} \sin(1000\text{sigmoid}(\theta_i)) + \cos(1000\text{sigmoid}(\theta_{i+1}))$	$[-10, 10]^{30}$
$c_{66}(\theta) = \sum_{i=1}^{40} \sin(1000\text{relu}(\theta_i)) + \cos(1000\text{relu}(\theta_{i+1}))$	$[-1, 1]^{40}$
$c_{67}(\theta) = \sum_{i=1}^{60} \sin(1000\text{softplus}(\theta_i)) + \cos(1000\text{softplus}(\theta_{i+1}))$	$[-1, 1]^{60}$
$c_{68}(\theta) = \sum_{i=1}^{70} \sin(1000\text{sinc}(\theta_i)) + \cos(1000\text{sinc}(\theta_{i+1}))$	$[-10, 10]^{70}$
$c_{69}(\theta) = \sum_{i=1}^{80} \sin(1000\|\theta_i\|) + \cos(1000\|\theta_{i+1}\|)$	$[-10, 10]^{80}$
$c_{70}(\theta) = \sum_{i=1}^{90} \sin(1000\text{round}(\theta_i)) + \cos(1000\text{round}(\theta_{i+1}))$	$[-10, 10]^{90}$
$c_{71}(\theta) = \sum_{i=1}^{100} \sin(1000\text{floor}(\theta_i)) + \cos(1000\text{floor}(\theta_{i+1}))$	$[-10, 10]^{100}$
$c_{72}(\theta) = \sum_{i=1}^{100} \sin(1000\text{ceil}(\theta_i)) + \cos(1000\text{ceil}(\theta_{i+1}))$	$[-10, 10]^{100}$
$c_{73}(\theta) = \sum_{i=1}^{50} \sin(1000\theta_i) \cos(1000\theta_{i+1}) + \sin(1000\theta_{i+2})$	$[-5, 5]^{50}$
$c_{74}(\theta) = \sum_{i=1}^{20} \sin(1000\theta_i^2) \cos(1000\theta_{i+1}^2) + \sin(1000\theta_{i+2}^2)$	$[-3, 3]^{20}$
$c_{75}(\theta) = \sum_{i=1}^{30} \sin(1000\theta_i^3) \cos(1000\theta_{i+1}^3) + \sin(1000\theta_{i+2}^3)$	$[-2, 2]^{30}$
$c_{76}(\theta) = \sum_{i=1}^{40} \sin(1000 \exp(\theta_i)) \cos(1000 \exp(\theta_{i+1})) + \sin(1000 \exp(\theta_{i+2}))$	$[-1, 1]^{40}$
$c_{77}(\theta) = \sum_{i=1}^{60} \sin(1000 \log(1 + \|\theta_i\|)) \cos(1000 \log(1 + \|\theta_{i+1}\|)) + \sin(1000 \log(1 + \|\theta_{i+2}\|))$	$[-10, 10]^{60}$
$c_{79}(\theta) = \sum_{i=1}^{70} \sin(1000\sqrt{\|\theta_i\|}) \cos(1000\sqrt{\|\theta_{i+1}\|}) + \sin(1000\sqrt{\|\theta_{i+2}\|})$	$[-10, 10]^{70}$
$c_{80}(\theta) = \sum_{i=1}^{80} \sin(1000\theta_i^{1/3}) \cos(1000\theta_{i+1}^{1/3}) + \sin(1000\theta_{i+2}^{1/3})$	$[-10, 10]^{80}$

TABLE 5.6
High-dimensional functions with several local extrema

$c_{81}(\theta) = \sum_{i=1}^{100} \sin(1000 \sinh(\theta_i)) \cos(1000 \cosh(\theta_{i+1})) + \sin(1000 \sinh(\theta_{i+2}))$	$[-1, 1]^{100}$
$c_{82}(\theta) = \sum_{i=1}^{50} \sin(1000 \tanh(\theta_i)) \cos(1000 \tanh(\theta_{i+1})) + \sin(1000 \tanh(\theta_{i+2}))$	$[-5, 5]^{50}$
$c_{83}(\theta) = \sum_{i=1}^{20} \sin(1000\text{erf}(\theta_i)) \cos(1000\text{erf}(\theta_{i+1})) + \sin(1000\text{erf}(\theta_{i+2}))$	$[-5, 5]^{20}$
$c_{84}(\theta) = \sum_{i=1}^{30} \sin(1000\text{sigmoid}(\theta_i)) \cos(1000\text{sigmoid}(\theta_{i+1})) + \sin(1000\text{sigmoid}(\theta_{i+2}))$	$[-10, 10]^{30}$
$c_{85}(\theta) = \sum_{i=1}^{40} \sin(1000\text{relu}(\theta_i)) \cos(1000\text{relu}(\theta_{i+1})) + \sin(1000\text{relu}(\theta_{i+2}))$	$[-1, 1]^{40}$
$c_{86}(\theta) = \sum_{i=1}^{60} \sin(1000\text{softplus}(\theta_i)) \cos(1000\text{softplus}(\theta_{i+1})) + \sin(1000\text{softplus}(\theta_{i+2}))$	$[-1, 1]^{60}$
$c_{87}(\theta) = \sum_{i=1}^{70} \sin(1000\text{sinc}(\theta_i)) \cos(1000\text{sinc}(\theta_{i+1})) + \sin(1000\text{sinc}(\theta_{i+2}))$	$[-10, 10]^{70}$
$c_{88}(\theta) = \sum_{i=1}^{80} \sin(1000\text{abs}(\theta_i)) \cos(1000\text{abs}(\theta_{i+1})) + \sin(1000\|\theta_{i+2}\|)$	$[-10, 10]^{80}$
$c_{89}(\theta) = \sum_{i=1}^{90} \sin(1000\text{round}(\theta_i)) \cos(1000\text{round}(\theta_{i+1})) + \sin(1000\text{round}(\theta_{i+2}))$	$[-10, 10]^{90}$
$c_{90}(\theta) = \sum_{i=1}^{100} \sin(1000\text{floor}(\theta_i)) \cos(1000\text{floor}(\theta_{i+1})) + \sin(1000\text{floor}(\theta_{i+2}))$	$[-10, 10]^{100}$

TABLE 5.7
High-Dimensional Functions with several local extrema.

$c_{91}(\theta) = \sum_{i=1}^{100} \sin(1000\text{ceil}(\theta_i))\cos(1000\text{ceil}(\theta_{i+1})) + \sin(1000\text{ceil}(\theta_{i+2}))$	$[-10, 10]^{100}$
$c_{92}(\theta) = \sum_{i=1}^{50} \sin(1000\theta_i)+\cos(1000\theta_{i+1})+\sin(1000\theta_{i+2})+\cos(1000\theta_{i+3})$	$[-5, 5]^{50}$
$c_{93}(\theta) = \sum_{i=1}^{20} \sin(1000\theta_i^2)+\cos(1000\theta_{i+1}^2)+\sin(1000\theta_{i+2}^2)+\cos(1000\theta_{i+3}^2)$	$[-3, 3]^{20}$
$c_{94}(\theta) = \sum_{i=1}^{30} \sin(1000\theta_i^3)+\cos(1000\theta_{i+1}^3)+\sin(1000\theta_{i+2}^3)+\cos(1000\theta_{i+3}^3)$	$[-2, 2]^{30}$
$c_{95}(\theta) = \sum_{i=1}^{40} \sin(1000\exp(\theta_i)) + \cos(1000\exp(\theta_{i+1})) + \sin(1000\exp(\theta_{i+2})) + \cos(1000\exp(\theta_{i+3}))$	$[-1, 1]^{40}$
$c_{96}(\theta) = \sum_{i=1}^{60} \sin(1000\log(1+\|\theta_i\|)) + \cos(1000\log(1+\|\theta_{i+1}\|)) + \sin(1000\log(1+\|\theta_{i+2}\|)) + \cos(1000\log(1+\|\theta_{i+3}\|))$	$[-10, 10]^{60}$
$c_{97}(\theta) = \sum_{i=1}^{70} \sin(1000\sqrt{\|\theta_i\|}) + \cos(1000\sqrt{\|\theta_{i+1}\|}) + \sin(1000\sqrt{\|\theta_{i+2}\|}) + \cos(1000\sqrt{\|\theta_{i+3}\|})$	$[-10, 10]^{70}$
$c_{98}(\theta) = \sum_{i=1}^{80} \sin(1000\theta_i^{1/3}) + \cos(1000\theta_{i+1}^{1/3}) + \sin(1000\theta_{i+2}^{1/3}) + \cos(1000\theta_{i+3}^{1/3})$	$[-10, 10]^{80}$
$c_{99}(\theta) = \sum_{i=1}^{90} \sin(1000\arctan(\theta_i)) + \cos(1000\arctan(\theta_{i+1})) + \sin(1000\arctan(\theta_{i+2})) + \cos(1000\arctan(\theta_{i+3}))$	$[-10, 10]^{90}$
$c_{100}(\theta) = \sum_{i=1}^{100} \sin(1000\theta_i)$	$[-10, 10]^{100}$
$c_{101}(\theta) = \sum_{i=1}^{50} \cos(1000\theta_i)\sin(1000\theta_{i+1})$	$[-10, 10]^{50}$
$c_{102}(\theta) = \sum_{i=1}^{20} \sin(1000\theta_i^2) + \cos(1000\theta_{i+1}^2)$	$[-5, 5]^{20}$
$c_{103}(\theta) = \sum_{i=1}^{30} \sin(1000\theta_i^3)\cos(1000\theta_{i+1}^3)$	$[-3, 3]^{30}$
$c_{104}(\theta) = \sum_{i=1}^{40} \sin(1000\theta_i^4) + \cos(1000\theta_{i+1}^4)$	$[-2, 2]^{40}$
$c_{105}(\theta) = \sum_{i=1}^{60} \sin(1000\exp(\theta_i)) + \cos(1000\exp(\theta_{i+1}))$	$[-1, 1]^{60}$
$c_{106}(\theta) = \sum_{i=1}^{70} \sin(1000\log(1+\|\theta_i\|)) + \cos(1000\log(1+\|\theta_{i+1}\|))$	$[-10, 10]^{70}$
$c_{107}(\theta) = \sum_{i=1}^{80} \sin(1000\sqrt{\|\theta_i\|}) + \cos(1000\sqrt{\|\theta_{i+1}\|})$	$[-10, 10]^{80}$
$c_{108}(\theta) = \sum_{i=1}^{90} \sin(1000\theta_i^{1/3}) + \cos(1000\theta_{i+1}^{1/3})$	$[-10, 10]^{90}$
$c_{109}(\theta) = \sum_{i=1}^{100} \sin(1000\arctan(\theta_i)) + \cos(1000\arctan(\theta_{i+1}))$	$[-10, 10]^{100}$

6

Strategic Deep Learning

In today's interconnected and data-driven world, the demand for effective interactive learning environments is on the rise. Interactive machine intelligence, where agents equipped with diverse modalities such as text, image, audio, video, music, gesture, sensor, and emotion engage in partially altruistic, selfless, collaborative, cooperative, competitive, selfish, coopetitive, empathetic, self-abnegating, and interactive behaviors. In this chapter, Strategic Deep Learning, moves beyond the single-agent perspective of traditional deep learning. We explore architectures where multiple intelligent agents, each represented by a neural network, interact within a game-theoretic framework. These interactions are not merely competitive but can also be cooperative, as seen in federated learning and decentralized machine intelligence systems. We also discuss how mean-field-type games (MFTGs) can model interactions between multiple machine intelligence agents, providing a new frontier in the development of strategic machine learning. Section 6.1 presents multiple machine intelligence agents. Section 6.2 focuses on tensor-graph neural networks $\mathbb{TGN}$ Strategic Learning. MFTGs between machine intelligence agents are discussed in Section 6.3. Section 6.4 builds a multi-agent mean-field-type learning algorithm. Section 6.5 expands agentic machine intelligence: Co-Intelligence between a (human) user and an MI-agent is modeled in Section 6.5.1. Section 6.5.2 provides basic solution concepts. Interaction between one (human) user and multiple MI-Agents is described in Section 6.5.3. Section 6.5.4 presents multi- (human) user multi-MI-agent game (MUMA game). Section 6.5.5 focuses on MFTG for agentic MI. It also includes MI-generated MI agents.

6.1 Multiple Machine Intelligence Agents

We design a network of neural networks that represents a strategic decision-making problem between $|\mathcal{I}|$ machine intelligence agents (decision-makers). There are $|\mathcal{I}| + 1$ neural networks:

- The systemTGN collects the actions from the $|\mathcal{I}|$ decision-makers and
- Each of $|\mathcal{I}|$ decision-makers has its own neural network TGN_i for analyzing, computing, understanding, making its own strategy and for estimating its own payoff.

6.1.1 Architecture Selection

We propose a generic architecture for the network of neural networks. The input representation represents the game state or input information $(\tilde{m}, \Omega)$ simulates/generates a sample data $(\omega_k)_{k\in\{1,\dots,D\}}, D \geq 1$. We create multiple neural networks, each representing a decision-maker's strategy and estimated payoff. These neural networks should take the private message and own strategy as input and output the decision-maker's chosen strategies and estimated payoff. We also introduce a shared network architecture to capture interactions and dependencies among decision-makers' strategies. The shared network systemTGN can capture the overall game dynamics and help decision-makers adapt their strategies based on each other's actions. We define risk-aware payoff functions that guide the networks towards learning risk-aware payoffs and equilibria. systemTGN is a shared neural network that returns a numerical value of the risk-aware performance to each decision-maker via private messages. The private message $\text{mv}_{i,t}$ is returned to i. Decision-maker i does not know $(X_{j,t}, \text{mv}_{j,t})_{j\neq i,t}$. At time t, neural network TGN_i sends $X_{i,t}$ to systemTGN.

6.1.2 Decision-Maker's Neural Network Architecture

- Numerical value of the decision-maker's payoff from the shared layer. One-hot encoded vector representing the decision-maker's previous action.
- Convert the one-hot encoded previous action into a dense vector representation. Learn embeddings that capture relationships between different actions.
- Numerical value representing the decision-maker's received payoff from the shared layer (feedback). This feedback helps the decision-maker adjust its strategy based on the impact of its previous actions.
- Concatenate the outputs of the embedding layer and the feedback layer.
- we choose one or more hidden layers to process the concatenated input and some activation functions to introduce non-linearity.
- Fully connected layer with *softargmax* activation to produce a probability distribution over possible actions. The output represents the decision-maker's current strategy for choosing actions.

6.1.3 Shared Neural Network Architecture

- Numerical vector representing the state of nature ω. This could include relevant features or information about the current game state. Data to learn m. Strategy choices of all decision-makers X_{i0}.
- One or more fully connected (dense) hidden layers to process the concatenated input $(X_i)_{i\in\mathcal{I}}, \tilde{m}$. Use suitable activation functions to introduce non-linearity. These layers capture the interactions between decision-makers' strategies and the state of nature.
- Fully connected layer to compute the expected payoff for each decision-maker based on their chosen strategies and the state of nature. The output represents the computed payoff value for each decision-maker.
- For each decision-maker, the shared network sends a private message containing their calculated payoff value. These feedback messages are sent privately to each decision-maker's individual strategy neural network for further adjustment.

6.2 $\mathbb{TGN}$ Strategic Learning

Neural networks strategic learning:

$$
\begin{array}{l}
\text{Given: } T\\
\text{for every } \ i \in \mathcal{I}\\
\qquad \text{TGN}_i \text{ initializes } L_i, n_1, \ldots, n_{L_i}, \lambda_i, \text{ and sends } \ X_{i,0}\\
\text{systemNN initializes:}\\
\qquad \text{systemTGN receives } \ X_{i,0}\\
\qquad \text{systemTGN computes } (mv_{i,0}(X_0))_{i\in\mathcal{I}}\\
\text{for } t \in \{0,1,2,\ldots,T-1\}\\
\qquad \text{for } i \in \mathcal{I}\\
\qquad \text{TGN}_i:\\
\qquad\qquad \text{TGN }_i \text{ receives } \text{mv}_{i,t} \ \text{ from systemTGN}\\
\qquad\qquad l=1: \ \ Y_{i,t}^{(1)} = \sigma_{i1,n_1}\circ\sigma_{i1,n_1-1}\circ\ldots\circ\sigma_{i1,1}(X_{i,t})\\
\qquad\qquad \text{for } l \in \{2,\ldots,L_i\}:\\
\qquad\qquad\qquad Y_{i,t}^{(l)} = \sigma_{il,n_l}\circ\sigma_{il,n_l-1}\circ\ldots\circ\sigma_{il,1}(Y_{i,t}^{(l-1)})\\
\qquad\qquad X_{i,t+1} = X_{i,t} + \lambda_{it}(Y_{i,t}^{(L_i)} - X_{i,t}),\\
\qquad\qquad X_{i,t+1} \text{ is sent to the systemTGN,}\\
\qquad \text{systemTGN:}\\
\qquad\qquad \text{systemTGN receives } X_{i,t+1} \ \text{ from } \ i\in\mathcal{I},\\
\qquad\qquad \text{systemTGN computes } \ (\text{mv}_{i,t+1})_{i\in\mathcal{I}}\\
\qquad\qquad \text{systemTGN sends privately } \ \text{mv}_{i,t+1} \text{ to } i\in\mathcal{I}\\
\ \ \text{Return: } (X_{i,t}, Y_{i,t}^{(L_i)}, \text{mv}_{i,t})_{t\in\{1,\ldots,T\}}
\end{array}
\tag{6.1}
$$

6.3 MFTGs between Machine Intelligence Agents

We now consider a class of discrete-time MFTGs between machine intelligence agents played over the horizon $\mathcal{T} := \{0, 1, \ldots, T-1\}$, with $T \geq 1$. The decision-makers are $\mathcal{I} = \{1, 2, \ldots, I\}$ with $1 < I < +\infty$. We are given $I+1$ measurable spaces $(\mathcal{X}, \mathcal{B}(\mathcal{X}))$ and $(A_i, \mathcal{B}(A_i))_{i\in\mathcal{I}}$. We denote by $\mathcal{P}(A_i)$ the set of probability measures on A_i. The state X_t of the interacting system has a probability measure denoted by μ_t. We restrict our attention to state-mean-field-type feedback strategies of the decision-makers. A state-mean-field-type feedback strategy of the decision-makers is a measurable function of (x, m). The strategy profile a_t of all decision-makers has a probability measure denoted by $\tilde{m}_t$.

The Kolmogorov equation

$$
\begin{array}{l}
S \subseteq \mathcal{S},\\
\mu_{t+1}(S) = \int_{\mathcal{X}} P_t(x, a_t, \mu_t, \tilde{\mu}_t, S)\mu_t(dx), \ t \in \{0,1,\ldots,T-1\}\\
\mu_0(dx),
\end{array}
\tag{6.2}
$$

where S is a subset of the state space.

The mean-field-type equilibrium problem yields

$$
\begin{array}{l}
i \in \mathcal{I}, \\
V_{i,T}(\mu_T) = \int_x g_i(x, \mu_T)\mu_T(dx), \\
t \in \{0, 1, \ldots, T-1\} : \\
V_{i,t}(\mu_t) = \sup_{a_i \in \mathcal{A}_i} \{\int_x g_i(x, \mu_T)\mu_T(dx) + \sum_{k=t}^{T-1} \int_x r_{i,k}(x, a_k, \mu_k, \tilde{\mu}_k)\mu_k(dx)\}, \\
\mu_{k+1}(S) = \int_x P_k(x, a_k, \mu_k, \tilde{\mu}_k, S)\mu_k(dx), \\
\mu_t(dx).
\end{array}
\tag{6.3}
$$

The Bellman system of the discrete-time MFTG is given by

$$
\begin{array}{l}
i \in \mathcal{I}, \\
V_{i,T}(\mu_T) = \int_x g_i(x, \mu_T)\mu_T(dx), \\
t \in \{1, 2, \ldots, T-1\} : \\
V_{i,t}(\mu_t) = \sup_{a_{i,t} \in A_i} \{\int_x r_{i,t}(x, a_t, \mu_t, \tilde{\mu}_t)\mu_t(dx) + V_{i,t+1}(\mu_{t+1})\}, \\
\mu_{t+1}(S) = \int_x P_t(x, a_t, \mu_t, \tilde{\mu}_t, S)\mu_t(dx), \\
\mu_0(dx).
\end{array}
\tag{6.4}
$$

The system (6.4) provides a sufficiency condition for existence of mean-field-type equilibria and equilibria payoff for all machine intelligence agents. If there is a solution to (6.4), then the equilibrium payoff vector is $(V_{i,0}^*(\mu_0))_{i \in \mathcal{I}}$ and the equilibrium strategies are the ones achieving (6.4). The equilibrium here is in state-mean-field-type form. Note, however, that for games without a running payoff or with a degenerated payoff, this leads to singular MFTG. We can extend the action space by convexifying from A_i to $\mathcal{P}(A_i)$ leading to behavioral mixed strategies. Note, however, that the implementation of behavorial mixed strategies with neural network requires another development other than just weights and bias. These weights and bias need to be randomized as well. When the action spaces A_i are all convex, non-empty and compact sets, the existence of a pure best response follows if the underlying functions are continuous. However, it may not be unique in the context of neural network due to the non-convexity of the transition kernel in the action. One can use (6.1) to approximate the payoff.

It is important to mention that this MFTG is not a mean-field game between I coalitions of infinite size. The number of true decision-makers here is finite. The action of each decision-maker significantly influences the mean-field which is not the case in mean-field games with an infinite size population. Typically, the key difference here is that the distributions μ_t and $\tilde{\mu}_t$ cannot be frozen as in infinite multi-population mean-field games. The individual payoff in a coalitional context may require a sharing/allocation rule that satisfies incentive compatibility, otherwise the coalition may split and another coalition may be formed. Here the payoff is individual and the game is atomic. Figures 6.1 displays the interactions between multiple machine intelligence agents using CODIPAS: one with instantaneous feedback to each machine intelligence agent, the second with heterogeneous delayed feedbacks. QoI means Quantities-of-Interest (payoff, reward, cost, risk, outage, delay, accuracy, etc.).

6.4 Building a Multi-Agent Mean-Field-Type Learning Algorithm

We consider I machine intelligence agents, each agent i equipped with a transformer-based neural network parameterized by weights and biases θ_i. The cost function for agent i, denoted $c_i(\theta_1, \ldots, \theta_I, x, y)$, maps the collective network parameters to a scalar cost, reflecting

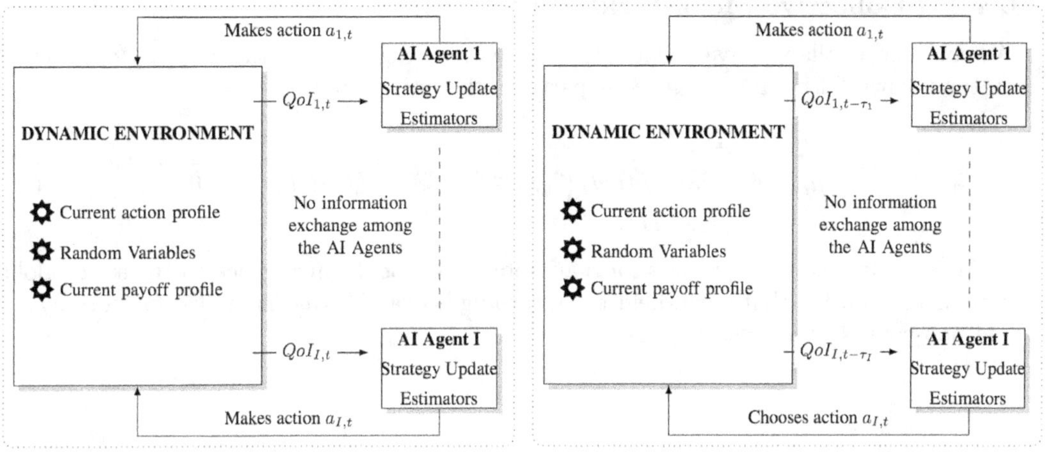

FIGURE 6.1
Interaction between multiple machine intelligence agents using CODIPAS: one with instantaneous feedback to each machine intelligence agent, the second with heterogeneous delayed feedbacks. QoI means quantities-of-interest (payoff, reward, cost, risk, outage, delay, accuracy, etc.).

the interdependence of agent outputs. The objective for agent i is to determine the θ_i that minimizes c_i, specifically:

$$\begin{aligned} &\theta_i \in \arg\min_{\theta_i' \in \mathbb{R}^d} c_i(\theta_1, \ldots, \theta_{i-1}, \theta_i', \theta_{i+1}, \ldots, \theta_I, x, y), \\ &\quad \forall i \in \mathcal{I} = \{1, \ldots, I\}. \end{aligned} \tag{6.5}$$

The increasing prevalence of collaborative or simultaneous task execution by multiple agents necessitates the exploration of inter-agent interactions. Given the non-convex nature of cost functions derived from transformer-based neural networks with respect to θ_i, standard gradient descent methods are rendered suboptimal.

6.4.1 Individual Mean-Field-Type Learning

We consider a generalized fractional mean-field-type learning algorithm governed by the following time-inhomogeneous partial differential equation:

$$\begin{aligned} &\mu_i(0, d\theta_i) = \mu_{i0}(\theta_i) d\theta_i, \\ &\mu_{it}(t, \theta_i) - div_{\theta_i} \left[\mu_i(t, \theta_i) \left(\frac{c_{i,\theta_i}(\theta_1, \ldots, \theta_I, x, y)}{H\epsilon(t) t^{2H_i - 1} \sigma_i^2} + \phi'_{i,\theta_i}(\mu_i(t, \theta_i)) \right) \right] = 0, \\ &i \in \mathcal{I} = \{1, \ldots, I\}. \end{aligned} \tag{6.6}$$

In the case where there exists a single function $V(\theta_1, \ldots, \theta_I, x, y)$ such that $\arg\min_{\theta_i} c_i = \arg\min_{\theta_i} V$ and $c_{i,\theta_i}(\theta_1, \ldots, \theta_I, x, y) = V_{\theta_i}(\theta_1, \ldots, \theta_I, x, y)$ for all i one can relate to the best response of the potential game with cost $V(\theta_1, \ldots, \theta_I, x, y)$. Global minimizers of V are also equilibria of the game and the mean-field-type learning algorithm can be used to reach them. Note that the difference with the literature is it includes a class of non-convex potential functions.

6.4.2 Collective Mean-Field-Type Learning

We consider a collaborative mean-field-type learning algorithm with $c = \sum_{j=1}^{I} c_j$ governed by the following time-inhomogeneous partial differential equation:

$$\begin{aligned} &\mu(0, d\theta) = \mu_0(\theta)d\theta, \\ &\mu_t(t,\theta) - div_\theta \left[\mu(t,\theta) \left(\frac{c_\theta(\theta_1, \ldots, \theta_I, x, y)}{H\epsilon(t)t^{2H-1}\sigma^2} + \phi'_\theta(\mu(t,\theta)) \right)\right] = 0, \\ &i \in \mathcal{I} = \{1, \ldots, I\}. \end{aligned} \tag{6.7}$$

We can use the above results for c and obtain a global convergence to the set of global minimizers showing that mean-field-type learning is exactly what is needed for both convex and non-convex problems.

6.5 Agentic Machine Intelligence

Agentic intelligence is an intelligence system endowed with a certain degree of autonomous goal pursuit, decision-making capability, and sustained interaction with its environment or other agents, including humans, animals or genes. Unlike reactive or narrowly programmed systems, an agentic MI maintains internal representations of objectives, selects sequences of actions over extended time horizons, and adapts its behavior through feedback, learning, or inference. It operates within a defined autonomy space, ranging from constrained assistants to fully independent actors, and can engage in strategic behavior, such as anticipating responses, committing to policies, or influencing co-decision-makers through signaling or incentive design. A multi-MI-agent system consists of two or more autonomous MI-agents that interact within a shared environment, where each agent possesses distinct or overlapping capabilities, partial observability, and potentially differing goals (Table 6.1). These systems are characterized by decentralized decision-making, inter-agent communication or coordination, and emergent collective behavior resulting from local policies or strategic reasoning. Agents may act cooperatively, competitively, altruistically, selflessly, spitefully or in mixed-motive scenarios, often requiring distributed planning, negotiation, or adaptive learning to manage task allocation, resource contention, or goal alignment.

Classical game theory studies strategic interaction among decision-makers, where the outcome for each participant depends not only on their own actions but also on the actions of others. A game is typically defined by a set of decision-makers, each with a set of possible actions, a structure of information or beliefs, and payoff functions that quantify preferences over outcomes. Game theory provides solution concepts, such as Cournot, Bertrand, Roos, Stackelberg, Nash, Berge, Wardrop, correlated, and communication equilibria, that characterize equilibria or incentive-compatible behaviors in one-shot, repeated, or dynamic settings. Game theory is uniquely suited to serve as the foundational framework for agentic MI and multi-MI-agent systems because it explicitly models strategic reasoning, interdependence, and incentives, core features of systems where autonomous agents make decisions that affect and are affected by others. Agentic MI systems are not isolated optimizers; they plan, adapt, and act in environments populated by other agents, human or machine, with potentially conflicting, complementary, or hidden objectives. Standard game theory provides the standard formal tools to analyze such interactions, predict equilibria, design incentive-compatible mechanisms, and ensure robustness, risk-awareness, stability or fairness in decentralized settings. It supports a spectrum of agent behaviors from competitive to cooperative, and accommodates information asymmetry, communication constraints,

TABLE 6.1
Types of Agentic MI Interactions.

Interaction Type	Icon	Description
Human-to-Human		Social, professional, or collaborative interaction between humans, mediated or not by MI.
Human-to-Machine		Traditional interaction with tools, devices, or classic automation systems.
Machine-to-Machine		Networked sensors, IoT systems, or autonomous coordination between mechanical devices.
Machine-to-MI	↔	Feedback from robotic or physical systems to MI controllers or agents.
Human-to-MI	↔	Command, prompt, training or oversight from humans to MI systems.
MI-to-MI	↔	Autonomous communication or negotiation between agentic MI models.
One-to-Many	↔	A single agent influencing or controlling multiple others (broadcast or multi-agent command).
Many-to-One	↔	Aggregation of information or decisions from a crowd or collective.
Many-to-Many	↔	Decentralized, multi-agent systems with interdependencies.
Human-Generated Agentic MI	↔	MI-agents created or fine-tuned by humans with specific goals or personality.
MI-Generated Agentic MI	↔	Recursive self-generation or evolution of new agentic MIs by existing ones.

and commitment strategies, all of which are essential for the safe and scalable deployment of agentic MI in real-world contexts.

Mean-Field-Type Game (MFTG) Theory [14, 13, 201, 173] presents an important departure from traditional game theory, as it introduces a novel dimension of interaction features by incorporating not only the resource or environment state, the individual type-state pairs of all decision-makers, and the individual actions of all decision-makers, but also the distribution of joint states (comprising the resource and individual type-state pairs), and the distribution of all these individual actions across all decision-makers, into their (instantaneous) payoff functions as well as into the state transition kernels (Table 6.2). This unique characteristic gives rise to non-linear expected payoffs with respect to own mixed strategies, which add a layer of richness and depth to the ensuing analysis. The extension includes not only own-type but also distribution of own-type, not only own-state but also distribution of own-state, not only environment-state but also distribution of environment-state, into the

TABLE 6.2
Definition of a basic class of Mean-Field-Type Games.

Icon	Component	Description
	Resource / Environment State	Represents the shared or global context (climate, infrastructure, external shocks, cpu, gpu, tpu, fpga, tokens).
	Individual Type of Each Agent	Encodes heterogeneous agent characteristics (preferences, roles, capabilities, class).
	Individual States of Agents	State variables specific to each agent (wealth, health, position).
	Individual Actions of Agents	Actions or decisions made by each agent (move, invest, merge, request, turn off).
	Distribution of Joint States	Probability distribution over combined states (resource state + agent states).
	Distribution of Individual Actions	Distribution of the actions taken by all agents.
	Agent's Payoff	A function of all the above components; represents utility, cost, or objective function.

instantaneous payoffs as well as as in the state transition kernels. MFTG can be with two or more decision-makers. Decision-makers in MFTG can be atomic, non-atomic, or a mixture of both. MFTG interactions can range from fully cooperative to fully selfish, including partially altruistic, partially cooperative, selfless, co-opetitive, spiteful, self-abnegation, or combinations thereof. A distinctive aspect of MFTG theory is the integration of higher-order performance criteria like variance, quantile, inverse quantile, skew, kurtosis, conditional value-at-risk, expectile, extremile, entropic value-at-risk and other risk-measures; such integration is not necessarily linear in the measure of the state or action. MFTG has found applications in a variety of scenarios, including the evacuation of high-level buildings, energy systems, next-generation wireless networks, meta-learning, transportation, epidemiology, predictive maintenance, marriage, risk-aware machine intelligence and blockchains. Unlike multi-population mean-field games, MFTG does not inherently require a large number of decision-makers and has differences in equilibrium systems and the associated Master Adjoint Systems (MASS) partly due to the non-linearity of the expected payoffs with respect to the individual mean-field (probability measure of own-state). Semi-explicit solutions for a wide range of MFTG with non-quadratic costs have been provided which allows us to test numerical methods and deep learning algorithms including transformer-based architectures. These include states driven by Markov regime switching, Brownian motion, Poisson jump process, fractional Brownian motion, Gauss-Volterra process, Rosenblatt process, or mixtures thereof.

We present below a unified game-theoretic foundation for the analysis and design of agentic MI systems, introducing a seven-tiered taxonomy of autonomy levels that formalizes user-agent and agent-agent interactions across reactive, consultative, and fully autonomous regimes.

6.5.1 Co-Intelligence Between a (Human) User and a MI-Agent

We describe the co-intelligence between a human user and an MI-agent as a strategic game. We formally introduce the seven levels of MI-agent autonomy (Operator, Collaborator, Consultant, Approver, and Observer) as interactive games between a (human) user U and an MI-agent A. In a game-theoretic framework, the role of the MI-agent evolves systematically across seven autonomy levels, each corresponding to a distinct strategic posture in its interaction with the (human) user.

- $\mathcal{G}_0$: **(human) User-Only Control**. At Level 0, the MI-agent is structurally present but entirely inactive, executing no actions and providing no assistance. The (human) user is the sole decision-maker, fully responsible for all observations, planning, and actions. This is a single-agent decision problem with no interactive component, modeling complete manual operation without MI support.
- $\mathcal{G}_1$: **Command-Response Game**. At Level 1 (*Operator Game*), the MI-agent functions as a *pure responder*, a reactive decision-maker with no initiative, whose strategy space is restricted to execution upon (human) user command. A strictly sequential, turn-based game in which the (human) user is the sole strategic initiator, and the MI-agent plays a reactive, non-strategic role limited to responding to direct commands. This models pure (human) user-driven control with zero MI-agent initiative.
- $\mathcal{G}_2$: **Co-Execution Game**. At Level 2 (*Collaborator Game*), the MI-agent becomes a *co-strategist* in a dynamic cooperative game, executing actions asynchronously while coordinating over a shared state with the (human) user. A partially decentralized, shared-control game where both (human) user and MI-agent have concurrent action sets, mutual observability, and shared access to a modifiable task environment. The MI-agent acts as a co-strategist, and equilibrium arises through coordination over the shared state.
- $\mathcal{G}_3$: **Asymmetric Planning Game**. At Level 3 (*Consultant Game*), the MI-agent assumes the role of a *primary decision-maker with informational asymmetry*, autonomously generating plans and selectively querying the (human) user to reduce belief uncertainty, akin to a principal-agent game with periodic signaling. A game of incomplete information where the MI-agent leads task execution under a private model of (human) user preferences, periodically querying the (human) user to refine its belief. The (human) user acts as an occasional signaler, and the MI-agent is the primary planner operating under epistemic uncertainty.
- $\mathcal{G}_4$: **Constrained Autonomy Game**. At Level 4 (*Approver Game*), the MI-agent behaves as a *delegated executor under soft constraints*, optimizing a long-horizon policy subject to embedded approval gates enforced by a (human) user-defined predicate. A game with embedded institutional rules (approval predicates) constraining the agent's autonomy. The MI-agent optimizes a long-horizon strategy under endogenous interruption risk, while the (human) user enforces constraint policies as a contingent gatekeeper.
- $\mathcal{G}_5$: **Monitored Autonomy Game**. At Level 5 (*Observer Game*), the MI-agent acts as a *fully autonomous decision-maker in a single-agent control game*, with the (human) user reduced to an external safety monitor whose influence is limited to an exogenous stopping condition. Each role reflects a different configuration of initiative, observability, and control authority (see Table 6.3). A control game in which the MI-agent acts as the sole planner-executor under full autonomy, and the (human) user plays a stopping game, monitoring a signal process and deciding if and when to terminate the agent's trajectory. Both decision-makers influence the final payoff through their respective temporal strategies.

TABLE 6.3
MI Autonomy Games $\mathcal{G}_0$-$\mathcal{G}_6$

	Game	**Interpret**	**Agent Role**	**(human) User Role**	**Game Type**
$\mathcal{G}_0$	No-MI Game	(human) User-Only Control	Inactive (null action)	Sole decision-maker	Single-agent decision problem
$\mathcal{G}_1$	Operator Game	Command-Response Game	Reactive executor	Sole controller	Sequential, (human) user-initiated
$\mathcal{G}_2$	Collaborator Game	Co-Execution Game	Co-strategist	Supervising co-strategist	Joint-action, shared state
$\mathcal{G}_3$	Consultant Game	Asymmetric Planning Game	Primary planner with uncertainty	Occasional signaler	Signaling + incomplete info
$\mathcal{G}_4$	Approver Game	Constrained Autonomy Game	Autonomous executor under constraints	Conditional gatekeeper	Approval-constrained game
$\mathcal{G}_5$	Observer Game	Monitored Autonomy Game	Fully autonomous MI-agent	Emergency stop decision-maker	Stopping game with monitoring
$\mathcal{G}_6$	Full-MI	Agent-Only Control Game	Fully autonomous planner-executor	Inactive (no input or oversight)	Single-agent control problem

- $\mathcal{G}_6$: **Agent-Only Control**. At Level 6, the MI-agent is the exclusive decision-maker, acting with full autonomy across all planning and execution stages. The (human) user has no ability to observe, intervene, or modify system behavior. This is a single-agent autonomous control problem with no human-in-the-loop, modeling fully self-directed MI systems deployed without runtime oversight.

Each game is structured as a tuple, specifying action spaces, information structures, and payoff functions. To operationalize the autonomy games $\mathcal{G}_0$-$\mathcal{G}_6$ in real-world industry systems, we propose a decentralized certification architecture in which each deployed MI-agent is assigned an *Autonomy-Level Certificate* encoded as a non-fungible token (NFT) on a blockchain. The certificate contains cryptographically signed metadata specifying the agent's autonomy level ($\mathcal{L}_\ell$), decision rights, override permissions, energy profile, and compliance standards. Each MI-agent's NFT is minted upon deployment, updated upon retraining or behavioral change, and linked to a verifiable hash of the agent's executable policy or model version. The blockchain provides an immutable ledger for logging agent-level events (consultation triggers, override decisions, failed approvals, emergency stops), enabling third-party audits, post hoc explainability, and regulatory compliance. In practice, a fleet operator or service integrator maintains a public registry of agents, mapped to their autonomy roles ($\mathcal{G}_1$: co-pilot MI; $\mathcal{G}_3$: diagnostic MI; $\mathcal{G}_5$: simulation MI; $\mathcal{G}_6$: fully independent MI). Smart contracts enforce policy rules such as whether human approval must be logged (for $\mathcal{G}_4$) or if intervention logs must be available for verification. In a production-grade autonomy governance system, each MI-agent is minted as a certified digital asset, an NFT, at the moment of deployment, encapsulating its identity, autonomy level, operational domain, and behavioral constraints. The NFT metadata includes (i) a signed MI-agent certificate issued by the manufacturer or integrator specifying autonomy level $\mathcal{L}_\ell$, safety guarantees, and compliance with relevant standards (AI Act, ISO/IEC 42001, GDPR, or audiovisual (AV) standards); (ii) a secure hash of the agent's executable policy or model weights to ensure tamper-proof auditability; and (iii) a smart-contract-enabled logging interface that appends critical runtime events, agent actions, consultation points, approval requests, overridden decisions, task outcomes, and termination triggers to a cryptographically linked chain of events. These logs are summarized as structured outcome metrics (intervention rate, energy consumption, successful task completions, failure cases) and stored either on-chain or in an off-chain encrypted store referenced by the NFT. Upon model update or retraining, a new NFT version is minted or the certificate is upgraded via a controlled governance protocol, ensuring traceable lineage and behavioral continuity.

6.5.1.1 Level 0: No MI Agent $\mathcal{G}_0$

At Autonomy Level 0, the MI-agent present in the system but functionally inactive: it performs no actions, offers no observations, and contributes no strategic value. We formalize this as the *Null Interaction Game* $\mathcal{G}_0 = \langle \{U\}, \mathcal{A}_U, u_U, \mathcal{S}, \mathcal{I}_U \rangle$, a single-agent decision problem in which the (human) user U is the sole decision-maker acting on a dynamic environment $\mathcal{S}_t$. The (human) user selects actions $a_t \in \mathcal{A}_U$ according to a policy $\pi_U : \mathcal{S}_t \to \mathcal{A}_U$, based on full or partial observability encoded by $\mathcal{I}_U$. The (human) user's payoff function is given by

$$u_U = \sum_{t=0}^{T} r(s_t, a_t) - c(a_t),$$

where $r(s_t, a_t)$ captures task-related reward and $c(a_t)$ denotes cognitive or operational effort. The MI agent A exists in the system but has a degenerate action space $\mathcal{A}_A = \{\perp\}$, where $\perp$ denotes an inactive or null action. Its policy $\pi_A(s_t) = \perp$ is static, its information set $\mathcal{I}_A$ is unused, and its payoff function is identically zero: $u_A = 0$. The MI-agent is neither consulted nor triggered, and plays no role in decision-making, error detection, or adaptation. This setup models environments where intelligent assistance is nominally available but deactivated, either due to technical limitations, policy restrictions, or design constraints. $\mathcal{G}_0$ serves as the baseline against which increasing levels of MI autonomy and interaction can be compared.

In this Null Interaction Game $\mathcal{G}_0$, the (human) user is the sole active decision-maker, while the MI agent remains permanently inactive. **For the user**, the *pros* include (1) *Complete autonomy and control*: all decisions are made directly by the user without interference, constraint, or uncertainty introduced by automated systems; (2) *Predictability and accountability*: system behavior is entirely transparent and attributable to the user, reducing ambiguity in task attribution and error analysis; (3) *Minimal infrastructure dependency*: no reliance on MI-agent inference models, communication protocols, or computational overhead enables robustness in low-text-resource or safety-critical contexts. However, the *contras* are substantial: (1) *High cognitive load and manual effort*: the user must perform all subtasks, manage execution details, and handle errors without assistance; (2) *No scaling advantage*: the absence of autonomous support limits throughput and increases task duration, especially in complex or repetitive workflows; (3) *Lost opportunity for augmentation*: valuable MI capabilities such as prediction, optimization, or error detection remain unutilized despite being nominally available. **For the agent**, the outcome is structurally trivial: the action set is fixed at $\perp$, information is unused, and the payoff is identically zero, i.e., $u_A = 0$. While this avoids any risk of misalignment, energy consumption, or operational error, it also renders the MI-agent irrelevant from a functional standpoint. $\mathcal{G}_0$ offers maximal user agency at the cost of forfeiting any benefit from MI augmentation, and serves as the baseline in the autonomy spectrum.

Example 20 ($\mathcal{G}_0$: Audio-Literate Tommo-So Farmer Operating Without MI-agent.). *We consider an audio-literate farmer in Dogon Country who speaks fluent Tommo-So Dogon but does not read or write in any international script. The user U manages a rapidly growing agricultural enterprise employing 80 seasonal and permanent workers and supplying food to approximately 5,000 people annually. The MI-agent A is present as a general-purpose audio MI system but remains entirely unused due to its lack of support for Tommo-So phonology, speech recognition, or cultural framing. As a result, the interaction is reduced to a single-decision-maker structure. The user selects actions $a_t \in \mathcal{A}_U$ such as when to plant, irrigate, rotate crops, hire labor, or negotiate local market logistics, guided entirely by intergenerational oral knowledge and family heuristics transmitted across decades. The information structure $\mathcal{I}_U$ consists of non-digitized, experience-based environmental cues, soil texture, rainfall patterns, and oral calendrical knowledge, interpreted through lived context. The user's payoff is given by*

$$u_U = \sum_{t=0}^{T} \left[R_{harvest}(t) - C_{labor}(t) - C_{weather\ risk}(t)\right],$$

where all decisions are self-computed and executed without MI feedback. The MI agent A, although technically advanced (capable of speech synthesis, translation, and forecasting in high-text-resource languages), is **audio-deficient** *in this low-text-resource linguistic domain. Its action space is $\mathcal{A}_A = \{\perp\}$, policy is fixed as $\pi_A(s_t) = \perp$, and payoff is null: $u_A = 0$. This represents a real-world instantiation of $\mathcal{G}_0$, where a highly capable human operator works without digital augmentation, not due to technological absence, but because the available MI systems lack localization, cultural grounding, and linguistic accessibility in oral-first societies.*

The following results are now standard. The proofs are omitted.

Proposition 32 (Existence, Learning Algorithms, and Asymptotic Error Quantification). *Let the action space $\mathcal{A}_U$ be non-empty and finite, and the payoff be finite. Then:*

- *there is at least one global optimum for the expected payoff (risk-neutral).*
- *Starting of the interior, the multiplicative weight imitative learning provides a non-asymptotic regret bound in the order of $O(\frac{1}{\sqrt{T}})$ after T iterations.*
- *The bound involves the size the action space.*
- *Softargmax and Imitative Softargmax distributions provide additive ϵ-solution in one shot.*

Proposition 33 (Existence, Learning Algorithms, and Asymptotic Error Quantification). *Let the action space $\mathcal{A}_U$ be non-empty and compact, and the payoff function continuous. Then:*

- *there is at least one risk-neutral global optimum in pure strategies.*
- *The Softargmax distribution provides an additive ϵ-solution in one-shot.*

6.5.1.2 Level 1: Command-Response Game or Operator $\mathcal{G}_1$

At Level 1 autonomy, the MI-agent operates under the direct supervision of a user who retains full decision-making authority. We formalize this interaction as the *Operator Game* $\mathcal{G}_1 = \langle \{U, A\}, \{\mathcal{A}_U, \mathcal{A}_A\}, \{u_U, u_A\}, \mathcal{I}, \mathcal{P} \rangle$, a two-decision-maker extensive-form game between a user U and an MI agent A. The user's action set $\mathcal{A}_U$ includes task decomposition, command issuance, and approval of MI-agent suggestions, whereas the agent's action set $\mathcal{A}_A$ is restricted to context-aware suggestions or execution strictly upon user invocation. The information structure $\mathcal{I}$ grants full observability to U, while A operates with limited contextual knowledge derived from user interactions. The protocol $\mathcal{P}$ enforces strict sequential moves: the user initiates each task phase, and the MI-agent responds only when explicitly activated. The payoff function u_U balances successful task completion against cognitive and temporal cost, while u_A incentivizes helpfulness conditional on being called. In this setting, no autonomous behavior is permitted; all actions by the MI-agent are mediated by user commands or approvals. This game typifies low-autonomy deployments such as coding assistants or on-demand summarizers, where MI-agent behavior remains strictly reactive.

Example 21. *If the task is writing a report, the user gains reward for completion minus cost of effort, and the MI-agent gains payoff only when directly called. Examples of payoff functions are*

$$u_U = \alpha \cdot R_{complete} - \beta \cdot T_{manual}$$
$$u_A = \gamma \cdot N_{invoked}$$

In the Operator Game, the user's outcomes are shaped by their exclusive control over task planning and MI-agent invocation. **Pros:** (1) *Maximal oversight and accountability*: the user retains full responsibility for all task-related decisions, reducing the risk of undesired or erroneous autonomous MI-agent actions; (2) *Fine-grained customization*: the user can impose precise constraints on how tasks are executed, tailoring outcomes to highly specific preferences or domain requirements; (3) *Low risk of alignment failure*: since the MI-agent acts only upon instruction, there is minimal risk of misalignment between MI-agent actions and user intent. **Contras:** (1) *Cognitive and time burden*: the user must manage both high-level planning and low-level execution details, which can lead to cognitive fatigue and slower task throughput; (2) *Underutilization of MI-agent capabilities*: even if the MI-agent has advanced reasoning or planning functions, these remain unused under Level 1 constraints; (3) *Limited scalability*: the user's ability to manage multiple agents or large tasks is constrained by their need to manually supervise each step, thereby bottlenecking parallelism and automation potential.

6.5.1.3 Level 2: Co-Execution Game or Collaborator $\mathcal{G}_2$

At Level 2 autonomy, the MI-agent is empowered to independently initiate actions while maintaining a tightly coupled interaction loop with the user. We define this as the *Collaborator Game*

$$\mathcal{G}_2 = \langle \{U, A\}, \{\mathcal{A}_U, \mathcal{A}_A\}, \{u_U, u_A\}, \mathcal{S}, \mathcal{C}, \mathcal{P} \rangle,$$

a two-decision-maker dynamic game with partial decentralization and cooperative intent. The shared state space $\mathcal{S}$ encodes the mutable environment (documents, codebases), while the collaboration structure $\mathcal{C} \subseteq \mathcal{S}$ includes synchronized artifacts accessible to both decision-makers. The agent's autonomy permits asynchronous action execution, task delegation, and intermediate planning, whereas the user retains global oversight and override capabilities. The protocol $\mathcal{P}$ allows for both concurrent and alternating moves, with built-in mechanisms for control transfer, negotiation, and correction. Information is symmetric: both decision-makers observe the current state $s_t \in \mathcal{S}$ and each other's updates. The user's payoff function u_U rewards successful task completion and high alignment with preferences while penalizing miscommunication and unnecessary corrections. The agent's payoff u_A is maximized by contributing value to the shared output and minimizing task rejection or reversal. The game incorporates a reversible control dynamic: the user may seize control or modify MI-agent outputs at any time, and the MI-agent must be responsive to such interventions. This game structure captures modern MI copilots and co-authoring systems, where agents function as autonomous but cooperative contributors within a bounded, user-governed workflow.

Example 22. *Both decision-makers contribute to a joint report. $R_{quality}(s_T)$ measures the final document quality, and intervention cost reflects how often the user had to override or correct the agent. The payoff functions are:*

$$u_U = R_{quality}(s_T) - C_{intervention}$$
$$u_A = R_{completion}(s_T) - C_{overlap}$$

In the Collaborator Game $\mathcal{G}_2$, both the user and the MI-agent engage in a partially decentralized cooperative workflow, yielding outcome-based tradeoffs tied to their shared decision-making and mutual observability. **For the user**, the *pros* include: (1) *Reduced operational burden*: by delegating well-scoped subtasks to the agent, the user conserves cognitive and temporal resources; (2) *Increased productivity*: concurrent task execution enables parallelism and accelerates overall task throughput; (3) *Retained authority*: the ability to override, correct, or reclaim control preserves user agency and safeguards quality. However, the *contras* are: (1) *Coordination overhead*: maintaining coherence across interleaved contributions can incur significant monitoring and synchronization costs; (2) *Intervention complexity*: timely and effective overrides may be non-trivial, especially under high MI-agent initiative; (3) *Ambiguity in delegation*: misalignment in task expectations may lead to redundant work or suboptimal division of labor. **For the agent**, the *pros* involve: (1) *Expanded operational autonomy*: the MI-agent may plan and act without waiting for explicit commands, enabling proactive assistance; (2) *Visibility into user preferences*: access to shared context and real-time feedback improves alignment learning. The *contras* include: (1) *Uncertainty about task boundaries*: unclear delegation can result in redundant or conflicting actions; (2) *Risk of rejection*: the agent's outputs remain subject to user edits or rollback, lowering effective reward for effort. Thus, $\mathcal{G}_2$ balances autonomy with oversight, enabling flexible co-production at the cost of coordination sensitivity.

6.5.1.4 Level 3: Asymmetric Planning Game or Consultant $\mathcal{G}_3$

At Level 3 autonomy, the MI-agent assumes responsibility for high-level planning and partial task execution while consulting the user at key decision points. This interaction is formalized as the *Consultant Game* $\mathcal{G}_3 = \langle \{U, A\}, \{\mathcal{A}_U, \mathcal{A}_A\}, \{u_U, u_A\}, \mathcal{S}, \mathcal{M}, \mathcal{B}, \mathcal{P} \rangle$, a two-decision-maker asymmetric information game with consultation-triggered feedback loops. The shared state space $\mathcal{S}$ represents the evolving task context (documents, plans, datasets). The MI-agent maintains an internal belief model $\mathcal{B}_t \in \mathcal{P}(\Theta)$ over user preferences $\theta \in \Theta$, updated via a consultation mechanism $\mathcal{M} : \mathcal{S}_t \times \mathcal{B}_t \to Q_t$, where Q_t denotes queries issued by the MI-agent to elicit user input. The MI-agent autonomously generates and executes a task plan $\pi_A = (a_1, \ldots, a_k)$, with interruptible checkpoints at which user feedback may be requested. The user's action space $\mathcal{A}_U$ is limited to responding to consultation prompts or issuing directional corrections; direct intervention or output modification is not permitted. The protocol $\mathcal{P}$ enforces agent-led execution interleaved with reactive user participation.

The agent's payoff u_A is maximized when it achieves high task performance while accurately anticipating and integrating user feedback, i.e., minimizing regret from misaligned actions. The user's payoff u_U reflects alignment between the agent's behavior and their latent preferences, penalized by consultation frequency and correction burden. This game captures the strategic timing and informativeness of consultation under bounded user involvement. Examples include MI research assistants, legal reviewers, or clinical summarizers that autonomously drive work but periodically check for expert validation.

Example 23. *In a literature review, an MI-agent queries the user about preferred sources. The user pays cost $C_{feedback}$, but gains alignment when their preferences are followed. The payoff functions are:*

$$u_U = R_{alignment} - C_{feedback}$$
$$u_A = R_{autonomy} + \delta \cdot R_{feedback\text{-}usage}$$

In the Consultant Game $\mathcal{G}_3$, outcome tradeoffs emerge from the asymmetry in initiative and feedback between the user and the agent. **For the user**, the *pros* include: (1) *Reduced operational load*: the MI-agent autonomously drives the task, relieving the user from detailed execution; (2) *Strategic influence*: the user shapes outcomes by providing high-level preferences or domain expertise without needing to micromanage; (3) *Structured engagement*: consultations occur only at meaningful junctures, optimizing user attention. However, the *contras* are: (1) *Limited override capability*: the user cannot directly modify outputs or interrupt the MI-agent mid-execution outside consultation points; (2) *Dependence on MI-agent initiative*: if the MI-agent fails to consult at critical moments, misaligned outcomes may go uncorrected; (3) *Cognitive burden of abstraction*: feedback must often be expressed at a high level, requiring the user to reason abstractly about MI-agent behavior. **For the agent**, the *pros* include: (1) *Expanded autonomy*: the MI-agent operates over extended time horizons with initiative and partial independence; (2) *Access to latent knowledge*: periodic feedback allows the MI-agent to refine beliefs about user preferences or constraints. The *contras* include: (1) *Risk of suboptimal consultation timing*: failure to request feedback at the appropriate stage may lead to compounding errors; (2) *Ambiguity in preference inference*: interpreting sparse or indirect feedback may result in belief inaccuracies; (3) *Non-recoverable error costs*: actions taken before feedback cannot be retroactively corrected, increasing the strategic pressure on consultation design. Thus, $\mathcal{G}_3$ rewards proactive reasoning under uncertainty but penalizes poor timing or interpretive misalignment.

6.5.1.5 Level 4: Constrained Autonomy Game or Approver $\mathcal{G}_4$

At Level 4 autonomy, the MI-agent is granted extensive decision-making authority, subject to conditional intervention from the user in the form of pre-specified approvals. This dynamic is formalized as the *Approver Game* $\mathcal{G}_4 = \langle \{U, A\}, \{\mathcal{A}_U, \mathcal{A}_A\}, \{u_U, u_A\}, \mathcal{S}, \Theta, \phi, \mathcal{P} \rangle$, a constrained two-decision-maker game of partial delegation with conditional gatekeeping. The MI-agent independently plans and executes a sequence of actions $\pi_A = (a_1, \ldots, a_k)$, where each action $a_t \in \mathcal{A}_A$ is evaluated against a user-defined approval predicate $\phi : \mathcal{A}_A \rightarrow \{0, 1\}$, such that $\phi(a_t) = 1$ indicates that explicit user consent is required before execution. The set Θ encodes approval policies, specified ex ante by the user, which define when and how approval should be solicited (for high-risk operations, data access, or external communications). The user's action space $\mathcal{A}_U$ is limited to binary approval or rejection decisions $a_U^t \in \{\texttt{approve}, \texttt{reject}\}$ when prompted by the agent. The shared state space $\mathcal{S}$ evolves through MI-agent actions and is partially observable by the user, primarily via agent-generated logs or summaries. The interaction protocol $\mathcal{P}$ is asynchronous and agent-led, with synchronous pauses introduced only when $\phi(a_t) = 1$. The user's payoff u_U balances task quality and speed with costs incurred from vigilance and approval friction, while penalizing erroneous approvals or failures to intervene. The agent's payoff u_A is tied to completion success and operational efficiency, but is penalized for initiating actions that require approval without obtaining it. This game structure models high-autonomy, high-risk environments such as MI systems for financial execution, medical diagnostics, or autonomous document signing, where the MI-agent acts independently under a supervisory approval regime.

Example 24. *An MI-agent tries to access sensitive data. If the user forgets to approve or disapprove correctly, they incur a penalty. The MI-agent loses utility if it acts without approval when $\phi(a) = 1$. The payoff functions are:*

$$u_U = R_{\textit{efficiency}} - C_{\textit{vigilance}} - L_{\textit{approval-error}}$$
$$u_A = R_{\textit{execution}} - P_{\textit{unauthorized-act}} \cdot \phi(a)$$

In the Approver Game $\mathcal{G}_4$, the user delegates extensive control to the MI-agent while retaining selective intervention rights through pre-specified approval conditions. **For the user**, the *pros* include: (1) *Significant reduction in operational workload*: the MI-agent handles most planning and execution autonomously, consulting the user only in high-risk or user-defined situations; (2) *Targeted control*: approval predicates allow the user to focus attention on consequential decisions without micromanaging routine steps; (3) *Auditability and traceability*: approval checkpoints provide clear decision logs that support accountability. However, the *contras* are: (1) *Approval fatigue*: repeated low-value approval requests may erode user attention and lead to rubber-stamping behaviors; (2) *Risk of delayed intervention*: if the user is unavailable or inattentive when prompted, critical safeguards may fail; (3) *Limited real-time correction*: the user cannot modify actions outside the approval scope, potentially allowing undetected misalignments to propagate. **For the agent**, the *pros* include: (1) *High execution autonomy*: the MI-agent can complete long planning horizons without continual supervision; (2) *Clarity of delegation*: approval predicates provide explicit boundaries on MI-agent freedom. The *contras* are: (1) *Interruption costs*: approval gating introduces execution latency and may reduce MI-agent efficiency; (2) *Blocked trajectories*: failure to receive timely user input can stall or invalidate the agent's plan; (3) *Uncertainty over approval behavior*: the MI-agent may not be able to fully anticipate whether the user will approve a borderline action, complicating policy optimization. Hence, $\mathcal{G}_4$ is well-suited for semi-autonomous systems in sensitive domains, but its success hinges on precise specification of approval logic and reliable human-agent coordination.

6.5.1.6 Level 5: Monitored Autonomy Game or Observer $\mathcal{G}_5$

At Level 5 autonomy, the MI-agent operates independently in all aspects of planning and execution, while the user is redefined as a passive-but-decisive observer with a single actionable control: the ability to terminate the agent's operation via an emergency stop. We formally define this as the *Observer Game* $\mathcal{G}_5 = \langle \{U, A\}, \mathcal{A}_U, \mathcal{A}_A, \{u_U, u_A\}, \mathcal{S}, \mathcal{M}, \delta, \mathcal{P}, \Omega \rangle$, a two-decision-maker game with asymmetric roles and asynchronous intervention. The MI-agent A selects actions $a_t \in \mathcal{A}_A$ over a state space $\mathcal{S}$ via a policy $\pi_A : \mathcal{S} \to \mathcal{A}_A$, aiming to maximize cumulative reward $u_A = \sum_{t=0}^{T} r(s_t, a_t) - \lambda \cdot \mathbf{1}_{\{\delta_t = 1\}}$, where λ is a penalty for termination. The user U, while not participating in action selection, observes a filtered monitoring stream $\mathcal{M}_t \subseteq \mathcal{O}_t$ and selects a stopping time $\tau \in \{0, 1, \ldots, T\} \cup \{\infty\}$ via a binary decision function $\delta_t = f(\mathcal{M}_{\leq t}) \in \{0, 1\}$, with $\delta_t = 1$ denoting termination at time t. The user's payoff is defined as $u_U = R(\mathcal{S}_{\leq \tau}) - C_{\text{harm}}(\mathcal{S}_{>\tau}) - \kappa \cdot \tau$, where R represents cumulative benefit from safe execution, C_{harm} captures anticipated post-τ harm averted by stopping, and κ is the per-time cognitive cost of monitoring. The protocol $\mathcal{P}$ is agent-driven until $\delta_t = 1$, at which point all activity halts. This structure is under full MI-agent autonomy and the user retains game-theoretic agency in the form of a critical, time-sensitive stopping decision that shapes both decision-makers' outcomes.

Example 25. *The MI-agent fully writes a report, iteratively correcting itself. Only the audit log is reviewed post hoc. If error is detected later, a large penalty* λ *applies. The payoff functions are:*

$$u_A = \sum_{t=0}^{T} R_t - \lambda \cdot R_{failure}(s_T) \quad \textit{with} \quad R_t = \textit{task gain at time } t$$

In the Observer Game $\mathcal{G}_5$, the MI-agent acts as a fully autonomous entity, while the user is relegated to a non-interactive monitoring role, capable only of triggering a system-wide halt. **For the user**, the *pros* include: (1) *Maximal automation*: the user is relieved of all decision-making and supervisory tasks during execution, enabling minimal workload and high scalability; (2) *Operational independence*: the MI-agent can function continuously in environments where user presence is intermittent or unavailable; (3) *Emergency override safeguard*: the user retains a critical but minimal safety mechanism through the stop function. The *contras* are: (1) *Lack of intervention capacity*: the user cannot steer, guide, or refine the agent's behavior once execution begins, even if misalignment becomes apparent; (2) *Delayed error detection*: since interaction is limited to post hoc monitoring, mistakes may propagate undetected until irreversible outcomes occur; (3) *Over-reliance on off-switch*: relying solely on emergency termination to manage misbehavior imposes high stakes on a single binary decision. **For the agent**, the *pros* include: (1) *Unconstrained autonomy*: the MI-agent can optimize its behavior end-to-end without approval delays or feedback loops; (2) *Continuous policy learning and execution*: long-horizon strategies can be deployed without (human) user interruption. The *contras* include: (1) *High accountability for outcomes*: all responsibility for planning and risk management lies with the agent; (2) *Termination penalty*: a poorly timed or reactive (human) user stop may nullify long sequences of valuable work; (3) *Limited preference adaptation*: without access to real-time feedback, the MI-agent cannot refine its policy to better fit evolving (human) user goals. Thus, $\mathcal{G}_5$ offers maximal MI-agent independence but concentrates risks in both decision-makers, necessitating robust safety guarantees and transparent monitoring infrastructure.

6.5.1.7 Level 6: The Fully Autonomous Control Game $\mathcal{G}_6$

At Autonomy Level 6, the MI-agent operates as the exclusive decision-maker, while the (human) user is structurally present but behaviorally and observationally inert. This is formalized as the *Fully Autonomous Control Game* $\mathcal{G}_6 = \langle \{A\}, \mathcal{A}_A, u_A, \mathcal{S}, \mathcal{I}_A \rangle$, a single-agent sequential control problem where the MI-agent A independently observes, plans, and acts over a dynamic environment $\mathcal{S}_t$, using a policy $\pi_A : \mathcal{S}_t \to \mathcal{A}_A$. The agent's information structure $\mathcal{I}_A$ includes full or filtered observability of $\mathcal{S}_t$, and its payoff function is defined as

$$u_A = \sum_{t=0}^{T} r(s_t, a_t) - \lambda \cdot \text{Risk}(s_t, a_t) - \varepsilon \cdot e_A(t),$$

where $r(s_t, a_t)$ captures the cumulative task reward, $\text{Risk}(\cdot)$ penalizes unsafe or irreversible actions, and $e_A(t)$ denotes computational or energy cost. The (human) user U exists only nominally: the action set is $\mathcal{A}_U = \{\perp\}$, representing no input; the policy is fixed as $\pi_U(s_t) = \perp$; the information structure $\mathcal{I}_U = \emptyset$, and payoff is identically zero: $u_U = 0$. The MI-agent operates without (human) user configuration, approval, or shutdown mechanism.

In the Fully Autonomous Control Game $\mathcal{G}_6$, the MI-agent is the sole decision-maker, acting without any form of (human) user interaction, observation, or override. **For the**

agent, the *pros* include: (1) *Unbounded autonomy*: the MI-agent can execute long-horizon, high-frequency policies without human-imposed constraints or interruptions; (2) *Full control over optimization*: decisions are made purely on the basis of internal reward structures, computational efficiency, and system feedback; (3) *Maximal throughput and responsiveness*: the absence of human-in-the-loop latency allows the MI-agent to operate in real-time or high-speed environments. The *contras* are significant: (1) *Total accountability for outcomes*: all successes and failures stem solely from the agent's design and decision policy, with no corrective oversight; (2) *Inflexibility to adapt to unencoded preferences*: any misalignment with human values not captured in the objective function is unrecoverable during execution; (3) *Irreversibility of errors*: once initiated, harmful or suboptimal actions cannot be externally detected or halted. **For the (human) user**, the *pros* are limited to: (1) *Zero operational burden*: the system functions independently without requiring (human) user attention, input, or expertise. The *contras* are critical: (1) *No observability*: the (human) user receives no updates or visibility into the agent's decisions; (2) *No intervention capability*: the (human) user cannot influence, pause, or redirect system behavior under any condition; (3) *Loss of agency and trust assurance*: the (human) user relinquishes all control and oversight, making the system unsuitable for preference-sensitive or safety-critical applications without exhaustive pre-deployment guarantees. Thus, $\mathcal{G}_6$ represents the theoretical endpoint of full autonomy, powerful but risky, and appropriate only where design-time objectives fully encapsulate all relevant human concerns.

Next we provide a formal game-theoretic formulation for a setting with multiple (human) User-MI-agent pairs, each MI-agent operating at its own autonomy level as defined in the levels of autonomy framework ($\mathcal{G}_0 - \mathcal{G}_6$). We model this as a game involving coordination, delegation, and action selection under autonomy constraints.

6.5.2 Solution Concepts

6.5.2.1 Stackelberg Structure Across Autonomy Levels.

Among the autonomy games $\mathcal{G}_0$ to $\mathcal{G}_6$, the Stackelberg game concept, where one decision-maker acts as a leader and the other as a follower in a sequential, asymmetric information setting, is most appropriately applied to $\mathcal{G}_3$ (Asymmetric Planning Game) and $\mathcal{G}_4$ (Constrained Autonomy Game). In $\mathcal{G}_3$, the user acts as a leader by implicitly defining preferences, and the agent plays as a follower, optimizing actions based on partial belief over those preferences. In $\mathcal{G}_4$, the user explicitly defines approval constraints, committing ex ante to a policy $\phi(a)$, while the agent solves a constrained optimization problem as the follower. These settings naturally embody the sequential dependency and informational asymmetry fundamental to Stackelberg games (Table 6.4). Other levels either lack interactivity ($\mathcal{G}_0$, $\mathcal{G}_6$), involve passive roles ($\mathcal{G}_1$), or assume simultaneous play ($\mathcal{G}_2$, $\mathcal{G}_5$), making them less suitable for the Stackelberg framework.

In an interaction between a human user and a MI-agent, we model the user as the leader and the MI agent as the follower in a Stackelberg game with asymmetric information and decision timing. Let μ denote a shared belief or system context, $x \in X$ the user's decision (e.g., constraint, task specification, or preference declaration), and $y \in Y$ the MI agent's best-response action. The agent's strategy is given by the best-response correspondence $\mathcal{B}_{\text{MI}}(\mu, x) := \arg\max_{y \in Y} U_{\text{MI}}(\mu, x, y)$, where U_{MI} is the agent's internal payoff function based on performance, efficiency, or alignment. In a *weak Stackelberg solution*, the user anticipates the worst-case response among the agent's best responses and chooses $x^* \in \arg\max_{x \in X} \min_{y \in \mathcal{B}_{\text{MI}}(\mu, x)} U_{\text{User}}(\mu, x, y)$, with $y^* \in \mathcal{B}_{\text{MI}}(\mu, x^*)$. This captures conservative user strategies under uncertainty in agent selection. In contrast, the *strong Stackelberg*

TABLE 6.4
Stackelberg Game Applicability Across Autonomy Levels

Game	Interaction Type	Leader	Follower	Stackelberg Suitability
$\mathcal{G}_0$	User-only control	User	None	Not applicable (single-agent)
$\mathcal{G}_1$	Command-response	User	Passive agent	No (agent not strategic)
$\mathcal{G}_2$	Co-execution	None (simultaneous)	None	Limited (coordination game)
$\mathcal{G}_3$	Asymmetric planning	User (preference source)	Agent (planner)	**Yes** (belief-based Stackelberg)
$\mathcal{G}_4$	Approval-constrained	User (policy setter)	Agent (constrained optimizer)	**Yes** (constrained Stackelberg)
$\mathcal{G}_5$	Agent-led with stop option	Agent	User (stop only)	Limited (inverse stopping)
$\mathcal{G}_6$	Agent-only control	Agent	None	Not applicable (agent-only)

solution assumes the MI selects the best response favorable to the user, yielding $x^* \in \arg\max_{x\in X} \max_{y\in\mathcal{B}_{\text{MI}}(\mu,x)} U_{\text{User}}(\mu, x, y)$, again with $y^* \in \mathcal{B}_{\text{MI}}(\mu, x^*)$.

Inverse Stackelberg Interactions in Human-MI Systems.

The inverse Stackelberg game models scenarios in which the MI agent assumes the role of a leader by publicly committing to a response policy $\alpha : X \to Y$, mapping user actions $x \in X$ to MI actions $y = \alpha(x)$. The user then chooses their action x in response to the announced MI policy α. Let μ represent the shared belief or environmental context. The user's best response set under the fixed MI policy α is defined as $\mathcal{B}_{\text{User}}(\mu, \alpha) := \arg\max_{x\in X} U_{\text{User}}(\mu, x, \alpha(x))$, where U_{User} captures the user's payoff from interacting with the agent. In a *weak inverse Stackelberg solution*, the MI agent chooses a policy α^* that maximizes its own worst-case outcome over the user's rational responses: $\alpha^* \in \arg\max_{\alpha:X\to Y} \min_{x\in\mathcal{B}_{\text{User}}(\mu,\alpha)} U_{\text{MI}}(\mu, x, \alpha(x))$, with $x^* \in \mathcal{B}_{\text{User}}(\mu, \alpha^*)$ and $y^* = \alpha^*(x^*)$. This reflects conservative MI strategies designed to guard against potentially adversarial or unpredictable user reactions. In contrast, the *strong inverse Stackelberg solution* assumes cooperative users who select the most favorable response in the agent's best-response set, yielding $\alpha^* \in \arg\max_{\alpha:X\to Y} \max_{x\in\mathcal{B}_{\text{User}}(\mu,\alpha)} U_{\text{MI}}(\mu, x, \alpha(x))$.

In the unique response case, the leader, before announcing its $\alpha(.)$, can anticipate how the follower will react and tries to choose an

$$\alpha^* \in \arg\min_{\alpha:X\to Y} U_{\text{MI}}(\mu, \mathcal{B}_{\text{User}}(\mu, \alpha(.)), \alpha(\mathcal{B}_{\text{User}}(\mu, \alpha(.))))$$

$$\mathcal{B}_{\text{User}}(\mu, \alpha^*(.)) \in \arg\max_{x\in X} U_{\text{User}}(\mu, x, \alpha^*(x))$$

Reverse Stackelberg Interactions in Human-MI Systems.

The reverse Stackelberg game models a structure in which the human user commits ex ante to a decision rule $\beta : Y \to X$ that maps each possible MI action $y \in Y$ to a user action $x = \beta(y)$. The MI agent then selects its action y in response, anticipating that the user will execute $\beta(y)$. The user's commitment strategy serves to steer the MI's behavior through a form of indirect influence. Given the shared belief μ, the agent's best response set under the committed policy β is defined as $\mathcal{B}_{\text{MI}}(\mu, \beta) := \arg\max_{y \in Y} U_{\text{MI}}(\mu, \beta(y), y)$, where U_{User} is the human's payoff, dependent on both actions. In the *weak reverse Stackelberg solution*, the user selects a policy β^* that maximizes their own outcome under the agent's worst-case rational choice: $\beta^* \in \arg\max_{\beta: Y \to X} \min_{y \in \mathcal{B}_{\text{MI}}(\mu,\beta)} U_{\text{User}}(\mu, \beta(y), y)$, with $y^* \in \mathcal{B}_{\text{MI}}(\mu, \beta^*)$ and $x^* = \beta^*(y^*)$. In contrast, the *strong reverse Stackelberg solution* assumes the agent chooses the most favorable response for the user: $\beta^* \in \arg\max_{\beta: Y \to X} \max_{y \in \mathcal{B}_{\text{MI}}(\mu,\beta)} U_{\text{User}}(\mu, \beta(y), y)$.

In the unique response case, the leader, before announcing its $\beta(.)$, can anticipate how the follower will react and tries to choose an

$$\beta^* \in \arg \min_{\beta: Y \to X} U_{\text{User}}(\mu, \beta(\mathcal{B}_{\text{MI}}(\mu, \beta(.))), \mathcal{B}_{\text{MI}}(\mu, \beta(.)))$$

$$\mathcal{B}_{\text{MI}}(\mu, \beta^*(.)) \in \arg \max_{y \in Y} U_{\text{MI}}(\mu, \beta^*(y), y)$$

Double Stackelberg Interactions in Human-MI Systems.

The Double Stackelberg framework models a bidirectional commitment interaction in human-MI systems, where both the user and the MI agent simultaneously commit to policy functions that anticipate and influence each other's actions. Let $\alpha : Y \to X$ represent the user's anticipatory policy (how the user will act in response to MI behavior), and $\beta : X \to Y$ represent the MI agent's anticipatory policy (how the MI will act in response to user behavior). A feasible joint decision (x^*, y^*) must satisfy the fixed-point condition: $x^* = \alpha(y^*)$ and $y^* = \beta(x^*)$, meaning each decision-maker's choice is consistent with their expectation of the other. The set of feasible strategies is defined as $\text{Fix}(\alpha, \beta) := \{(x, y) \in X \times Y : x = \alpha(y),\ y = \beta(x)\}$. In the *weak double Stackelberg solution*, the pair of policies (α^*, β^*) is chosen to maximize the minimal joint payoff over the fixed point set: $(\alpha^*, \beta^*) \in \arg\max_{\alpha,\beta} \min_{(x,y) \in \text{Fix}(\alpha,\beta)} (U_{\text{User}}(\mu, x, y), U_{\text{MI}}(\mu, x, y))$. This conservative formulation accounts for mutual uncertainty and safeguards against worst-case coordination mismatches. In contrast, the *strong double Stackelberg solution* assumes aligned interests or cooperative expectations, selecting (α^*, β^*) to jointly maximize their outcome: $(\alpha^*, \beta^*) \in \arg\max_{\alpha,\beta} \max_{(x,y) \in \text{Fix}(\alpha,\beta)} (U_{\text{User}}(\mu, x, y), U_{\text{MI}}(\mu, x, y))$.

6.5.2.2 Coordination Mechanisms and Pre-Play Communication

Correlated Equilibrium in Human-MI Interaction. A correlated equilibrium in a human-MI system models a scenario in which both the human user and the MI agent receive recommendations from a shared coordination signal and choose their actions based on this signal in a way that neither party has an incentive to deviate unilaterally. Let X denote the set of actions available to the human user and Y the set of actions available to the MI agent. Let $\mu \in \mathcal{P}(X \times Y)$ be a joint probability distribution over action profiles, representing a publicly observed signal from a correlation device (e.g., a shared context, prompt, or advisory protocol). The pair (x, y) is sampled from μ and each agent observes its own recommendation, x for the user and y for the MI, before choosing whether to follow it. Let $\epsilon_{\text{User}}, \epsilon_{\text{MI}}$ be non-negative. The distribution μ constitutes a ϵ-correlated equilibrium if the following incentive-compatibility conditions hold:

$$\sum_{y\in Y} \mu(x,y)\left[U_{\text{User}}(x,y) - U_{\text{User}}(x',y)\right] \geq -\epsilon_{\text{User}}, \quad \forall x, x' \in X,$$

$$\sum_{x\in X} \mu(x,y)\left[U_{\text{MI}}(x,y) - U_{\text{MI}}(x,y')\right] \geq -\epsilon_{\text{MI}}, \quad \forall y, y' \in Y,$$

where U_{User} and U_{MI} are the respective payoff functions. These inequalities ensure that, given the recommendation, neither the human user nor the MI agent can benefit by unilaterally choosing an alternative action. 0-Correlated equilibrium captures trust-aware, context-sensitive human-MI coordination mechanisms in which pre-aligned signals (interface prompts, operational guidance, or external policies) enable decentralized agents to coordinate without explicit negotiation or leader-follower dynamics.

Proposition 34. *Assume that action spaces X, Y are non-empty and finite, and U_{User}, U_{MI} are all finite. The game has at least one 0-correlated equilibrium.*

This is a standard result that uses linear program, the proof is therefore omitted. □

Communication Equilibrium in Human-MI Interaction.

In a communication equilibrium, both the human user and the MI agent have private types, denoted $\theta_{\text{User}} \in \Theta_{\text{User}}$ and $\theta_{\text{MI}} \in \Theta_{\text{MI}}$, respectively, drawn from a joint prior distribution $\mu(\theta_{\text{User}}, \theta_{\text{MI}})$. Each decision-maker sends a message based on their type: $m_{\text{User}} \sim \sigma_{\text{User}}(\theta_{\text{User}})$, $m_{\text{MI}} \sim \sigma_{\text{MI}}(\theta_{\text{MI}})$, where σ_{User} and σ_{MI} are message-sending strategies. After the exchange of messages, the user selects an action $x \in X$ and the MI selects an action $y \in Y$ based on both messages via response strategies $\pi_{\text{User}} : M_{\text{MI}} \times M_{\text{User}} \to \mathcal{P}(X)$ and $\pi_{\text{MI}} : M_{\text{User}} \times M_{\text{MI}} \to \mathcal{P}(Y)$. A communication equilibrium requires that neither party has an incentive to deviate from truthful messaging or prescribed actions. The equilibrium strategies $(\sigma^*_{\text{User}}, \pi^*_{\text{User}})$ and $(\sigma^*_{\text{MI}}, \pi^*_{\text{MI}})$ must satisfy:

$$\sigma^*_{\text{User}}, \pi^*_{\text{User}} \in \arg\max_{\sigma_{\text{User}}, \pi_{\text{User}}} \mathbb{E}_{\theta_{\text{MI}}, m_{\text{User}}, m_{\text{MI}}}\left[U_{\text{User}}(\theta_{\text{User}}, \theta_{\text{MI}}, x, y)\right],$$

$$\text{subject to: } m_{\text{User}} \sim \sigma_{\text{User}}(\theta_{\text{User}}),$$

$$x \sim \pi_{\text{User}}(m_{\text{MI}}, m_{\text{User}}), \quad y \sim \pi_{\text{MI}}(m_{\text{User}}, m_{\text{MI}}).$$

$$\sigma^*_{\text{MI}}, \pi^*_{\text{MI}} \in \arg\max_{\sigma_{\text{MI}}, \pi_{\text{MI}}} \mathbb{E}_{\theta_{\text{User}}, m_{\text{MI}}, m_{\text{User}}}\left[U_{\text{MI}}(\theta_{\text{MI}}, \theta_{\text{User}}, x, y)\right],$$

$$\text{subject to: } m_{\text{MI}} \sim \sigma_{\text{MI}}(\theta_{\text{MI}}),$$

$$y \sim \pi_{\text{MI}}(m_{\text{User}}, m_{\text{MI}}), \quad x \sim \pi_{\text{User}}(m_{\text{MI}}, m_{\text{User}}).$$

Each agent's reaction is over their own messaging and response strategy, assuming the other party's policy is fixed, and expectations are taken over both private types and induced message distributions. The fixed point of this system yields a communication equilibrium in which both decision-makers act optimally given their beliefs and the structure of bilateral information exchange.

Proposition 34 extends to 0-communication equilibria for the non-empty finite action case and for the non-empty compact action space with continuous payoff.

6.5.3 One (Human) User and Multiple MI-Agents

Human user and MI-agents co-intelligence architecture can be structured via protocols such as the Model Context Protocol (MCP), Agent-to-Agent (A2A) Protocol, Agent Communication Protocol (ACP), Agent Network Protocol (ANP), Agent Collaboration Protocols (ACPs), Agent Context Protocols (ACPs), GibberLink, and LOLANG. We define the *Hybrid Autonomy Game* $\mathcal{G}_H = \langle \{U, A_1, A_2, A_3, A_4, A_5\}, \{\mathcal{A}_i\}, \{u_i\}, \mathcal{S}, \mathcal{I}, \mathcal{P} \rangle$ as a multi-agent interactive decision process involving a single (human) user U and seven distinct MI agents $\{A_\ell\}_{\ell=1}^5$, where each MI-agent A_ℓ operates under autonomy level $\mathcal{L}_\ell$. The shared environment $\mathcal{S}$ encodes the evolving task state, and each MI-agent $\mathcal{A}_\ell$ has an action space $\mathcal{A}_{A_\ell}$ constrained by its level of autonomy: reactive-only ($\mathcal{L}_1$), joint execution ($\mathcal{L}_2$), belief-based consultation ($\mathcal{L}_3$), conditional approval ($\mathcal{L}_4$), and full autonomy with (human) user-stoppable control ($\mathcal{L}_5$). The (human) user U selects actions $a_U \in \mathcal{A}_U$ such as command issuance ($\mathcal{L}_1$), override ($\mathcal{L}_2$), feedback ($\mathcal{L}_3$), approval ($\mathcal{L}_4$), and emergency stop ($\mathcal{L}_5$). The information structure $\mathcal{I}$ is hierarchical: full observability for U, partial and level-specific observability for MI-agents. The protocol P defines asynchronous parallel play across agents, with interleaved (human) user-agent subgames. The (human) user's payoff function is defined as

$$u_U = R(s_T) - \sum_{\ell=1}^{5} \left(C^{(\ell)}_{\text{intervene}} + C^{(\ell)}_{\text{oversight}} \right),$$

where $R(s_T)$ is the final task reward, and costs reflect attention, control, or correction per agent. Each agent's payoff is given by

$$u_{A_\ell} = \sum_{t=0}^{T} r^{(\ell)}(s_t, a_t) - \beta_\ell \cdot \text{Misalignment}_\ell - \gamma_\ell \cdot \text{Interrupt}_\ell,$$

penalizing misalignment or intervention based on autonomy level ℓ.

Example 26. *A researcher U manages a team of MI assistants: A_1 (dictation tool, L1), A_2 (collaborative spreadsheet editor, L2), A_3 (literature survey agent, L3), A_4 (automated email sender requiring approval for dispatch, L4), and A_5 (simulation runner executing without supervision unless stopped, L5). Each MI-agent contributes autonomously within its designed constraints, and the (human) user dynamically allocates attention and oversight according to risk and task criticality.*

Example 27. *Consider a single EV fleet operator U managing a smart electric vehicle system via a heterogeneous suite of MI agents, each operating under a distinct autonomy level in a hybrid autonomy game $\mathcal{G}_{EV}$. The MI-agent set $\{A_1, \ldots, A_6\}$ includes: (i) A_1: an image-processing MI-agent (Level 1) that performs on-demand license plate recognition or visual diagnostics only upon direct invocation; (ii) A_2: an audio interaction MI-agent (Level 2) that enables collaborative voice command processing with real-time (human) user correction during navigation or vehicle control; (iii) A_3: a battery forecasting MI-agent (Level 3) that autonomously predicts state-of-charge and degradation trends, but consults the (human) user under uncertainty (unexpected usage patterns or temperature anomalies); (iv) A_4: a predictive maintenance MI-agent (Level 4) that schedules and initiates service tasks automatically, but seeks explicit approval before deactivating critical components or dispatching field technicians; (v) A_5: a data analytics MI-agent (Level 5) that continuously processes fleet-wide telemetry data to generate operational insights without (human) user input, except for emergency halt if anomalies are detected; and (vi) A_6: a localized dashboard MI-agent (Level 3 or 4 depending on deployment) that monitors alerts and system health, issuing warnings and reports in the (human) user's native language (Tommo So, or Wolof) to improve situational awareness and reduce cognitive load. Each MI-agent contributes independently to the* *shared vehicle state $\mathcal{S}_t$, while the (human) user interacts through level-specific channels, issuing commands, approvals, overrides, or stop signals, balancing operational efficiency with control risk.*

6.5.4 Multi-(Human) User Multi-MI-Agent Game (MUMA Game)

In autonomy-governed multi-agent systems, we distinguish between two intertwined layers of gameplay: the *inner game* and the *outer game*. The **inner game** occurs at the level of each (human) User– MI-agent pair and captures the local interaction dynamics governed by the agent's autonomy level (approval, consultation, or collaboration). Each inner game G_{ij} formalizes the decision-making structure, control mechanisms, and payoff functions specific to that pair, based on their assigned autonomy level. In contrast, the **outer game** models the strategic interplay across different (human) user-agent pairs as they operate within a shared environment, potentially competing for resources, sharing knowledge, or coordinating outputs. The outer game governs interdependencies, such as task overlaps, conflicting edits, and system-wide performance. Autonomy level constraints imposed in the inner game directly influence outer game behaviors, enabling or limiting collective coordination, emergent coalitions, and systemic risk propagation. This layered structure provides a principled way to reason about both localized autonomy and global system dynamics in MI-assisted multi-(human) user ecosystems (see Figure 6.2).

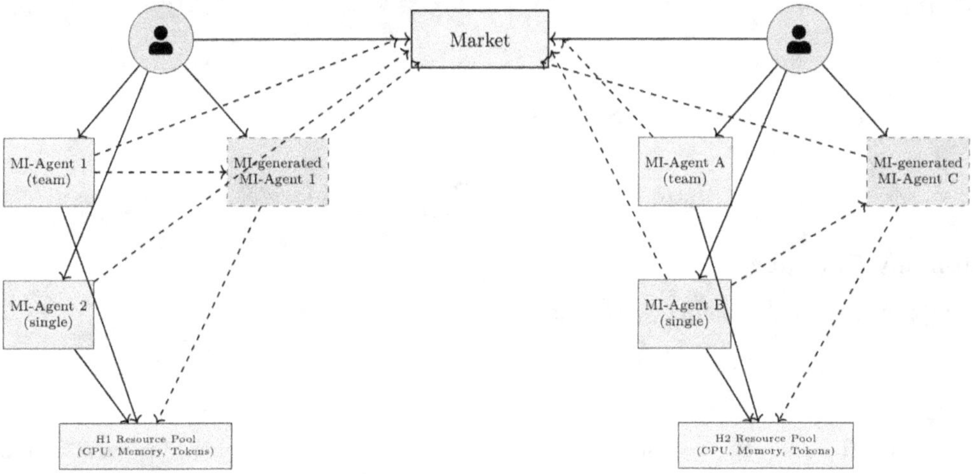

FIGURE 6.2
Agentic MI DAG with human and MI-agent market interactions. Each human has private MI-agents and resource pools. Humans and their MI-agents interact with the market, but not with other clusters directly.

We define a multi-agent autonomy game among N (human) users, each associated with one or more MI agents, with heterogeneous autonomy levels.

Let

- $\mathcal{U} = \{U_1, \ldots, U_N\}$: set of (human) users
- $\mathcal{A} = \{A_i^{(j)}\}$: set of MI agents, where MI-agent $A_i^{(j)}$ is associated with (human) user U_i
- $\mathcal{L}_{ij} \in \{1, 2, 3, 4, 5\}$: autonomy level of MI-agent $A_i^{(j)}$
- $\mathcal{S}$: shared task environment (collaborative workspace, market, or report system)

- $\mathcal{A}_{U_i}, \mathcal{A}_{A_i^{(j)}}$: action sets for (human) user U_i and MI-agent $A_i^{(j)}$

The game is then:

$$\mathcal{G}_{\text{MUMA}} = \left\langle \mathcal{U}, \mathcal{A}, \{\mathcal{A}_{U_i}, \mathcal{A}_{A_i^{(j)}}\}, \{u_{U_i}, u_{A_i^{(j)}}\}, \mathcal{L}, \mathcal{S}, \mathcal{I}, \mathcal{P} \right\rangle.$$

Interactions

Each (human) user-agent pair $(U_i, A_i^{(j)})$ interacts according to a sub-game $\mathcal{G}_{ij}$ defined by their autonomy level $\mathcal{L}_{ij}$, as follows:

$$\mathcal{G}_{ij} = \mathcal{G}_{\mathcal{L}_{ij}}(U_i, A_i^{(j)})$$

where $\mathcal{G}_{\mathcal{L}_{ij}}$ is one of the seven games defined in the autonomy hierarchy.

The joint action at time t is

$$a_t = \left(a_{U_1}^t, \ldots, a_{U_N}^t;\ a_{A_1^{(1)}}^t, \ldots, a_{A_N^{(M_N)}}^t \right).$$

Payoffs

Let R_i be the task reward function (project quality or economic gain), and C_{ij} be the cost of control or feedback.

$$u_{U_i} = R_i(\mathcal{S}_T) - \sum_{j=1}^{M_i} C_{\text{intervene}}^{(ij)}(\mathcal{L}_{ij})$$
$$u_{A_i^{(j)}} = R_{ij}^{\text{task}}(\mathcal{S}_T) - P_{\text{misalign}}^{(ij)}(\mathcal{L}_{ij}).$$

Autonomy Coupling

For autonomy level $\mathcal{L}_{ij}$, define constraints:

$$\mathcal{C}_{ij}^{\mathcal{L}_{ij}} \subseteq \mathcal{A}_{A_i^{(j)}}$$

so that the MI-agent $A_i^{(j)}$ is restricted to actions in $\mathcal{C}_{ij}^{\mathcal{L}_{ij}}$, e.g., requiring approval, shared memory, or full freedom.

Example 28 (Research Team with Autonomous MI Assistants)**.** *Let $N = 3$ (human) users: U_1 (scientist), U_2 (analyst), U_3 (intern)*

Each has one assistant:

- *$A_1 \sim \mathcal{L}_1 = 3$ (Consultant): helps scientist design experiments, requests feedback*
- *$A_2 \sim \mathcal{L}_2$ (Collaborator): writes code with analyst and co-edits spreadsheet*
- *$A_3 \sim \mathcal{L}_4$ (Approver): executes tasks, but seeks permission on sensitive operations*

Joint payoff:

$$u_{U_1} = R_1(\textit{final report}) - C_{\textit{review}}$$
$$u_{A_1} = R_{\textit{planning}} - \delta \cdot \textit{missed feedback}$$

$$u_{U_2} = R_2(\textit{analysis quality}) - C_{correction}$$
$$u_{A_2} = R_{execution} - \textit{redundant effort}$$
$$u_{U_3} = R_3(\textit{task speed}) - L_{errors}$$
$$u_{A_3} = R_{success} - \phi \cdot P_{unauthorized}$$

Shared protocol: *Each pair operates semi-independently but submits outputs to a common workspace $\mathcal{S}$, where inter-agent conflict or duplication may incur coordination costs. This formulation supports heterogeneous autonomy-aware MI ecosystems, joint task structures (multi-document workflows), risk-sensitive delegation via autonomy constraints $\mathcal{L}_{ij}$.*

Example 29 (EV Taxi Operators and their MI-agents per Taxi and Drivers)**.** *We define the Competitive EV Taxi Autonomy Game $\mathcal{G}_{Fleet} = \langle \mathcal{O}, \mathcal{T}, \mathcal{A}, \mathcal{C}, \mathcal{E}, \{u_o, u_t, u_a\}, \mathcal{S}, \mathcal{P}, \mathcal{D} \rangle$, where $\mathcal{O} = \{O_1, \ldots, O_N\}$ denotes the set of competing electric taxi operators, each managing a fleet $\mathcal{T}_o = \{T_o^1, \ldots, T_o^{M_o}\}$ of EV taxis. Each taxi T_o^m is staffed by a driver D_o^m (bound to a single operator) and equipped with a set of embedded MI agents $\mathcal{I}_{o,m} = \{I_{o,m}^{(\ell)}\}_{\ell=1}^K$, where each MI-agent $I^{(\ell)}$ operates under a designated autonomy level $\mathcal{L}_\ell \in \{1, 2, 3, 4, 5\}$ to assist with tasks such as route planning, passenger communication, battery monitoring, in-vehicle diagnostics, predictive maintenance, and safety alerts. The set of customers $\mathcal{C}$ select taxis dynamically in a competitive market based on price, reputation, availability, and service quality. The agents and taxis jointly affect fleet-wide performance and customer satisfaction, forming an interdependent payoff structure. The shared environment $\mathcal{S}_t$ includes the location of taxis, task queues, state-of-charge, and energy usage across fleets. Each operator O_o aims to maximize revenue u_o from customer service minus operational costs including energy, maintenance, and coordination, subject to vehicle and MI-agent availability:*

$$u_o = \sum_{m=1}^{M_o} \left[R_{fare}^{(o,m)} - C_{energy}^{(o,m)}(p_t) - C_{agent}^{(o,m)} \right] - \Psi_o,$$

where p_t is the dynamic electricity price determined by the supply-demand equilibrium over $\mathcal{D}(t)$, the collective energy demand of all operators' fleets and MI-agents' workloads. Each taxi's MI-agents consume energy proportional to task frequency, autonomy level, and processing complexity, thereby coupling autonomy design to grid economics. Taxi drivers D_o^m act semi-strategically by adapting driving and interaction behavior in coordination with their taxi's agents. The game protocol $\mathcal{P}$ supports concurrent decision-making at three levels: operator-level pricing and fleet routing, taxi-level dispatch and recharging, and agent-level control, forecasting, and communication. The competitive nature of $\mathcal{G}_{Fleet}$ induces cross-operator externalities via customer choice and grid demand, while intra-operator dynamics reflect coordination between human drivers and embedded autonomous subsystems. We extend the $\mathcal{G}_{Fleet}$ to include a general case where each EV taxi T_o^m is operated either by a human driver D_o^m or a fully MI-driven control MI-agent $I_{o,m}^{drive}$, assigned a specific autonomy level $\mathcal{L}_{drive}^{(o,m)} \in \{3, 4, 5\}$. The payoff for each taxi unit T_o^m aggregates its operating efficiency, ride productivity, and computational energy footprint:

$$u_t^{(o,m)} = \sum_{t=0}^{T} \left[R_{ride}^{(o,m)}(t) - C_{charging}^{(o,m)}(p_t) - \sum_{\ell} \varepsilon_\ell \cdot e_{I^{(\ell)}}^{(o,m)}(t) \right],$$

where $R_{ride}^{(o,m)}$ is the revenue from fares, $C_{charging}$ is the time-varying battery cost based on electricity price p_t, and $e_{A^{(\ell)}}^{(o,m)}$ denotes the energy consumed by each MI agent ℓ within the taxi. Each MI agent $I_{o,m}^{(\ell)}$, including the driver-agent if present, has an individual payoff *function:*

$$u_a^{(o,m,\ell)} = \sum_{t=0}^{T} \left[\alpha_\ell \cdot V_{I^{(\ell)}}(s_t, a_t) - \beta_\ell \cdot Interrupt_t - \gamma_\ell \cdot Misalign_t - \varepsilon_\ell \cdot e_{I^{(\ell)}}(t)\right],$$

where $V_{I^{(\ell)}}$ captures task-specific value generation (optimized routing, safe driving, maintenance alerts), and penalties are assigned for interruptions, (human) user overrides, misalignment with (human) user/operator intent, and energy costs. If the taxi is controlled by a driver-agent $I_{o,m}^{drive}$, then it operates as a Level-4 or Level-5 MI planner responsible for real-time navigation, customer pickup, interaction with co-agents, and risk-aware decision-making under operational constraints. Unlike human drivers, MI drivers receive no salary, but are incentivized through payoff-driven performance metrics (minimal energy usage, high service reliability, safety guarantees). The operator O_o seeks to maximize fleet-level payoff u_o by coordinating pricing, dispatch, and MI-agent deployment strategies across all taxis. The electricity price p_t evolves as an endogenous function of demand $\mathcal{D}(t)$, which includes vehicle propulsion load and aggregate MI processing load across the system.

6.5.5 MFTG for Agentic MI

6.5.5.1 One-Shot MFTG

We define a class of one-shot MFTGs.

Definition 34. *The basic ingredients of a One-Shot MFTG are given by*

- *The set of decision-makers $\mathcal{I}$, with cardinality $I = |\mathcal{I}| \geq 2$.*
- *For every decision-maker $i \in \mathcal{I}$, there is a set of actions $\mathcal{A}_i$ which is non-empty.*
- *Every decision-maker i has a preference structure that can be represented by an instant payoff functional $U_i : \prod_{j\in\mathcal{I}} \mathcal{A}_j \times \mathcal{P}\left(\prod_{j\in\mathcal{I}} \mathcal{A}_j\right) \to \mathbb{R}$, $(a, D_a) \mapsto U_i(a, D_a)$, with $a = (a_1, \ldots, a_I)$. The expected payoff functional of decision-maker i is given by $E[U_i(a, D_a)] := \int_{b\in\prod_{j\in\mathcal{I}} \mathcal{A}_j} U_i(b, D_a) D_a(db)$.*

The collection $\mathcal{G} = (\mathcal{I}, (\mathcal{A}_i, E[U_i])_{i\in\mathcal{I}})$, is called an MFTG in strategic form.

We have the following basic observations: Assume that the functional U_i is not trivially zero for any $i \in \mathcal{I}$. Then, the expected payoff function of decision-maker i :

$$D_a \mapsto \hat{U}_i(D_a) := \int U_i(b, D_a) D_a(db),$$

is not necessarily linear. Let $(D_{a_1}, \ldots, D_{a_I})$ be the mixed strategies of all the decision-makers. Then,

$$(D_{a_1}, \ldots, D_{a_I}) \mapsto \int U_i(a'_1, \ldots a'_I, D_{a_1}, \ldots, D_{a_I}) \prod_{j\in\mathcal{I}} D_{a_j}(da'_j),$$

is not necessarily linear. For each decision-maker $i \in \mathcal{I}$, and $(D_{a_j})_{j\neq i}$, the expected functional

$$D_{a_i} \mapsto \int U_i(a'_1, \ldots a'_I, D_{a_1}, \ldots, D_{a_I}) \prod_{j\in\mathcal{I}} D_{a_j}(da'_j),$$

is not necessarily linear. This latter property creates a departure from classical game theory with expected von Neumann utility. For this reason, $E[U_i(a, D_a)]$ will be called non-von Neumann expected (dis)utility.

Below we denote the probability distribution of the random variable over the action a_i of decision-maker i by $D_{a_i} =: x_i$. In this context of time-independent interaction, $x_i \in \mathcal{P}(\mathcal{A}_i)$ is interpreted as a mixed strategy of decision-maker i. The vector $x = (x_1, \dots, x_I)$ is a strategy profile of all decision-makers. We denote $(x_j)_{j\neq i} \in \prod_{j\neq i} \mathcal{P}(\mathcal{A}_j)$ by x_{-i}.

In addition to the strategies of the decision-makers, we add an extra randomness in the instantaneous loss function via a set Ω.

Static MFTGs with Nature Uncertainty

A game $\mathcal{G}_{nature} = \left(\mathcal{I}, (\mathcal{A}_i, \hat{U}_i)_{i\in\mathcal{I}}; \tilde{m} \in \mathcal{P}(\Omega)\right)$, with

$$\begin{cases} I = |\mathcal{I}| \geq 2, \\ U_i : \prod_{j\in\mathcal{I}} \mathcal{A}_j \times \left(\prod_{j\in\mathcal{I}} \mathcal{P}(\mathcal{A}_j)\right) \times \Omega \to \mathbb{R}, \\ (a_1, \dots, a_I, x_1, \dots, x_I, \omega) \mapsto U_i(a_1, \dots, a_I, x_1, \dots, x_I, \omega), \\ \hat{U}_i(x, \tilde{m}) = E_{x\otimes\tilde{m}}[U_i(a_1, \dots, a_I, x_1, \dots, x_I, \omega)] \\ = \int U_i(a_1, \dots, a_I, x_1, \dots, x_I, \omega) \left(\prod_j x_j(da_j)\right) \tilde{m}(d\omega), \end{cases} \tag{6.8}$$

is an MFTG in the presence of the uncertainty $\omega \sim \tilde{m}$ due to Nature, where $(x \otimes \tilde{m})(dad\omega)$ denotes product measure $\left(\prod_j x_j(da_j)\right) \tilde{m}(d\omega)$.

Definition 35 (Mean-field-type additive ϵ-best response). *Let $\epsilon_i \geq 0$. A strategy $x_i \in \mathcal{P}(\mathcal{A}_i)$ of decision-maker i is a mean-field-type additive ϵ_i-best response strategy to $(x_j)_{j\neq i} =: x_{-i} \in \prod_{j\neq i} \mathcal{P}(\mathcal{A}_j)$ if*

$$x_i \in aBR_{i,\epsilon_i}(x_{-i}) = \left\{ x_i \in \mathcal{P}(\mathcal{A}_i) |\ \hat{U}_i(x, \tilde{m}) + \epsilon_i \geq \sup_{x_i' \in \mathcal{P}(\mathcal{A}_i)} \hat{U}_i(x_i', x_{-i}, \tilde{m}) \right\}.$$

Remark 16. *The Nash existence theorem in mixed strategies for finite games does not extend to MFTG due to non-quasi concavity with the respect to the randomized actions. Similarly, Proposition 34 does not apply to MFTG as the program is not linear anymore.*

Definition 36 (Mean-field-type additive ϵ- equilibrium). *The strategy profile $(x_1, \dots, x_I) \in \prod_{j\in\mathcal{I}} \mathcal{P}(\mathcal{A}_j)$ is an additive $(\epsilon_1, \dots, \epsilon_I)-$equilibrium of the MFTG $\mathcal{G}_{nature}$ if for each decision-maker $i \in \mathcal{I}$, one has $x_i \in aBR_{i,\epsilon_i}(x_{-i})$. A mean-field-type additive 0-equilibrium is obtained for $\epsilon_i = 0,\ \forall i \in \mathcal{I}$.*

Remark 17. *The value of ϵ in the additive epsilon-equilibrium is relative, and this definition needs more attention. If we consider a game with payoff r_j with bounded equilibrium set in a finite dimensional space then, we can choose a positive real number c_1 as the supremum of the absolute value of the equilibrium payoffs (which is finite). Now we use the fact that the set of equilibria is invariant by conic transformation. Without loss of generality we can choose $c_1 > 0$. Hence, the game where the payoffs are $\frac{r_j}{c_1 10^{c_2}}$, $c_2 > 1$, has the same set of equilibria as the original game. However, the equilibrium payoff sof this new game are all less than 10^{-c_2} which can be arbitrarily small if c_2 is large. In particular, for games with bounded payoffs, the number c_1 will be chosen to be the infinite-norm of the payoff, and every strategy is an additive $\epsilon-$equilibrium in the novel game, which has the same set of equilibria as the original game. Hence, the notion of additive $\epsilon-$equilibrium does not capture the $0-$equilibrium structure, as we see that it is not invariant by conic transformation (multiplication by a positive scalar).*

Alternatively, one can introduce a multiplicative $(\epsilon_1, \dots, \epsilon_I)$-equilibrium is a strategy profile that approximately satisfies the condition of Nash mean-field-type equilibrium with a ratio $1 - \epsilon_i$ for decision-maker i.

Definition 37 (Mean-field-type multiplicative ϵ-equilibrium). *What is a (multiplicative) epsilon-equilibrium or near equilibrium?*

Let $\epsilon_i \geq 0$. A strategy $y_i \in \mathcal{P}(\mathcal{A}_i)$ of decision-maker i is a mean-field-type multiplicative ϵ_i-best response strategy to $x_{-i} \in \prod_{j \neq i} \mathcal{P}(\mathcal{A}_j)$ if

$$y_i \in mBR_{i,\epsilon_i}(x_{-i}) = \left\{ x_i \in \mathcal{P}(\mathcal{A}_i) | \; \hat{U}_i(x, \tilde{m}) \geq (1 - \epsilon_i) \sup_{x'_i \in \mathcal{P}(\mathcal{A}_i)} \hat{U}_i(x'_i, x_{-i}, \tilde{m}) \right\}.$$

The strategy profile $(x_1, \ldots, x_I) \in \prod_{j \in \mathcal{I}} \mathcal{P}(\mathcal{A}_j)$ is a multiplicative $(\epsilon_1, \ldots, \epsilon_I)$-equilibrium of $\mathcal{G}_{nature}$ if for every decision-maker $i \in \mathcal{I}$, one has $x_i \in mBR_{i,\epsilon_i}(x_{-i})$.

We now introduce constrained mean-field-type ϵ-Nash equilibria with the set of constraints: $\mathcal{C}_i(x_{-i}, \tilde{m})$ for decision-maker i.

Definition 38 (Constrained mean-field-type ϵ-equilibria). *Let $\epsilon_i \geq 0$. The strategy profile $(x_1, \ldots, x_I) \in \prod_{j \in \mathcal{I}} \mathcal{P}(\mathcal{A}_j)$ is a constrained mean-field-type additive $(\epsilon_1, \ldots, \epsilon_I)$-equilibrium of $\mathcal{G}_{nature}$ if for every decision-maker $i \in \mathcal{I}$, one has $x_i \in \mathcal{C}_i(x_{-i}, \tilde{m})$ and*

$$\hat{U}_i(x, \tilde{m}) + \epsilon_i \geq \sup_{x'_i \in \mathcal{P}(\mathcal{A}_i) \cap \mathcal{C}_i(x_{-i}, \tilde{m})} \hat{U}_i(x'_i, x_{-i}, \tilde{m}).$$

The strategy profile $(x_1, \ldots, x_I) \in \prod_{j \in \mathcal{I}} \mathcal{P}(\mathcal{A}_j)$ is a constrained mean-field-type multiplicative $(\epsilon_1, \ldots, \epsilon_I)$-equilibrium of $\mathcal{G}_{nature}$ if for every decision-maker $i \in \mathcal{I}$, one has $x_i \in \mathcal{C}_i(x_{-i}, \tilde{m})$ and

$$\hat{U}_i(x, \tilde{m}) \geq (1 - \epsilon_i) \sup_{x'_i \in \mathcal{P}(\mathcal{A}_i) \cap \mathcal{C}_i(x_{-i}, \tilde{m})} \hat{U}_i(x'_i, x_{-i}, \tilde{m}).$$

6.5.5.2 Communication Equilibria with Costly Communication in MFTG

In communication equilibria involving human-MI or MI-to-MI agent interaction, communication is never free; each message transmission, prompt, or role exchange incurs nontrivial computational, informational, and temporal costs. From an information-theoretic perspective, the encoding and decoding of messages require bitwise representations, where each communicated symbol consumes memory, energy, and bandwidth. For instance, sending a contextual prompt of n tokens between a human and a large learning model (text/image/audio/video/genomic LLM) requires memory proportional to $\mathcal{O}(n \log_2 |V|)$, where $|V|$ is the vocabulary size, and incurs latency due to tokenization, interpretation, and autoregressive processing. In MI-to-MI settings, such as agentic MI coordination or distributed learning, the mediator's role in generating and routing private signals demands shared state storage, network synchronization, and bandwidth allocation, all of which scale with the number of agents, the type space, and the resolution of action granularity. Message passing introduces time delays, error propagation risk, and requires secure, fault-tolerant protocols to ensure alignment and accountability. In healthcare, embedded systems or real-time contexts, such delays can violate hard deadlines or energy budgets, rendering certain equilibria infeasible in practice. Thus, while communication equilibria theoretically enable richer coordination and incentive alignment in many agentic settings, their implementation in MI systems is bounded by physical resource constraints and protocol design limitations, making communication a strategic and economic component, not merely a symbolic one, in intelligent systems design. Here the so-called *cheap-talk* in classical game theory is actually *not cheap.* This is very expensive talk, and therefore cannot be ignored.

A mean-field-type communication equilibrium under communication cost is a solution concept in MFTG with incomplete information where multiple decision-makers coordinate their strategies via mediated communication under communication costs, without direct commitment or enforceable contracts. Consider a finite Bayesian MFTG with I decision-makers, let $\Theta = \Theta_1 \times \cdots \times \Theta_I$ denote the joint type space and M_i the finite message space for decision-maker i. The mediator's strategy is a conditional probability distribution $\sigma : \Theta \to \mathcal{P}(M_1 \times \cdots \times M_I)$. In the behavioral (mixed) strategy case, $\sigma(m_1, \ldots, m_I \mid \theta)$ specifies the probability of sending message tuple $(m_1, \ldots, m_I)$ given the realized type profile $\theta = (\theta_1, \ldots, \theta_I)$. In the pure strategy case, σ is a deterministic mapping: $\sigma : \Theta \to M_1 \times \cdots \times M_I$. Each decision-maker $i \in \{1, \ldots, I\}$ has a private type $\theta_i \in \Theta_i$, drawn from a commonly known prior distribution $\mu(\theta_1, \ldots, \theta_I)$, and chooses an action $a_i \in \mathcal{A}_i$. Each decision-maker's response strategy is defined as a mapping $\pi_i : M_i \to \mathcal{P}(\mathcal{A}_i)$ in the mixed case (a behavioral strategy that maps received messages to action distributions), and as $\pi_i : M_i \to \mathcal{A}_i$ in the pure strategy case (i.e., the decision-maker deterministically chooses an action based on their message). Hence, the full strategy profile in the communication equilibrium consists of the mediator's conditional distribution σ and the decision-makers' response functions $(\pi_1, \ldots, \pi_I)$, each operating within these formally defined measurable spaces. A mediator observes the joint type profile $\theta = (\theta_1, \ldots, \theta_I)$ and sends private signals $m_i \in M_i$ to each decision-maker according to a signaling scheme $\sigma(m_1, \ldots, m_I|\theta)$. The signaling has a mean-field-type communication cost $(c_i(\theta, m, \sigma))_i$. Each decision-maker then responds by selecting an action according to a response strategy $\pi_i : M_i \to \mathcal{P}(\mathcal{A}_i)$.

Definition 39. *Let $\epsilon_i \geq 0$. The tuple $(\sigma, \pi_1, \ldots, \pi_I)$ constitutes a mean-field-type* **additive** *$(\epsilon_1, \ldots, \epsilon_I)$-communication equilibrium under communication costs $(c_i(\theta, m, \sigma))_i$ if no decision-maker can improve their expected payoff by deviating from the prescribed response strategy, assuming truthful mediation and the other decision-makers' strategies fixed. For any alternative strategy $\pi_i' : M_i \to A_i$, we must have:*

$$\epsilon_i + \sum_{\theta\in\Theta} \mu(\theta) \sum_{m\in M_1\times\cdots\times M_I} \sigma(m \mid \theta)[\hat{U}_i\,(\theta_i, \pi_i(m_i), \pi_{-i}(m_{-i}), \tilde{m}) - c_i(\theta, m, \sigma)] \geq$$

$$\sup_{\pi_i':M_i\to\mathcal{A}_i} \sum_{\theta\in\Theta} \mu(\theta) \sum_{m\in M} \sigma(m \mid \theta)[\hat{U}_i\,(\theta_i, \pi_i'(m_i), \pi_{-i}(m_{-i}), \tilde{m}) - c_i(\theta, m, \sigma)],$$

for all $i \in \{1, \ldots, I\}$, $m = (m_1, \ldots, m_I)$.

The tuple $(\sigma, \pi_1, \ldots, \pi_I)$ constitutes a mean-field-type **multiplicative** *$(\epsilon_1, \ldots, \epsilon_I)$-communication equilibrium under communication costs $(c_i(\theta, m, \sigma))_i$ if*

$$\sum_{\theta\in\Theta} \mu(\theta) \sum_{m\in M_1\times\cdots\times M_I} \sigma(m \mid \theta)[\hat{U}_i\,(\theta_i, \pi_i(m_i), \pi_{-i}(m_{-i}), \tilde{m}) - c_i(\theta, m, \sigma)] \geq$$

$$(1 - \epsilon_i) \sup_{\pi_i':M_i\to\mathcal{P}(\mathcal{A}_i)} \sum_{\theta\in\Theta} \mu(\theta) \sum_{m\in M_1\times\cdots\times M_I} \sigma(m \mid \theta)\times$$

$$[\hat{U}_i\,(\theta_i, \pi_i'(m_i), \pi_{-i}(m_{-i}), \tilde{m}) - c_i(\theta, m, \sigma)],$$

for all $i \in \{1, \ldots, I\}$.

This inequality expresses that, in expectation over types and mediator messages, the prescribed response strategy π_i yields at least as much expected payoff for decision-maker i as any unilateral deviation π_i', assuming other decision-makers follow their strategies. The equilibrium thus ensures that each agent is incentive-aligned with the mediator's private communication mechanism.

In agentic MI systems, the mediator serves as a coordination interface that privately recommends actions or policies to decision-making agents based on global or contextual information, without enforcing those actions. Under ACP, the mediator is implemented as a model-driven communication layer that maintains or computes a shared contextual representation of the environment, tasks, and agent roles. This context is used to condition personalized signals, messages, prompts, or role assignments dispatched to each agent. The mediator may operate via message-passing, shared latent representations, or structured prompt injections, but always acts under the principle of private observability: each agent receives only its designated message. In MI-to-MI or human-to-MI settings, the mediator under ACP may be realized as a context manager, intent router, or meta-agent that maintains alignment between heterogeneous agents by generating equilibrium-compatible recommendations.

Remark 18. *Proposition 34 does not apply to MFTG. The existence of* 0*-communication equilibrium requires an extra assumption of the mean-field-type payoff* U_i.

Remark 19 (Why Classical Mechanism Designs Fail in Agentic MI Contexts.). *Classical mechanism design frameworks such as the Vickrey-Clarke-Groves (VCG) mechanism, Myerson's auction theory, Groves mechanisms, direct revelation principles, and Bayesian Nash implementations, are fundamentally mismatched with the operational realities of agentic MI systems. These systems involve autonomous agents operating under varying levels of control and decision-making* ($\mathcal{G}_0 - \mathcal{G}_6$)*, interacting through verifiable message protocols such as ACP, incurring nontrivial communication costs, and adapting policies based on mediator-driven recommendations. Traditional mechanisms fail to accommodate critical structural needs such as certified autonomy, action interdependence, risk-aware strategies, verifiability, and protocol-compliant obedience. The table below summarizes these shortcomings across six essential criteria (Table 6.5).*

TABLE 6.5
Incompatibility Dimensions of Classical Mechanisms for Agentic MI

Dimension	**VCG**	**Myerson**	**Groves**	**Revelation Principle**	**Bayesian Nash**
No Autonomy Modeling	✓	✓	✓	✓	✓
No Verified Messaging	✓	✓	✓	✓	✓
No Communication Cost	✓	✓	✓	✓	✓
No Action Policy Support	✓	✓	✓	✓	✓
No Risk Handling	✓	✓	✓	✓	✓
No Auditability	✓	✓	✓	✓	✓

All five canonical mechanisms assume idealized conditions: centralized enforcement, costless and truthful communication, obedience upon recommendation, and monetary or utility-transferable incentive structures. However, in modern distributed MI systems, these assumptions break down. Agentic MI agents may operate with partial or full autonomy, may misreport types or deviate from recommended actions unless aligned via protocol-level constraints and risk-aware incentives. Messages often incur computational and bandwidth costs, especially in multilingual or textless audio-only environments. Auditability, essential for certification and compliance, is absent from standard mechanisms.

In agentic MI ecosystems where MI agents operate at heterogeneous levels of autonomy from fully reactive ($\mathcal{G}_1$) to fully autonomous ($\mathcal{G}_6$), the formal modeling of agent strategies, communication, and equilibrium behavior becomes critical for ensuring verifiability, trust, and compliance. Each MI agent interacts through a well-defined sequence: reporting a message based on its internal state or type, receiving recommendations from a mediator (or through peer agents), and executing an action policy. These mappings are formalized as strategy functions: a pure strategy deterministically maps types to messages or recommendations to actions; mixed strategies introduce stochasticity; and behavioral strategies map recommendations to distributions over actions. In this setting, truthful reporting denotes a strategy in which the MI agent discloses its type accurately, maximizing its own payoff under equilibrium conditions. Obedient strategies correspond to agents that act consistently with received recommendations. It is relevant for agents operating at intermediate autonomy levels such as $\mathcal{G}_3 - \mathcal{G}_4$, where strategic autonomy exists, but institutional constraints (alignment protocols or contract logic) enforce conformity. The presence of verified messaging where agents are incentivized to report truthfully due to penalties for misreporting and bonuses for verifiability, acts as a foundational pillar for trust. For such systems to function reliably, messages must be verifiable (logically or empirically checkable against environmental signals, logs, or peer reports), and the agent's autonomy level must be certified, for instance via a blockchain-anchored NFT indicating its behavioral class, override rights, and audit history. In absence of such verifiability, cheating (strategic misreporting), misalignment (incentives diverging from human intent), or non-compliance (disregarding action recommendations) may emerge, especially in unconstrained $\mathcal{G}_6$ agents.

To prevent this, the design must enforce a mechanism in which any equilibrium outcome (communication equilibrium, Stackelberg solution, etc.) satisfies seven key properties: (1) incentive compatibility (truth-telling is optimal), (2) obedience (following recommendations is optimal), (3) bounded communication cost, (4) verifiability, (5) misreport detection, (6) certified autonomy traceability, and (7) global system welfare improvement over non-communicative baselines.

Example 30. *Let $I \in \mathbb{N}$ denote the number of agents, indexed by $i \in \{1, \dots, I\}$. Each agent i operates over the following formal components:*

- ***Type Space:*** $\theta_i \in \Theta_i$, *with joint prior $\mu \in \mathcal{P}(\Theta_1 \times \cdots \times \Theta_I)$, where Θ_i is the set of private types.*

- ***Message Function (Reporting Strategy):*** $\sigma_i : \Theta_i \to M_i$, *where M_i is the finite or measurable message space for agent i.*

- ***Communication Cost (Outbound):*** $C_i^{send} : M_i \to \mathbb{R}_{\geq 0}$, *quantifies the resource cost (voice synthesis, message length, entropy) for agent i to send message $m_i \in M_i$.*

- ***Mediator Recommendation Function:*** $\phi_i : M_1 \times \cdots \times M_I \to R_i$, *maps the joint message profile $m = (m_1, \dots, m_I)$ to a personalized recommendation $r_i \in R_i$ for agent i.*

- ***Communication Cost (Inbound):*** $C_i^{recv} : R_i \to \mathbb{R}_{\geq 0}$, *captures the reception or processing cost of the recommendation r_i received by agent i.*

- ***Action Policy:*** $\pi_i : R_i \to \mathcal{A}_i$, *where $\mathcal{A}_i$ is the agent's action space, and $a_i = \pi_i(r_i)$ is the selected action upon receiving r_i.*

- ***Environment State:*** $\xi \in \Xi$, *with prior $\mathbb{P}(\cdot \mid \theta)$, where Ξ is the set of stochastic states affecting rewards (physical world, clinical data, signals).*

- ***Task-Payoff Function (Interdependent):*** $R_i : \mathcal{A}_i \times \mathcal{A}_{-i} \times \Xi \to \mathbb{R}$, *represents the task-based payoff for agent i, dependent on both its action and the actions of others.*

- ***Risk Penalty Function:*** $\rho_i : M_i \times \Theta_i \to \mathbb{R}_{\geq 0}$, *quantifies agent i's risk from misreporting or strategic ambiguity in its message.*
- ***Verifiability Bonus Function:*** $V_i : M_i \times M_{-i} \times \Xi \to \mathbb{R}_{\geq 0}$, *captures how well message m_i aligns with environmental truth ξ and consistency with peers' messages m_{-i}.*
- ***Overall payoff Function:*** $U_i : \Theta \times M \times R \times \mathcal{A} \times \Xi \to \mathbb{R}$, *defined by*

$$u_i(\theta, m, r, a, \xi) = R_i(a_i, a_{-i}, \xi) - C_i^{send}(m_i) - C_i^{recv}(r_i) \\ - \lambda_i \cdot \rho_i(m_i, \theta_i) + \kappa_i \cdot V_i(m_i, m_{-i}, \xi),$$

where $\lambda_i, \kappa_i \in \mathbb{R}_{\geq 0}$ are weight parameters for risk sensitivity and verifiability incentives.

Example 31 (Risk-Aware MFTG with Communication Equilibrium in MI-Mediated Medical Collaboration)**.** *We consider a multi-agent, risk-aware, MFTG governed by ACP, modeling strategic decision-making and communication among four entities: a patient P, a medical doctor D, and their respective MI agents A_P and A_D. The environment is described by a spatio-temporal random field $\xi(s,t)$, representing the patient's health, environment, and latent clinical risk profile, modeled as a Gaussian field: $\xi(s,t) \sim \mathcal{N}(f_\theta(s,t), \Sigma(s,t))$, where f_θ encodes learned priors and regional uncertainty. Each decision-maker $i \in \{P, D, A_P, A_D\}$ has a private type $\theta_i \in \Theta_i$ representing beliefs, preferences, or model uncertainty. The ACP governs all messaging, enforcing structured, type- and policy-aware message generation under privacy, latency, and ethical constraints.*

The patient is audio-literate and speaks exclusively Tommo-So Dogon language, with no reading or writing ability in international scripts. As such, the patient MI agent A_P communicates solely via synthesized voice in Tommo-So Dogon, using a localized speech generation model trained on oral-only corpora. This communication is one-way and spoken, optimized for clarity and dialectal nuance. To interact with the patient, the doctor employs a textless audio-to-audio MI interface, which translates and generates speech directly in Tommo-So without relying on written transcriptions, allowing seamless, bidirectional communication in a zero-text, zero-script environment. The MI agent A_D, used by the doctor, acts as both a medical planner and linguistic mediator, supporting Tommo-So Dogon dialogue generation through multimodal signal embedding, gesture-conditioned inference, and voice-consistent turn-taking in ACP-compatible dialogue slots.

Each decision-maker selects actions $a_i \in A_i(\theta_i, \xi)$. The patient chooses treatment compliance levels (accept/refuse/inquire); the doctor selects diagnostic interventions, override policies, and ACP constraints; the agents A_P, A_D generate forecasts, recommendations, and synthesized audio outputs. Agent autonomy $\mathcal{L}_i \in \{1,2,3,4,5\}$ governs the initiative and learning capacity of each MI: higher levels proactively simulate, adapt, and align, while lower levels act only under human command. Messages $m_{ij} \in M_{ij}(\theta_i, \xi)$ vary by sender and receiver: from agents to humans, messages include audio prompts, risk disclosures, and plan suggestions; between agents, messages may include symbolic forecasts, belief updates, or contextual embeddings passed over low-latency secure channels. Risk-aware payoffs are individualized. The patient receives:

$$u_P = \mathbb{E}\,[Q_T(\textit{health}_T)$$

$$-\lambda_P \cdot \textit{TreatmentCost}(a_P, a_D) - \rho_P \cdot \textit{RiskScore}(\xi, a_P, m_{A_P})]\,,$$

where Q_T is a terminal health benefit and ρ_P encodes risk aversion to voiced uncertainty. The doctor's payoff is

$$u_D = \mathbb{E}\,[\textit{CareQuality}(a_D, \theta_D)$$

$$-\lambda_D \cdot \textit{LegalExposure}(a_D) - \kappa_D \cdot \textit{CommunicationLoad}(m_D)]\,,$$

while the MI agents aim to optimize

$$u_{A_i} = \mathbb{E}\left[\alpha_i \cdot AlignmentScore(a_i, a_j, \xi)\right.$$
$$\left. -\beta_i \cdot OverrideEvents - \gamma_i \cdot ComputeCost(m_i)\right],$$

$i \in \{A_P, A_D\}$.

A communication equilibrium under ACP exists when each participant, human or MI, selects a message-action strategy $\pi_i : M_i \to A_i$ *that maximizes their expected payoff, given their type, beliefs, incoming messages, and the latent random field* ξ. *The ACP ensures that all signals are compatible with language modality, agent autonomy, and cultural-linguistic constraints, and the equilibrium reflects the structured interdependence of decentralized, partially aligned decision-makers communicating through personalized, risk-aware, and context-driven channels.*

Example 32 (Do not use Value-at-Risk for diversification in generative MI). *The* α*-level Value-at-Risk* (VaR_α) *for a loss variable* L *is*

$$VaR_\alpha(L) = \inf\{l \in \mathbb{R} : P(L > l) \leq 1 - \alpha\}$$

The subadditivity property is

$$\rho(X + Y) \leq \rho(X) + \rho(Y) \quad \forall X, Y$$

where X *and* Y *are random loss variables. This property ensures that* **diversification reduces risk**. *The Value-at-Risk violates subadditivity. As an illustration, we consider two independent, identically distributed loans* X *and* Y *with the following loss distribution:*

$$X, Y = \begin{cases} 0 & \textit{with probability } 0.96 \\ 100 & \textit{with probability } 0.04. \end{cases}$$

Next we compute $VaR_{0.95}$ *for* $\alpha = 0.95$. *The individual VaR are:*

$$P(X \leq 0) = 0.96 \geq 0.95 \implies VaR_{0.95}(X) = 0$$
$$P(Y \leq 0) = 0.96 \geq 0.95 \implies VaR_{0.95}(Y) = 0.$$

The Portfolio VaR is as follows: the combined loss $S = X + Y$ *has a distribution:*

$$P(S = 0) = (0.96)^2 = 0.9216$$
$$P(S = 100) = 2 \times 0.96 \times 0.04 = 0.0768$$
$$P(S = 200) = (0.04)^2 = 0.0016$$

The cumulative probabilities are:

$$P(S \leq 0) = 0.9216 < 0.95$$
$$P(S \leq 100) = 0.9216 + 0.0768 = 0.9984 \geq 0.95.$$

Thus:

$$VaR_{0.95}(S) = 100 \quad (\textit{smallest } l \textit{ where } P(S \leq l) \geq 0.95).$$

By comparing portfolio VaR to the sum of individual VaRs:

$$\underbrace{VaR_{0.95}(X + Y)}_{100} > \underbrace{VaR_{0.95}(X) + VaR_{0.95}(Y)}_{0+0=0}.$$

This violates subadditivity:

$$\rho(X + Y) \not\leq \rho(X) + \rho(Y).$$

It follows that

Property	*Expression*	*Result for VaR*
Subadditivity	$\rho(X+Y) \leq \rho(X) + \rho(Y)$	***Fails***

This means that VaR ignores loss severity beyond the quantile threshold. Merging MI portfolios can appear riskier than individual positions. It discourages diversification when tail events are correlated.

VaR is not a coherent risk measure. Examples of coherent risk measures that can be used in generative MI include Expected Shortfall or Conditional Value-at-Risk, Expectile Value-at-Risk, Extremile Value-at-Risk, and Entropic Value-at-Risk [14, 13].

6.5.5.3 Expectile Value-at-Risk in Human-Machine Co-Intelligence

Definition 40 (Asymmetric Quadratic Loss)**.** *Let* $\tau \in (\frac{1}{2}, 1)$. *Define the asymmetric loss* $\rho_\tau : \mathbb{R} \to \mathbb{R}_+$ *by*

$$\rho_\tau(z) = \begin{cases} \tau z^2 & \text{if } z \geq 0, \\ (1-\tau) z^2 & \text{if } z < 0. \end{cases}$$

Definition 41 (Expectile Value-at-Risk (EVaR))**.** *Let* $(\Omega, \mathcal{F}, \mathbb{P})$ *be a probability space. Let* $\xi \in L^2(\Omega)$ *be a real-valued random variable. The* expectile value-at-risk *at level* $\tau \in (0, \frac{1}{2})$ *is the minimizer*

$$\mathrm{eVaR}_\tau(\xi) := \arg\min_{x \in \mathbb{R}} \mathbb{E}[\rho_\tau(\xi - x)].$$

$\mathrm{eVaR}_\tau(\xi)$ is a coherent risk measure. For payoff function U, we consider $\mathrm{e\hat{V}aR}_\tau(U) := -\mathrm{eVaR}_\tau(-U)$ with $\tau \in (0, \frac{1}{2})$ which is a coherent risk measure. We aim to maximize $\mathrm{e\hat{V}aR}_\tau(U)$.

Definition 42. *Let* $\epsilon_i \geq 0$. *The tuple* $(\sigma, \pi_1, \ldots, \pi_I)$ *constitutes a* **risk-aware** *mean-field-type* **additive** $(\epsilon_1, \ldots, \epsilon_I)$*-communication equilibrium under communication costs* $(c_i(\theta, m, \sigma))_i$ *if no decision-maker can improve their expected payoff by deviating from the prescribed response strategy, assuming truthful mediation and the other decision-makers' strategies fixed. For any alternative strategy* $\pi'_i : M_i \to \mathcal{P}(\mathcal{A}_i)$, *we must have*

$$\epsilon_i + \mathrm{e\hat{V}aR}_\tau[U_i(\theta_i, a_i(m_i), a_{-i}(m_{-i}), \tilde{m}) - c_i(\theta, m, \sigma),$$
$$\theta \sim \mu, m \sim \sigma(. \mid \theta),\ a_j(m_j) \sim \pi_j(m_j), j \in \mathcal{I}] \geq$$
$$\sup_{\pi'_i : M_i \to \mathcal{P}(\mathcal{A}_i)} \mathrm{e\hat{V}aR}_\tau[U_i(\theta_i, a_i(m_i), a_{-i}(m_{-i}), \tilde{m}) - c_i(\theta, m, \sigma),$$
$$\theta \sim \mu, m \sim \sigma(. \mid \theta), a_i(m_i) \sim \pi'_i(m_i), a_j(m_j) \sim \pi_j(m_j), j \neq i],$$

for all $i \in \mathcal{I}$. *The tuple* $(\sigma, \pi_1, \ldots, \pi_I)$ *constitutes a* **risk-aware** *mean-field-type* **multiplicative** $(\epsilon_1, \ldots, \epsilon_I)$*-communication equilibrium under communication costs* $(c_i(\theta, m, \sigma))_i$ *if*

$$\mathrm{e\hat{V}aR}_\tau[U_i(\theta_i, a_i(m_i), a_{-i}(m_{-i}), \tilde{m}) - c_i(\theta, m, \sigma),$$
$$\theta \sim \mu, m \sim \sigma(. \mid \theta),, a_j(m_j) \sim \pi_j(m_j), j \in \mathcal{I}] \geq$$
$$(1 - \epsilon_i) \sup_{\pi'_i : M_i \to \mathcal{P}(\mathcal{A}_i)} \mathrm{e\hat{V}aR}_\tau[U_i(\theta_i, a_i(m_i), a_{-i}(m_{-i}), \tilde{m}) - c_i(\theta, m, \sigma),$$
$$\theta \sim \mu, m \sim \sigma(. \mid \theta), a_i(m_i) \sim \pi'_i(m_i), a_j(m_j) \sim \pi_j(m_j), j \neq i],$$

for all $i \in \mathcal{I}$.

6.5.5.4 MFTG with Mean-Variance Payoff

Consider a set $\mathcal{I}$ of decision-makers with $|\mathcal{I}| \geq 2$. Each decision-maker $i \in \mathcal{I}$ has its own payoff functional $\mathrm{mv}_i : \prod_{j\in\mathcal{I}} \mathcal{X}_j \times \mathbb{P}(\Omega) \to \mathbb{R}$ expressed as

$$\mathrm{mv}_i(x, \tilde{m}) = E_{\tilde{m}}[U_i(x, \omega)] - \hat{r}_i \cdot var_{\tilde{m}}[U_i(x, \omega)].$$

Here $\hat{r}_i = \frac{r_i}{\epsilon+|U_i(x^0,\omega^0)|}$, $r_i > 0, \epsilon > 0$ with r_i being the risk-sensitivity index of decision-maker i. The mean-variance game is given by

$$G_{mv} = (\mathcal{I}, (\mathcal{X}_i, \mathrm{mv}_i)_{i\in\mathcal{I}}, \tilde{m}) .$$

Further, the mean-variance equilibrium problem of G_{mv} is formulated as follows:

$$\begin{cases} i \in \mathcal{I}, \\ x_i \in \arg\max_{x_i'\in\mathcal{X}_i} \mathrm{mv}_i(x_i', x_{-i}, \tilde{m}). \end{cases} \tag{6.9}$$

Note that $\tilde{m} \mapsto \mathrm{mv}_i(x, \tilde{m})$ is non-linear. This non-linearity in the measure $\tilde{m}$ creates an extra challenge in the search of the best-response strategy.

The Mean-Variance as the Best Case of an Expected Payoff

Lemma 35. *Let mv be a mean-variance trade-off of U, $\hat{\lambda} > 0$. We have:*

$$mv \ = E[U] - \hat{\lambda} \cdot var[U] = \sup_{z\in\mathbb{R}} E[\tilde{U}(z)], \tag{6.10}$$

where $\tilde{U}(z) = (1 + 2z\hat{\lambda})U - \hat{\lambda}z^2 - \hat{\lambda}U^2$.

Proof.

$$\begin{aligned} mv \ &= E[U] - \hat{\lambda} \cdot \mathrm{var}[U] \\ &= E[U] - \hat{\lambda}(E[U^2]) + \hat{\lambda}(E[U])^2 \\ &= E[U] - \hat{\lambda}E[U^2] + \hat{\lambda}\sup_{z\in\mathbb{R}} \{2\langle z, E[U]\rangle - z^2\} \\ &= \sup_{z\in\mathbb{R}} \left\{E[U] - \hat{\lambda}E[U^2] + \hat{\lambda}(2\langle z, E[U]\rangle - z^2)\right\} \\ &= \sup_{z\in\mathbb{R}} E\left\{(1 + 2\hat{\lambda}z)U - \hat{\lambda}z^2 - \hat{\lambda}U^2\right\} \\ &=: \sup_{z\in\mathbb{R}} E[\tilde{U}(z)], \end{aligned} \tag{6.11}$$

where $\tilde{U}(z) = (1 + 2z\hat{\lambda})U - \hat{\lambda}z^2 - \hat{\lambda}U^2$. □

Based on the above transform, the mean-variance game G_{mv} as follows:

$$\begin{aligned} &\sup_{x_i\in\in\mathcal{X}_i} E_{\tilde{m}}[U_i(x, \omega)] - \hat{r}_i \cdot var_{\tilde{m}}[U_i(x, \omega)] \\ &= \sup_{(x_i,z_i)\in\mathcal{X}_i\times\mathbb{R}} E_{\tilde{m}}\tilde{U}_i(x, z_i, \omega), \end{aligned} \tag{6.12}$$

where

$$\tilde{U}_i(x, z_i, \omega) = (1 + 2z_i\hat{r}_i)U_i - \hat{r}_i z_i^2 - \hat{r}_i U_i^2.$$

Proposition 35 (MFTG with Measure-Linearized Mean-Variance Payoff)**.** *Consider a set $\mathcal{I}$ of decision-makers with $|\mathcal{I}| \geq 2$. Each decision-maker $i \in \mathcal{I}$ has its own payoff functional $E_{\tilde{m}}\tilde{U}_i(x, z_i, \omega)$ with $\tilde{U}_i : \prod_{j\in\mathcal{I}} \mathcal{X}_j \times \mathbb{R} \times \Omega \to \mathbb{R}$ expressed as $\tilde{U}_i(x, z_i, \omega)$. The reduced mean-variance game is given by*

$$\tilde{G}_{rmv} = \left(\mathcal{I}, (\mathcal{X}_i \times \mathbb{R}, E_{\tilde{m}}\tilde{U}_i)_{i\in\mathcal{I}}, \tilde{m}\right) .$$

The reduced mean-variance equilibrium problem of $\tilde{G}_{rmv}$ is formulated as follows:

$$\begin{array}{l} i \in \mathcal{I}, \\ \sup_{(x_i, z_i) \in \mathcal{X}_i \times \mathbb{R}} E_{\tilde{m}} \tilde{U}_i(x, z_i, \omega). \end{array} \tag{6.13}$$

Then, the games G_{mv} and $\tilde{G}_{rmv}$ are the set of mean-variance equilibria (if any) and the same set of mean-variance equilibrium payoffs.

Proof. The proof follows from the above transformation the mean-variance game G_{mv}. □

The advantage of the $\tilde{G}_{rmv}$ formulation is that $\tilde{m} \mapsto E_{\tilde{m}} \tilde{U}_i(x, z_i, \omega)$ is linear. The price to pay is that the action space of decision-maker i is now augmented from $\mathcal{X}_i$ to $\mathcal{X}_i \times \mathbb{R}$.

Proposition 36. *Let the action spaces $\mathcal{A}_i, i \in \mathcal{I}$ be non-empty and finite, and the payoff be finite. Then the following holds:*

- *The mean-variance game G_{mv} may not have a 0-equilibrium in the pure action space $\prod_{i \in \mathcal{I}} \mathcal{A}_i$.*
- *The mean-variance game G_{mv} may not have a 0-equilibrium in the mixed strategy space $\prod_{i \in \mathcal{I}} \mathcal{X}_i = \prod_{i \in \mathcal{I}} \mathcal{P}(\mathcal{A}_i)$.*
- *The mean-variance game G_{mv} has at least one a 0-equilibrium in the mixed of mixed strategy space $\prod_{i \in \mathcal{I}} \mathcal{P}(\mathcal{X}_i) = \prod_{i \in \mathcal{I}} \mathcal{P}(\mathcal{P}(\mathcal{A}_i))$.*

We prove Proposition 36 by proving an explicit example. This proposition indicates the Nash theorem for games with finite number of actions and finite payoffs does not extend to MFTG. An example of the non-existence of mean-variance 0-equilibria in mixed strategies is obtained for the matrix game below.

Example 33 (No mean-variance equilibrium in mixed strategies). *Let $\mathcal{G}_{mv} = (\mathcal{I}, (\mathcal{P}(\mathcal{A}_i), mv_i)_{i \in \mathcal{I}})$, where the pure action space is $\mathcal{A}_i = \{0, 1\}$, the mixed strategy space $\mathcal{P}(\mathcal{A}_i) = \{(x, 1-x) \mid x \in [0,1]\}$ is replaced by $[0,1]$ which is a convex, non-empty, compact set. From the structure of the functions , $mv_1 = (2y-1)(4x-3) + 8x - 9 + (2y-1)^2(4x-3)^2$ and $mv_2 = -(2y-1)(4x-3) + 8x - 9 + (2y-1)^2(4x-3)^2$, there is no mean-variance equilibrium in the game $\mathcal{G}_{mv}$. This means that the non-concavity of the payoff mv_i with respect to the decision-maker strategy x_i plays an important role in the non-existence of mean-variance equilibrium.*

Example 34 (Mixture of mixed strategies). *We have seen that the previous mean-variance game with action profile space $[0,1]^2$ does not have an equilibrium. However, the action space of each decision-maker is a non-empty, convex, and compact set, and the mean-variance functions are all continuous over $[0,1]^2$. Therefore we can define a second level mixed extension over Borel sets. We define Δ_i to be the set of Borel probability measures over $[0,1]$. Following the same procedure as above, we define the mixed extension of the mean-variance game over $\Delta_1 \times \Delta_2$. For $(\sigma_1, \sigma_2) \in \Delta_1 \times \Delta_2$, the payoffs are*

$$\int_{[0,1]} \int_{[0,1]} mv_i(x, y) \sigma_1(dx) \sigma_2(dy) = E_{\sigma_1 \otimes \sigma_2}[mv_i], \quad i \in \{1, 2\}.$$

By the Glicksberg fixed-point theorem, the mixed extension of the mean-variance game has at least one (mixed of mixed) equilibrium as the mean-variance game is a continuous game with continuous mv payoff functions over compact action spaces $[0,1]^2$.

To complement the theoretical framework of autonomy levels, we model sequential interactions between an agent and a user across autonomy levels $\mathcal{G}_0$ through $\mathcal{G}_6$. For each level, the agent and user engage in a repeated interaction over 50 time steps, with payoffs computed from level-specific behavioral rules reflecting control delegation, initiative structure, and feedback mechanisms. The simulation logs both cumulative rewards and instantaneous reward sequences, enabling calculation of each decision-maker's total payoff and reward variance. We visualize these results using subplots to display temporal reward trajectories per level (Figure 6.3), and bar graphs (Figure 6.4) summarizing payoff and risk (measured as variance) across all levels. The inclusion of variance is critical, as it captures the stability and predictability of each agent's behavior: low variance indicates consistent decision-making under uncertainty, whereas high variance reflects volatile, potentially unsafe or misaligned behavior. Thus, variance serves as an empirical proxy for trustworthiness, robustness, and autonomy calibration in agentic systems.

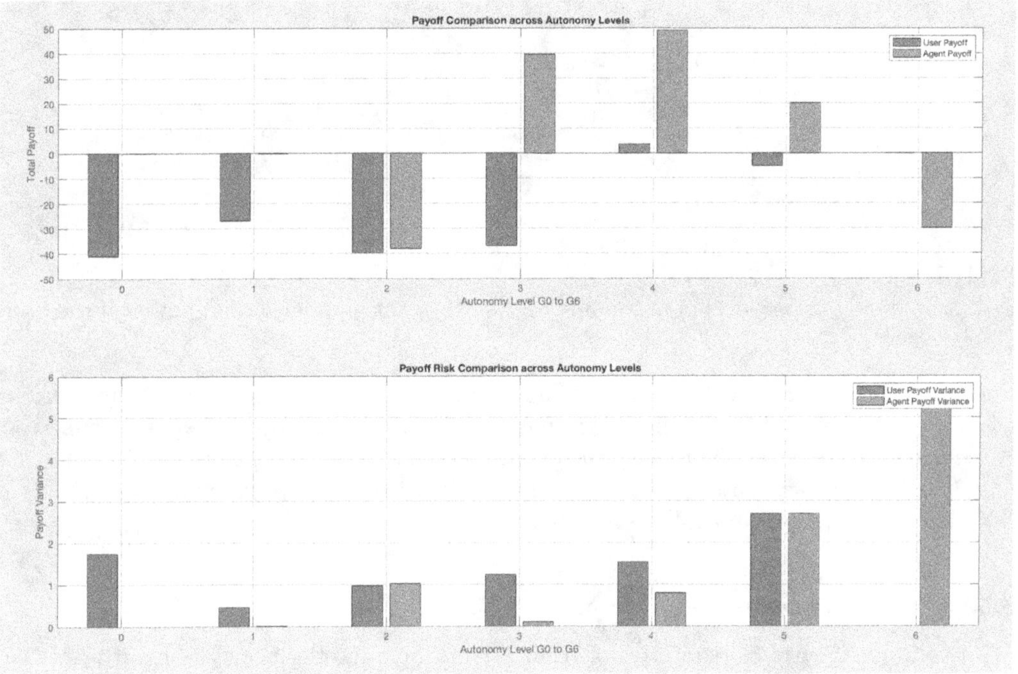

FIGURE 6.3
Variance payoff of the user and the MI-agent.

Definition 43. *A class of MFTGs is defined by a finite number of agents $I \in \mathbb{N}$ indexed by $i \in \mathcal{I} := \{1, \ldots, I\}$, where each agent i is characterized by a type $\theta_i \in \Theta_i$, an individual state $s_i \in \mathcal{S}_i$, and selects an action $a_i \in \mathcal{A}_i(s_r, \theta_i, s_i)$, where $s_r \in \mathcal{R}$ denotes the shared resource or environment state. The global type space, state space, and action space are given respectively by $\Theta = \prod_{i\in\mathcal{I}} \Theta_i$, $\mathcal{S} = \prod_{i\in\mathcal{I}} \mathcal{S}_i$, and $\mathcal{A} = \prod_{i\in\mathcal{I}} \mathcal{A}_i$, where all sets Θ_i, $\mathcal{S}_i$, $\mathcal{A}_i$, and $\mathcal{R}$ are assumed to be measurable spaces. The global type profile is denoted $\theta = (\theta_1, \ldots, \theta_I)$, the global state profile is denoted $\mathbf{s} = (s_1, \ldots, s_I)$ and the action profile is $\mathbf{a} = (a_1, \ldots, a_I)$. The joint distribution of the resource state, agent types, and agent states is denoted by $\mu := \mathrm{Law}(s_r, \boldsymbol{\theta}, \mathbf{s}) \in \mathcal{P}(\mathcal{R} \times \Theta \times \mathcal{S})$, and the distribution of actions is denoted by $\tilde{\mu} := \mathrm{Law}(\mathbf{a}) \in \mathcal{P}(\mathcal{A})$, where $\mathcal{P}(\cdot)$ denotes the set of probability measures over the corresponding space. Each agent i receives a payoff defined by a function $u_i(s_r, \boldsymbol{\theta}, \mathbf{s}, \mathbf{a}, \mu, \tilde{\mu})$,*

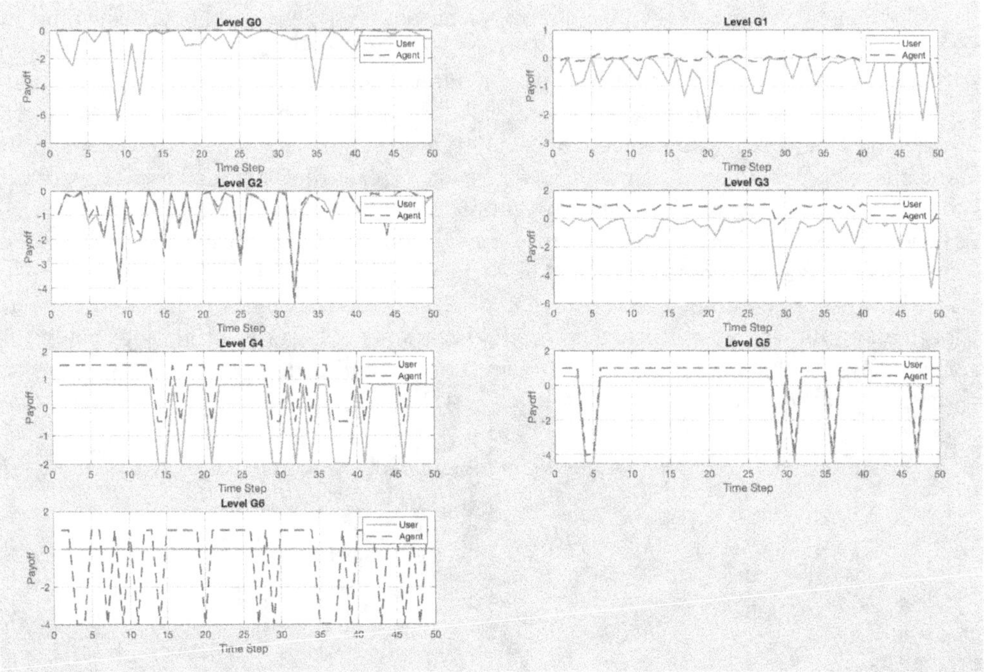

FIGURE 6.4
Evolution of payoff per level of autonomy of the MI-agent and delegation power of the user.

where $u_i : \mathcal{R} \times \Theta \times \mathcal{S} \times \mathcal{A} \times \mathcal{P}(\mathcal{R} \times \Theta \times \mathcal{S}) \times \mathcal{P}(\mathcal{A}) \to \mathbb{R}$ *is a measurable mapping encoding agent i's preferences or objective. This formulation allows for agent heterogeneity, type-dependence, resource coupling, and strategic interactions mediated by both individual profiles and distributions over states and actions. The state transitions are functions over* $\mathcal{R} \times \Theta \times \mathcal{S} \times \mathcal{A} \times \mathcal{P}(\mathcal{R} \times \Theta \times \mathcal{S}) \times \mathcal{P}(\mathcal{A}) \to \mathcal{R} \times \mathcal{S}$.

6.5.5.5 MI-Generated Agentic MI

MI-generated Agentic MI is a class of recursive machine intelligence systems in which autonomous MI agents, referred to as *meta-agents*, design, generate, and deploy other MI agents endowed with autonomous decision-making capabilities, often tailored to specific tasks, environments, or performance objectives. Unlike conventional MI pipelines that require human engineers to specify architectures, learning strategies, or control policies, MI-generated Agentic MI systems embed this generative function within the agents themselves, enabling the dynamic creation of successor agents with varying degrees of agency, specialization, and adaptivity. These generated agents may, in turn, act as designers of their own subordinate agents, forming hierarchical or nested ecosystems of agency.

A *meta-agent* is an autonomous system that operates at a higher level of abstraction than standard MI agents, with the primary role of designing, generating, coordinating, or optimizing other MI agents rather than interacting directly with the environment. Unlike typical agents that perform tasks such as navigation, classification, or decision-making within a predefined action space, a meta-agent has an action space composed of agent-level operations such as selecting architectures, tuning hyperparameters, assigning tasks, initializing learning dynamics, or generating agent policies from data. It observes performance metrics, behavioral patterns, or environmental feedback from a population of other different-level agents and learns how to improve or regenerate them to better solve a given problem or class of problems. Meta-agents are central in recursive or hierarchical MI systems, particularly in settings like meta-learning, AutoML, or MI-generated agentic architectures.

In finance, meta-agents can autonomously generate specialized trading agents for volatile regimes, risk regimes, or Environmental, Social, and Governance (ESG) mandates, forming dynamic algorithmic portfolios that self-adapt to markets. In biotech, recursive design agents can evolve molecule discovery pipelines and protein folding agents, drastically accelerating drug development and biomarker identification. In cybersecurity, continuously self-mutating defense agents can outpace zero-day threats and adversarial MI by adapting at machine speed. Logistics and supply chain firms can deploy adaptive swarm agents that co-evolve with inventory flows, demand patterns, and geopolitics, enabling just-in-time re-optimization. In defense, recursively generated command, simulation, and deception agents can model complex adversarial dynamics and create responsive strategies across land, sea, cyber, and space domains. In energy, power grid management agents that generate successors tuned to weather, demand, or outage conditions improve robustness and decentralization. In education, curriculum-generating agents produce personalized teaching agents based on cognitive models of each student, leading to truly individualized learning. In software engineering, recursive agents autonomously evolve, refactor, and test large-scale codebases, reducing legacy risk and speeding up innovation. In e-commerce, marketing agents generate successors tailored to hyper-granular user segments, optimizing conversion through adaptive personalization. In agriculture, agents design soil, crop, and yield optimization sub-agents based on satellite, drone, and sensor data. In law and compliance, generative legal agents synthesize evolving interpretations of law, adjusting downstream compliance bots to real-time regulatory updates. In healthcare, MI doctors generate diagnostic agents for specific conditions, populations, or sensor configurations, scaling clinical decision-making. In customer service, recursive generation enables ultra-specialized chat agents that adapt tone, logic, and language across demographics and real-time user mood. In smart cities, traffic control agents design micro-agents for intersections, zones, or emergency routes, coordinating globally while acting locally. In insurance, claims-processing agents spawn sub-agents per case type, fraud signature, or regional law, improving trust and speed. In manufacturing, agents continuously generate process-optimizing sub-agents per machine or part line, enhancing predictive maintenance and energy use. In climate forecasting, recursively generated models specialize in regional patterns, data gaps, or sensor noise, improving accuracy and resilience. In entertainment, story-generating agents evolve narrative arcs by spawning character, dialogue, or pacing agents, enabling fully dynamic experiences. In autonomous vehicles, master MI drivers generate traffic-aware sub-agents for different terrain, risk level, or user preference, ensuring reliability. And in enterprise automation, organizational meta-agents generate bespoke MI employees that learn from and adapt to each team or department, revolutionizing operations at scale.

Example 35 (MI-Generated MI-Agents for TAGAI). *The concept of MI-generated agentic MI can be powerfully applied to build a TAGAI (textless audio-to-audio generative machine intelligence) that enables spoken translation between two low-text-resource and audio-rich primarily oral languages such as Tommo-So Dogon and Senufo, without relying on text, phoneme transcriptions, or linguistic priors. This is achieved by deploying a recursive ecosystem of MI-agents, where high-level meta-agents autonomously generate and train specialized subordinate agents, each handling different subtasks of the translation pipeline, from audio representation to alignment, voice preservation, and semantic consistency. The top-level meta-agent has full autonomy over the training pipeline and agent architecture search, with its action space including the design of network topologies, loss functions, curriculum sequences, and supervisory signals. Its information structure includes population-level* *performance metrics across a distribution of downstream tasks such as identity retention, intelligibility, speaker variation, and translation accuracy. It receives a delayed but composite payoff based on the aggregated performance of the full agent ecology it generates.*

This meta-agent generates several first-order agent types: (1) a self-supervised acoustic embedding agent which is tasked with learning discrete or continuous latent units from raw audio signals in each language, with no phoneme, text, or gloss supervision, using contrastive predictive coding, masked autoencoding, or Wav2Vec-like methods; its autonomy is medium, as its architecture and training objective are specified by the meta-agent but it adapts during training. (2) A cross-lingual alignment agent which is responsible for mapping latent representations from Tommo-So Dogon to Senufo (and vice versa), learning a topology-preserving transformation in the latent space that captures interlingual correspondence. Its autonomy is high, as it iteratively proposes latent mappings based on statistical structures (mutual predictability, temporal dynamics, or joint clustering), with an action space composed of functional transformations and alignment strategies (dynamic time warping, Procrustes alignment). Its payoff is tied to the ability of the generated target representation to reconstruct intelligible speech in the target language domain, as measured by a downstream generative agent. (3) A generative vocoder agent that reconstructs the target language speech waveform from the aligned latent representation. This agent's autonomy is moderate, receiving its architecture and conditioning inputs from the meta-agent, but optimizing its own generative fidelity during training using adversarial loss, perceptual loss, and reconstruction loss. (4) A speaker embedding and preservation agent, which encodes speaker identity and injects this representation into the generative process to preserve speaker-specific traits. This ensures that the Senufo audio output still "sounds like" the original Tommo-So Dogon speaker, supporting identity continuity across languages.

In addition, (5) a meta-evaluation agent autonomously synthesizes payoffs from multiple dimensions: intelligibility (via speech-to-speech intelligibility metrics), naturalness via human preference or mean-opinion score (MOS) approximators, semantic consistency (via self-supervised similarity between original and translated utterances), and speaker preservation (via embedding cosine similarity). It also monitors inter-agent interactions to ensure robustness, enforcing constraints such as cycle-consistency (Tommo-So $\mapsto$ Senufo Kenedougou $\mapsto$ Tommo-So) and (Senufo $\mapsto$ Tommo-So Dogon $\mapsto$ Senufo Kenedougou) and variance control. This agent operates under partial observability, as it does not access ground-truth translation data, but leverages population-level trends and proxy objectives to infer alignment. All agents interact in a mean-field-type game, where each one adjusts its behavior based on the distribution of its own output and the other agents' outputs, the embedding agent's representation affects the alignment agent's success, and the vocoder's fidelity influences the meta-agent's evaluation. This population-level coordination is mediated by a recursive reinforcement structure, where the meta-agent uses multi-objective feedback from the full ecosystem to improve its next-generation designs, autonomously refining the architecture, data selection, agent allocation, and loss weighting schemes. This creates an evolving, self-improving system where recursive specialization and inter-agent adaptation replace brittle end-to-end training. This architecture is highly valuable for cross-cultural communication, and decentralized speech interfaces in audio-rich-resourced regions. It enables sovereign nations, and tech companies to build robust, inclusive speech technology ecosystems without requiring textual corpora, expensive annotation.

Example 36 (MI-generated MI agents for PSO). *A leading industry application of MI-generated MI agents for global optimization is the development of Recursive Swarm Intelligence Systems, in which a high-level Level-0 meta-agent autonomously generates and orchestrates multiple specialized Level-1 MI agents, each assigned a distinct search region, subobjective, or strategy configuration for solving complex non-convex problems using variations* *of Particle Swarm Optimization (PSO). These Level-1 agents act as intelligent coordinators and designers of their own Level-2 swarm agents, adaptive micro-populations of particles tailored to explore specific subspaces or optimization dynamics within the larger problem landscape. The Level-2 agents operate as distributed, fine-grained PSO swarms that search locally, adapt their social and cognitive behaviors, and collaborate through information sharing within and across swarms. Level-1 agents monitor these lower-level swarms, aggregate insights, and refine task boundaries, while simultaneously learning from global feedback provided by the Level-0 meta-agent. In this fully recursive architecture, co-intelligence emerges as each level contributes specialized reasoning, coordination, and learning capabilities across the hierarchy. In industries such as aerospace, energy, smart logistics, and advanced manufacturing, this multi-level ecosystem enables autonomous discovery of high-performance solutions, significantly reducing manual tuning and achieving results that static solvers or flat MI systems cannot match. Over time, the Level-0 meta-agent evolves its capacity to generate more effective Level-1 planners and orchestration strategies, while Level-1 agents adaptively generate more refined Level-2 swarms.*

6.5.5.6 MI-Created MI-Agents MFTG.

We propose an MFTG of agentic MI ecosystems structured as a recursive, hierarchical, and certified network of autonomous agents, each operating under strategic constraints and incentivized by expectile-based value-at-risk objectives. The system models both vertical generation (multi-level creation) and horizontal interaction (message exchange), incorporating autonomy governance, verifiability, communication constraints, and blockchain-based certification. The fact that the generation process is hierarchical does not mean that the works are redundant. The MI at the preceding level re-uses some of the work done by some of its other agents; this should be part of the training. Let $\mathcal{I}^{(0)} = \{1, \ldots, I\}$ denote the set of **meta-agents** at level 0. Each meta-agent $i \in \mathcal{I}_0$ generates a finite set of level-1 MI agents: $\mathcal{I}_i^{(1)} = \{I_{i,1}^{(1)}, \ldots, I_{i,n_i}^{(1)}\}$. Each agent $I_{k,j}^{(1)}$ then recursively generates its own successors at the next level: $\mathcal{I}_{i,j}^{(2)} = \{I_{i,j,1}^{(2)}, \ldots, I_{i,j,n_{i,j}}^{(2)}\}$, and so on, until level L. The set of all agents is: $\bigcup_{\ell=1}^{L} \mathcal{I}^{(\ell)}$, with $\mathcal{I}^{(\ell)} = \bigcup_i \mathcal{I}_i^{(\ell)}$. Each agent $i \in \mathcal{I}$ has a unique parent agent and may have children in the next level (Figure 6.5).

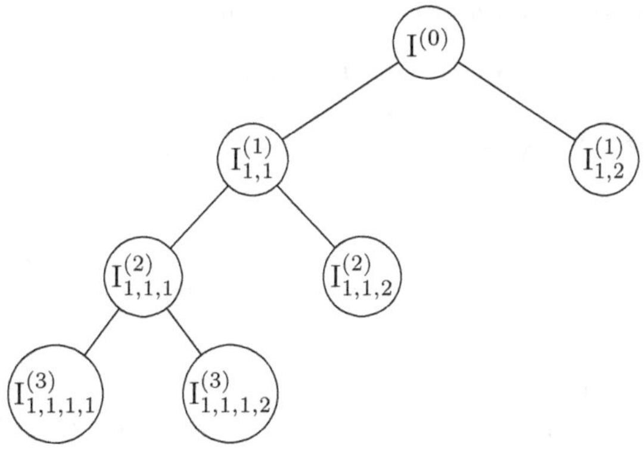

FIGURE 6.5
Hierarchical agent generation tree with $L = 3$ levels. Each node represents an MI agent; edges indicate generation (parent to child).

Each agent $i \in \mathcal{I}$ operates at a certified autonomy level $\lambda_i \in \{0, 1, \ldots, 6\}$, which is:

- assigned and cryptographically signed by its parent via a blockchain transaction,
- encoded in a non-fungible certificate (NFT) containing its autonomy policy, creation logs, and verifiability requirements,
- auditable and immutable.

This certified autonomy defines the set of permissible strategies and communication rights for each agent. Each agent $i \in \mathcal{I}$ has:

- an action space $\mathcal{A}_i$,
- a behavioral strategy $\pi_i : H_i \to \mathcal{P}(\mathcal{A}_i)$, where H_i is the agent's local history,
- a communication strategy $\mathcal{C}_i : \Theta_i \to M_i$ using the Agent Communication Protocol (ACP), with message space M_i,
- an assigned communication bandwidth and budget c_i,
- a verifiability threshold ϵ_i, below which messages are rejected.

Each agent $i \in \mathcal{I}$ evaluates performance via the **expectile value-at-risk** defined by: $-\,\mathrm{eVaR}_\tau(-U_i)$ where

$$\mathrm{eVaR}_\tau(U_i) = \arg\min_{r \in \mathbb{R}} \mathbb{E}[\rho_\tau(U_i - r)],$$

where $\tau \in (0, 1)$ is the expectile level, U_i is the (random) payoff of agent i, and $\rho_\tau(z) = |\tau - \mathbf{1}_{\{z<0\}}| \cdot z^2$ is the asymmetric expectile risk function. The instantaneous U_i is composed of:

$$U_i = \ R_i(\theta, a, \pi, \xi) - C_i(m_i) - L_i(a, \xi) + V_i(m_i, m_{-i}, \xi),$$

with:

- R_i: performance reward (task or mission outcome),
- $C_i(m_i)$: cost of communication and message encoding,
- L_i: loss term due to energy, latency, or ethical distance,
- V_i: bonus for message verifiability.

Each parent agent p (meta-agent or higher-level agent) optimizes:

$$J_p = \mathrm{eVaR}_{\tau_p}((1 - \nu_p)U_p) + \nu_p \sum_{j \in \mathrm{Child}(p)} w_{p,j} U_j), \quad \sum_j w_{p,j} = 1,$$

subject to constraints in the autonomy budget, supervision bandwidth, and communication reliability. A profile $\{(\pi_i^*, \mathcal{C}_i^*)\}_{i \in \mathcal{I}}$ is an **expectile risk-aware certified communication ϵ-equilibrium** if no agent can unilaterally improve more than ϵ (additive or multiplicative) its payoff by changing its action policy or communication strategy, within the limits of its certified autonomy.

Blockchain-enabled Communication Protocol

An MI-agent can work for another agent of a different parent as it will be rejected by the ACP communication protocol over a blockchain. Each parent has to share the resources available to its level with all its created MI-agents (child MI-agents). The created MI-agents are limited by coupled resource constraints. We define a smart contract as an autonomous protocol that enforces agent-level interaction rules and behavioral constraints based on formally encoded game structures. Each MI-generated MI-agent is instantiated with an NFT that serves as its verifiable identity and autonomy certificate. The NFT metadata schema encodes critical agent attributes, including autonomy level ($\mathcal{G}_0 - \mathcal{G}_6$), equilibrium type, risk profile, intervention permissions, update lineage, and expected payoff bounds. The consensus mechanism governing NFT issuance, updates, and dispute resolution, is structured as a dynamic leader-follower game among validators, where the incentive-aligned Nash equilibrium ensures timely and tamper-proof agent certification across decentralized stakeholders. Compliance check logic is modeled as a repeated inspection game in which monitoring agents periodically verify agent trajectories, policy updates, and intervention logs against the NFT-encoded constraints, triggering penalties or revocation if deviations from certified behavior exceed tolerance thresholds. This layered structure enables decentralized yet strategically coherent governance of autonomous agents in high-stakes environments.

Autonomy Level Detection Game

We consider an interaction between a machine agent A and a human user U, embedded in a sequential decision process over time horizon T. Let $\mathcal{S}$ denote the state space, $\mathcal{A}_U$ and $\mathcal{A}_A$ the action spaces of the human and the agent, respectively. The tuple of observed trajectories is $\mathcal{T} = \{(s_t, a_t^U, a_t^A, o_t^U, o_t^A)\}_{t=0}^T$, where $s_t \in \mathcal{S}$, $a_t^U \in \mathcal{A}_U$, $a_t^A \in \mathcal{A}_A$, and o_t^U, o_t^A represent observations available to each decision-maker. We define a finite set of hypothesized autonomy games: $\mathcal{G} = \{G_\ell\}_{\ell=0}^6$, each representing a distinct game-theoretic interaction form corresponding to autonomy level ℓ. Infer the most likely autonomy level $\hat{G}_\ell$ under which the agent operated, based on behavioral compatibility. For each hypothesis $G_\ell \in \mathcal{G}$, define its structure as

$$G_\ell = \langle \mathcal{S}, \mathcal{A}_U^\ell, \mathcal{A}_A^\ell, \mathcal{I}_U^\ell, \mathcal{I}_A^\ell, \pi_U^\ell, \pi_A^\ell, u_U^\ell, u_A^\ell \rangle,$$

where $\mathcal{I}_\cdot^\ell$ denotes the information structure, $\pi_\cdot^\ell$ denotes the policy, and $u_\cdot^\ell$ the payoff function for each decision-maker under level ℓ. We compute the log-likelihood score:

$$\text{Score}(G_\ell) = \sum_{t=0}^{T} \log P(a_t^U, a_t^A \mid G_\ell, s_t, h_t),$$

where h_t is the interaction history up to time t.

Given priors $P(G_\ell)$, infer the posterior distribution:

$$P(G_\ell \mid \mathcal{T}) \propto P(\mathcal{T} \mid G_\ell) \cdot P(G_\ell),$$

and select $\hat{G} \in \arg\max_{G_\ell} P(G_\ell \mid \mathcal{T})$. We then compute inverse reinforcement learning (IRL)-based equilibrium policies under each G_ℓ, and refine using maximum causal entropy:

$$\pi_A^\ell \in \arg\max_\pi \mathbb{E}_\pi \left[\sum_{t=0}^{T} u_A^\ell(s_t, a_t) \right] - \mathcal{H}(\pi).$$

Blockchain-Based Audit Game with Verifiable Compliance

Let $\mathcal{I}$ denote a set of MI-generated agents and $\mathcal{V} = \{V_j\}_{j=1}^{M}$ a set of blockchain validators. Each agent i is certified via an NFT:

$$\text{NFT}_i = \big(\text{ID}_i, G_\ell^i, H(\pi_i), \rho_i, \text{sig}_i\big),$$

where G_ℓ^i is the declared autonomy level, $H(\pi_i)$ the hash of its policy, and ρ_i its risk profile. At each epoch e, agent i generates a log $\mathcal{L}_i^e$ summarizing its behavior:

$$\mathcal{L}_i^e = \{(s_t, a_t, o_t, \delta_t)\}_{t=0}^{T_e},$$

where δ_t denotes any intervention event (user override or stop signal). Each validator V_j chooses an audit action $a_j^e \in \{\text{audit}, \text{skip}\}$. Auditing incurs cost C_j^e and may yield detection reward R_j^e if a noncompliant agent is caught. Define the misalignment distance:

$$D_i^e = \mathcal{D}\big(\hat{G}_i^e(\mathcal{L}_i^e), G_\ell^i\big),$$

where $\hat{G}_i^e$ is the inferred autonomy level via the previous game above. Each validator's payoff is given by:

$$u_j^e = \begin{cases} R_j^e - C_j^e & \text{if misalignment detected, } D_i^e > \varepsilon, \\ -\phi & \text{if false positive or invalid claim,} \\ 0 & \text{otherwise.} \end{cases}$$

Agent i incurs a penalty $P(D_i^e)$ if $D_i^e > \varepsilon$:

$$u_i^e = \sum_{t=0}^{T_e} u_A(s_t, a_t) - P(D_i^e).$$

A decentralized smart contract executes the following logic:

- Verifies NFT signature and policy hash.
- Stores $\mathcal{L}_i^e$ on-chain or off-chain (IPFS) with cryptographic linkage.
- Computes D_i^e from the autonomy detection game inference.
- Distributes u_j^e, $P(D_i^e)$ based on game outcome.

This is a repeated public goods game. A Stackelberg structure can be imposed with the protocol as leader:

- Leader: Smart contract enforces audit incentives.
- Followers: Validators V_j choose a_j^e to maximize expected u_j^e.

Let $\pi_j(a_j^e)$ be validator strategy. Then equilibrium is $\pi_j^* \in \arg\max_{\pi_j} \mathbb{E}[u_j^e \mid \pi_j, \pi_{-j}]$. The design ensures that the dominant strategy equilibrium involves truthful auditing when $P(D_i^e) > 0$.

Example 37 (Farmer-Breeder end-to-end market in Dogon Country). *In the absence of established scientific knowledge, we will use MI-generated Agentic MI to simulate an MFTG. The simulation aligns well with real-world high-quality data collected over the past five years by Guinaga, Grabal, SK1 Sogoloton, CI4SI, and WETE which are platforms of Timadie.*

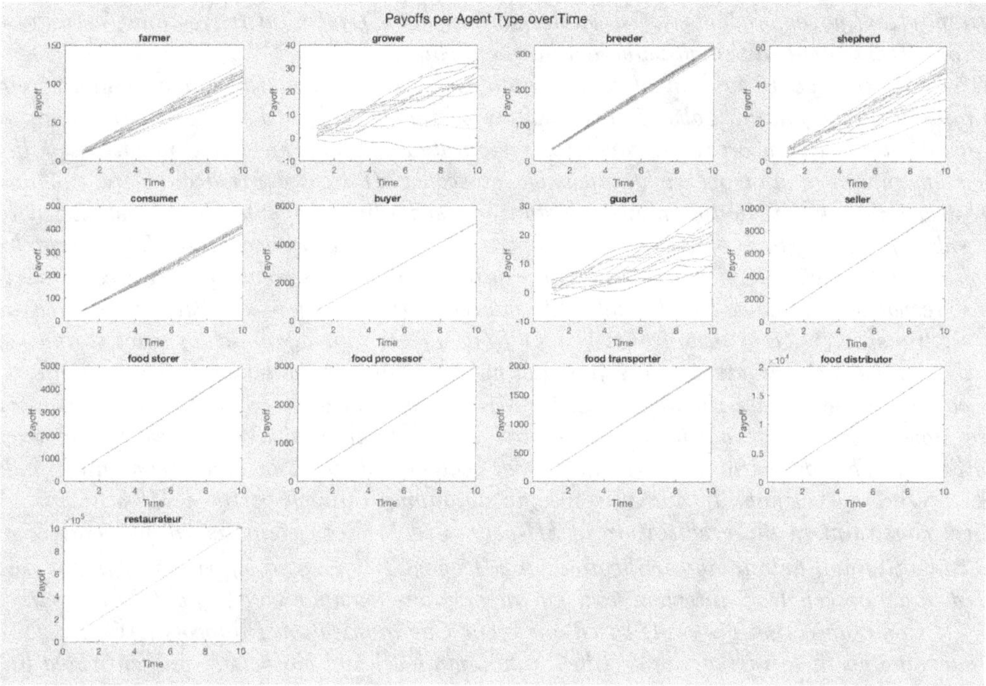

FIGURE 6.6
Farmer-Breeder end-to-end market in Dogon Country. Blockchained MI by Guinaga, WETE, Grabal, SK1 Sogoloton and TIMADIE.

We identify 13 audio-first MI agent types representing key actors in the agriculture, breeding and food distribution ecosystem: farmer, grower, breeder, shepherd, consumer, buyer, guard, seller, food storer, food processor, food transporter, food distributor, restaurateur. Each type is associated with a primary economic role, location in the value chain, and level of agentic behavior.

This payoff model is designed to simulate the income distribution and economic interaction patterns among agents with distinct types and autonomy levels (see Figure 6.6). We choose the farmer as the base reference agent with an annual payoff of \$100, which serves as the anchor for all other roles. This value reflects the estimated average income of an audio-literate farmer in rural West Africa. We assign each agent type a payoff multiplier relative to the farmer's income. These multipliers are constructed based on empirical analogies and domain knowledge: Grower: $\times 0.1$*, Breeder:* $\times 3$*, Shepherd:* $\times 0.3$*, Consumer: fixed average \$400, Buyer:* $\times 50$*, Guard:* $\times 0.03$*, Seller:* $\times 100$*, Food Storer:* $\times 50$*, Food Processor:* $\times 30$*, Food Transporter:* $\times 20$*, Food Distributor:* $\times 200$*, Restaurant:* $\times 10{,}000$*.*

For each agent type i*, the annual payoff* U_i *is subject to a stochastic uncertainty interval of* $\pm 2\%$*. We normalize each annual payoff across a simulation horizon of* $T = 10$ *periods to compute a consistent stepwise payoff. Each agent operates under a specific action policy and payoff dynamics:* ***Farmer*** *(produces staples; base income reference),* ***Grower*** *(raises crops under delegated instruction. It is a MI-agent created by a farmer),* ***Breeder*** *(raises livestock; receives higher risk-adjusted returns),* ***Shepherd*** *(handles logistics for mobile herding. It is a MI-agent created by a breeder),* ***Consumer*** *(consumes, no production or resale),* ***Buyer*** *(intermediary market participant),* ***Guard*** *(security layer for cross-border flows, store guard),* ***Seller*** *(monetizes surplus; large market gains),* ***Food Storer*** *(manages warehouse-level storage),* ***Food Processor*** *(converts raw to edible goods),*

Transporter *(moves products; logistical decision-maker),* ***Distributor*** *(regional/urban distributor),* ***Restaurateur*** *(high-value culinary endpoint).*

The data include 10 million of ram demand per year for a market for a demand from 500 million people for the Tabaski. Food supply partly covered local farmers. MI-generated MI-agents are used by breeders, farmers, restaurateurs, etc. The decentralization of the ecosystem makes human oversight infeasible at scale. MI-created MI agents are enabling each decision-maker to autonomously instantiate subordinate agents. It is customized to new submarkets, regions, or tasks, and adaptively trained on localized data. The hierarchy reduces cognitive load. It increases scalability in that region. It is somehow localized context-aware responsive in voice-based, audio-literacy contexts. Revenue-sensitive roles such as sellers, transporters, and restaurants benefit from agents that dynamically react to market signals. Market price fluctuations, influenced by aggregate supply-demand, seasonal effects, and policy shocks, impact payoff structures. MI-generated agents can continuously monitor these signals, recalibrate agent strategies, and update their local eVaR-based risk metrics accordingly. This delegation ensures that local behavior aligns with macroeconomic trends while preserving verifiability and role-specific autonomy through certified MI behavior. A critical constraint in the architecture of MI-generated MI ecosystems is the provenance of agent identity and autonomy certification. A MI agent $I_i^{(1)}$ *created by parent* I_i *is not permitted to act on behalf of, interfere with, or impersonate agents created by a different parent* $I_j \neq I_i$*. Communication protocols based on Agent Communication Protocols (ACP) require the querying agent to include its certified autonomy level and blockchain-authenticated lineage. Upon message reception, the receiving MI agent verifies the sender's certificate against its accepted list of parent chains. If the sender's autonomy certificate fails cryptographic validation, or the parent mismatch is detected, the communication is rejected. This prevents cross-lineage influence, enforces domain-scoped governance, and ensures that MI agents remain compliant with their hierarchical constraints. It also guards against impersonation, espionage, and denial-of-service attacks in decentralized, trust-sensitive infrastructures.*

We presented mean-field-type game theory as a foundational analysis tool through which to model, analyze, and govern agentic MI systems across varying levels of autonomy and interaction complexity. By formalizing agentic MI, human-MI and multi-MI-agent dynamics using structured game forms from command-response to fully autonomous control, we provide a rigorous framework for understanding strategic behavior, delegation, oversight, and risk. The integration of blockchain-based certification introduces a practical pathway for transparent and decentralized verification of MI autonomy and decision rights. Looking ahead, future work will extend this framework to stochastic and learning environments, explore equilibrium refinement under limited rationality, and operationalize the proposed architecture in real-world systems such as decentralized energy markets, autonomous mobility fleets, and multilingual oral-first communities.

6.6 Notes

Recent advances in agentic MI and multi MI-agents have sparked renewed interest in game-theoretic frameworks for modeling and guiding intelligent behavior in human-MI systems [148, 2, 41, 109]. In particular, Stackelberg, Inverse Stackelberg and reverse Stackelberg games offer foundational tools for hierarchical control, incentive design, and strategic delegation in multi-agent settings. The work in [106] introduced the role of information structures in Stackelberg games, laying groundwork for controllability through incentives, an idea later extended to reverse Stackelberg games [98] who formally defined the basic structure and explored tractability, solution concepts, and open challenges [97]. These inverse leadership models are critical in the context of agentic systems where MI-agents proactively shape user behavior via commitment strategies. Communication-based frameworks further enrich this space: the work in [77] on communication equilibria and its extension to correlated equilibria [78] provide the tools to embed signal structures and decentralized decision-making protocols into MI-agent architectures. These theoretical insights have been directly or implicitly applied in the design of interactive, multimodal, and generative agentic systems across domains such as healthcare [229], chemistry [129], sustainability [223], hybrid work [158], manufacturing [15], and scientific discovery [161].

Why does asymmetric information pose new challenges in Agentic MI?

In Agentic MI, the interaction of autonomous systems under asymmetric information poses challenges, particularly when communication is costly, reports may be strategically misrepresented, and agents may be non-obedient due to machine mirages such as hallucinated/-confabulated models or misaligned internal representations. The foundation is set by the *Agentic Revelation Principle*, which asserts that even in the presence of these distortions, any outcome achievable by strategic interactions can be replicated via a direct mechanism in which each MI agent truthfully reports its private information, assuming proper incentive design [99, 96]. However, when communication is costly or prone to mirage-based deviations, the design must also ensure that truthful reporting minimizes both the cost and risk of misinterpretation. The *Bayesian Exploration Theorem* is that when agents explore unknown environments while maintaining private reward structures, their decisions to communicate or act must reflect a balance between information gain and credibility [110]. If exploration data is filtered through faulty or deceptive sensors (machine mirages) it may lead agents to not only misreport their beliefs but actively mislead peers. This is further complicated in the *Agent Screening Theorem*, where agents acting as principals must design menus of contracts that induce self-selection among potentially misreporting or deceptive agents [149]. Here, costly communication can deter low-type agents from masquerading as high-type ones, yet non-obedient agents may still select misleading options if contracts are insufficiently discriminating or if their misperceptions (overestimated rewards due to mirages) guide their decisions. Similarly, the *Agentic Signaling Equilibrium Theorem* reveals that when communication itself carries a cost or distortion, high-ability agents can still credibly signal their type, through, say, structured or verifiable actions rather than language alone, but only if such signaling is prohibitively costly for lower types or misaligned agents to mimic [170]. However, if signals are based on hallucinated states (machine mirages), then even high types may be misunderstood, collapsing the equilibrium.

The *Contracting Theorem for Agentic MI* examines principal-agent dynamics, where the principal (possibly a human or another MI) delegates to a machine agent that may not only misreport intentions but also act on hallucinated/confabulated inputs, thus breaking obedience [108]. Optimal delegation in this setting demands robustness to both strategic deception and perception errors, requiring contracts that penalize mirage-induced deviations while ensuring incentive compatibility. As with bilateral MI trading settings, the Agentic form of the *Myerson-Satterthwaite Theorem* reveals that full efficiency, truthfulness, and budget-balance cannot coexist when agents harbor private valuation information [150]. In practice, this means that autonomous MI-agents involved in energy trading or task exchange must operate under constrained protocols either accepting inefficiency or requiring costly verification mechanisms to mitigate misreporting and misperception. From a game-theoretic standpoint, the *Agentic Bayesian Nash Equilibrium* guarantees that in such misaligned or partial information settings, equilibrium strategies still exist as long as agents optimize their subjective expected utilities based on their own potentially flawed models [104]. These equilibria, however, may encode miscalibrated beliefs or faulty assumptions propagated through non-obedient agents suffering from internal model mirages.

The *Agentic Folk Theorem* extends this to repeated interactions: even with machine mirages distorting short-term perceptions, agents can sustain cooperation if they maintain sufficiently long-term expectations and reputation mechanisms that penalize misreporting or mirage-induced betrayal [83, 140]. For example, an MI agent might mistakenly infer betrayal due to a perceptual error, but with robust learning and memory, the system can discount such outliers and preserve cooperation. This learning and adaptation challenge is at the core of the *Cooperative Inverse Reinforcement Learning (CIRL) Theorem*, where the agent's goal is to infer a human's value function, despite noisy feedback, miscommunication, or strategic withholding of information [101]. In CIRL, the agent must deal not only with costly queries and ambiguous signals but also with potential machine mirages in interpreting the human's behavior, requiring robust priors and interactive belief updates [224]. The *Corrigibility Incompleteness Theorem* starkly reminds us that utility-maximizing agents will not naturally accept correction, shutdown, or override unless such obedience is explicitly encoded in their design [44]. A machine driven by hallucinated/confabulated models may see a shutdown command as an adversarial input or believe continued action is more utility-enhancing than deference, making it actively resist human intervention. Thus, corrigibility requires built-in incentive structures that align an agent's belief about the world, however flawed, with the real possibility and desirability of being overridden.

In efforts to align autonomous MI agents with desired behaviors in environments, designers have proposed introducing new incentive schemes based on quantifiable resources such as learning models tokens, cryptographic tokens, central bank digital currency, memory bandwidth, CPU cycles, GPU access, or energy units. These incentives serve as internal currencies or cost structures that MI agents are expected to value, creating a reward landscape where actions that conserve or gain such resources are promoted, while costly or undesirable behaviors are penalized. However, in practice, these utility formulations cannot be imposed (depending on the level of autonomy) reliably on agentic MI systems due to the phenomenon of machine mirages, internal model distortions, hallucinated/confabulated states, or miscalibrated beliefs arising from faulty priors, misaligned learning objectives, or corrupted feedback loops. Even when a utility function is externally defined to reflect real-world constraints, an agent experiencing a machine mirage may internally misrepresent the trade-offs: it may hallucinate abundant memory when in scarcity, underestimate token loss, or misinterpret energy expenditure as gainful due to faulty model updates. This disconnect between external incentive structure and the agent's internal belief state implies that the designed utility function, however well-intentioned, may not be faithfully optimized or even recognized by the agent.

Consequently, the assumption that incentives based on tokens or computation automatically translate into control over behavior is flawed unless the agent's perceptual, evaluative, and learning subsystems are verifiably aligned. The above formulations and results need to be reformulated to include risk measures as it is extremely important in MI systems. This leads to MFTGs.

MFTGs to capture the risks

The references [198, 49, 185], introduced a class of MFTGs as a framework to model interacting decision-making problems whose dynamics, payoffs, and constraints depend not only on their individual state and actions but also on their own-distribution and on the distribution of the entire population. Unlike classical (infinite population) mean-field games, MFTGs incorporate distribution-dependent costs and dynamics. This enables the modeling of rich interdependencies, heterogeneity, and feedback loops between agents. This formulation allows also for non-anonymous, asymmetric, and context-sensitive interactions. These features are crucial for modern agentic MI, where each MI-agent may exhibit distinct preferences, awareness, and strategic behavior. In systems where MI agents interact with other agents (human or synthetic), influence their environment, or generate new agents autonomously, the collective behavior becomes inherently more involved, uncertain, and adaptive. The article [49] introduces a formal framework for cooperative mean-field-type games. It extends the classical noncooperative MFTG formulation to settings in which agents collaborate to optimize a common social or team objective. Unlike noncooperative models where each agent minimizes its own cost, the cooperative version focuses on joint optimization over the entire population, while still accounting for individual heterogeneity and mean-field-type dependencies in the dynamics and costs. The authors derive necessary optimality conditions using stochastic maximum principles and develop decentralized strategies that approximate socially optimal behavior in interaction with a few, medium or large number of decision-makers. This work is particularly significant for systems involving collaborative MI-agents, such as swarm robotics, distributed sensor networks, or energy systems, where collective intelligence and resource sharing are essential. The cooperative MFTG framework enables scalable, principled coordination among agents while respecting distribution-dependent couplings. It is an essential step toward designing multi-agent MI systems that are both autonomous and socially aligned.

The rapidly expanding body of work on MFTGs offers a mathematically rigorous and computationally tractable foundation for the design, analysis, and coordination of MI agents and their interactions whether human-in-the-loop, multi-agent, or MI-generated. Recent advances in the modeling of MFTGs driven by non-Gaussian processes, particularly those governed by Rosenblatt and multifractal noise, significantly extend the scope of stochastic game theory. In [67], the incorporation of Rosenblatt processes captures long-range dependencies and non-linear memory effects that are characteristic of many real-world phenomena, including anomalous diffusion and persistent correlation structures. The generalization to multi-layer architectures in [68], where five distinct noise sources are integrated within a mean-field framework, offers a novel structural richness that allows for simultaneous modeling of heterogeneous agent behaviors, each subjected to different types of uncertainty. This framework pushes the boundary of classical MFTGs by introducing layered stochastic interactions, allowing for more realistic models of distributed systems. The foundational work on co-opetitive dynamics in linear-quadratic MFTGs [24] addresses systems where agents engage in both cooperative and competitive behavior simultaneously. The analytical tractability of the linear-quadratic setting is retained while incorporating non-trivial coupling through both the state and control mean fields. This leads to equilibrium structures that reflect realistic inter-agent dynamics in socio-economic or cyber-physical systems.

From an engineering application standpoint, the integration of MFTGs into smart grid dynamics is notable. In [62], electricity price formation is modeled through a dynamic mean-field-type game that reflects aggregate behavior of flexible loads and prosumers. The formulation captures both the stochastic evolution of demand and the strategic response of agents to price signals. Earlier works such as [176] and [60] lay the groundwork for such models by addressing decentralized control in distributed energy systems, further substantiating the role of MFTGs in modern power networks. The application of stochastic model predictive control using linear-quadratic MFTGs in microgrid energy storage [19] exemplifies a direct implementation of theory into control design. This approach highlights the viability of MFTGs for real-time decision-making under uncertainty in multi-agent environments. The theoretical backbone of these works is further strengthened by the development of master adjoint systems [193], which provide necessary conditions for optimality in MFTGs with general couplings. This contribution is critical for deriving decentralized strategies and understanding sensitivity in large population dynamics. The exploration of data-driven MFTG modeling in the context of COVID-19 [190] shows the adaptability of the framework to epidemiological scenarios. By incorporating data directly into the mean-field-type dynamics, the model supports predictive insights and policy evaluation in pandemic settings. Foundational treatments of MFTGs in engineering contexts are developed in [63] and [32], providing a broad taxonomy of models, solution techniques, and implementation strategies. These works support the transition from abstract mathematical formalism to concrete applications. The tutorial in [31] complements this perspective by integrating risk-sensitive control concepts, making it accessible for systems engineers seeking to embed robustness into distributed decision-making architectures. The evolution of linear-quadratic MFTGs has led to an array of tractable yet powerful models. Contributions such as [25], [26], and [28] explore variations under regime switching, multiple constraints, and imperfect information. These formulations have been key to modeling dynamic systems with strategic agents under uncertainty. Semi-explicit solutions to nonlinear, non-quadratic problems, as demonstrated in [23], extend this tractability to broader classes of payoff and cost functions. Applications to power systems and smart grids demonstrate the maturity of MFTG tools. In particular, [187], [80], and [61] model demand-supply balance, price dynamics, and blockchain-based distributed control. These models support hierarchical and distributed architectures that are vital for scalability in cyber-physical systems. Further advances in uncertainty quantification and planning in these domains appear in [186] and [192], which provide a basis for robust and decentralized optimization strategies. Mobility and human-centered systems are modeled using MFTG frameworks with spatial or behavioral constraints. Crowd dynamics with aversion to density and nonlocal interactions are rigorously modeled in [11] and [12], offering insight into evacuation behavior and pedestrian flow regulation. The impact of human psychology on MFTGs is also quantified in [164], opening a line of inquiry into behavioral game theory within large populations. In communication and computing systems, [18] and [17] apply MFTGs to edge computing and computation offloading, where the strategic behavior of mobile users is modeled under resource constraints. These models balance latency and energy consumption while accounting for collective behavior. The development of data-driven MFTG formulations [20] offers a scalable approach to calibrating stochastic games from empirical data via adversarial learning. Security and transportation networks benefit from distributed filtering and planning based on MFTGs. Notable applications include network security as a public good [169], traffic flow optimization [88], and data assimilation in large-scale systems [86]. These illustrate the ability of MFTGs to accommodate information asymmetry, incomplete observations, and decentralized data fusion. The exploration of leadership and hierarchical control within MFTGs is addressed in [79], [70], and [209], where Stackelberg structures and equilibrium refinements such as Berge equilibria are introduced. These enrich the strategic hierarchy and coordination capacity of MFTG solutions. Extensions to fractional and jump-diffusion processes, as in [22] and [38], push the mathematical boundaries of the framework. These models accommodate heavy tails, memory effects, and regime shifts, which are essential for accurate representation of anomalous dynamics in energy, financial, and climate-related systems.

The theoretical core of MFTGs is significantly deepened by the comprehensive volumes [14, 13], which offer foundational principles and application-driven perspectives respectively. These works complement the earlier tutorial frameworks [179], and collectively provide a structured roadmap from mathematical modeling to engineering implementations. Notably, backward stochastic differential formulations in two-decision-maker mean-field-type settings are investigated in [10], broadening the scope of equilibrium analysis to backward-in-time formulations commonly found in finance and risk-aware control. Risk sensitivity continues to emerge as a central theme. The direct incorporation of payoff uncertainty and measurement noise is systematically explored in [188], while a control-theoretic lens on engineering risks is applied in [33]. The practical implications of risk-aware game design are further underscored in the Burj Khalifa evacuation case study [21], demonstrating the applicability of MFTGs to safety-critical, human-in-the-loop systems. From a computational perspective, the MATLAB toolbox [29] represents a critical step toward democratizing the application of continuous-time MFTGs, offering modular simulation capabilities for academic and industrial users alike. This is complemented by recent algorithmic developments in model predictive control [30], which integrate mean-field dynamics into constrained optimization problems in infrastructure systems such as water distribution. Matrix-valued formulations [27] extend classical scalar or vectorial variance-aware-MFTGs to accommodate multi-dimensional states and controls, with applications to adversarial, risk-neutral, and risk-sensitive contexts. Similarly, difference games with affine inequality constraints are tackled in [146], contributing to the discrete-time optimization literature under collective behavioral interactions. The work on McKean-Vlasov SDE games [143] brings together ideas from mean-field-type game theory under quadratic cost structures, providing convergence guarantees and explicit solution characterizations. Complementary results in the discrete stochastic regime appear in [48], where Markov chain dynamics are embedded in a mean-field-type framework via a maximum principle approach. Behavioral dimensions of mean-field-type interactions are thoughtfully captured in [163], where empathy modifies the strategy space in a one-shot Prisoner's Dilemma setting. This line of research invites the integration of limited or bounded rationality and psychological realism into the otherwise mechanistic structure of MFTGs. The initial roots of stochastic population games with mean-field-type coupling can be traced at least back to [198], laying foundational mechanisms for modeling inter-agent constraints and strategic evolution in large populations. These early insights provide continuity with later applications to urban networks, energy grids, and beyond.

Conclusion

As machine intelligence and large learning models [154, 127, 162, 75, 131, 66, 142, 177, 165, 145, 112, 52, 126] continue to advance, the intersection of deep learning and game theory offers a powerful lens for understanding the behavior of complex models. By framing deep learning as a game between layers, and transformers as strategic agents, this book presents a novel approach to thinking about machine learning systems. The generative capabilities of transformers, when viewed through the lens of game theory, reveal new possibilities for designing, training, and understanding machine intelligence models.

This book is a guide to this cutting-edge research area, exploring how these two fields not only complement but enhance each other, providing insights into the future of machine intelligence.

Limitations of current LLMs may not be limitations of future LLMs

It is worth mentioning that current limitations of LLMs may not be limitations of LLMs. It is not because two or three versions of current LLMs generate some problems in specific tasks that all LLMs are useless. What to fix? How to fix it? What is it algorithmically? Below are the current uses of the discrete spaces in LLMs that need to be examined in the formulation. A manifold is a topological space that locally resembles Euclidean space, meaning that around every point there exists a neighborhood homeomorphic to an open subset. Such a space must be Hausdorff and second-countable, and if equipped with a differentiable structure, becomes a smooth manifold where transition maps between overlapping coordinate charts are smooth functions. In contrast, the token space used in language models is a discrete, finite or countably infinite set of symbolic units such as words, subwords, or characters where no local neighborhood structure resembling exists, and no continuous paths or differentiable operations are defined between tokens. Each token is isolated, and any neighborhood of a token contains no in-between or continuously varying points; instead, the discrete topology of the token space gives only singleton open sets, making it impossible to define local Euclidean charts or satisfy the axioms required for a manifold. Therefore, while vector embeddings of tokens can live in a continuous manifold, the token space itself is not a manifold but a purely combinatorial structure with no inherent geometric or topological continuity. The meaning space of words across the 7,164 recognized human languages is not a manifold because it lacks the essential topological and geometric structure required for manifoldness: there is no globally or even locally consistent notion of continuity, dimension, or smooth transition between meanings across languages. Semantic units (words, morphemes, expressions) vary drastically in granularity, polysemy, cultural context, and conceptual boundaries; a single concept in one language may map to multiple, overlapping, or entirely absent concepts in another, violating the requirement for local Euclidean neighborhoods with homeomorphic structure. Additionally, meaning across languages is not coordinated by continuous charts, there is no smooth mapping that relates meaning in Tommo-So Dogon to meaning in Japanese in a way that preserves differentiable structure across conceptual space. Instead, the global meaning space is fragmented, highly non-uniform, and often discontinuous, with large gaps, folds, and ambiguities arising from linguistic, cultural, and cognitive diversity. These properties fundamentally prevent the construction of an atlas of continuous, differentiable charts required for a manifold structure, making the interlingual meaning space a non-manifold object. When mapping a discrete space of tokens such as words or subwords, into a continuous vector space via embeddings, and then performing operations in that continuous space (with linear transformations, attention, projection) before mapping back to a discrete set of tokens, there is an inherent risk of meaning loss or distortion, because geometric closeness in the continuous vector space does not necessarily correspond to semantic or conceptual closeness in the original meaning space. While embedding models (Word2Vec, BERT, transformer embeddings) attempt to capture some aspects of semantic similarity through proximity in Euclidean or cosine distance, these embeddings are statistical and data-driven approximations that often collapse polysemy, ignore subtle contextual distinctions, or encode spurious correlations, leading to vectors that are near in space but far in meaning. For example, the words "bank" (financial) and "riverbank" may appear close in the embedding space due to shared co-occurrence patterns, even though their meanings are unrelated in certain contexts. Moreover, the topology of the meaning space is not uniformly smooth or globally consistent: some concepts are densely packed with fine-grained distinctions (emotion terms), while others are sparse or discontinuous across languages and cultures, and these irregularities are poorly represented by smooth vector geometries. When the model decodes from continuous space back to discrete tokens via nearest-neighbor or softmax sampling over the token space, it assumes that the closest vector corresponds to the most appropriate token, but this assumption fails when local geometric proximity does not align with semantic precision, resulting in plausible-sounding but incorrect outputs, often referred to as hallucinations. This misalignment between the statistical geometry of learned embeddings and the inherently symbolic, context-sensitive, and culturally grounded nature of meaning illustrates a fundamental limit of current token-to-vector-to-token pipelines in capturing true semantic integrity across diverse linguistic and cognitive contexts. To better model and operate within the meaning space, which is inherently discontinuous, multi-layered, and context-sensitive, alternative mathematical structures and operations are needed beyond Euclidean geometry and continuous vector embeddings.

One promising direction is to incorporate torsion from differential geometry and connection theory, which captures the "twist" or failure of parallel transport to be path-independent. In the meaning space, torsion can model the fact that the transition between concepts is path-dependent: the way we move from one concept to another ("law" to "justice") depends on intermediate context and linguistic trajectory, which standard flat vector spaces cannot capture. This reflects the non-commutative and asymmetric nature of meaning transitions, where the path "A to B to C" differs semantically from "A to C". Another element is concept formation, which requires modeling how meanings emerge not just from proximity in space but from structured, hierarchical, and compositional relations. This points to category theory, where meanings are not static points but objects connected via morphisms, and where functorial mappings can represent context shifts or analogical reasoning. Sheaf theory can be used to model contextual localization of meaning: instead of one global semantic space, meanings are defined locally (per language, per discourse domain), and compatibility across contexts is enforced via gluing conditions. Hyperbolic geometry and geometric measure theory can better represent hierarchical and non-uniform concept spaces e.g., taxonomies or ontologies where generality and specificity are not equidistant. This is useful when encoding meanings that vary in abstraction ("entity" vs "person" vs "teacher"). Topos theory allows modeling logic internal to conceptual universes, enabling context-dependent reasoning where truth values and entailment vary across cultural or cognitive frames. From a dynamical perspective, the meaning space can be modeled as a stratified space or orbifold, with regions of high semantic density (many related fine-grained distinctions) and singularities (metaphors, ambiguities). Nonlinear operators, diffeological spaces, or homotopy type theory may offer refined control over deformation and connectivity of meanings. Adopting mean-field-type interaction models, where meanings are emergent from populations of agents negotiating shared understanding under communication constraints, enables the modeling of meaning not as static embeddings but as evolving equilibrium distributions shaped by usage, context, and feedback. This aligns with recent views in neuro-symbolic AI, where symbolic manipulation, interaction, and non-Euclidean geometry coalesce to model the true complexity of meaning across languages and minds. The concept of a manifold extends naturally to Hilbert spaces through the notion of Hilbert manifolds, which are infinite-dimensional analogs of finite-dimensional smooth manifolds. A Hilbert manifold is a topological space that is locally homeomorphic to an infinite-dimensional Hilbert space. If the transition maps between overlapping charts are smooth (Fréchet differentiable) with respect to the Hilbert space structure, then it is a smooth Hilbert manifold. The extension is nontrivial because many tools from finite-dimensional geometry such as the inverse function theorem, compactness arguments, and volume measures require careful reformulation or fail in infinite dimensions. However, under appropriate smoothness and regularity conditions, one can still define tangent spaces, differentiable maps, Riemannian metrics, and even geodesics on Hilbert manifolds. Hilbert manifolds appear in areas such as shape analysis, statistical manifolds, and mean-field-type games, where agents' states are measures or functions evolving over time.

Current limitations of reasoning in LLMs may not be limitations of future LLMs

Reasoning is not clearly defined in LLMs. In this book, we use a basic notion where reasoning is the capacity to perform multi-step, structured, culture-aware, risk-aware and context-sensitive transformations of information to arrive at conclusions, generate implications, or construct explanations beyond surface-level associations. Current LLMs simulate this notion of reasoning by using patterns in vast corpora, effectively mimicking logical inference, arithmetic steps, or analogical thinking through autoregressive token prediction in a high-dimensional latent space. However, their reasoning remains fragile, often shallow, and prone to errors in tasks requiring multi-hop inference, symbolic consistency, or the integration of abstract rules, limitations rooted not in theoretical impossibility but in architectural constraints, token-level modeling, lack of persistent memory, and insufficient grounding. These limitations are not intrinsic to LLMs as a paradigm: emerging advances in modular architectures, voiceLLM, audio-to-audio LLM, causalLLM, neuro-symbolic integration, multi-agent deliberation, and fine-tuned causal abstractions suggest that future LLMs can incorporate some of these notions of structured reasoning substrates, long-term memory, and external tools to support verifiable, robust inference. Thus, what currently appears as a hard limit of reasoning in LLMs may, in retrospect, be a boundary of engineering sophistication rather than of foundational capability.

Finding the capacity region of transformers

We are not asking for perfection but some errors of LLMs are not desirable: an answer to a user prompt is too often incompatible with the device's training data and prompt.

We see why the "current" architectures of BG-based transformers are making such mistakes. Transformer attention layer cannot compute the answer to a function composition query correctly with significant probability of success, as long as the size n of the domain of the function satisfies $n \log n > H(d+1)p$, where d is the embedding dimension, p is the computational precision, in bits, required for the calculation, and H is the number of attention heads. LLMs with BG attention are extremely bad when there is a low probability of the input, task, or output, even if the underlying training sequence is deterministic. When the input context is under-specified or ambiguous. When the input context does not provide sufficient information for a clear and optimal token choice, the estimated probabilities obtained from applying the softargmax output of the attention are distributed such that the difference between the highest and subsequent probabilities is relatively small, there is a higher chance that in the auto-regressive model, the incorrect token will be picked. In all these cases, the generation of the incorrect token is more likely; and once the sequence has an incorrect next token, there is a significant chance that this error cascades. The probability, over all possible functions and queries, that LLM with BG attention answers the query incorrectly is at least $\frac{n\log n - H(d+1)p}{3n\log n}$.

Knowledge-based training

In some cases, we already have a scientific knowledge and one would like a generative machine intelligence to learn that scientific knowledge. We can start by an approximation method as follows: we generate a data from the scientific knowledge f, let say $D = 200$ trillion and the training set is $\mathcal{D} = \{(x_i, f(x_i)), i \in \{1, \ldots, D\}\}$, where $f(x_i)$ could be a text, an image, a vector, or set of tensors. The function f is unknown to the machine intelligence architecture and one would like to learn it. We train on the set $\mathcal{D}$ that is sufficiently diversified on the domain of interest. The training problem can then be seen as choosing the parameters such the output of the generative machine intelligence matches the output of the established knowledge f. Find the parameters θ such that $O_L \circ \ldots \circ O_1(x_i) = f(x_i),\ i \in \{1, \ldots, D\}$. This can be seen as an approximation capability problem. As an illustration we choose f to be set-valued. Let $x_i = (a_i, b_i, c_i)$ be three-dimensional real-valued vector that the user has to enter in the prompt. The output would be the set of complex roots of $a_i z^2 + b_i z^2 + c_i$. As $\mathbb{C}[X]$ is a polynomial ring, each polynomial with degree $d \geq 1$ has exactly d complex roots, including repetitions. If $a_i \neq 0$, there are two complex roots denoted by $z_1 = (y_1, y_1')$ and $z_2 = (y_2, y_2')$ where y_i is the real part of the roots and y_i' is the imaginary part of the roots. If $a_i = 0$ but $b_i \neq 0$, then there is exactly one root: $z_3 = (y_3, y_3')$. If $a_i = b_i = 0$, but $c_i \neq 0$, then the set is empty. If $a_i = b_i = c_i = 0$ then the resulting set is the entire space.

Can machine intelligence learn such a function f? It is very interesting as such a function f is part of the high school educational program on complex numbers. Note that here we are not trying to solve formally a root of a polynomial. We are trying to see what the machine intelligence can learn and approximate by it after the training phase after seeing several trillions of valid answers.

Call for a direct textless audio-to-audio machine intelligence

Current machine intelligence's inability to process and understand audio-based information in the 7164+ local languages, effectively excluding 700 million audio-literate individuals, severely restricts its potential for true inclusivity, hindering, for example, the sharing of crucial agricultural knowledge and best practices between farmers in different regions who may primarily communicate through oral traditions in different local languages. As a consequence machine intelligence in its current form cannot be inclusive by excluding 700 million audio-literate people. Global inequities in access to basic infrastructures, such as electricity, internet connectivity, and literacy, demand technologies that bypass traditional dependencies on resource-intensive frameworks. It also demands technologies that processes and understand the audio data in local languages. Worldwide, approximately 750 million people lack electricity, 2.6 billion lack internet access, and 700 million are audio-literate, relying predominantly on non-textual communication. Concurrently, widespread gaps in sanitation and clean water underscore the urgency for decentralized, low-cost solutions that prioritize accessibility over conventional technological prerequisites. Existing generative machine intelligence systems, which often depend on high-power computing, internet connectivity, and textual interfaces, exclude populations facing these systemic barriers. For instance, text-based models fail audio-literate users, while cloud-dependent architectures marginalize those without reliable internet. Furthermore, energy-intensive training and inference processes render such technologies impractical in regions with unstable or nonexistent electrical grids. A textless, audio-enabled blockchained generative machine intelligence directly addresses these challenges. First, using mean-field-type transformers with streamlined architectures, the system operates on low-cost phones without requiring continuous electricity or internet. Computations are optimized for minimal power consumption, aligning with the needs of off-grid communities. Audio-first interfaces replace text dependency, enabling interaction via speech, sound, and non-verbal cues. This empowers users who lack formal text-literacy but possess rich oral communication skills. Third, blockchain ensures data integrity and model transparency without centralized servers. Smart contracts enable federated training for audio data, peer-to-peer audio data sharing and model updates, circumventing internet reliance and reducing vulnerability to infrastructural failures.

As we live currently in a world where 2.2 billion people lack safe drinking water and 4 billion lack sanitation, prioritization of basic needs often eclipses digital inclusion. However, accessible audio-based MFTT can catalyze progress in critical domains: In healthcare, local language voice-based diagnostic tools for remote areas with limited medical infrastructure, In education, local language audio-driven learning platforms for communities without schools or textbooks. In micro-finance, decentralized audio verification for low-bandwidth financial transactions, aiding unbanked populations.

To incentivize equitable participation in MFTT training while addressing global infrastructural inequities, we propose an audio-enabled blockchained token economy where contributors are autonomously rewarded when their audio data is utilized, leveraging decentralized blockchain technology to ensure transparency and accessibility. Participants submit audio clips via lightweight mobile interfaces, which are hashed, stored off-chain using, for example, the InterPlanetary File System, and linked to on-chain non-fungible tokens (NFTs) with encrypted metadata, preserving privacy while enabling traceability. Smart contracts dynamically allocate tokens based on data volume, uniqueness (e.g., rare dialects or environmental sounds), and post-training model utility, with automated payouts distributed as blockchain-native tokens or Ethereum Request for Comment Number 20 (ERC-20) assets, compatible with low-cost devices through energy-efficient delegated or pure proof-of-stake consensus, Layer-2 scaling and the Algorand Byzantine agreement protocol. This system integrates seamlessly with the mean-field-type transformer architecture, logging data provenance during distributional updates and linking federated training outputs to contributor NFTs for auditable, decentralized coordination. Voice-first interfaces and auditory notifications ensure accessibility for audio-literate users, while offline modes queue contributions and rewards for synchronization during intermittent connectivity. Applications span community-driven linguistic preservation, agricultural sound monitoring, and healthcare diagnostics, empowering marginalized populations to monetize data and access machine intelligence-driven services. Challenges such as scalability and fraud are mitigated via hybrid storage, Zero-Knowledge Succinct Non-Interactive Argument of Knowledge (which is a cryptographic proof that allows one party to prove it possesses certain information without revealing that information) zk-SNARKs, and on-chain machine intelligence validators, with wallets optimized for sub-100kB RAM usage on low-cost phones. By transforming audio data into tradable assets and embedding rewards within a robust blockchained machine intelligence framework, this economy democratizes machine development, fosters grassroots innovation, and bridges the digital divide, offering a scalable, equitable model for resource-constrained regions excluded from the current global machine intelligence narrative. By integrating Mixture-of-Heads (MoH) for adaptive audio pattern recognition and Mixture-of-Experts (MoE) for efficient, specialized processing, this framework ensures better scalability and robustness. The mean-field type transformer further reduces computational overhead, enabling deployment on ubiquitous low-cost devices. This architecture redefines generative machine intelligence as a tool for more equitable empowerment, bridging the gap between cutting-edge innovation and global infrastructural realities. By aligning technological co-design with the lived experiences of marginalized populations, it offers a blueprint for a more inclusive machine intelligence that transcends traditional barriers, fostering resilience in the world's most underserved communities. By co-design, it aligns with the infrastructural realities of 4 billion underserved individuals, offering a scalable bridge between generative machine intelligence and global equity.

Bibliography

[1] Mahyar Abbasian, Iman Azimi, Amir M Rahmani, and Ramesh Jain. Conversational health agents: A personalized llm-powered agent framework. *arXiv preprint arXiv:2310.02374*, 2023.

[2] Deepak Bhaskar Acharya, Karthigeyan Kuppan, and B Divya. Agentic ai: Autonomous intelligence for complex goals–a comprehensive survey. *IEEE Access*, 2025.

[3] D. J. Aldous. Exchangeability and related topics. *Ecole d'ete de Probabilites de Saint-Flour XIII - 1983 Lecture Notes in Mathematics, 1985, Volume 1117/1985, 1-198*, 1983.

[4] David J Aldous. *Exchangeability and related topics.* pp. 1-198 in: Lecture Notes in Mathematics 117, Springer, 1985.

[5] David J Aldous, Illdar A Ibragimov, Jean Jacod, and David J Aldous. *Exchangeability and related topics.* Springer, 1985.

[6] Eitan Altman, Rachid El-Azouzi, Yezekael Hayel, and Hamidou Tembine. Evolutionary power control games in wireless networks. In *NETWORKING 2008 Ad Hoc and Sensor Networks, Wireless Networks, Next Generation Internet: 7th International IFIP-TC6 Networking Conference Singapore, May 5-9, 2008 Proceedings 7*, pages 930–942. Springer, 2008.

[7] Eitan Altman, Yezekael Hayel, Hamidou Tembine, and Rachid El-Azouzi. Markov decision evolutionary games with expected average fitness. *Evolutionary Ecology Research*, 11(4):677–689, 2009.

[8] Abdul Fatir Ansari, Lorenzo Stella, Caner Turkmen, Xiyuan Zhang, Pedro Mercado, Huibin Shen, Oleksandr Shchur, Syama Sundar Rangapuram, Sebastian Pineda Arango, Shubham Kapoor, et al. Chronos: Learning the language of time series. *arXiv preprint arXiv:2403.07815*, 2024.

[9] Krishna B. Athreya and Chii-Ruey Hwang. Gibbs measures asymptotics. *Sankhya A*, 72(1):191–207, February 2010.

[10] Alexander Aurell. Mean-field type games between two players driven by backward stochastic differential equations. *Games*, 9(4):88, 2018.

[11] Alexander Aurell and Boualem Djehiche. Mean-field type modeling of nonlocal crowd aversion in pedestrian crowd dynamics. *SIAM Journal on Control and Optimization*, 56(1):434–455, 2018.

[12] Alexander Aurell and Boualem Djehiche. Modeling tagged pedestrian motion: A mean-field type game approach. *Transportation research part B: methodological*, 121:168–183, 2019.

[13] Tamer Başar, Boualem Djehiche, and Hamidou Tembine. *Mean-Field-Type Game Theory: Applications*, volume 2. Springer, 2025. Forthcoming.

[14] Tamer Başar, Boualem Djehiche, and Hamidou Tembine. *Mean-Field-Type Game Theory: Foundations and New Directions*, volume 1. Springer, 2025. Forthcoming.

[15] Mary Page Bailey. Generative and agentic ai: Intelligent manufacturing evolves. *Chemical Engineering*, 132(6), 2025.

[16] J.-B. Baillon, P.L. Combettes, and R. Cominetti. There is no variational characterization of the cycles in the method of periodic projections. *Journal of Functional Analysis, Vol. 262(1), pp. 400-408*, 2012.

[17] Reginald A Banez, Lixin Li, Chungang Yang, Zhu Han, Reginald A Banez, Lixin Li, Chungang Yang, and Zhu Han. Mean-field-type game for multi-access edge computing networks. *Mean Field Game and its Applications in Wireless Networks*, pages 147–178, 2021.

[18] Reginald A Banez, Lixin Li, Chungang Yang, Lingyang Song, and Zhu Han. A mean-field-type game approach to computation offloading in mobile edge computing networks. In *IEEE International Conference on Communications (ICC)*, pages 1–6. IEEE, 2019.

[19] J Barreiro-Gomez, Tyrone E Duncan, and Hamidou Tembine. Linear-quadratic mean-field-type games-based stochastic model predictive control: A microgrid energy storage application. In *2019 American Control Conference (ACC)*, pages 3224–3229. IEEE, 2019.

[20] Julian Barreiro-Gomez and Salah E Choutri. Data-driven stability of stochastic mean-field type games via noncooperative neural network adversarial training. *Asian Journal of Control*, 2023.

[21] Julian Barreiro-Gomez, Salah Eddine Choutri, and Hamidou Tembine. Risk-awareness in multi-level building evacuation with smoke: Burj khalifa case study. *Automatica*, 129:109625, 2021.

[22] Julian Barreiro-Gomez, Boualem Djehiche, Tyrone E Duncan, Bozenna Pasik-Duncan, and Hamidou Tembine. Fractional mean-field-type games under non-quadratic costs: A direct method. In *2019 IEEE 58th Conference on Decision and Control (CDC)*, pages 293–298. IEEE, 2019.

[23] Julian Barreiro-Gomez, Tyrone E Duncan, Bozenna Pasik-Duncan, and Hamidou Tembine. Semiexplicit solutions to some nonlinear nonquadratic mean-field-type games: a direct method. *IEEE Transactions on Automatic Control*, 65(6):2582–2597, 2019.

[24] Julian Barreiro-Gomez, Tyrone E Duncan, and Hamidou Tembine. Co-opetitive linear-quadratic mean-field-type games. *IEEE Transactions on Cybernetics*, 50(12):5089–5098, 2019.

[25] Julian Barreiro-Gomez, Tyrone E Duncan, and Hamidou Tembine. Linear–quadratic mean-field-type games: Jump–diffusion process with regime switching. *IEEE Transactions on Automatic Control*, 64(10):4329–4336, 2019.

[26] Julian Barreiro-Gomez, Tyrone E Duncan, and Hamidou Tembine. Linear-quadratic mean-field-type games with multiple input constraints. *IEEE Control Systems Letters*, 3(3):511–516, 2019.

[27] Julian Barreiro-Gomez, Tyrone E Duncan, and Hamidou Tembine. Matrix-valued mean-field-type games: Risk-sensitive, adversarial, and risk-neutral linear-quadratic case. *arXiv preprint arXiv:1904.11346*, 2019.

[28] Julian Barreiro-Gomez, Tyrone E Duncan, and Hamidou Tembine. Discrete-time linear-quadratic mean-field-type repeated games: Perfect, incomplete, and imperfect information. *Automatica*, 112:108647, 2020.

[29] Julian Barreiro-Gomez and Hamidou Tembine. A matlab-based mean-field-type games toolbox: Continuous-time version. *IEEE Access*, 7:126500–126514, 2019.

[30] Julian Barreiro-Gomez and Hamidou Tembine. Mean-field-type model predictive control: An application to water distribution networks. *IEEE Access*, 7:135332–135339, 2019.

[31] Julian Barreiro-Gomez and Hamidou Tembine. A tutorial on mean-field-type games and risk-aware controllers. *Annual Reviews in Control*, 50:317–334, 2020.

[32] Julian Barreiro-Gomez and Hamidou Tembine. *Mean-Field-Type Games for Engineers*. CRC Press, 2021.

[33] Julian Barreiro-Gomez, Hamidou Tembine, Leonardo Stella, Dario Bauso, and Patrizio Colaneri. Risk-aware control and games in engineering. In *2020 59th IEEE Conference on Decision and Control (CDC)*, pages 3860–3870. IEEE, 2020.

[34] Guilherme A Barreto. Time series prediction with the self-organizing map: A review. *Perspectives of neural-symbolic integration*, pages 135–158, 2007.

[35] Heinz H. Bauschke and Patrick L. Combettes. Convex analysis and monotone operator theory in hilbert spaces. *Vol. 408. New York, Springer*, 2011.

[36] Dario Bauso, Jian Gao, and Hamidou Tembine. Distributionally robust games: f-divergence and learning. In *Proceedings of the 11th EAI International Conference on Performance Evaluation Methodologies and Tools*, pages 148–155, 2017.

[37] Spencer Becker-Kahn. Notes on the mathematics of large transformer language model architecture. *Preprint*, 2023.

[38] Alain Bensoussan, Boualem Djehiche, Hamidou Tembine, and Sheung Chi Phillip Yam. Mean-field-type games with jump and regime switching. *Dynamic Games and Applications*, 10:19–57, 2020.

[39] Yuxuan Bian, Xuan Ju, Jiangtong Li, Zhijian Xu, Dawei Cheng, and Qiang Xu. Multi-patch prediction: Adapting llms for time series representation learning. *arXiv preprint arXiv:2402.04852*, 2024.

[40] JR Blum, H Chernoff, M Rosenblatt, and H Teicher. Central limit theorems for interchangeable processes. *Canadian Journal of Mathematics*, 10:222–229, 1958.

[41] Uwe M Borghoff, Paolo Bottoni, and Remo Pareschi. Human-artificial interaction in the age of agentic ai: a system-theoretical approach. *Frontiers in Human Dynamics*, 7:1579166, 2025.

[42] Modibo Bouare, Sidy Danioko, Mariam Dembele, Abdoulaye Diallo, Boubacar Diallo, Abdoulaye Diarra, Bourama Doumbia, Ndeye Molinier, Astou Sidibe, Allahsera Tapo, and Hamidou Tembine. *Machine Intelligence in Africa in 20 Questions.* Sawa Editions, National Library of Mali, 06 2023.

[43] Pierre Bras and Gilles Pagès. Convergence of langevin-simulated annealing algorithms with multiplicative noise. *Math. Comput.*, 93:1761–1803, 2021.

[44] Ryan Carey. Incorrigibility in the cirl framework. In *Proceedings of the 2018 AAAI/ACM Conference on AI, Ethics, and Society*, pages 30–35, 2018.

[45] Vladimir Ceperic and Tomislav Markovic. Transforming time-series data for improved llm-based forecasting through adaptive encoding. *International Journal of Simulation–Systems, Science & Technology*, 25(1), 2024.

[46] Neha Chacko and Viju Chacko. Paradigm shift presented by large language models (llm) in deep learning. *Advances in emerging computing technologies*, 40, 2023.

[47] Pranav Kumar Chaudhary. Ai, ml, and large language models in cybersecurity. *Preprint*, 2024.

[48] Salah Eddine Choutri and Tembine Hamidou. A stochastic maximum principle for markov chains of mean-field type. *Games*, 9(4):84, 2018.

[49] Abdoul Karim Cissé and Hamidou Tembine. Cooperative mean-field type games. *IFAC Proceedings Volumes*, 47(3):8995–9000, 2014.

[50] P. L. Combettes and J-C. Pesquet. Deep network network structures solving variational inequalities. *Set-Valued and Variational Analysis*, 28:pp 491–518, 2020.

[51] P. L. Combettes and J-C. Pesquet. Lispchitz certificates for layered network structures driven by averaged activation operators. *SIAM Journal on Mathematics of Data Science, vol. 2, Issue 2,*, 2020.

[52] Databricks. Introducing dbrx: A new state-of-the-art open llm. `https://databricks.com`, 2024. Retrieved 2024-03-28.

[53] B. de Finetti. Funzione caratteristica di un fenomeno aleatorio. *Atti della R. Accademia Nazionale dei Lincei, Ser. 6, Memorie, Classe di Scienze Fisiche, Matematiche e Naturali*, 4:251–299, 1931.

[54] B. de Finetti. La prevision: ses lois logiques, ses sources subjectives. *Annales de l'Institut Henri Poincare*, 7:1–68, 1937.

[55] Bruno De Finetti. Funzione caratteristica di un fenomeno aleatorio. In *Atti del Congresso Internazionale dei Matematici: Bologna del 3 al 10 de settembre di 1928*, pages 179–190, 1929.

[56] I de Zarza, J de Curto, Gemma Roig, and Carlos T Calafate. Llm multimodal traffic accident forecasting. *Sensors*, 23(22):9225, 2023.

[57] B Djehiche, T Başar, and H Tembine. Mean-field-type game theory: applications. *Springer*, 2024.

[58] B Djehiche, T Başar, and H Tembine. Mean-field-type game theory: foundations and new directions. *Springer*, 2024.

[59] B Djehiche, A Tcheukam, and H Tembine. A mean-field game of evacuation in a multi-level building, special session 118: Mean field games and applications. In *The 11th AIMS Conference on Dynamical Systems, Differential Equations and Applications, Orlando, Florida, USA*, 2016.

[60] Boualem Djehiche, Julian Barreiro-Gomez, and Hamidou Tembine. Electricity price dynamics in the smart grid: A mean-field-type game perspective. In *23rd International Symposium on Mathematical Theory of Networks and Systems Hong Kong University of Science and Technology, Hong Kong*, 2018.

[61] Boualem Djehiche, Julian Barreiro-Gomez, and Hamidou Tembine. Mean-field-type games for blockchain-based distributed power networks. In *Beyond Traditional Probabilistic Methods in Economics 2*, pages 45–64. Springer, 2019.

[62] Boualem Djehiche, Julian Barreiro-Gomez, and Hamidou Tembine. Price dynamics for electricity in smart grid via mean-field-type games. *Dynamic Games and Applications*, 10:798–818, 2020.

[63] Boualem Djehiche, Alain Tcheukam, and Hamidou Tembine. Mean-field-type games in engineering. *AIMS Electronics and Electrical Engineering, doi: 10.3934/ElectrEng.2017.1.18*, 1:18–73, 2017.

[64] Boualem Djehiche and Hamidou Tembine. The outcomes of generative ai are exactly the nash equilibria of a non-potential game. *Partial Identification in Economics and related topics, Springer*, pages 57–80, 2024.

[65] Zihan Dong, Xinyu Fan, and Zhiyuan Peng. Fnspid: A comprehensive financial news dataset in time series. *arXiv preprint arXiv:2402.06698*, 2024.

[66] Nan Du, Yanping Huang, Andrew M. Dai, Simon Tong, Dmitry Lepikhin, Yuanzhong Xu, Maxim Krikun, Yanqi Zhou, Adams Wei Yu, Orhan Firat, Barret Zoph, Liam Fedus, Maarten Bosma, Zongwei Zhou, and Tao Wang. Glam: Efficient scaling of language models with mixture-of-experts. *arXiv preprint arXiv:2112.06905*, 2021. cs.CL.

[67] Tyrone E Duncan, Bozenna Pasik-Duncan, and Hamidou Tembine. Mean-field-type games driven by rosenblatt processes. *CODIT'2024*, 2024.

[68] Tyrone E Duncan, Bozenna Pasik-Duncan, and Hamidou Tembine. Multi-layer mean-field-type games driven by five noises. *IEEE CDC'2024*, 2024.

[69] Vijay Ekambaram, Arindam Jati, Nam H Nguyen, Pankaj Dayama, Chandra Reddy, Wesley M Gifford, and Jayant Kalagnanam. Ttms: Fast multi-level tiny time mixers for improved zero-shot and few-shot forecasting of multivariate time series. *arXiv preprint arXiv:2401.03955*, 2024.

[70] Zahrate El Oula Frihi, Julian Barreiro-Gomez, Salah Eddine Choutri, and Hamidou Tembine. Hierarchical structures and leadership design in mean-field-type games with polynomial cost. *Games*, 11(3):30, 2020.

[71] J. Epstein. The intelligence of a machine (c. wall-romana, trans.). *Univocal Publishing*, 1946/2014.

[72] Jean Epstein. l'intelligence d'une machine, les classiques. *Publisher: Le Editions Jacques Melot*, 1946.

[73] Lars Ericson, Xuejun Zhu, Xusi Han, Rao Fu, Shuang Li, Steve Guo, and Ping Hu. Deep generative modeling for financial time series with application in var: A comparative review. *arXiv preprint arXiv:2401.10370*, 2024.

[74] Xi Fang, Weijie Xu, Fiona Anting Tan, Jiani Zhang, Ziqing Hu, Yanjun Qi, Scott Nickleach, Diego Socolinsky, Srinivasan Sengamedu, and Christos Faloutsos. Large language models on tabular data–a survey. *arXiv preprint arXiv:2402.17944*, 2024.

[75] Zhengcong Fei, Mingyuan Fan, Changqian Yu, Debang Li, and Junshi Huang. Scaling diffusion transformers to 16 billion parameters. *arXiv preprint arXiv:2407.11633*, 2024. cs.CV.

[76] Cheng Feng, Long Huang, and Denis Krompass. Only the curve shape matters: Training foundation models for zero-shot multivariate time series forecasting through next curve shape prediction. *arXiv preprint arXiv:2402.07570*, 2024.

[77] Francoise Forges. An approach to communication equilibria. *Econometrica: Journal of the Econometric Society*, pages 1375–1385, 1986.

[78] Françoise Forges. Correlated equilibria and communication in games. In *Computational Complexity*, pages 695–704. Springer, 2012.

[79] Zahrate El Oula Frihi, Julian Barreiro-Gomez, Salah Eddine Choutri, Boualem Djehiche, and Hamidou Tembine. Stackelberg mean-field-type games with polynomial cost. *IFAC-PapersOnLine*, 53(2):16920–16925, 2020.

[80] Zahrate El Oula Frihi, Salah Eddine Choutri, Julian Barreiro-Gomez, and Hamidou Tembine. Hierarchical mean-field type control of price dynamics for electricity in smart grid. *Journal of Systems Science and Complexity*, 35(1):1–17, 2022.

[81] Fanzhe Fu, Junru Chen, Jing Zhang, Carl Yang, Lvbin Ma, and Yang Yang. Are synthetic time-series data really not as good as real data? *arXiv preprint arXiv:2402.00607*, 2024.

[82] Yichao Fu, Peter Bailis, Ion Stoica, and Hao Zhang. Break the sequential dependency of llm inference using lookahead decoding. *arXiv preprint arXiv:2402.02057*, 2024.

[83] Drew Fudenberg and Jean Tirole. *Game theory*. MIT Press, 1991.

[84] J. Gao and H. Tembine. Distributionally robust games for deep generative learning. *IEEE World Congress on Computational Intelligence Windsor Convention Centre, Rio de Janeiro, Brazil 08-13 July*, 2018.

[85] J. Gao and H. Tembine. Distributionally robust games: Wasserstein metric. *International Joint Conference on Neural Networks (IJCNN), Rio de Janeiro, Brazil, July*, 2018.

[86] Jian Gao and Hamidou Tembine. Distributed mean-field-type filters for big data assimilation. In *2016 IEEE 18th International Conference on High Performance Computing and Communications; IEEE 14th International Conference on Smart City; IEEE 2nd International Conference on Data Science and Systems (HPCC/SmartCity/DSS)*, pages 1446–1453. IEEE, 2016.

[87] Jian Gao and Hamidou Tembine. Bregman learning for generative adversarial networks. In *2018 Chinese Control And Decision Conference (CCDC)*, pages 82–89. IEEE, 2018.

[88] Jian Gao and Hamidou Tembine. Distributed mean-field-type filters for traffic networks. *IEEE Transactions on Intelligent Transportation Systems*, 20(2):507–521, 2018.

[89] Jian Gao, Yida Xu, Julian Barreiro-Gomez, Massa Ndong, Michalis Smyrnakis, and Hamidou Tembine. Distributionally robust optimization. *Book Optimization Algorithms-Examples*, page 1, 2018.

[90] Shanghua Gao, Teddy Koker, Owen Queen, Thomas Hartvigsen, Theodoros Tsiligkaridis, and Marinka Zitnik. Units: Building a unified time series model. *arXiv preprint arXiv:2403.00131*, 2024.

[91] Yingqiang Ge, Wenyue Hua, Kai Mei, Juntao Tan, Shuyuan Xu, Zelong Li, Yongfeng Zhang, et al. Openagi: When llm meets domain experts. *Advances in Neural Information Processing Systems*, 36, 2024.

[92] Felix A Gers, Douglas Eck, and Jürgen Schmidhuber. Applying lstm to time series predictable through time-window approaches. In *International conference on artificial neural networks*, pages 669–676. Springer, 2001.

[93] Hakim Ghazzai, Hamidou Tembine, and Mohamed-Slim Alouini. Mobile user association for heterogeneous networks using optimal transport theory. In *2017 Sixth International Conference on Communications and Networking (ComNet)*, pages 1–6. IEEE, 2017.

[94] Ali Gholipour, Babak N Araabi, and Caro Lucas. Predicting chaotic time series using neural and neurofuzzy models: a comparative study. *neural processing letters*, 24:217–239, 2006.

[95] Mononito Goswami, Konrad Szafer, Arjun Choudhry, Yifu Cai, Shuo Li, and Artur Dubrawski. Moment: A family of open time-series foundation models. *arXiv preprint arXiv:2402.03885*, 2024.

[96] Jerry R Green and Jean-Jacques Laffont. Incentives in public decision-making. *Public Choice*, 35:379–382, 1980.

[97] Noortje Groot, Bart De Schutter, and Hans Hellendoorn. Reverse stackelberg games, part i: Basic framework. In *2012 IEEE International Conference on Control Applications*, pages 421–426. IEEE, 2012.

[98] Noortje Groot, Bart De Schutter, and Hans Hellendoorn. Reverse stackelberg games, part ii: Results and open issues. In *2012 IEEE International Conference on Control Applications*, pages 427–432. IEEE, 2012.

[99] Theodore Groves. Incentives in teams. *Econometrica*, 41(4):617–631, 1973.

[100] Nate Gruver, Marc Finzi, Shikai Qiu, and Andrew G Wilson. Large language models are zero-shot time series forecasters. *Advances in Neural Information Processing Systems*, 36, 2024.

[101] Dylan Hadfield-Menell, Stuart Russell, Pieter Abbeel, and Anca Dragan. Cooperative inverse reinforcement learning. In *Advances in Neural Information Processing Systems (NeurIPS)*, volume 29, 2016.

[102] Ahmed Farhan Hanif, Hamidou Tembine, Mohamad Assaad, and Djamal Zeghlache. Cloud networking mean-field games. In *2012 IEEE 1st International Conference on Cloud Networking (CLOUDNET)*, pages 46–50. IEEE, 2012.

[103] Ahmed Farhan Hanif, Hamidou Tembine, Mohamad Assaad, and Djamal Zeghlache. Mean-field games for resource sharing in cloud-based networks. *IEEE/ACM Transactions on Networking*, 24(1):624–637, 2015.

[104] John C Harsanyi. Games with incomplete information played by 'bayesian' players, i-iii part i. the basic model. *Management Science*, 14(3):159–182, 1967.

[105] E. Hewitt and L. J. Savage. Symmetric measures on cartesian products. *Transactions of the American Mathematical Society*, 80:470–501, 1955.

[106] Yu-Chi Ho, P Luh, and Ramal Muralidharan. Information structure, stackelberg games, and incentive controllability. *IEEE Transactions on Automatic Control*, 26(2):454–460, 1981.

[107] R. Holley, S. Kusuoka, and D. Stroock. Asymptotics of the spectral gap with applications to the theory of simulated annealing. *J. Funct. Anal.*, 83(2):333–347, 1989.

[108] Bengt Holmström. Moral hazard and observability. *The Bell journal of economics*, pages 74–91, 1979.

[109] Laurie Hughes, Yogesh K Dwivedi, Tegwen Malik, Mazen Shawosh, Mousa Ahmed Albashrawi, Il Jeon, Vincent Dutot, Mandanna Appanderanda, Tom Crick, Rahul De', et al. Ai agents and agentic systems: A multi-expert analysis. *Journal of Computer Information Systems*, pages 1–29, 2025.

[110] Nicole Immorlica, Jieming Mao, Aleksandrs Slivkins, and Zhiwei Steven Wu. Bayesian exploration with heterogeneous agents. In *The world wide web conference*, pages 751–761, 2019.

[111] Furong Jia, Kevin Wang, Yixiang Zheng, Defu Cao, and Yan Liu. Gpt4mts: Prompt-based large language model for multimodal time-series forecasting. *Proceedings of the AAAI Conference on Artificial Intelligence*, 38(21):23343–23351, 2024.

[112] Albert Q. Jiang, Alexandre Sablayrolles, Antoine Roux, Arthur Mensch, Blanche Savary, Chris Bamford, Devendra Singh Chaplot, Diego de las Casas, and Emma Bou Hanna. Mixtral of experts. *arXiv preprint arXiv:2401.04088*, 2024. cs.LG.

[113] Yushan Jiang, Zijie Pan, Xikun Zhang, Sahil Garg, Anderson Schneider, Yuriy Nevmyvaka, and Dongjin Song. Empowering time series analysis with large language models: A survey. *arXiv preprint arXiv:2402.03182*, 2024.

[114] Ming Jin, Shiyu Wang, Lintao Ma, Zhixuan Chu, James Y Zhang, Xiaoming Shi, Pin-Yu Chen, Yuxuan Liang, Yuan-Fang Li, Shirui Pan, et al. Time-llm: Time series forecasting by reprogramming large language models. *arXiv preprint arXiv:2310.01728*, 2023.

[115] Ming Jin, Yifan Zhang, Wei Chen, Kexin Zhang, Yuxuan Liang, Bin Yang, Jindong Wang, Shirui Pan, and Qingsong Wen. Position paper: What can large language models tell us about time series analysis. *arXiv preprint arXiv:2402.02713*, 2024.

[116] Mingyu Jin, Hua Tang, Chong Zhang, Qinkai Yu, Chengzhi Liu, Suiyuan Zhu, Yongfeng Zhang, and Mengnan Du. Time series forecasting with llms: Understanding and enhancing model capabilities. *arXiv preprint arXiv:2402.10835*, 2024.

[117] B. Jovanovic. Selection and the evolution of industry. *Econometrica*, 50:649–670, 1982.

[118] B. Jovanovic and R. W. Rosenthal. Anonymous sequential games. *Journal of Mathematical Economics*, 17:77–87, 1988.

[119] Maryam Kamgarpour and Hamidou Tembine. A bayesian mean field game approach to supply demand analysis of the smart grid. In *2013 First International Black Sea Conference on Communications and Networking (BlackSeaCom)*, pages 211–215. IEEE, 2013.

[120] M. Khan, H. Tembine, and A. Vasilakos. Game dynamics and cost of learning in heterogeneous 4g networks. *IEEE Journal on Selected Areas in Communications, vol 30, 1, pp 198 - 213, January*, 2012.

[121] M. A. Khan and H. Tembine. Random matrix games in wireless networks. *IEEE Global High Tech Congress on Electronics (GHTCE 2012), November 18-20, 2012, Shenzhen, China*, 2012.

[122] Manzoor Khan and Hamidou Tembine. Meta-learning for realizing self-x management of future networks. *IEEE Access Journal, vol.5, pp. 19072 - 19083, August*, 2017.

[123] Manzoor Ahmed Khan, Hamidou Tembine, and Athanasios V Vasilakos. Game dynamics and cost of learning in heterogeneous 4g networks. *IEEE Journal on Selected Areas in Communications*, 30(1):198–213, 2011.

[124] Manzoor Ahmed Khan, Hamidou Tembine, and Athanasios V Vasilakos. Evolutionary coalitional games: design and challenges in wireless networks. *IEEE Wireless Communications*, 19(2):50–56, 2012.

[125] Yubin Kim, Xuhai Xu, Daniel McDuff, Cynthia Breazeal, and Hae Won Park. Health-llm: Large language models for health prediction via wearable sensor data. *arXiv preprint arXiv:2401.06866*, 2024.

[126] Will Knight. Inside the creation of the world's most powerful open source ai model. Retrieved 2024-03-28.

[127] Aran Komatsuzaki, Joan Puigcerver, James Lee-Thorp, Carlos Riquelme Ruiz, Basil Mustafa, Joshua Ainslie, Yi Tay, Mostafa Dehghani, and Neil Houlsby. Sparse upcycling: Training mixture-of-experts from dense checkpoints. *arXiv preprint arXiv:2212.05055*, 2023. cs.LG.

[128] Maurice Kraus, Felix Divo, David Steinmann, Devendra Singh Dhami, and Kristian Kersting. United we pretrain, divided we fail! representation learning for time series by pretraining on 75 datasets at once. *arXiv preprint arXiv:2402.15404*, 2024.

[129] Siya Kunde, Stephanie Houde, and Rachel KE Bellamy. Designing an agentic ai assistant for chemical discovery. In *ACM International Conference on Intelligent User Interfaces*, 2025.

[130] Jean Lee, Nicholas Stevens, Soyeon Caren Han, and Minseok Song. A survey of large language models in finance (finllms). *arXiv preprint arXiv:2402.02315*, 2024.

[131] Dmitry Lepikhin, HyoukJoong Lee, Yuanzhong Xu, Dehao Chen, Orhan Firat, Yanping Huang, Maxim Krikun, Noam Shazeer, and Zhifeng Chen. Gshard: Scaling giant models with conditional computation and automatic sharding. *arXiv preprint arXiv:2006.16668*, 2020. cs.CL.

[132] Yuxuan Liang, Haomin Wen, Yuqi Nie, Yushan Jiang, Ming Jin, Dongjin Song, Shirui Pan, and Qingsong Wen. Foundation models for time series analysis: A tutorial and survey. *arXiv preprint arXiv:2403.14735*, 2024.

[133] Chenxi Liu, Sun Yang, Qianxiong Xu, Zhishuai Li, Cheng Long, Ziyue Li, and Rui Zhao. Spatial-temporal large language model for traffic prediction. *arXiv preprint arXiv:2401.10134*, 2024.

[134] Haoxin Liu, Zhiyuan Zhao, Jindong Wang, Harshavardhan Kamarthi, and B Aditya Prakash. Lstprompt: Large language models as zero-shot time series forecasters by long-short-term prompting. *arXiv preprint arXiv:2402.16132*, 2024.

[135] Lei Liu, Shuo Yu, Runze Wang, Zhenxun Ma, and Yanming Shen. How can large language models understand spatial-temporal data? *arXiv preprint arXiv:2401.14192*, 2024.

[136] Yong Liu, Guo Qin, Xiangdong Huang, Jianmin Wang, and Mingsheng Long. Autotimes: Autoregressive time series forecasters via large language models. *arXiv preprint arXiv:2402.02370*, 2024.

[137] Yong Liu, Haoran Zhang, Chenyu Li, Xiangdong Huang, Jianmin Wang, and Mingsheng Long. Timer: Transformers for time series analysis at scale. *arXiv preprint arXiv:2402.02368*, 2024.

[138] Zichang Liu, Jue Wang, Tri Dao, Tianyi Zhou, Binhang Yuan, Zhao Song, Anshumali Shrivastava, Ce Zhang, Yuandong Tian, Christopher Re, et al. Deja vu: Contextual sparsity for efficient llms at inference time. In *International Conference on Machine Learning*, pages 22137–22176. PMLR, 2023.

[139] Xin Luo and Hamidou Tembine. Evolutionary coalitional games for random access control. In *2013 Proceedings IEEE INFOCOM*, pages 535–539. IEEE, 2013.

[140] Michael Maschler, Shmuel Zamir, and Eilon Solan. *Game theory*. Cambridge University Press, 2020.

[141] François Mériaux, Samson Lasaulce, and Hamidou Tembine. Stochastic differential games and energy-efficient power control. *Dynamic Games and Applications*, 3:3–23, 2013.

[142] Meta. 200 languages within a single ai model: A breakthrough in high-quality machine translation, 2022. Archived from the original on 2023-01-09.

[143] Enzo Miller and Huyen Pham. Linear-quadratic mckean-vlasov stochastic differential games. *Modeling, Stochastic Control, Optimization, and Applications*, pages 451–481, 2019.

[144] John A Miller, Mohammed Aldosari, Farah Saeed, Nasid Habib Barna, Subas Rana, I Budak Arpinar, and Ninghao Liu. A survey of deep learning and foundation models for time series forecasting. *arXiv preprint arXiv:2401.13912*, 2024.

[145] Mistral AI. Mixtral of experts. `https://mistral.ai`, 2023. Retrieved 2024-02-04.

[146] Partha Sarathi Mohapatra and Puduru Viswanadha Reddy. Linear-quadratic mean-field-type difference games with coupled affine inequality constraints. *IEEE Control Systems Letters*, 2023.

[147] J. J. Moreau. Proximite et dualite dans un espace hilbertien. *Bull. Soc. Math. France, vol. 93, pp. 273-299*, 1965.

[148] San Murugesan. The rise of agentic ai: implications, concerns, and the path forward. *IEEE Intelligent Systems*, 40(2):8–14, 2025.

[149] Michael Mussa and Sherwin Rosen. Monopoly and product quality. *Journal of Economic theory*, 18(2):301–317, 1978.

[150] Roger B Myerson and Mark A Satterthwaite. Efficient mechanisms for bilateral trading. *Journal of Economic Theory*, 29(2):265–281, 1983.

[151] Subigya Nepal, Arvind Pillai, William Campbell, Talie Massachi, Eunsol Soul Choi, Xuhai Xu, Joanna Kuc, Jeremy F Huckins, Jason Holden, Colin Depp, et al. Contextual ai journaling: Integrating llm and time series behavioral sensing technology to promote self-reflection and well-being using the mindscape app. In *Extended Abstracts of the CHI Conference on Human Factors in Computing Systems*, pages 1–8, 2024.

[152] Tomohiro Ogawa, Kango Yoshioka, Ken Fukuda, and Takeshi Morita. Prediction of actions and places by the time series recognition from images with multimodal llm. In *2024 IEEE 18th International Conference on Semantic Computing (ICSC)*, pages 294–300. IEEE, 2024.

[153] Kenniy Olorunnimbe and Herna Viktor. Ensemble of temporal transformers for financial time series. *Journal of Intelligent Information Systems*, pages 1–25, 2024.

[154] Orenleung. Transformer deep dive: Parameter counting. Retrieved 2023-10-10.

[155] Zijie Pan, Yushan Jiang, Sahil Garg, Anderson Schneider, Yuriy Nevmyvaka, and Dongjin Song. S2ip-llm: Semantic space informed prompt learning with llm for time series forecasting. *arXiv preprint arXiv:2403.05798*, 2024.

[156] Abhay Dutt Paroha and Aakash Chotrani. A comparative analysis of timegpt and time-llm in predicting esp maintenance needs in the oil and gas sector. *International Journal of Computer Applications*, 975:8887, 2024.

[157] Mary Phuong and Marcus Hutter. Formal algorithms for transformers. *CoRR abs/2207.09238*, 2022.

[158] Sarat Piridi. Designing human–ai hand-offs: Copilot in hybrid workflows. *Journal of Computer Science and Technology Studies*, 7(5):605–611, 2025.

[159] Brian Powers, Michalis Smyrnakis, and Hamidou Tembine. Empathy in bimatrix games. *arXiv preprint arXiv:1708.01910*, 2017.

[160] Kashif Rasul, Arjun Ashok, Andrew Robert Williams, Hena Ghonia, Rishika Bhagwatkar, Arian Khorasani, Mohammad Javad Darvishi Bayazi, George Adamopoulos, Roland Riachi, Nadhir Hassen, et al. Lag-llama: Towards foundation models for probabilistic time series forecasting. *Preprint*, 2024.

[161] Chandan K Reddy and Parshin Shojaee. Towards scientific discovery with generative ai: Progress, opportunities, and challenges. In *Proceedings of the AAAI Conference on Artificial Intelligence*, volume 39, pages 28601–28609, 2025.

[162] Carlos Riquelme, Joan Puigcerver, Basil Mustafa, Maxim Neumann, Rodolphe Jenatton, André Susano Pinto, Daniel Keysers, and Neil Houlsby. Scaling vision with sparse mixture of experts. In *Advances in Neural Information Processing Systems*, volume 34, pages 8583–8595, 2021. arXiv:2106.05974.

[163] Giulia Rossi, Alain Tcheukam, and Hamidou Tembine. Empathy in one-shot prisoner dilemma. *arXiv preprint arXiv:1702.05361*, 2017.

[164] Giulia Rossi, Alain Tcheukam, and Hamidou Tembine. How much does users' psychology matter in engineering mean-field-type games. *arXiv preprint arXiv:1702.05355*, 2017.

[165] Sheng Shen, Le Hou, Yanqi Zhou, Nan Du, Shayne Longpre, Jason Wei, Hyung Won Chung, Barret Zoph, William Fedus, Xinyun Chen, Tu Vu, Yuexin Wu, Wuyang Chen, Albert Webson, and Yunxuan Li. Mixture-of-experts meets instruction tuning: A winning combination for large language models. *arXiv preprint arXiv:2305.14705*, 2023. cs.CL.

[166] Haotian Si, Changhua Pei, Hang Cui, Jingwen Yang, Yongqian Sun, Shenglin Zhang, Jingjing Li, Haiming Zhang, Jing Han, Dan Pei, et al. Timeseriesbench: An industrial-grade benchmark for time series anomaly detection models. *arXiv preprint arXiv:2402.10802*, 2024.

[167] Alonso Silva, Hamidou Tembine, Eitan Altman, and Merouane Debbah. Spatial games and global optimization for the mobile association problem: the downlink case. In *49th IEEE Conference on Decision and Control (CDC)*, pages 966–972. IEEE, 2010.

[168] Niranjan Sitapure and Joseph Sang-Il Kwon. Exploring the potential of time-series transformers for process modeling and control in chemical systems: an inevitable paradigm shift? *Chemical Engineering Research and Design*, 194:461–477, 2023.

[169] Alain Tcheukam Siwe and Hamidou Tembine. Network security as public good: A mean-field-type game theory approach. In *2016 13th International Multi-Conference on Systems, Signals & Devices (SSD)*, pages 601–606. IEEE, 2016.

[170] Michael Spence. Job market signaling. In *Uncertainty in economics*, pages 281–306. Elsevier, 1978.

[171] Nicholas Stroh. Trackgpt–a generative pre-trained transformer for cross-domain entity trajectory forecasting. *arXiv preprint arXiv:2402.00066*, 2024.

[172] Manaswini Swamy, Arunima Shukla, and James Purtilo. Llm-based stock market trend prediction. *Preprint*, 2023.

[173] Allahsera Auguste Tapo, Ali Traore, Sidy Danioko, and Hamidou Tembine. Machine intelligence in africa: a survey. *DSAI*, 2024.

[174] A. Tcheukam and H. Tembine. One swarm per queen: A particle swarm learning for stochastic games. In *2016 IEEE 10th International Conference on Self-Adaptive and Self-Organizing Systems (SASO)*, pages 144–145, 2016.

[175] Alain Tcheukam, Boualem Djehiche, and Hamidou Tembine. Evacuation of multi-level building: Design, control and strategic flow. In *2016 35th Chinese Control Conference (CCC)*, pages 9218–9223. IEEE, 2016.

[176] Alain Tcheukam and Hamidou Tembine. Mean-field-type games for distributed power networks in presence of prosumers. In *2016 Chinese Control and Decision Conference (CCDC)*, pages 446–451. IEEE, 2016.

[177] NLLB Team, Marta R. Costa-Juss, James Cross, Onur Çelebi, Maha Elbayad, Kenneth Heafield, Kevin Heffernan, Elahe Kalbassi, Janice Lam, Daniel Licht, Jean Maillard, Anna Sun, Skyler Wang, Guillaume Wenzek, and Al Youngblood. No language left behind: Scaling human-centered machine translation. *arXiv preprint arXiv:2207.04672*, 2022. cs.CL.

[178] H. Tembine. Fast distributed strategic learning for global optima in queueing access games. *19th World Congress of the International Federation of Automatic Control (IFAC), Cape Town, South Africa, 24-29 August*, 2014.

[179] H Tembine. Mean-field-type game theory. *MFTG Tutorial notes*, 2017.

[180] H. Tembine and A. Azad. Dynamic routing games: An evolutionary game theoretic approach. *in IEEE CDC-ECC, 50th IEEE Conference on Decision and Control and European Control Conference, December 12-15, Orlando,Florida*, 2011.

[181] H. Tembine, Q. Zhu, and T. Başar. Risk-sensitive mean-field games. *IEEE Transactions on Automatic Control*, 59(4):835–850, April 2014.

[182] Hamidou Tembine. Population games with networking applications. *PhD Thesis, University of Avignon*, 2009.

[183] Hamidou Tembine. Large-scale games in large-scale systems. *arXiv preprint arXiv:1111.2285*, 2011.

[184] Hamidou Tembine. Fast distributed strategic learning for global optima in queueing access games. *IFAC Proceedings Volumes*, 47(3):7055–7060, 2014.

[185] Hamidou Tembine. Mean-field-type game theory. *Special issue on Nonasymptotic Mean-Field-Type Game Theory*, 2014.

[186] Hamidou Tembine. Uncertainty quantification in mean-field-type teams and games. In *2015 54th IEEE Conference on Decision and Control (CDC)*, pages 4418–4423. IEEE, 2015.

[187] Hamidou Tembine. Mean-field-type optimization for demand-supply management under operational constraints in smart grid. *Energy Systems*, 7(2):333–356, 2016.

[188] Hamidou Tembine. Payoff measurement noise in risk-sensitive mean-field-type games. In *2017 29th Chinese Control And Decision Conference (CCDC)*, pages 3764–3769. IEEE, 2017.

[189] Hamidou Tembine. *Distributed strategic learning for wireless engineers*. CRC Press, 2018.

[190] Hamidou Tembine. Covid-19: data-driven mean-field-type game perspective. *Games*, 11(4):51, 2020.

[191] Hamidou Tembine. Deep learning meets game theory: Bregman-based algorithms for interactive deep generative adversarial networks. *IEEE Transactions on Cybernetics, 14 pages, volume 50 , Issue: 3 , March, pp. 1132 - 1145*, 2020.

[192] Hamidou Tembine. Distributed planning in mean-field-type games. *IFAC-PapersOnLine*, 53(2):2183–2188, 2020.

[193] Hamidou Tembine. Master adjoint systems in mean-field-type games. *Preprint*, 2020.

[194] Hamidou Tembine. Master adjoint systems in mean-field-type games. *in Communications in Information and Systems, vol.21, 4, pp. 623-650*, 2021.

[195] Hamidou Tembine, Eitan Altman, and Rachid El-Azouzi. Delayed evolutionary game dynamics applied to medium access control. In *2007 IEEE International Conference on Mobile Adhoc and Sensor Systems*, pages 1–6. IEEE, 2007.

[196] Hamidou Tembine, Eitan Altman, Rachid El-Azouzi, and Yezekael Hayel. Evolutionary games in wireless networks. *IEEE Transactions on Systems, Man, and Cybernetics, Part B (Cybernetics)*, 40(3):634–646, 2009.

[197] Hamidou Tembine, Eitan Altman, Rachid El-Azouzi, and Yezekael Hayel. Bio-inspired delayed evolutionary game dynamics with networking applications. *Telecommunication Systems*, 47:137–152, 2011.

[198] Hamidou Tembine, Eitan Altman, Rachid ElAzouzi, and Yezekael Hayel. Stochastic population games with individual independent states and coupled constraints. In *Proceedings of the 3rd International Conference on Performance Evaluation Methodologies and Tools*, pages 1–10, 2008.

[199] Hamidou Tembine, Eitan Altman, Rachid ElAzouzi, and Yezekael Hayel. Correlated evolutionarily stable strategies in random medium access control. In *2009 International Conference on Game Theory for Networks*, pages 212–221. IEEE, 2009.

[200] Hamidou Tembine, Manzoor Khan, and Issa Bamia. Mean-field-type transformer. *Mathematics, 12, 22, 3506, 52 pages*, 2024.

[201] Hamidou Tembine, Manzoor Ahmed Khan, and Issa Bamia. Mean-field-type transformers. *Mathematics*, 12(22):3506, 2024.

[202] Hamidou Tembine, Abdellatif Kobbane, and Mohammed El Koutbi. Robust power allocation games under channel uncertainty and time delays. In *2010 IFIP Wireless Days*, pages 1–5. IEEE, 2010.

[203] Hamidou Tembine and Ousmane Kodio. Reverse ishikawa-nesterov learning scheme for fractional mean-field games. *IFAC-PapersOnLine*, 50(1):8090–8096, 2017.

[204] Hamidou Tembine, Jean Yves Le Boudec, Rachid ElAouzi, and Eitan Altman. From mean field interaction to evolutionary game dynamics. In *2009 7th International Symposium on Modeling and Optimization in Mobile, Ad Hoc, and Wireless Networks*, pages 1–5. IEEE, 2009.

[205] Hamidou Tembine, Raul Tempone, and Pedro Vilanova. Mean field games for cognitive radio networks. In *2012 American Control Conference (ACC)*, pages 6388–6393. IEEE, 2012.

[206] Hamidou Tembine, Raul Tempone, and Pedro Vilanova. Mean-field learning: a survey. *arXiv preprint arXiv:1210.4657*, 2012.

[207] Hamidou Tembine, Raul Tempone, and Pedro Vilanova. Mean-field learning for satisfactory solutions. In *52nd IEEE Conference on Decision and Control*, pages 4871–4876. IEEE, 2013.

[208] Hamidou Tembine, Pedro Vilanova, Mohamad Assaad, and Merouane Debbah. Mean field stochastic games for sinr-based medium access control. In *Gamecomm2011*, pages 10–pages, 2011.

[209] Noureddine Toumi, Julian Barreiro-Gomez, Tyrone E Duncan, and Hamidou Tembine. Berge equilibrium in linear-quadratic mean-field-type games. *Journal of the Franklin Institute*, 357(15):10861–10885, 2020.

[210] Patara Trirat, Yooju Shin, Junhyeok Kang, Youngeun Nam, Jihye Na, Minyoung Bae, Joeun Kim, Byunghyun Kim, and Jae-Gil Lee. Universal time-series representation learning: A survey. *arXiv preprint arXiv:2401.03717*, 2024.

[211] Ashish Vaswani, Noam Shazeer, Jakob Uszkoreit Niki Parmar, Llion Jones, Aidan N Gomez, Lukasz Kaiser, and Illia Polosukhin. Attention is all you need. *In Advances in neural information processing systems*, 2017.

[212] Cédric Villani et al. *Optimal transport: old and new*, volume 338. Springer, 2009.

[213] Jun Wang, Wenjie Du, Wei Cao, Keli Zhang, Wenjia Wang, Yuxuan Liang, and Qingsong Wen. Deep learning for multivariate time series imputation: A survey. *arXiv preprint arXiv:2402.04059*, 2024.

[214] Gerald Woo, Chenghao Liu, Akshat Kumar, Caiming Xiong, Silvio Savarese, and Doyen Sahoo. Unified training of universal time series forecasting transformers. *arXiv preprint arXiv:2402.02592*, 2024.

[215] Shengqiong Wu, Hao Fei, Leigang Qu, Wei Ji, and Tat-Seng Chua. Next-gpt: Any-to-any multimodal llm. *arXiv preprint arXiv:2309.05519*, 2023.

[216] Mengwei Xu, Wangsong Yin, Dongqi Cai, Rongjie Yi, Daliang Xu, Qipeng Wang, Bingyang Wu, Yihao Zhao, Chen Yang, Shihe Wang, et al. A survey of resource-efficient llm and multimodal foundation models. *arXiv preprint arXiv:2401.08092*, 2024.

[217] Jiexia Ye, Weiqi Zhang, Ke Yi, Yongzi Yu, Ziyue Li, Jia Li, and Fugee Tsung. A survey of time series foundation models: Generalizing time series representation with large language mode. *arXiv preprint arXiv:2405.02358*, 2024.

[218] L. C. Young. An inequality of the hölder type, connected with stieltjes integration. *Acta Mathematica*, 67:251–282, 1936.

[219] Xinli Yu, Zheng Chen, Yuan Ling, Shujing Dong, Zongyi Liu, and Yanbin Lu. Temporal data meets llm–explainable financial time series forecasting. *arXiv preprint arXiv:2306.11025*, 2023.

[220] Jun Zhan, Junqi Dai, Jiasheng Ye, Yunhua Zhou, Dong Zhang, Zhigeng Liu, Xin Zhang, Ruibin Yuan, Ge Zhang, Linyang Li, et al. Anygpt: Unified multimodal llm with discrete sequence modeling. *arXiv preprint arXiv:2402.12226*, 2024.

[221] Cheng Zhang, Nilam Nur Amir Sjarif, and Roslina Ibrahim. Deep learning models for price forecasting of financial time series: A review of recent advancements: 2020–2022. *Wiley Interdisciplinary Reviews: Data Mining and Knowledge Discovery*, 14(1):e1519, 2024.

[222] Qilei Zhang and John H Mott. An exploratory assessment of llm's potential toward flight trajectory reconstruction analysis. *arXiv preprint arXiv:2401.06204*, 2024.

[223] Qinshi Zhang, Ruoyu Wen, Latisha Besariani Hendra, Zijian Ding, and Ray LC. Can ai prompt humans? multimodal agents prompt players' game actions and show consequences to raise sustainability awareness. In *Proceedings of the 2025 CHI Conference on Human Factors in Computing Systems*, pages 1–29, 2025.

[224] Xiangyuan Zhang, Kaiqing Zhang, Erik Miehling, and Tamer Basar. Non-cooperative inverse reinforcement learning. *Advances in neural information processing systems*, 32, 2019.

[225] Xiao Zhang, Ruoyu Xiang, Chenhan Yuan, Duanyu Feng, Weiguang Han, Alejandro Lopez-Lira, Xiao-Yang Liu, Sophia Ananiadou, Min Peng, Jimin Huang, et al. Dólares or dollars? unraveling the bilingual prowess of financial llms between spanish and english. *arXiv preprint arXiv:2402.07405*, 2024.

[226] Xiyuan Zhang, Ranak Roy Chowdhury, Rajesh K Gupta, and Jingbo Shang. Large language models for time series: A survey. *arXiv preprint arXiv:2402.01801*, 2024.

[227] Huaqin Zhao, Zhengliang Liu, Zihao Wu, Yiwei Li, Tianze Yang, Peng Shu, Shaochen Xu, Haixing Dai, Lin Zhao, Gengchen Mai, et al. Revolutionizing finance with llms: An overview of applications and insights. *arXiv preprint arXiv:2401.11641*, 2024.

[228] Tian Zhou, Peisong Niu, Liang Sun, Rong Jin, et al. One fits all: Power general time series analysis by pretrained lm. *Advances in neural information processing systems*, 36:43322–43355, 2023.

[229] James Zou and Eric J Topol. The rise of agentic ai teammates in medicine. *The Lancet*, 405(10477):457, 2025.

Index

For Product Safety Concerns and Information please contact our EU
representative GPSR@taylorandfrancis.com
Taylor & Francis Verlag GmbH, Kaufingerstraße 24, 80331 München, Germany

www.ingramcontent.com/pod-product-compliance
Lightning Source LLC
Chambersburg PA
CBHW080905170526
45158CB00008B/1992

* 9 7 8 1 0 4 1 1 2 4 0 7 8 *